Theory of Nonregular Factorial Designs

Nonregular factorial designs are a class of factorial designs that enable researchers to study simultaneously the effects of many explanatory variables on a response variable of interest. Factorial designs have been in the mainstream of design research for decades, but interest in nonregular factorial designs has been increasing significantly in the last twenty years or so. These designs are practically useful as they offer much richer and more flexible choices for experimenters to contemplate than regular designs. Theoretically, they present new challenges and raise new issues that require new ideas and methods to properly address. This book is the first to synthesize research on this topic into a consistent statistical design framework. It is focused on the theoretical underpinnings of nonregular factorial designs but emphasizes the ideas and methods of statistical designs, rather than mathematical rigor. Some background is provided in an appendix, and the book also includes lots of worked examples to illustrate the ideas and methods.

Features:

- A unified and comprehensive overview of the theory and application of nonregular factorial designs
- Self-contained, with an appendix of necessary background knowledge in factorial designs
- Mathematics is kept to a minimum, with proofs only included if simple or insightful, or is otherwise in chapter appendices
- Great attention has been paid to the narrative flow
- Lots of worked examples to illustrate the ideas and methods

This book is primarily aimed at researchers and practitioners working on the design of experiments. It will be a perfect entry into the field for a junior researcher and a useful reference for all researchers already in the field. It could be used as a textbook for graduate students who have taken a first design course and have a passion for design ideas and research. Scientists from other disciplines will find the book useful as well, if their research involves multi-factor experiments.

MONOGRAPHS ON STATISTICS AND APPLIED PROBABILITY

Recently Published Titles

Sufficient Dimension Reduction
Methods and Applications with R
Bing Li

Large Covariance and Autocovariance Matrices
Arup Bose and Monika Bhattacharjee

The Statistical Analysis of Multivariate Failure Time Data: A Marginal Modeling Approach
Ross L. Prentice and Shanshan Zhao

Dynamic Treatment Regimes
Statistical Methods for Precision Medicine
Anastasios A. Tsiatis, Marie Davidian, Shannon T. Holloway, and Eric B. Laber

Sequential Change Detection and Hypothesis Testing
General Non-i.i.d. Stochastic Models and Asymptotically Optimal Rules
Alexander Tartakovsky

Introduction to Time Series Modeling
Genshiro Kitigawa

Replication and Evidence Factors in Observational Studies
Paul R. Rosenbaum

Introduction to High-Dimensional Statistics, Second Edition
Christophe Giraud

Object Oriented Data Analysis
J.S. Marron and Ian L. Dryden

Martingale Methods in Statistics
Yoichi Nishiyama

The Energy of Data and Distance Correlation
Gabor J. Szekely and Maria L. Rizzo

Sparse Graphical Modeling for High Dimensional Data: Sparse Graphical Modeling for High Dimensional Data
Faming Liang and Bochao Jia

Bayesian Nonparametric Methods for Missing Data and Causal Inference
Michael J. Daniels, Antonio Linero, and Jason Roy

Functional Data Analysis with R
Ciprian M. Crainiceanu, Jeff Goldsmith, Andrew Leroux, and Erjia Cui

Mixture Models: Parametric, Semiparametric, and New Directions
Weixin Yao and Sijia Xiang

Seminal Ideas and Controversies in Statistics
Roderick J.A. Little

Theory of Nonregular Factorial Designs
Ching-Shui Cheng and Boxin Tang

For more information about this series please visit: https://www.crcpress.com/Chapman--HallCRC-Monographs-on-Statistics--Applied-Probability/book-series/CHMONSTAAPP

Theory of Nonregular Factorial Designs

Ching-Shui Cheng and Boxin Tang

CRC Press
Taylor & Francis Group
Boca Raton London New York

CRC Press is an imprint of the
Taylor & Francis Group, an **informa** business
A CHAPMAN & HALL BOOK

First edition published 2025
by CRC Press
2385 NW Executive Center Drive, Suite 320, Boca Raton FL 33431

and by CRC Press
4 Park Square, Milton Park, Abingdon, Oxon, OX14 4RN

CRC Press is an imprint of Taylor & Francis Group, LLC

ISBN: 978-1-032-44379-9 (hbk)
ISBN: 978-1-032-44382-9 (pbk)
ISBN: 978-1-003-37186-1 (ebk)

DOI: 10.1201/9781003371861

Typeset in CMR10 font
by KnowledgeWorks Global Ltd.

Publisher's note: This book has been prepared from camera-ready copy provided by the authors.

To Suzanne, Adelaide, and in Memory of Leonard, CSC

To Joan, Bradley, and my parents, BT

Contents

Preface

This book is aimed at presenting a unified and comprehensive treatment of the theory of nonregular factorial designs. It is intended to provide interested statisticians and data scientists one place to find and systematically study key ideas and concepts without the need to labor over numerous research articles scattered around in the literature.

Nonregular factorial designs are a class of factorial designs, which are designs that enable researchers to study simultaneously the effects of many explanatory variables on a response variable of interest. Factorial designs are in the mainstream of design research for at least three reasons. Theoretical elegance alone is an irresistible force that draws design researchers to the field. Various connections with other types of designs make the studies of factorial designs interesting and rewarding. Most important is their practical utility, for no other types of designs have seen so much success in applications to science and technology.

Many books have been written on factorial designs but almost exclusively on regular factorial designs, except for a few recent books that have a chapter or two on nonregular factorial designs. Although nonregular designs were touched upon long ago, serious research on them really burst onto the scene in the early 1990s. It gradually picked up the pace, and to date has been developed into a full-fledged theory that is still expanding. Nonregular factorial designs are practically useful as they offer much richer and more flexible choices for experimenters to contemplate than regular designs do. Theoretically, they present new challenges and raise new issues that require new ideas and methods to properly address. An easily overlooked but important side benefit from studying nonregular designs is new insights into regular designs.

This is the first book that is devoted entirely to nonregular designs. It brings together various results and material into a single volume, and presents them in a unified, coherent, and comprehensive fashion. The book can be used not only as a resource for design researchers but also as a textbook for graduate students who have taken a first design course and have a passion for design ideas or research.

We intended to write a theoretical book that emphasizes the ideas and methods of statistical designs rather than mathematical rigor. As a result, the level of mathematics is kept as low as possible without jeopardizing the understanding of concepts. The proofs, if simple or insightful, are given right after the corresponding results. Those that are lengthy, highly technical, or not of much help in understanding the results are either omitted or given in proof-only sections, which can be skipped by all readers unless their research requires such specialized techniques. The purpose of this strategy is to maintain the flow of main ideas and concepts at a reasonably low mathematical level.

The readers will need basic design knowledge such as that taught in a first design course. More exposure to regular designs will be a plus but not required. To make the book more or less self-contained, we provide some statistical and mathematical background in an appendix chapter.

The book would not be possible without the help from many people. We were fortunate to have opportunities to collaborate with numerous researchers. Research results from these collaborations are an integral part of this book. To our coauthors, we thank you all. Several people read parts of this book, and offered comments and identified mistakes. They all deserve our gratitude. Special thanks go to Guanzhou Chen and Chunfang D. Lin, who painstakingly read the whole book and provided insightful comments. We also want to thank Rob Calver and Sherry Thomas of Chapman & Hall/CRC Press. Although the idea of writing this book was conceived long ago, it was Rob's persistent encouragement that really got us started. The book would not be completed on schedule without Sherry's friendly reminders and prompt responses to our questions. Finally, support from NSERC is gratefully acknowledged.

Author

Ching-Shui Cheng is a Professor Emeritus at the University of California, Berkeley. **Boxin Tang** is a Professor of Statistics at Simon Fraser University. Both have their main research interests in the area of experimental design.

Professor Cheng is a Fellow of the Institute of Mathematical Statistics and a Fellow of the American Statistical Association. He is a former Chair-Editor for *Statistica Sinica*, and also served as an Associate Editor for the *Annals of Statistics*, *Biometrika*, *Journal of Statistical Planning and Inference*, *Statistica Sinica*, and *Technometrics*.

Professor Tang is a Fellow of the Institute of Mathematical Statistics and a Fellow of the American Statistical Association. He has served on the editorial boards of the *Annals of Statistics*, *Bernoulli Journal*, *Electronic Journal of Statistics*, *Journal of the Royal Statistical Society: Series B*, *Journal of Statistical Theory and Practice*, and *Statistica Sinica*.

1

Introduction

We introduce nonregular designs by examples and highlight their major differences with regular designs. These distinct features lead to some key ideas and issues that will serve as useful guides for our studies of nonregular designs in later chapters. The second part of the chapter outlines a plan for the book.

1.1 Nonregular Designs by Examples

Consider an experiment for studying the effects of nine factors on a response variable, where two levels, denoted by ± 1, are chosen for each factor. A full factorial experiment would require $2^9 = 512$ runs. Suppose that limited monetary and other resources including time constraints allow the experimenter to use a design with no more than 20 runs. One possibility is to consider a regular design of 16 runs generated by:

$$\boldsymbol{d}_5 = \boldsymbol{d}_1\boldsymbol{d}_2, \quad \boldsymbol{d}_6 = \boldsymbol{d}_1\boldsymbol{d}_2\boldsymbol{d}_3, \quad \boldsymbol{d}_7 = \boldsymbol{d}_1\boldsymbol{d}_2\boldsymbol{d}_4, \quad \boldsymbol{d}_8 = \boldsymbol{d}_1\boldsymbol{d}_3\boldsymbol{d}_4, \quad \boldsymbol{d}_9 = \boldsymbol{d}_2\boldsymbol{d}_3\boldsymbol{d}_4.$$

As displayed in Table 1.1, the design matrix has 16 rows and 9 columns with each row representing a run and each column corresponding to a factor, and it is obtained as follows. The first four columns $\boldsymbol{d}_1, \boldsymbol{d}_2, \boldsymbol{d}_3$, and $\boldsymbol{d}_4$ are given by a full factorial design for four factors. The columns for the remaining factors are then generated using the first four columns. For example, column $\boldsymbol{d}_5$ is the componentwise product of columns $\boldsymbol{d}_1$ and $\boldsymbol{d}_2$, which is expressed symbolically as $\boldsymbol{d}_5 = \boldsymbol{d}_1\boldsymbol{d}_2$. Similarly, $\boldsymbol{d}_6 = \boldsymbol{d}_1\boldsymbol{d}_2\boldsymbol{d}_3$ means that column $\boldsymbol{d}_6$ is the componentwise product of columns $\boldsymbol{d}_1$, $\boldsymbol{d}_2$, and $\boldsymbol{d}_3$. The reader is referred to Section A.1 of the appendix for an introduction to two-level regular factorial designs.

A consequence of using this fractional factorial design instead of a full factorial design is aliasing among factorial effects. For example, the relation $\boldsymbol{d}_5 = \boldsymbol{d}_1\boldsymbol{d}_2$ implies that the main effect of factor 5 is fully aliased with the two-factor interaction between factors 1 and 2. In fact, one can work out a complete aliasing pattern of the design. It turns out that the aliasing pattern divides the collection of all $2^9 = 512$ factorial effects, which, for convenience, includes the grand mean corresponding to the all-ones column, into 16 groups, each containing 32 effects. The effects from the same group are fully aliased with one another, and those from different groups are mutually orthogonal.

Let us now look at a nonregular design of 12 runs, as displayed in Table 1.2.

This design is obtained by deleting two columns from a Plackett-Burman design of 12 runs. The nine main-effect columns, $\boldsymbol{d}_1, \ldots, \boldsymbol{d}_9$, are mutually orthogonal just as those of the 16-run regular design are. Beyond this similarity, the relationships among the factorial effects are now much more complicated. For example, the interaction column $\boldsymbol{d}_1\boldsymbol{d}_2$, the componentwise product of columns $\boldsymbol{d}_1$ and $\boldsymbol{d}_2$, is neither orthogonal to nor fully aliased with the main-effect column $\boldsymbol{d}_3$. In fact, $\boldsymbol{d}_1\boldsymbol{d}_2$ and $\boldsymbol{d}_3$ are partially aliased with each other, as can be seen by their correlation, which is $-1/3$. One can check that column $\boldsymbol{d}_1\boldsymbol{d}_2$ is

DOI: 10.1201/9781003371861-1

TABLE 1.1
A regular design of 16 runs

d_1	d_2	d_3	d_4	d_5	d_6	d_7	d_8	d_9
−1	−1	−1	−1	1	−1	−1	−1	−1
−1	−1	−1	1	1	−1	1	1	1
−1	−1	1	−1	1	1	−1	1	1
−1	−1	1	1	1	1	1	−1	−1
−1	1	−1	−1	−1	1	1	−1	1
−1	1	−1	1	−1	1	−1	1	−1
−1	1	1	−1	−1	−1	1	1	−1
−1	1	1	1	−1	−1	−1	−1	1
1	−1	−1	−1	−1	1	1	1	−1
1	−1	−1	1	−1	1	−1	−1	1
1	−1	1	−1	−1	−1	1	−1	1
1	−1	1	1	−1	−1	−1	1	−1
1	1	−1	−1	1	−1	−1	1	1
1	1	−1	1	1	−1	1	−1	−1
1	1	1	−1	1	1	−1	−1	−1
1	1	1	1	1	1	1	1	1

TABLE 1.2
A nonregular design of 12 runs

d_1	d_2	d_3	d_4	d_5	d_6	d_7	d_8	d_9
−1	−1	−1	−1	−1	−1	−1	−1	−1
1	−1	1	−1	−1	−1	1	1	1
1	1	−1	1	−1	−1	−1	1	1
−1	1	1	−1	1	−1	−1	−1	1
1	−1	1	1	−1	1	−1	−1	−1
1	1	−1	1	1	−1	1	−1	−1
1	1	1	−1	1	1	−1	1	−1
−1	1	1	1	−1	1	1	−1	1
−1	−1	1	1	1	−1	1	1	−1
−1	−1	−1	1	1	1	−1	1	1
1	−1	−1	−1	1	1	1	−1	1
−1	1	−1	−1	−1	1	1	1	−1

orthogonal to columns $\boldsymbol{d}_1$ and $\boldsymbol{d}_2$ but has a correlation of $\pm 1/3$ with any other main-effect column. Similar phenomena can be observed for other two-factor interactions.

This complex aliasing structure among factorial effects is the hallmark of nonregular designs. It is a curse because it causes difficulties in ascertaining significant effects when conducting data analysis. It is also a blessing as we now have an opportunity of estimating both the interaction effect between factors 1 and 2 and the main effect of factor 3 because they are not fully aliased with each other. When two effects are fully aliased, there is no way one can separate them without further experimentation.

The differences in aliasing structure between regular and nonregular designs are front and center. Less apparent but more interesting are their differences in projection properties. Focusing on any three factors, we see that the 12-run nonregular design contains a full 2^3 factorial. This remarkable property implies that one can estimate all factorial effects produced by a set of three factors, no matter which three factors turn out to be active if it is reasonable to assume that no more than three factors can be active. In contrast, the

16-run regular design lacks this property. A three-factor projection of a regular design does not contain a full 2^3 factorial if the three factors form a defining word, meaning that the componentwise product of the three corresponding design columns is a column of all 1's or all -1's. An example of such situations is the projection design onto factors 1, 2, and 5. If a regular design is to contain a full 2^3 factorial in all of its three-factor projection designs, then it cannot have any defining word of length three, and therefore must have resolution IV. With 16 runs, resolution IV designs are available only for up to eight factors.

Even less apparent and in fact quite hidden is the property of the 12-run nonregular design when projected onto four factors. A projection design onto any four factors, though not containing a full 2^4 factorial, permits the joint estimation of all main effects and all two-factor interactions, which can easily be verified if one wishes to do so. This hidden projection property, as appropriately named in the literature, is simply striking. For a regular design to have the same property, it must have resolution V. With 16 runs, we can obtain a resolution V design only for five factors.

Last, we consider two 20-run nonregular designs in Tables 1.3 and 1.4.

TABLE 1.3
Nonregular 20-run design 1

−1	1	1	1	1	1	1	1	1
1	−1	1	1	−1	−1	1	−1	1
1	1	−1	1	1	−1	−1	−1	−1
1	1	1	−1	−1	1	−1	1	−1
1	−1	1	−1	−1	1	1	−1	−1
1	−1	−1	1	1	−1	1	1	−1
1	1	−1	−1	1	1	−1	−1	1
1	−1	−1	1	−1	1	−1	−1	1
1	1	−1	−1	−1	−1	1	1	−1
1	−1	1	−1	1	−1	−1	1	1
1	1	1	1	1	1	1	1	1
−1	−1	−1	−1	1	1	−1	1	−1
−1	−1	−1	−1	−1	1	1	1	1
−1	−1	−1	−1	1	−1	1	−1	1
−1	1	−1	1	−1	−1	−1	1	1
−1	1	1	−1	−1	−1	−1	−1	1
−1	−1	1	1	−1	−1	−1	1	−1
−1	1	1	−1	1	−1	1	−1	−1
−1	−1	1	1	1	1	−1	−1	−1
−1	1	−1	1	−1	1	1	1	1

These two nonregular designs are similar in that they both enjoy the projection property and hidden projection property of the 12-run design as discussed earlier but are dissimilar otherwise. For design 1, any of its three-factor projection designs contains two copies of a full 2^3 factorial. The story is different for design 2. While some of the three-factor projection designs of design 2 contain two copies of a full 2^3 factorial, the others contain just one copy of a full 2^3 factorial. For example, the projection design onto factors 1, 2, and 4 contains two copies of a full 2^3 factorial whereas the projection design onto factors 1, 2, and 3 contains only one copy of a full 2^3 factorial. Common sense would suggest that design 1 is better than design 2.

The example designs presented in this section highlight some of the major differences between regular and nonregular designs, which in turn raise a number of principal issues that will guide us through the studies of nonregular designs. The aliasing, projection, and hidden projection properties are some of the central themes in later developments. Investigations

TABLE 1.4
Nonregular 20-run design 2

1	1	1	1	1	1	1	1	1
1	1	1	1	1	1	1	−1	−1
1	1	1	1	−1	−1	−1	1	1
1	1	1	−1	1	−1	−1	−1	−1
1	1	−1	−1	−1	1	1	1	1
1	−1	1	−1	−1	1	−1	1	−1
1	−1	−1	1	1	−1	−1	−1	−1
1	−1	−1	1	−1	1	−1	−1	1
1	−1	−1	−1	1	−1	1	−1	1
1	−1	−1	−1	−1	−1	1	1	−1
−1	1	1	−1	−1	−1	−1	−1	1
−1	1	−1	1	1	−1	1	1	−1
−1	1	−1	1	−1	−1	−1	1	−1
−1	1	−1	−1	1	1	−1	−1	1
−1	1	−1	−1	−1	1	1	−1	−1
−1	−1	1	1	−1	1	1	−1	−1
−1	−1	1	1	−1	−1	1	−1	1
−1	−1	1	−1	1	1	−1	1	−1
−1	−1	1	−1	1	−1	1	1	1
−1	−1	−1	1	1	1	−1	1	1

of these and other design properties lead us further to the formulation of suitable criteria for the assessment and selection of nonregular designs.

1.2 Plan for the Book

The book comprises two parts with Chapters 2–9 allocated to two-level designs and Chapters 10–12 to multi-level and mixed-level designs. We have chosen this structure for the book, mainly because two-level designs are the most useful and research results for them are much richer, and partly because even if general results are available, the two-level versions are simpler and more intuitive, hence more helpful for the understanding of concepts and results.

In Chapter 2, we provide some fundamentals for factorial designs, examining such issues as models and analysis, aliasing, estimability, optimality, and bias. Orthogonal arrays, forming a most useful class of factorial designs, are introduced and their estimation capacity and optimality are discussed. Regular and nonregular designs are distinguished by their distinct aliasing features. We also devote a section to Hadamard matrices due to their close relationship with orthogonal arrays.

In Chapter 3, we develop a theory of J-characteristics, which provides us with a powerful machinery for studying nonregular designs just like the defining relation for studying regular designs. Making use of J-characteristics, we then investigate the projection and hidden projection properties of nonregular designs. The chapter is concluded with an application of J-characteristics to the enumeration of orthogonal arrays.

The generalized resolution is the topic of Chapter 4. We first introduce the generalized resolution and then justify it as a selection criterion for nonregular designs from two different angles. We next embark on the construction of nonregular designs with maximum generalized resolutions. Designs from Paley constructions of Hadamard matrices are instrumental in this endeavor. Good Hadamard matrices and quaternary codes also have roles to play and thus deserve their fair share of treatment. We conclude the chapter with a discussion on the minimum G-aberration criterion, an extension of generalized resolution, and a survey of theoretical and computational results on the construction of minimum G-aberration designs.

Chapter 5 is devoted to the minimum G_2-aberration criterion, its various statistical justifications, and the construction of minimum G_2-aberration designs. There are two key ideas in the construction results: a complementary design theory and minimum moment aberration. The minimum moment aberration criterion is equivalent to the minimum G_2-aberration criterion but has a form that significantly eases the effort of both theoretical construction and computer search of minimum G_2-aberration designs.

Requirement sets, clear effects, and second-order saturated designs are treated in Chapter 6. A general contamination framework is introduced to facilitate the studies of the requirement sets and clear effects. Second-order saturated designs are useful in their own right, and are important also because of their connections with clear effects and space-filling designs.

In Chapter 7, we examine minimum G_2-aberration under complex settings. The minimum G_2-aberration criterion introduced in Chapter 5 is for the simple setting of unstructured experimental units. In Chapter 7, a Bayesian framework is used to derive appropriate minimum G_2-aberration criteria for several complex settings where the experimental units have structures as those given by block designs, split-plot designs, and row-column designs.

Chapter 8 is on supersaturated designs. A central theme is the $E(s^2)$-optimality. There has been a rich literature with an aim at deriving lower bounds on the $E(s^2)$-criterion and developing associated methods for constructing $E(s^2)$-optimal designs. We will present some of the most important results. The chapter also allocates adequate space for discussing alternative criteria such as those for controlling the maximum correlations, $UE(s^2)$-optimality, and MDS-resolution. These criteria are useful in their own right or can be used in combination with the $E(s^2)$-criterion. The chapter is concluded with a brief discussion on group screening designs.

Chapter 9 deals with miscellaneous topics that are not covered in previous chapters. We first examine nonorthogonal designs of resolution III and IV as well as Rechtschaffner's designs, and then turn our attention to fractional factorial designs under a baseline parametrization and their relationships with those under the usual orthogonal parametrization. A brief discussion is given to search designs and parallel-flat designs.

In Chapter 10, we commence our studies on multi-level and mixed-level nonregular designs. After introducing factorial effects and orthogonal arrays, we define and justify the minimum G_2-aberration criterion. We next discuss an equivalent criterion, the minimum moment aberration criterion, which provides a powerful machinery for both theoretical studies and computational searches of minimum G_2-aberration designs. We also examine the use of minimum G-aberration for further design selection and of generalized resolution as a concise summary of a design's properties. Lastly, fair shares of treatment are given to the topics of supersaturated designs, three-level designs for quantitative factors, and definitive screening designs.

In Chapter 11, mixed-level designs under a broad class of structures of experimental units are investigated, and a general theory of minimum G_2-aberration is developed. This chapter can be thought of as a continuation of Chapter 7 in the most general settings. Here the treatments have a mixed-level factorial structure and the experimental units are allowed to have more general structures than those in Chapter 7.

Space-filling designs are for computer experiments and they are the subject of Chapter 12. Orthogonality is an attractive design property, and we present some results on the construction of orthogonal designs and orthogonal Latin hypercubes. Some extensive discussion is given to designs based on orthogonal arrays and strong orthogonal arrays; designs so constructed enjoy certain guaranteed low-dimensional space-filling properties. We also give an account of maximin distance designs, nested space-filling designs, and designs suitable for experiments with both quantitative and qualitative factors.

The book has an appendix chapter, aiming at providing background knowledge in design theory; topics covered here include regular designs, Galois fields, projective geometries, and coding theory. The purpose of this appendix is to make the book more or less self-contained so that the reader does not have to reach very far to find necessary knowledge whenever such need arises.

A few words need to be said about notation. We strive to achieve a balance between rigor and frugality in our usage of notation. Boldface fonts are used for vectors and matrices. By a vector, we mean a column vector unless specified otherwise. For two vectors $\boldsymbol{x}_1$ and $\boldsymbol{x}_2$ of length n, their componentwise product, also called Hadamard product, is another vector of length n. We will use the simple notation $\boldsymbol{x}_1\boldsymbol{x}_2$ to denote the componentwise product of $\boldsymbol{x}_1$ and $\boldsymbol{x}_2$. This does not risk confusion with matrix products because $\boldsymbol{x}_1\boldsymbol{x}_2$ would not be meaningful as a matrix product. The meaningful matrix products are $\boldsymbol{x}_1^T\boldsymbol{x}_2$, which is a scalar, and $\boldsymbol{x}_1\boldsymbol{x}_2^T$, which is an $n \times n$ matrix.

2

Basics of Factorial Designs

We give in this chapter a fairly broad account of two-level factorial designs. Linear models are introduced for factorial effects, thereby laying a foundation for discussing several fundamental design concepts: estimability, optimality, aliasing, and contamination. As seen throughout, orthogonality offers a key to the understanding of these concepts. We introduce orthogonal arrays and discuss their statistical properties. Hadamard matrices also receive a treatment due to their close connections with orthogonal arrays. Through aliasing structures, nonregular factorial designs, the focus of this book, are distinguished from regular factorial designs.

2.1 Introduction

Suppose that in a scientific or technological investigation, several variables, called factors, are to be studied to see how they may affect a variable of ultimate interest, called a response variable. Two levels are often chosen for each factor, which is especially the case where a large number of factors are examined at a preliminary stage of an investigation. A convenient choice is to code the two levels as ± 1, with -1 called the low level and $+1$ the high level. The low and high levels are used purely for easy referencing and have no bearing on the physical meaning of magnitude for a factor.

When m factors are studied at two levels, there is a total of 2^m level combinations, also called treatment combinations. Ideally, one would like to conduct one experiment, called a run, at each level combination, and the resulting design is a complete 2^m factorial. Even better is to have several replicates of a complete 2^m factorial. A complete factorial is also called a full factorial, and we use the two terms interchangeably. Unless m is very small, however, a complete 2^m factorial is prohibitively expensive as the number 2^m of runs required quickly becomes exceedingly large as m increases. This necessitates the use of fractional factorials. Any design is a fractional factorial if it does not contain all 2^m level combinations.

Formally, a two-level factorial design can be represented by an $n \times m$ matrix $\boldsymbol{D} = (d_{ij})_{n \times m}$ with entries $d_{ij} = \pm 1$. While a column of $\boldsymbol{D}$ corresponds to a factor, a row of $\boldsymbol{D}$ specifies a run. Thus, $\boldsymbol{D}$ is a two-level factorial design of n runs for m factors. Design $\boldsymbol{D}$ is a complete 2^m factorial if it has all the 2^m level combinations exactly once as its rows and is a fractional factorial if at least one level combination is missing from the rows of $\boldsymbol{D}$. A row of $\boldsymbol{D}$ may occur more than once and is called a repeated run if this happens. Two runs are mirror images of each other if one can be obtained by switching the signs of the other. For the two levels, we sometimes use $\pm$ in place of ± 1 to save space, especially when designs are displayed as matrices.

DOI: 10.1201/9781003371861-2

Example 2.1. *Consider three factorial designs:*

$$\boldsymbol{D}_1 = \begin{bmatrix} -1 & -1 & -1 \\ -1 & -1 & 1 \\ -1 & 1 & -1 \\ -1 & 1 & 1 \\ 1 & -1 & -1 \\ 1 & -1 & 1 \\ 1 & 1 & -1 \\ 1 & 1 & 1 \end{bmatrix}, \quad \boldsymbol{D}_2 = \begin{bmatrix} -1 & -1 & -1 & -1 \\ -1 & -1 & 1 & 1 \\ -1 & 1 & -1 & 1 \\ -1 & 1 & 1 & -1 \\ 1 & -1 & -1 & 1 \\ 1 & -1 & 1 & -1 \\ 1 & 1 & -1 & -1 \\ 1 & 1 & 1 & 1 \end{bmatrix}, \quad \boldsymbol{D}_3 = \begin{bmatrix} -1 & -1 & 1 \\ -1 & -1 & 1 \\ -1 & 1 & -1 \\ -1 & 1 & -1 \\ 1 & -1 & -1 \\ 1 & -1 & 1 \\ 1 & 1 & 1 \end{bmatrix}.$$

We see that design $\boldsymbol{D}_1$ is a full factorial with $n = 8$ runs for $m = 3$ factors, and $\boldsymbol{D}_2$ is a fractional factorial of 8 runs for four factors, and $\boldsymbol{D}_3$ is a fractional factorial of 7 runs for three factors. Whereas designs $\boldsymbol{D}_1$ and $\boldsymbol{D}_2$ do not have repeated runs, design $\boldsymbol{D}_3$ has a repeated run $(-1, -1, 1)$. Both $\boldsymbol{D}_2$ and $\boldsymbol{D}_3$ contain runs that are mirror images of each other. For example, in design $\boldsymbol{D}_2$, the first run is a mirror image of the last run and vice versa.

2.2 Linear Models

Consider a factorial design of n runs for m factors, $\boldsymbol{D} = (d_{ij})_{n\times m}$ with $d_{ij} = \pm 1$. Let $\boldsymbol{d}_j$ be the jth column of $\boldsymbol{D}$. Then $\boldsymbol{D} = (\boldsymbol{d}_1, \ldots, \boldsymbol{d}_m)$. Data collected using this design can be analyzed using a linear model:

$$\boldsymbol{Y} = \boldsymbol{X}\boldsymbol{\beta} + \boldsymbol{\epsilon}, \tag{2.1}$$

where $\boldsymbol{Y} = (Y_1, \ldots, Y_n)^T$ denotes the vector of observations on a response variable, $\boldsymbol{X} = (x_{ij})_{n\times p}$ is the model matrix, $\boldsymbol{\beta} = (\beta_1, \ldots, \beta_p)^T$ is the vector of unknown parameters which may consist of an intercept and various factorial effects, and $\boldsymbol{\epsilon} = (\epsilon_1, \ldots, \epsilon_n)^T$ is the vector of random errors assumed to be uncorrelated and have a zero mean and a constant variance σ^2. The assumptions on $\boldsymbol{\epsilon}$ can be written in matrix form as

$$\mathrm{E}(\boldsymbol{\epsilon}) = \boldsymbol{0}_n \text{ and } \mathrm{Var}(\boldsymbol{\epsilon}) = \sigma^2 \boldsymbol{I}_n, \tag{2.2}$$

where $\boldsymbol{0}_n$ is the vector of n zeros, $\boldsymbol{I}_n$ is the identity matrix of order n, and $\mathrm{Var}(\boldsymbol{\epsilon})$ denotes the variance-covariance matrix of $\boldsymbol{\epsilon}$.

The variance-covariance matrix of a random vector $\boldsymbol{x}$ is also called the covariance matrix of $\boldsymbol{x}$. Besides $\mathrm{Var}(\boldsymbol{x})$, another commonly used notation is $\mathrm{cov}(\boldsymbol{x})$, which we will use in Chapters 5, 7, and 11.

Using the case of $m = 3$, we illustrate various possibilities for (2.1). If the main-effect model is considered, then

$$\boldsymbol{X} = (\boldsymbol{1}_n, \boldsymbol{d}_1, \boldsymbol{d}_2, \boldsymbol{d}_3) \text{ and } \boldsymbol{\beta} = (\beta_0, \beta_1, \beta_2, \beta_3)^T,$$

with β_0 being the intercept and β_1, β_2, and β_3 denoting the main effects of the three factors. In the above, column $\boldsymbol{1}_n$ is a vector of n ones.

If all the factorial effects are of interest, we obtain the full model in which case

$$\boldsymbol{X} = (\boldsymbol{1}_n, \boldsymbol{d}_1, \boldsymbol{d}_2, \boldsymbol{d}_3, \boldsymbol{d}_1\boldsymbol{d}_2, \boldsymbol{d}_1\boldsymbol{d}_3, \boldsymbol{d}_2\boldsymbol{d}_3, \boldsymbol{d}_1\boldsymbol{d}_2\boldsymbol{d}_3)$$

and

$$\boldsymbol{\beta} = (\beta_0, \beta_1, \beta_2, \beta_3, \beta_{12}, \beta_{13}, \beta_{23}, \beta_{123})^T,$$

where, for example, $\boldsymbol{d}_1\boldsymbol{d}_2$ is the componentwise product of columns $\boldsymbol{d}_1$ and $\boldsymbol{d}_2$, and β_{12} represents the interaction effect between factors 1 and 2.

There are many possibilities intermediate between the main-effect model and the full model. For example, one may consider a model consisting of the intercept, all the main effects, and one two-factor interaction β_{13}. Then we have that

$$\boldsymbol{X} = (\boldsymbol{1}_n, \boldsymbol{d}_1, \boldsymbol{d}_2, \boldsymbol{d}_3, \boldsymbol{d}_1\boldsymbol{d}_3) \text{ and } \boldsymbol{\beta} = (\beta_0, \beta_1, \beta_2, \beta_3, \beta_{13})^T.$$

Using design $\boldsymbol{D}_1$ in Example 2.1, we give a full display of model (2.1) for this case:

$$\begin{bmatrix} Y_1 \\ Y_2 \\ Y_3 \\ Y_4 \\ Y_5 \\ Y_6 \\ Y_7 \\ Y_8 \end{bmatrix} = \begin{bmatrix} 1 & -1 & -1 & -1 & 1 \\ 1 & -1 & -1 & 1 & -1 \\ 1 & -1 & 1 & -1 & 1 \\ 1 & -1 & 1 & 1 & -1 \\ 1 & 1 & -1 & -1 & -1 \\ 1 & 1 & -1 & 1 & 1 \\ 1 & 1 & 1 & -1 & -1 \\ 1 & 1 & 1 & 1 & 1 \end{bmatrix} \begin{bmatrix} \beta_0 \\ \beta_1 \\ \beta_2 \\ \beta_3 \\ \beta_{13} \end{bmatrix} + \begin{bmatrix} \epsilon_1 \\ \epsilon_2 \\ \epsilon_3 \\ \epsilon_4 \\ \epsilon_5 \\ \epsilon_6 \\ \epsilon_7 \\ \epsilon_8 \end{bmatrix}.$$

For $\boldsymbol{\beta}$ in (2.1) to be estimable, $\boldsymbol{X}^T\boldsymbol{X}$ needs to be invertible. Suppose this is the case. Then the well-known Gauss-Markov theorem says that the best linear unbiased estimator of $\boldsymbol{\beta}$ is the least squares estimator, given by $\hat{\boldsymbol{\beta}} = (\boldsymbol{X}^T\boldsymbol{X})^{-1}\boldsymbol{X}^T\boldsymbol{Y}$, which is unbiased and has a variance-covariance matrix $\text{Var}(\hat{\boldsymbol{\beta}}) = \sigma^2(\boldsymbol{X}^T\boldsymbol{X})^{-1}$. We present this standard result from the theory of linear models in a theorem.

Theorem 2.1. *Consider the linear model in* (2.1) *under the assumptions in* (2.2). *Suppose that* $\boldsymbol{X}^T\boldsymbol{X}$ *is invertible. Then the least squares estimator* $\hat{\boldsymbol{\beta}} = (\boldsymbol{X}^T\boldsymbol{X})^{-1}\boldsymbol{X}^T\boldsymbol{Y}$ *is the best linear unbiased estimator for* $\boldsymbol{\beta}$ *and*

$$\text{Var}(\hat{\boldsymbol{\beta}}) = \sigma^2(\boldsymbol{X}^T\boldsymbol{X})^{-1}. \tag{2.3}$$

The error variance σ^2 is generally unknown and has to be estimated. The standard estimator of σ^2 is the mean residual sum of squares

$$\hat{\sigma}^2 = (\boldsymbol{Y} - \boldsymbol{X}\hat{\boldsymbol{\beta}})^T(\boldsymbol{Y} - \boldsymbol{X}\hat{\boldsymbol{\beta}})/(n-p).$$

Inferences about $\boldsymbol{\beta}$ or, more generally, linear functions of $\boldsymbol{\beta}$ can then be carried out using standard theory in a straightforward manner. Inferences are exact if normality is assumed for the response variable and approximate otherwise.

When $n = p$, the model in (2.1) becomes saturated and there is no degree of freedom left for estimating σ^2. However, one can still obtain a reasonable estimate for σ^2 if effect sparsity is assumed, meaning that only a few out of a large number of effects are actually active. One example of such situation is a single replicate of a full factorial design when one is uncomfortable with assuming negligibility of any specific set of factorial effects but feels fine with effect sparsity. In this case, an estimate of σ^2 can be obtained by employing half-normal probability plots (Daniel, 1959) or, alternatively, by using Lenth's method (Lenth, 1989).

When $\boldsymbol{X}^T\boldsymbol{X}$ is singular, $\boldsymbol{\beta}$ is not estimable. Although not all effects in $\boldsymbol{\beta}$ can be estimated, certain linear functions of $\boldsymbol{\beta}$ are estimable. There exists a simple characterization for estimable linear functions, which states that a linear function $\boldsymbol{a}^T\boldsymbol{\beta}$ is estimable if and only if $\boldsymbol{a}^T$ is a linear combination of the row vectors of $\boldsymbol{X}$. A formal theory for making inferences about estimable linear functions can be developed using generalized inverses. See Cheng (2014a, Chapter 2) for a reference.

Factorial effects are introduced as coefficients in the linear model (2.1). The reader should be fine with most of the developments with this intuitive understanding of factorial effects. To have a full appreciation of the meanings of factorial effects, the reader is referred to Section 2.8 where a formal definition is given. Factorial effects are in fact a special set of mutually orthogonal contrasts of the 2^m treatment means, independent of a factorial design $\boldsymbol{D}$ chosen to use and a specific linear model postulated in (2.1).

In model (2.1) under the assumptions in (2.2), $\boldsymbol{\beta}$ consists of the factorial effects of interest. This model also entails the negligibility of all factorial effects that are not in $\boldsymbol{\beta}$. The following situation, however, is more common in practice: we are interested in estimating the effects in $\boldsymbol{\beta}_1$ in the presence of nonnegligible effects in $\boldsymbol{\beta}_2$, assuming that all other effects are negligible. This gives rise to

$$\boldsymbol{Y} = \boldsymbol{X}_1\boldsymbol{\beta}_1 + \boldsymbol{X}_2\boldsymbol{\beta}_2 + \boldsymbol{\epsilon}, \tag{2.4}$$

where $\boldsymbol{X}_1$ and $\boldsymbol{X}_2$ are model matrices of $\boldsymbol{\beta}_1$ and $\boldsymbol{\beta}_2$, respectively, and $\boldsymbol{\epsilon}$ still satisfies the assumptions in (2.2). Here we want to estimate $\boldsymbol{\beta}_1$, given that the nuisance parameters $\boldsymbol{\beta}_2$ are needed to validate the model (2.4). In the present situation, Theorem 2.1 has the following counterpart.

Theorem 2.2. *Consider the linear model in* (2.4) *under the assumptions in* (2.2). *Then* $\boldsymbol{\beta}_1$ *is estimable if and only if* $\boldsymbol{X}_1^T\boldsymbol{X}_1 - \boldsymbol{X}_1^T\boldsymbol{P}_{\boldsymbol{X}_2}\boldsymbol{X}_1$ *is invertible, where* $\boldsymbol{P}_{\boldsymbol{X}_2}$ *is the projection matrix onto the linear subspace spanned by the columns of* $\boldsymbol{X}_2$. *In this case the best linear unbiased estimator for* $\boldsymbol{\beta}_1$ *is given by*

$$\hat{\boldsymbol{\beta}}_1 = (\boldsymbol{X}_1^T\boldsymbol{X}_1 - \boldsymbol{X}_1^T\boldsymbol{P}_{\boldsymbol{X}_2}\boldsymbol{X}_1)^{-1}(\boldsymbol{X}_1 - \boldsymbol{P}_{\boldsymbol{X}_2}\boldsymbol{X}_1)^T\boldsymbol{Y},$$

which has

$$\mathrm{Var}(\hat{\boldsymbol{\beta}}_1) = \sigma^2(\boldsymbol{X}_1^T\boldsymbol{X}_1 - \boldsymbol{X}_1^T\boldsymbol{P}_{\boldsymbol{X}_2}\boldsymbol{X}_1)^{-1}. \tag{2.5}$$

The projection matrix $\boldsymbol{P}_{\boldsymbol{X}_2}$ has an explicit form of $\boldsymbol{P}_{\boldsymbol{X}_2} = \boldsymbol{X}_2(\boldsymbol{X}_2^T\boldsymbol{X}_2)^{-}\boldsymbol{X}_2^T$, often called a hat matrix in a linear regression course. In the above expression for $\boldsymbol{P}_{\boldsymbol{X}_2}$, $\boldsymbol{A}^{-}$ denotes a generalized inverse of $\boldsymbol{A}$, which is defined as any matrix $\boldsymbol{B}$ that satisfies $\boldsymbol{ABA} = \boldsymbol{A}$.

2.3 Optimality Criteria

Consider the linear model in (2.1). The least squares estimator $\hat{\boldsymbol{\beta}}$ of $\boldsymbol{\beta}$ has a variance-covariance matrix $\mathrm{Var}(\hat{\boldsymbol{\beta}}) = \sigma^2(\boldsymbol{X}^T\boldsymbol{X})^{-1}$ as given in (2.3), which depends on a factorial design $\boldsymbol{D}$ that is used to collect the data. The optimal design problem is to choose $\boldsymbol{D}$ among all competing designs to minimize a certain meaningful functional of $\mathrm{Var}(\hat{\boldsymbol{\beta}})$.

The most popular is the D-optimality criterion that aims at minimizing $\det[\mathrm{Var}(\hat{\boldsymbol{\beta}})]$, the determinant of $\mathrm{Var}(\hat{\boldsymbol{\beta}})$, which is proportional to the volume of a confidence ellipsoid for $\boldsymbol{\beta}$ under the normality assumption. Alternatively, one can use the A-optimality criterion requiring the minimization of $\mathrm{tr}[\mathrm{Var}(\hat{\boldsymbol{\beta}})]$, the trace of $\mathrm{Var}(\hat{\boldsymbol{\beta}})$, which is proportional to the average of the variances of the individual estimators in the vector $\hat{\boldsymbol{\beta}}$. Another often-used criterion is the E-optimality criterion, which minimizes the largest eigenvalue of $\mathrm{Var}(\hat{\boldsymbol{\beta}})$. The largest eigenvalue of $\mathrm{Var}(\hat{\boldsymbol{\beta}})$ has an interpretation of $\max_{\|\mathbf{c}\|=1} \mathrm{Var}(\boldsymbol{c}^T\hat{\boldsymbol{\beta}})$, which is the variance of the estimator of a linear parameter in the worst direction.

These three criteria can also be viewed as functionals of the information matrix $\boldsymbol{M} = \sigma^2[\mathrm{Var}(\hat{\boldsymbol{\beta}})]^{-1}$. In the current setting, we have $\boldsymbol{M} = \boldsymbol{X}^T\boldsymbol{X}$. Let $\lambda_1, \ldots, \lambda_p$ be the eigenvalues of the information matrix $\boldsymbol{M}$. Then the eigenvalues of $\boldsymbol{M}^{-1}$ are $\lambda_1^{-1}, \ldots, \lambda_p^{-1}$. Equivalently, the three criteria can all be described in terms of the eigenvalues $\lambda_1, \ldots, \lambda_p$ of $\boldsymbol{M}$. The D-criterion is to maximize $\lambda_1 \cdots \lambda_p$, the A-criterion is to minimize $\lambda_1^{-1} + \cdots + \lambda_p^{-1}$, and the E-criterion is to maximize $\min(\lambda_1, \ldots, \lambda_p)$.

Kiefer (1974) introduced a class of Φ_s-criteria for $s \geq 0$, which is defined as

$$\Phi_s(\boldsymbol{M}) = \begin{cases} [p^{-1}\text{tr}(\boldsymbol{M}^{-s})]^{1/s} & \text{for } 0 < s < \infty, \\ (\det(\boldsymbol{M}^{-1}))^{1/p} & \text{for } s = 0, \\ \text{the largest eigenvalue of } \boldsymbol{M}^{-1} & \text{for } s = \infty. \end{cases}$$

The Φ_s-criteria include the commonly used D-, A- and E-criteria as special cases, since they are equivalent to $\Phi_0 = \lim_{s\to 0}\Phi_s$, Φ_1, and $\Phi_\infty = \lim_{s\to\infty}\Phi_s$, respectively. The equivalencies can easily be established by noting that

$$\text{tr}(\boldsymbol{M}^{-s}) = \lambda_1^{-s} + \cdots + \lambda_p^{-s} \text{ for } 0 < s < \infty$$

and then considering the limiting behaviors of Φ_s as $s \to 0$ and $s \to \infty$.

Kiefer (1975) proved the following general result that also applies to settings other than factorial designs.

Theorem 2.3. *If a design maximizes tr($\boldsymbol{M}$) among all competing designs and its information matrix is of the form $a\boldsymbol{I}$, where $\boldsymbol{I}$ is the identity matrix of order p, then it minimizes $\Phi(\boldsymbol{M})$ among all competing designs for all real-valued functions Φ such that*

(i) Φ is convex,

(ii) Φ is orthogonally invariant, i.e. $\Phi(\boldsymbol{M}) = \Phi(\boldsymbol{O}^T\boldsymbol{M}\boldsymbol{O})$ for all orthogonal matrices $\boldsymbol{O}$, and

(iii) For fixed $\boldsymbol{M}$, $\Phi(c\boldsymbol{M})$ is nonincreasing in positive scalars c.

Proof. Since Φ is orthogonally invariant, it depends on $\boldsymbol{M}$ through its eigenvalues. Without loss of generality, we may view Φ as a function of $(\lambda_1, \ldots, \lambda_p)$, where $\lambda_1, \ldots, \lambda_p$ are the eigenvalues of $\boldsymbol{M}$. Furthermore, $\Phi(\lambda_1, \ldots, \lambda_p) = \Phi(\lambda_{\pi(1)}, \ldots, \lambda_{\pi(p)})$ for all permutations π of $1, \ldots, p$. We are given a design such that $\boldsymbol{M} = a\boldsymbol{I}$ and that it maximizes tr($\boldsymbol{M}$). Let the eigenvalues of this information matrix be $\lambda_1^*, \ldots, \lambda_p^*$. Then $\lambda_1^* = \cdots = \lambda_p^* = a$, and $\sum_{i=1}^p \lambda_i \leq \sum_{i=1}^p \lambda_i^* = pa$, where $\lambda_1, \ldots, \lambda_p$ are the eigenvalues of the information matrix of any other design. We need to show that $\Phi(\lambda_1^*, \ldots, \lambda_p^*) \leq \Phi(\lambda_1, \ldots, \lambda_p)$. Let S_p be the set of all the $p!$ permutations of $1, \ldots, p$. Then

$$\begin{aligned} \Phi(\lambda_1, \ldots, \lambda_p) &= \tfrac{1}{p!}\textstyle\sum_{\pi\in S_p} \Phi(\lambda_{\pi(1)}, \ldots, \lambda_{\pi(p)}) \\ &\geq \Phi(\tfrac{1}{p!}\textstyle\sum_{\pi\in S_p} \lambda_{\pi(1)}, \ldots, \tfrac{1}{p!}\textstyle\sum_{\pi\in S_p} \lambda_{\pi(p)}) \\ &= \Phi(\tfrac{1}{p}\textstyle\sum_{i=1}^p \lambda_i, \ldots, \tfrac{1}{p}\textstyle\sum_{i=1}^p \lambda_i) \\ &\geq \Phi(a, \ldots, a), \end{aligned}$$

where the first equality and inequality follow from, respectively, the orthogonal invariance and convexity of Φ, and the last inequality follows from (iii) in Theorem 2.3 and the assumption that $\sum_{i=1}^p \lambda_i \leq pa$. □

In the optimal design literature, a design that is optimal with respect to all the criteria satisfying conditions (i)-(iii) in Theorem 2.3 is called *universally optimal.* Because all the Φ_s-criteria satisfy conditions (i)-(iii) in Theorem 2.3, which one can easily verify, a universally optimal design must also be Φ_s-optimal. Applying Theorem 2.3 to the current setting of two-level factorial designs, we have the following result.

Corollary 2.1. *If a factorial design $\boldsymbol{D}$ is such that the model matrix $\boldsymbol{X}$ has orthogonal columns, then it is universally optimal; in particular, it is Φ_s-optimal for all $s \geq 0$.*

Proof. Since the entries of $\boldsymbol{X}$ are 1 and -1, all the diagonal entries of the information matrix $\boldsymbol{M} = \boldsymbol{X}^T\boldsymbol{X}$ are equal to n. Thus $\text{tr}(\boldsymbol{M})$ is a constant. If a design $\boldsymbol{D}$ is such that the model matrix $\boldsymbol{X}$ has orthogonal columns, then $\boldsymbol{M} = n\boldsymbol{I}$. It follows from Theorem 2.3 that such a design $\boldsymbol{D}$ is universally optimal and, in particular, it is Φ_s-optimal for all $s \geq 0$. □

Remark 2.1. *For an arbitrary factorial design, since the diagonal entries of $\boldsymbol{M}$ are equal to n, by a result from linear algebra, all the diagonal entries of $\boldsymbol{M}^{-1}$ must be at least $1/n$. This lower bound is achieved by designs with $\boldsymbol{M} = n\boldsymbol{I}$, i.e. those whose model matrices have orthogonal columns. Thus the designs whose model matrices have orthogonal columns are not only universally optimal; they also minimize $\text{Var}(\hat{\beta}_j)$ for all $j = 1, \ldots, p$.*

Our discussion so far has been on estimating $\boldsymbol{\beta}$ in model (2.1). Now suppose we are interested only in estimating $\boldsymbol{\beta}_1$ in model (2.4) with $\boldsymbol{\beta}_2$ as nuisance parameters. By (2.5), the information matrix in this case is $\boldsymbol{M} = \sigma^2[\text{Var}(\hat{\boldsymbol{\beta}}_1)]^{-1} = \boldsymbol{X}_1^T\boldsymbol{X}_1 - \boldsymbol{X}_1^T\boldsymbol{P}_{\boldsymbol{X}_2}\boldsymbol{X}_1$. All the criteria introduced and Theorem 2.3 apply to this new information matrix without change. But Corollary 2.1 needs a modification.

Corollary 2.2. *Under the model $\boldsymbol{Y} = \boldsymbol{X}_1\boldsymbol{\beta}_1 + \boldsymbol{X}_2\boldsymbol{\beta}_2 + \boldsymbol{\epsilon}$ as in (2.4), a factorial design $\boldsymbol{D}$ is universally optimal for estimating $\boldsymbol{\beta}_1$ if its model matrix $(\boldsymbol{X}_1, \boldsymbol{X}_2)$ is such that $\boldsymbol{X}_1$ has orthogonal columns and the columns of $\boldsymbol{X}_1$ are orthogonal to those of $\boldsymbol{X}_2$.*

Proof. We apply Theorem 2.3 to the current information matrix $\boldsymbol{M} = \boldsymbol{X}_1^T\boldsymbol{X}_1 - \boldsymbol{X}_1^T\boldsymbol{P}_{\boldsymbol{X}_2}\boldsymbol{X}_1$. Since $\boldsymbol{X}_1$ has orthogonal columns and the columns of $\boldsymbol{X}_1$ are orthogonal to those of $\boldsymbol{X}_2$, we obtain that $\boldsymbol{M} = n\boldsymbol{I}$. Suppose $\boldsymbol{M}$ is $k \times k$. By Theorem 2.3, it suffices to show that $\text{tr}(\boldsymbol{M}') \leq nk$, where $\boldsymbol{M}'$ is the information matrix of an arbitrary design. Because $\boldsymbol{X}_1^T\boldsymbol{P}_{\boldsymbol{X}_2}\boldsymbol{X}_1$ is nonnegative definite, we must have that $\text{tr}(\boldsymbol{M}') \leq \text{tr}(\boldsymbol{X}_1^T\boldsymbol{X}_1) = nk$. □

Another optimality criterion, the (M.S)-criterion, plays a significant role in several of the later chapters. It will be introduced in due course. In discussing nonorthogonal designs in Chapter 9, we also need a class of criteria called type 1 criteria.

The family of Φ_s-criteria was extended by Pukelsheim (1980) to what he called j_a-criteria. For the information matrix $\boldsymbol{M}$ of a design, the j_a-criteria are defined for all a such that $a \in [-\infty, 1]$, and the objective is to maximize $j_a(\boldsymbol{M}) = \left(n^{-1}\sum_{i=1}^{n}\lambda_i^a\right)^{1/a}$. Specifically for $a \neq 0$ or $-\infty$, this is to minimize $\text{tr}(\boldsymbol{M}^a)$ for $a < 0$ and to maximize $\text{tr}(\boldsymbol{M}^a)$ for $a > 0$. Therefore for $a < 0$, the j_a criteria are equivalent to the Φ_{-a}-criteria. In particular, the j_{-1}-criterion is the same as the A-criterion. By taking limits, we see that for $a = 0$ and $-\infty$, the j_a criteria are equivalent to the D- and E-criteria, respectively. For $a = 1$, to maximize $j_1(\boldsymbol{M})$ is the same as to maximize $\text{tr}(\boldsymbol{M})$, one of the sufficient conditions for universal optimality. While the Φ_s-criteria are convex functions for all $s \geq 0$, the j_a criteria are concave functions for all $a \in [-\infty, 1]$. We will need the j_a-criteria in Section 9.6.

2.4 Orthogonal Arrays and Their Optimality

Orthogonal arrays are fascinating design objects. They were first introduced by Rao (1947) as generalizations of sets of mutually orthogonal Latin squares. An orthogonal array is not just more general than a set of mutually orthogonal Latin squares but also has a more concise and elegant form than the latter. This historical origin gives the first application of orthogonal arrays in comparative experiments as they can be utilized to block experimental material in several directions. Orthogonal arrays have enjoyed tremendous success in computer experiments for their usefulness in constructing space-filling designs (Owen, 1992; Tang, 1993), a topic we will look into in the last chapter. Otherwise, this book is concerned with the use of orthogonal arrays as factorial designs. Readers who wish to learn more about orthogonal arrays with an emphasis on their existence and construction should read a treatise on the subject by Hedayat, Sloane, and Stufken (1999). While multi-level and mixed-level orthogonal arrays will be covered in Chapters 10-12, we focus on two-level orthogonal arrays for the time being.

Definition 2.1. *A two-level factorial design* $\boldsymbol{D}$ *of* n *runs for* m *factors is said to be an orthogonal array of strength* t *if, for any of its submatrices of* t *columns, the* 2^t *level combinations occur as rows of the submatrix with the same frequency. This array is denoted by* $OA(n, 2^m, t)$.

If $\boldsymbol{D}$ is an $\mathrm{OA}(n, 2^m, t)$, then n must be a multiple of 2^t, meaning that $n = \lambda 2^t$ for some integer λ. The λ here is called the index of the array. From the definition, it is immediate that if $\boldsymbol{D}$ is an orthogonal array of strength t, it must also be of strength t' for any $t' < t$. The example designs in Chapter 1 are all orthogonal arrays of strength two. Consider the three designs in Example 2.1. Designs $\boldsymbol{D}_1$ and $\boldsymbol{D}_2$ are both orthogonal arrays of strength three. Design $\boldsymbol{D}_3$ is not an orthogonal array of strength t for any $t \geq 1$.

Regular two-level factorial designs are orthogonal arrays. In a resolution R design, any $R - 1$ columns are independent as there can be no defining words among them, and hence form a full factorial design or several replicates thereof. This says that a regular factorial design of resolution R is an orthogonal array of strength $R - 1$.

As factorial designs, orthogonal arrays enjoy many appealing statistical properties, one of which is their universal optimality. Throughout the book, we use $\lfloor x \rfloor$, the floor function, to denote the greatest integer not exceeding x.

Theorem 2.4. *If design* $\boldsymbol{D} = (d_{ij})_{n \times m}$ *is an orthogonal array of strength* t*, then it is universally optimal for any model consisting of factorial effects involving at most* $\lfloor t/2 \rfloor$ *factors.*

Proof. Consider a model consisting of factorial effects involving at most $\lfloor t/2 \rfloor$ factors. By Corollary 2.1, we need to show that the model matrix $\boldsymbol{X}$ has orthogonal columns. The column-orthogonality of $\boldsymbol{X}$ follows if we can establish that

$$\sum_{i=1}^{n} d_{ij_1} d_{ij_2} \cdots d_{ij_k} = 0, \tag{2.6}$$

for any $j_1 < j_2 < \cdots < j_k$ such that $k \leq t$. Equation (2.6) can easily be proved by an induction on k in the background of knowing that any t columns of $\boldsymbol{D}$ form a full factorial design or several replicates thereof. □

An orthogonal array of odd strength can accomplish more than just what Theorem 2.4 claims.

Theorem 2.5. *An orthogonal array of odd strength* t *is universally optimal for the estimation of factorial effects involving up to* $\lfloor t/2 \rfloor$ *factors under any model consisting of factorial effects involving at most* $\lfloor t/2 \rfloor + 1$ *factors.*

Proof. Let β_1 consist of the factorial effects involving up to $\lfloor t/2 \rfloor$ factors, β_2 consist of those involving $\lfloor t/2 \rfloor + 1$ factors, $\boldsymbol{X}_1$ consist of the columns corresponding to the effects in β_1, and $\boldsymbol{X}_2$ consist of those corresponding to the effects of β_2. By equation (2.6), we have that the columns of $\boldsymbol{X}_1$ are mutually orthogonal, and are also orthogonal to those of $\boldsymbol{X}_2$. Theorem 2.5 now follows from Corollary 2.2. □

We look at several cases of Theorems 2.4 and 2.5. If $\boldsymbol{D}$ is an orthogonal array of strength two or three, Theorem 2.4 says that it is universally optimal for the main-effect model, which consists of the intercept and main effects. If $\boldsymbol{D}$ has strength four, then it is universally optimal for the model containing the intercept, main effects, and all two-factor interactions. According to Theorem 2.5, an orthogonal array of strength three is also universally optimal for the estimation of the main effects even if the model contains two-factor interactions. In fact, even more can be said about orthogonal arrays of strength three. A strength-three orthogonal array is also universally optimal for a model comprising the intercept, main effects, and all those two-factor interactions involving one particular factor, say the first factor. This is because the columns of the model matrix

$$\boldsymbol{X} = (\boldsymbol{1}_n, \boldsymbol{d}_1, \ldots, \boldsymbol{d}_m, \boldsymbol{d}_1\boldsymbol{d}_2, \ldots, \boldsymbol{d}_1\boldsymbol{d}_m)$$

are mutually orthogonal, where $\boldsymbol{d}_j$ is the jth column of $\boldsymbol{D}$. This result deserves to be documented as a theorem.

Theorem 2.6. *An orthogonal array of strength three is universally optimal for a model comprising the intercept, main effects, and all those two-factor interactions involving any particular factor.*

When orthogonal arrays do not exist, the problem of determining optimal designs is usually difficult. There are only some limited results. For example, Cheng (1980) showed that if there exists an $\mathrm{OA}(n, 2^m, t)$ where t is even, then the design obtained by adding a run in which all the factors are at level 1 to an $\mathrm{OA}(n, 2^m, t)$ is optimal for the estimation of factorial effects involving up to $t/2$ factors among all possible designs of run size $n+1$ with respect to a large class of criteria that include all the Φ_s-criteria, although such designs are not universally optimal. In fact, such optimality property holds for any design obtained by adding an *arbitrary* run to an $\mathrm{OA}(n, 2^m, t)$. Whereas general theoretical results are scarce, designs whose information matrices are close to $n\boldsymbol{I}$ (say the columns of the model matrices are nearly orthogonal) are expected to be highly efficient.

We point out that the optimality criteria considered here are model-dependent. In practice, especially at the initial stage of experimentation, the model is often uncertain. This will be the main focus of the book. Appropriate criteria for choosing model-robust designs will be introduced and studied in later chapters.

2.5 Hadamard Matrices

Hadamard matrices are beautiful mathematical objects that have enormous applications in science and technology. They can be used to construct error-correcting codes, which help send stunning pictures of deep space back to the Earth. In survey sampling, the balanced repeated replication, a method of variance estimation, is based on Hadamard matrices. Hadamard matrices are particularly popular among the design community for their intimate relationships with various types of designs: weighing designs, balanced incomplete block designs, and orthogonal arrays, just to name a few. It is the connection with orthogonal arrays, and hence factorial designs, that the book needs this section to supply some rudimentary knowledge about Hadamard matrices.

We first provide a coverage of some basics of Hadamard matrices that include the definition, existence and isomorphism, and then review the most useful methods of construction. For more details, we refer to Dey and Mukerjee (1999), Hedayat, Sloane, and Stufken (1999), and Cheng (2014a). Highly recommended is a reading of Chapter 7 of Hedayat, Sloane, and Stufken (1999). For a treatise on Hadamard matrices and their generalizations, we refer to Horadam (2007).

A Hadamard matrix $\boldsymbol{H}$ of order n is an $n \times n$ square matrix with entries ± 1 such that $\boldsymbol{H}^T\boldsymbol{H} = \boldsymbol{H}\boldsymbol{H}^T = n\boldsymbol{I}$, where $\boldsymbol{I}$ is the identity matrix of order n. The column vectors of a Hadamard matrix are mutually orthogonal, which follows from $\boldsymbol{H}^T\boldsymbol{H} = n\boldsymbol{I}$, and so are its row vectors due to $\boldsymbol{H}\boldsymbol{H}^T = n\boldsymbol{I}$. It is obvious that the transpose of a Hadamard matrix is also a Hadamard matrix. The following are Hadamard matrices of the smallest orders $n = 1, 2, 4$:

$$\begin{bmatrix} 1 \end{bmatrix}, \quad \begin{bmatrix} 1 & 1 \\ 1 & -1 \end{bmatrix}, \quad \begin{bmatrix} 1 & -1 & -1 & 1 \\ 1 & -1 & 1 & -1 \\ 1 & 1 & -1 & -1 \\ 1 & 1 & 1 & 1 \end{bmatrix}.$$

Notably missing from the above is order 3. This is not a coincidence due to a fundamental existence result.

Theorem 2.7. *If a Hadamard matrix of order $n \geq 3$ exists, then n must be a multiple of 4.*

Proof. Suppose that a Hadamard matrix $\boldsymbol{H}$ of order $n \geq 3$ exists. Let $\boldsymbol{a}, \boldsymbol{b}, \boldsymbol{c}$ be its first three column vectors. We may assume that $\boldsymbol{a}$ has entries of all plus ones, for if $\boldsymbol{a}$ has an entry of -1, we can multiply the entire row of $\boldsymbol{H}$ by -1, which converts this entry to $+1$, and the resulting matrix is still a Hadamard matrix because the multiplication by -1 does not change the orthogonality of column vectors. Let n_1, n_2, n_3, n_4 be the numbers of times the four pairs $(-1, -1)$, $(-1, 1)$, $(1, -1)$, and $(1, 1)$ occur in the matrix consisting of two columns $\boldsymbol{b}$ and $\boldsymbol{c}$. Then the orthogonality of $\boldsymbol{a}$ and $\boldsymbol{b}$ implies that $n_3 + n_4 - n_1 - n_2 = 0$, the orthogonality of $\boldsymbol{a}$ and $\boldsymbol{c}$ implies that $n_2 - n_1 + n_4 - n_3 = 0$, and the orthogonality of $\boldsymbol{b}$ and $\boldsymbol{c}$ implies that $n_1 - n_2 - n_3 + n_4 = 0$. Combining the above three equations with the obvious one that $n_1 + n_2 + n_3 + n_4 = n$, we obtain $n_1 = n_2 = n_3 = n_4 = n/4$, which shows that n is a multiple of 4 as n_1 is an integer. □

One wonders what happens with the converse of Theorem 2.7. More precisely, given that n is a multiple of 4, does a Hadamard matrix of order n exist? Unfortunately, this is neither proved nor disproved and is hence called the Hadamard conjecture in the literature. Although many infinite families of Hadamard matrices have been constructed, we still do not know whether or not the conjecture is true. As of today, there are three unresolved cases for $n \leq 1000$, and they are $n = 668, 792$, and 868.

Starting with a Hadamard matrix, many other Hadamard matrices can easily be obtained. Any of the isomorphism operations (i) row permutation, (ii) column permutation, (iii) sign-switching a row, and (iv) sign-switching a column, either individually or in combination, on a Hadamard matrix still gives a Hadamard matrix. This gives rise to the notion of isomorphic Hadamard matrices. Two Hadamard matrices are said to be isomorphic if one can be obtained from the other through a combination of isomorphism operations.

Finding all nonisomorphic Hadamard matrices for a given order is an interesting but very challenging problem. So far, the results are available for $n \leq 32$. There is only one nonisomorphic Hadamard matrix for each of $n = 1, 2, 4, 8$, and 12. The number of nonisomorphic Hadamard matrices for $n = 16, 20, 24, 28$, and 32 is 5, 3, 60, 487, and 13,710,027, respectively (Kharaghani and Tayfeh-Rezaie, 2013).

The simplest method of constructing Hadamard matrices is the tensor or Kronecker product construction.

Theorem 2.8. *Let $\boldsymbol{F} = (f_{ij})$ and $\boldsymbol{G}$ be two Hadamard matrices of orders n_1 and n_2, respectively. Then the tensor product $\boldsymbol{F} \otimes \boldsymbol{G} = (f_{ij}\boldsymbol{G})$ is a Hadamard matrix of order $n_1 n_2$.*

The tensor product method allows the construction of Hadamard matrices of large orders from those of small orders. An application of this method produces a series of Hadamard matrices of Sylvester type. Using a Hadamard matrix of order 2 as both $\boldsymbol{F}$ and $\boldsymbol{G}$, we obtain a Hadamard matrix of order 4, as shown below:

$$\begin{bmatrix} 1 & 1 \\ 1 & -1 \end{bmatrix} \otimes \begin{bmatrix} 1 & 1 \\ 1 & -1 \end{bmatrix} = \begin{bmatrix} 1 & 1 & 1 & 1 \\ 1 & -1 & 1 & -1 \\ 1 & 1 & -1 & -1 \\ 1 & -1 & -1 & 1 \end{bmatrix}.$$

Using a Hadamard matrix of order 2 as $\boldsymbol{F}$ and a Hadamard matrix of order 4 as $\boldsymbol{G}$, we obtain a Hadamard matrix of order 8:

$$\begin{bmatrix} 1 & 1 \\ 1 & -1 \end{bmatrix} \otimes \begin{bmatrix} 1 & 1 & 1 & 1 \\ 1 & -1 & 1 & -1 \\ 1 & 1 & -1 & -1 \\ 1 & -1 & -1 & 1 \end{bmatrix} = \begin{bmatrix} 1 & 1 & 1 & 1 & 1 & 1 & 1 & 1 \\ 1 & -1 & 1 & -1 & 1 & -1 & 1 & -1 \\ 1 & 1 & -1 & -1 & 1 & 1 & -1 & -1 \\ 1 & -1 & -1 & 1 & 1 & -1 & -1 & 1 \\ 1 & 1 & 1 & 1 & -1 & -1 & -1 & -1 \\ 1 & -1 & 1 & -1 & -1 & 1 & -1 & 1 \\ 1 & 1 & -1 & -1 & -1 & -1 & 1 & 1 \\ 1 & -1 & -1 & 1 & -1 & 1 & 1 & -1 \end{bmatrix}.$$

Repeating this procedure yields a Hadamard matrix of order 2^m for any integer $m \geq 1$. An equivalent form of this Hadamard matrix is $\boldsymbol{H} = \left((-1)^{|u \cap v|}\right)$, with u and v running through all subsets of $Z_m = \{1, \ldots, m\}$, as will be seen in Section 2.8. When the column of all ones is removed from a Hadamard matrix of Sylvester type, a saturated regular factorial design is obtained.

Paley (1933) presented two constructions of Hadamard matrices, both based on a matrix $\boldsymbol{Q}$ to be introduced next. Let s be an odd prime power, and $GF(s) = \{\alpha_0, \alpha_1, \ldots, \alpha_{s-1}\}$ be a Galois field of order s. An element $\alpha \neq 0$ in $GF(s)$ is called a quadratic residue if α is a square, meaning that $\alpha = \beta^2$ for some β in $GF(s)$. Define

$$\chi(\alpha) = \begin{cases} 1 & \text{if } \alpha \text{ is a quadratic residue,} \\ 0 & \text{if } \alpha = 0, \\ -1 & \text{if } \alpha \text{ is not a quadratic residue,} \end{cases}$$

and then define the matrix

$$\boldsymbol{Q} = (q_{ij})_{s \times s}, \quad \text{where } q_{ij} = \chi(\alpha_i - \alpha_j) \tag{2.7}$$

for $i, j = 0, 1, \ldots, s-1$. It can be shown that $\boldsymbol{Q}$ has the following three properties:

(i) the columns of $\boldsymbol{Q}$ are all orthogonal to $\mathbf{1}_s$,

(ii) the inner product of any two columns of $\boldsymbol{Q}$ is -1, and

(iii) $\boldsymbol{Q}$ is symmetric if $s \equiv 1 \pmod 4$ and skew-symmetric if $s \equiv 3 \pmod 4$.

Example 2.2. *As an illustration, consider* $GF(11) = \{0, 1, 2, \ldots, 10\}$ *in which one can check that* $1, 3, 4, 5, 9$ *are quadratic residues and* $2, 6, 7, 8, 10$ *are not. The* $\boldsymbol{Q}$ *matrix is then given by*

$$\boldsymbol{Q} = \begin{bmatrix} 0 & -1 & 1 & -1 & -1 & -1 & 1 & 1 & 1 & -1 & 1 \\ 1 & 0 & -1 & 1 & -1 & -1 & -1 & 1 & 1 & 1 & -1 \\ -1 & 1 & 0 & -1 & 1 & -1 & -1 & -1 & 1 & 1 & 1 \\ 1 & -1 & 1 & 0 & -1 & 1 & -1 & -1 & -1 & 1 & 1 \\ 1 & 1 & -1 & 1 & 0 & -1 & 1 & -1 & -1 & -1 & 1 \\ 1 & 1 & 1 & -1 & 1 & 0 & -1 & 1 & -1 & -1 & -1 \\ -1 & 1 & 1 & 1 & -1 & 1 & 0 & -1 & 1 & -1 & -1 \\ -1 & -1 & 1 & 1 & 1 & -1 & 1 & 0 & -1 & 1 & -1 \\ -1 & -1 & -1 & 1 & 1 & 1 & -1 & 1 & 0 & -1 & 1 \\ 1 & -1 & -1 & -1 & 1 & 1 & 1 & -1 & 1 & 0 & -1 \\ -1 & 1 & -1 & -1 & -1 & 1 & 1 & 1 & -1 & 1 & 0 \end{bmatrix}. \tag{2.8}$$

Paley's first construction yields

$$\boldsymbol{H} = \begin{bmatrix} 1 & -\boldsymbol{1}_s^T \\ \boldsymbol{1}_s & \boldsymbol{Q} + \boldsymbol{I}_s \end{bmatrix}. \tag{2.9}$$

Theorem 2.9. *Let* n *be a multiple of* 4. *If* $s = n - 1$ *is a prime power, then the matrix* $\boldsymbol{H}$ *defined in* (2.9) *is a Hadamard matrix of order* n.

The validity of Theorem 2.9 follows straightforwardly from the three properties of $\boldsymbol{Q}$. To put it in words, $\boldsymbol{H}$ in (2.9) is obtained from $\boldsymbol{Q}$ in (2.7) by first replacing the zeros on the diagonal by $+1$s, then adding a row of -1s, and finally appending a column of $+1$s. While Hadamard matrices of Sylvester type are associated with regular designs, Hadamard matrices from Paley's first construction give rise to a class of appealing nonregular designs; see Section 4.2.

Applying Theorem 2.9 to the matrix $\boldsymbol{Q}$ in Example 2.2, we obtain a Hadamard matrix of order 12:

$$\boldsymbol{H} = \begin{bmatrix} 1 & -1 & -1 & -1 & -1 & -1 & -1 & -1 & -1 & -1 & -1 & -1 \\ 1 & 1 & -1 & 1 & -1 & -1 & -1 & 1 & 1 & 1 & -1 & 1 \\ 1 & 1 & 1 & -1 & 1 & -1 & -1 & -1 & 1 & 1 & 1 & -1 \\ 1 & -1 & 1 & 1 & -1 & 1 & -1 & -1 & -1 & 1 & 1 & 1 \\ 1 & 1 & -1 & 1 & 1 & -1 & 1 & -1 & -1 & -1 & 1 & 1 \\ 1 & 1 & 1 & -1 & 1 & 1 & -1 & 1 & -1 & -1 & -1 & 1 \\ 1 & 1 & 1 & 1 & -1 & 1 & 1 & -1 & 1 & -1 & -1 & -1 \\ 1 & -1 & 1 & 1 & 1 & -1 & 1 & 1 & -1 & 1 & -1 & -1 \\ 1 & -1 & -1 & 1 & 1 & 1 & -1 & 1 & 1 & -1 & 1 & -1 \\ 1 & -1 & -1 & -1 & 1 & 1 & 1 & -1 & 1 & 1 & -1 & 1 \\ 1 & 1 & -1 & -1 & -1 & 1 & 1 & 1 & -1 & 1 & 1 & -1 \\ 1 & -1 & 1 & -1 & -1 & -1 & 1 & 1 & 1 & -1 & 1 & 1 \end{bmatrix}. \tag{2.10}$$

Paley's second construction applies to the case where n is a multiple of 4 but not a multiple of 8 and requires that $s = n/2 - 1$ be a prime power. It uses matrix $\boldsymbol{Q}$ in (2.7) to construct a matrix of order n as follows

$$\boldsymbol{H} = \begin{bmatrix} 1 & \boldsymbol{1}_s^T & -1 & \boldsymbol{1}_s^T \\ \boldsymbol{1}_s & \boldsymbol{Q} + \boldsymbol{I}_s & \boldsymbol{1}_s & \boldsymbol{Q} - \boldsymbol{I}_s \\ -1 & \boldsymbol{1}_s^T & -1 & -\boldsymbol{1}_s^T \\ \boldsymbol{1}_s & \boldsymbol{Q} - \boldsymbol{I}_s & -\boldsymbol{1}_s & -\boldsymbol{Q} - \boldsymbol{I}_s \end{bmatrix}. \tag{2.11}$$

Theorem 2.10. *Let n be a multiple of 4 but not a multiple of 8. If $s = n/2 - 1$ is a prime power, then the $\boldsymbol{H}$ in (2.11) is a Hadamard matrix of order n.*

Hadamard matrices from Paley's second construction are also useful for finding good nonregular designs; see Section 4.2.

We conclude this section with a brief description of a construction due to Williamson (1944). Unlike those of Paley (1933), Williamson's construction only offers a template of the form

$$\begin{bmatrix} \boldsymbol{A} & \boldsymbol{B} & \boldsymbol{C} & \boldsymbol{D} \\ -\boldsymbol{B} & \boldsymbol{A} & -\boldsymbol{D} & \boldsymbol{C} \\ -\boldsymbol{C} & \boldsymbol{D} & \boldsymbol{A} & -\boldsymbol{B} \\ -\boldsymbol{D} & -\boldsymbol{C} & \boldsymbol{B} & \boldsymbol{A} \end{bmatrix},$$

which yields a Hadamard matrix if the component matrices $\boldsymbol{A}$, $\boldsymbol{B}$, $\boldsymbol{C}$, and $\boldsymbol{D}$ satisfy certain conditions. To apply Williamson's construction, one has to search for suitable $\boldsymbol{A}$, $\boldsymbol{B}$, $\boldsymbol{C}$, and $\boldsymbol{D}$. Fortunately, successes can often be achieved by limiting the search to symmetric circulant matrices.

2.6 Orthogonal Arrays from Hadamard Matrices

Two classes of two-level orthogonal arrays can readily be constructed from Hadamard matrices. In short, removing an all-one column from a Hadamard matrix produces an orthogonal array of strength two and folding-over a Hadamard matrix generates an orthogonal array of strength three. A detailed discussion follows.

Let $\boldsymbol{H}$ be a Hadamard matrix of order n. We may assume that $\boldsymbol{H}$ is normalized, meaning that its first column consists of all ones. This is because if $(\boldsymbol{h}_1, \boldsymbol{h}_2, \ldots, \boldsymbol{h}_n)$ is a Hadamard matrix, so is

$$(\boldsymbol{h}_1\boldsymbol{h}_1, \boldsymbol{h}_1\boldsymbol{h}_2, \ldots, \boldsymbol{h}_1\boldsymbol{h}_n) = (\boldsymbol{1}_n, \boldsymbol{h}_1\boldsymbol{h}_2, \ldots, \boldsymbol{h}_1\boldsymbol{h}_n).$$

We are ready for the next result.

Theorem 2.11. *Let $\boldsymbol{D}_1$ be obtained by deleting the all-one column from a normalized Hadamard matrix of order n. Then $\boldsymbol{D}_1$ is an orthogonal array of strength two. More precisely, $\boldsymbol{D}_1$ is an $OA(n, 2^{n-1}, 2)$.*

Proof. The proof will be brief as it is essentially that of Theorem 2.7. Consider the $n \times 2$ matrix $(\boldsymbol{b}, \boldsymbol{c})$ formed by two different columns $\boldsymbol{b}$ and $\boldsymbol{c}$ of $\boldsymbol{D}_1$. Let n_1, n_2, n_3, and n_4 be the numbers of times the pairs $(-1,-1)$, $(-1,1)$, $(1,-1)$, and $(1,1)$ appear as rows of matrix $(\boldsymbol{b}, \boldsymbol{c})$. We can then establish that $n_1 = n_2 = n_3 = n_4$ by using the fact that $\boldsymbol{1}_n, \boldsymbol{b}$, and $\boldsymbol{c}$ are mutually orthogonal. □

As an $\mathrm{OA}(n, 2^{n-1}, 2)$, design $\boldsymbol{D}_1$ in Theorem 2.11 is called saturated since $n - 1$ is the largest integer m for an $\mathrm{OA}(n, 2^m, 2)$ to exist. This is easily seen because the columns of an $\mathrm{OA}(n, 2^m, 2)$ are mutually orthogonal and are all orthogonal to $\boldsymbol{1}_n$. From this saturated $\mathrm{OA}(n, 2^{n-1}, 2)$, one can obtain an $\mathrm{OA}(n, 2^m, 2)$ for any $m \leq n - 1$ by dropping $n - 1 - m$ columns.

We have constructed a rich class of orthogonal arrays of strength two by choosing columns from a normalized Hadamard matrix. However, it should be pointed out that not all orthogonal arrays of strength two can be obtained this way. We will return to this point when we discuss the enumeration of orthogonal arrays in Chapter 3.

We next consider the construction of orthogonal arrays of strength three from Hadamard matrices.

Theorem 2.12. *Let* $\boldsymbol{H}$ *be a Hadamard matrix of order* n*, which need not be normalized. Then* $\boldsymbol{D}_2 = (\boldsymbol{H}^T, -\boldsymbol{H}^T)^T$ *is an orthogonal array of strength three. More precisely,* $\boldsymbol{D}_2$ *is an* $OA(2n, 2^n, 3)$.

Proof. Let $\boldsymbol{a}, \boldsymbol{b}, \boldsymbol{c}$ be any three columns of $\boldsymbol{H}$. We observe that $-\boldsymbol{H}$ is composed of the mirror images of the runs of $\boldsymbol{H}$. As a result, multiplying a row, say $\boldsymbol{r}$, of $\boldsymbol{H}$ by -1 only causes a position switch of the two runs $\boldsymbol{r}$ and $-\boldsymbol{r}$ in design $\boldsymbol{D}_2$ and has no effect on $\boldsymbol{D}_2$ otherwise. We can therefore assume that $\boldsymbol{a} = \boldsymbol{1}_n$.

We need to show that the matrix

$$\begin{pmatrix} \boldsymbol{1}_n & \boldsymbol{b} & \boldsymbol{c} \\ -\boldsymbol{1}_n & -\boldsymbol{b} & -\boldsymbol{c} \end{pmatrix}$$

contains the eight possible level combinations with the same frequency.

The arguments used in proving Theorem 2.7 show that $(\boldsymbol{1}_n, \boldsymbol{b}, \boldsymbol{c})$ contains $(1, -1, -1)$, $(1, -1, 1)$, $(1, 1, -1)$, and $(1, 1, 1)$ with the same number $n/4$ of times. Immediately, we see that $(-\boldsymbol{1}_n, -\boldsymbol{b}, -\boldsymbol{c})$ must contain $(-1, -1, -1)$, $(-1, -1, 1)$, $(-1, 1, -1)$, and $(-1, 1, 1)$ with the same number $n/4$ of times. □

An $\mathrm{OA}(2n, 2^n, 3)$, like design $\boldsymbol{D}_2$ in Theorem 2.12, is called a saturated orthogonal array of strength three since there does not exist a strength-three array with more than n factors in $2n$ runs, which is so because an $\mathrm{OA}(2n, 2^n, 3)$ allows joint estimation of the intercept, all main effects, and all two-factor interactions involving one particular factor. From $\boldsymbol{D}_2$, one can obtain an $\mathrm{OA}(2n, 2^m, 3)$ for any $m \leq n$ by dropping $n - m$ columns. Strength-three arrays so constructed all have a foldover structure—one-half design is made up of the mirror images of the other half. Non-foldover orthogonal arrays of strength three exist. Maximal resolution IV designs are important examples of this type of designs (Chen and Cheng, 2006). More discussion on foldover designs will be given in Chapter 3.

Remark 2.2. *Mathematically, Theorems 2.11 and 2.12 can be made stronger. The stronger version of Theorem 2.11 is that the existence of a Hadamard matrix of order n implies the existence of an* $OA(n, 2^{n-1}, 2)$ *and vice versa; and the stronger version of Theorem 2.12 is that the existence of a Hadamard matrix of order n implies the existence of an* $OA(2n, 2^n, 3)$ *and vice versa. The converses of Theorems 2.11 and 2.12 as stated in the stronger versions, though theoretically interesting, seldom have practical implications.*

We give an example to illustrate Theorems 2.11 and 2.12.

Example 2.3. *By using the Hadamard matrix of order 12 as in (2.10), one can apply Theorem 2.11 to produce an* $OA(12, 2^{11}, 2)$:

$$\begin{bmatrix}
-1 & -1 & -1 & -1 & -1 & -1 & -1 & -1 & -1 & -1 & -1 \\
1 & -1 & 1 & -1 & -1 & -1 & 1 & 1 & 1 & -1 & 1 \\
1 & 1 & -1 & 1 & -1 & -1 & -1 & 1 & 1 & 1 & -1 \\
-1 & 1 & 1 & -1 & 1 & -1 & -1 & -1 & 1 & 1 & 1 \\
1 & -1 & 1 & 1 & -1 & 1 & -1 & -1 & -1 & 1 & 1 \\
1 & 1 & -1 & 1 & 1 & -1 & 1 & -1 & -1 & -1 & 1 \\
1 & 1 & 1 & -1 & 1 & 1 & -1 & 1 & -1 & -1 & -1 \\
-1 & 1 & 1 & 1 & -1 & 1 & 1 & -1 & 1 & -1 & -1 \\
-1 & -1 & 1 & 1 & 1 & -1 & 1 & 1 & -1 & 1 & -1 \\
-1 & -1 & -1 & 1 & 1 & 1 & -1 & 1 & 1 & -1 & 1 \\
1 & -1 & -1 & -1 & 1 & 1 & 1 & -1 & 1 & 1 & -1 \\
-1 & 1 & -1 & -1 & -1 & 1 & 1 & 1 & -1 & 1 & 1
\end{bmatrix}.$$

The nonregular design of 12 runs for 9 factors in Table 1.2 is actually obtained from this $OA(12, 2^{11}, 2)$ *by deleting the last two columns. Theorem 2.12 can be used to generate an* $OA(24, 2^{12}, 3)$*:*

$$\begin{bmatrix}
1 & -1 & -1 & -1 & -1 & -1 & -1 & -1 & -1 & -1 & -1 & -1 \\
1 & 1 & -1 & 1 & -1 & -1 & -1 & 1 & 1 & 1 & -1 & 1 \\
1 & 1 & 1 & -1 & 1 & -1 & -1 & -1 & 1 & 1 & 1 & -1 \\
1 & -1 & 1 & 1 & -1 & 1 & -1 & -1 & -1 & 1 & 1 & 1 \\
1 & 1 & -1 & 1 & 1 & -1 & 1 & -1 & -1 & -1 & 1 & 1 \\
1 & 1 & 1 & -1 & 1 & 1 & -1 & 1 & -1 & -1 & -1 & 1 \\
1 & 1 & 1 & 1 & -1 & 1 & 1 & -1 & 1 & -1 & -1 & -1 \\
1 & -1 & 1 & 1 & 1 & -1 & 1 & 1 & -1 & 1 & -1 & -1 \\
1 & -1 & -1 & 1 & 1 & 1 & -1 & 1 & 1 & -1 & 1 & -1 \\
1 & -1 & -1 & -1 & 1 & 1 & 1 & -1 & 1 & 1 & -1 & 1 \\
1 & 1 & -1 & -1 & -1 & 1 & 1 & 1 & -1 & 1 & 1 & -1 \\
1 & -1 & 1 & -1 & -1 & -1 & 1 & 1 & 1 & -1 & 1 & 1 \\
-1 & 1 & 1 & 1 & 1 & 1 & 1 & 1 & 1 & 1 & 1 & 1 \\
-1 & -1 & 1 & -1 & 1 & 1 & 1 & -1 & -1 & -1 & 1 & -1 \\
-1 & -1 & -1 & 1 & -1 & 1 & 1 & 1 & -1 & -1 & -1 & 1 \\
-1 & 1 & -1 & -1 & 1 & -1 & 1 & 1 & 1 & -1 & -1 & -1 \\
-1 & -1 & 1 & -1 & -1 & 1 & -1 & 1 & 1 & 1 & -1 & -1 \\
-1 & -1 & -1 & 1 & -1 & -1 & 1 & -1 & 1 & 1 & 1 & -1 \\
-1 & -1 & -1 & -1 & 1 & -1 & -1 & 1 & -1 & 1 & 1 & 1 \\
-1 & 1 & -1 & -1 & -1 & 1 & -1 & -1 & 1 & -1 & 1 & 1 \\
-1 & 1 & 1 & -1 & -1 & -1 & 1 & -1 & -1 & 1 & -1 & 1 \\
-1 & 1 & 1 & 1 & -1 & -1 & -1 & 1 & -1 & -1 & 1 & -1 \\
-1 & -1 & 1 & 1 & 1 & -1 & -1 & -1 & 1 & -1 & -1 & 1 \\
-1 & 1 & -1 & 1 & 1 & 1 & -1 & -1 & -1 & 1 & -1 & -1
\end{bmatrix}.$$

2.7 Effect Aliasing in Factorial Designs

Two factorial effects β_1 and β_2 are said to be orthogonal if their corresponding column vectors $\boldsymbol{x}_1$ and $\boldsymbol{x}_2$ in the model matrix $\boldsymbol{X}$ are orthogonal, and are said to be fully aliased with each other if $\boldsymbol{x}_1 = \pm\boldsymbol{x}_2$. Two effects are said to be partially aliased with each other if they are neither orthogonal nor fully aliased. The extent of aliasing between two factorial effects can be measured using the *aliasing coefficient* defined by $\boldsymbol{x}_1^T\boldsymbol{x}_2/n$, which is exactly the same as the correlation coefficient unless neither of $\boldsymbol{x}_1$ and $\boldsymbol{x}_2$ is orthogonal to $\mathbf{1}_n$. This aliasing coefficient is equal to 0 if the two effects are orthogonal, ± 1 if they are fully aliased with each other, and has an absolute value between 0 and 1 if they are partially aliased with each other.

Aliasing causes problems in estimation. If two effects are fully aliased with each other, they cannot be simultaneously estimated. Only one of them can be included in the model if the model is to be estimable. Partial aliasing can also render a model nonestimable if it occurs between many pairs of factorial effects. Even if the model is estimable, partial aliasing between the effects in the model results in multicollinearity, thus causing variance inflation. As a result, the estimation efficiency decreases—the estimation efficiency can be measured by $1/\Phi$ using any optimality criterion Φ introduced in Section 2.3.

Another undesirable consequence of aliasing is the estimation bias. To make this precise, let $\boldsymbol{Y} = \boldsymbol{X}_1\boldsymbol{\beta}_1 + \boldsymbol{\epsilon}$ be the fitted model and $\boldsymbol{Y} = \boldsymbol{X}_1\boldsymbol{\beta}_1 + \boldsymbol{X}_2\boldsymbol{\beta}_2 + \boldsymbol{\epsilon}$ be the true model. In the fitted model, $\boldsymbol{\beta}_1$ collects all the factorial effects the experimenter wants to estimate but this model may not be adequate because the factorial effects in $\boldsymbol{\beta}_2$ may not be negligible. The least squares estimator $\hat{\boldsymbol{\beta}}_1$ of $\boldsymbol{\beta}_1$ from the fitted model has an expectation of $E(\hat{\boldsymbol{\beta}}_1) = \boldsymbol{\beta}_1 + (\boldsymbol{X}_1^T\boldsymbol{X}_1)^{-1}\boldsymbol{X}_1^T\boldsymbol{X}_2\boldsymbol{\beta}_2$ when evaluated under the true model. This means that the estimator of $\boldsymbol{\beta}_1$ is contaminated by the possibly nonnegligible $\boldsymbol{\beta}_2$. The magnitude of the contamination is the bias given by $(\boldsymbol{X}_1^T\boldsymbol{X}_1)^{-1}\boldsymbol{X}_1^T\boldsymbol{X}_2\boldsymbol{\beta}_2$.

We see that the aliasing between the effects in $\boldsymbol{\beta}_1$ contributes to the bias through $(\boldsymbol{X}_1^T\boldsymbol{X}_1)^{-1}$. When the effects in $\boldsymbol{\beta}_1$ are mutually orthogonal, the bias becomes $n^{-1}\boldsymbol{X}_1^T\boldsymbol{X}_2\boldsymbol{\beta}_2$, which depends only on the aliasing between the effects in $\boldsymbol{\beta}_1$ and those in $\boldsymbol{\beta}_2$. In particular, let β_1 be a component of $\boldsymbol{\beta}_1$ and β_2 a component of $\boldsymbol{\beta}_2$, and let the corresponding columns in $\boldsymbol{X}_1$ and $\boldsymbol{X}_2$ be $\boldsymbol{x}_1$ and $\boldsymbol{x}_2$, respectively. Then the contribution to the estimation bias of β_1 from β_2 is $n^{-1}\boldsymbol{x}_1^T\boldsymbol{x}_2\beta_2$, which is proportional to β_2 with the constant being precisely the aliasing coefficient $n^{-1}\boldsymbol{x}_1^T\boldsymbol{x}_2$.

Under a full factorial or several replicates thereof, all factorial effects are mutually orthogonal. But this is unrealistic in most applications. An inevitable consequence of using fractional factorial designs is effect aliasing. Analyzing the aliasing structure is therefore instrumental in understanding a factorial design.

Some factorial designs have a simple aliasing structure in that any two factorial effects are either orthogonal or fully aliased with each other, and others have a complex aliasing structure because there exist two factorial effects that are partially aliased with each other. We give a formal definition.

Definition 2.2. *A factorial design is called regular if any two factorial effects are either orthogonal or fully aliased with each other, and is called nonregular otherwise.*

Regular factorial designs are discussed in standard textbooks, and they are constructed as follows. Starting with a full factorial design for a set of basic factors, additional factors are then defined using selected interaction columns of the basic factors. These equations for defining additional factors are called generators. Combining the generators, one can then obtain the defining relation for the design. Based on the defining relation, the complete aliasing pattern can be obtained, which divides all factorial effects into several groups such that effects in the same group are fully aliased with each other and those from different groups are mutually orthogonal. Section A.1 of the appendix chapter provides a concise, yet fairly broad, account of regular two-level factorial designs. Here we give an example to illustrate.

Example 2.4. *We first obtain a full factorial for three factors. Let $\mathbf{d}_1, \mathbf{d}_2, \mathbf{d}_3$ denote columns of the three basic factors. We then define $\mathbf{d}_4 = \mathbf{d}_1\mathbf{d}_2$ and $\mathbf{d}_5 = \mathbf{d}_1\mathbf{d}_2\mathbf{d}_3$, which generates a design of 8 runs for five factors. The two generators $\mathbf{d}_4 = \mathbf{d}_1\mathbf{d}_2$ and $\mathbf{d}_5 = \mathbf{d}_1\mathbf{d}_2\mathbf{d}_3$ are equivalent to $\mathbf{1}_n = \mathbf{d}_1\mathbf{d}_2\mathbf{d}_4$ and $\mathbf{1}_n = \mathbf{d}_1\mathbf{d}_2\mathbf{d}_3\mathbf{d}_5$, where $\mathbf{1}_n$ is a column vector of $n = 8$ plus ones. Multiplying the two equations gives $\mathbf{1}_n = \mathbf{d}_3\mathbf{d}_4\mathbf{d}_5$. Together, we have $\mathbf{1}_n = \mathbf{d}_1\mathbf{d}_2\mathbf{d}_4 = \mathbf{d}_1\mathbf{d}_2\mathbf{d}_3\mathbf{d}_5 = \mathbf{d}_3\mathbf{d}_4\mathbf{d}_5$, which is called the defining relation of the design. From the definition relation, we obtain that*

$$
\begin{aligned}
&\mathbf{1}_n = \mathbf{d}_1\mathbf{d}_2\mathbf{d}_4 = \mathbf{d}_1\mathbf{d}_2\mathbf{d}_3\mathbf{d}_5 = \mathbf{d}_3\mathbf{d}_4\mathbf{d}_5, && \mathbf{d}_1 = \mathbf{d}_2\mathbf{d}_4 = \mathbf{d}_2\mathbf{d}_3\mathbf{d}_5 = \mathbf{d}_1\mathbf{d}_3\mathbf{d}_4\mathbf{d}_5\\
&\mathbf{d}_2 = \mathbf{d}_1\mathbf{d}_4 = \mathbf{d}_1\mathbf{d}_3\mathbf{d}_5 = \mathbf{d}_2\mathbf{d}_3\mathbf{d}_4\mathbf{d}_5, && \mathbf{d}_3 = \mathbf{d}_1\mathbf{d}_2\mathbf{d}_3\mathbf{d}_4 = \mathbf{d}_1\mathbf{d}_2\mathbf{d}_5 = \mathbf{d}_4\mathbf{d}_5\\
&\mathbf{d}_1\mathbf{d}_2 = \mathbf{d}_4 = \mathbf{d}_3\mathbf{d}_5 = \mathbf{d}_1\mathbf{d}_2\mathbf{d}_3\mathbf{d}_4\mathbf{d}_5, && \mathbf{d}_1\mathbf{d}_3 = \mathbf{d}_2\mathbf{d}_3\mathbf{d}_4 = \mathbf{d}_2\mathbf{d}_5 = \mathbf{d}_1\mathbf{d}_4\mathbf{d}_5\\
&\mathbf{d}_2\mathbf{d}_3 = \mathbf{d}_1\mathbf{d}_3\mathbf{d}_4 = \mathbf{d}_1\mathbf{d}_5 = \mathbf{d}_2\mathbf{d}_4\mathbf{d}_5, && \mathbf{d}_1\mathbf{d}_2\mathbf{d}_3 = \mathbf{d}_3\mathbf{d}_4 = \mathbf{d}_5 = \mathbf{d}_1\mathbf{d}_2\mathbf{d}_4\mathbf{d}_5.
\end{aligned}
$$

The above equations provide the complete aliasing pattern of the design. We see that the 32 factorial effects are divided into eight groups of 4 factorial effects. Any two factorial effects from the same group are fully aliased with each other. Any two factorial effects from different groups are mutually orthogonal. For convenience, we treat the grand mean, corresponding to $\mathbf{1}_n$*, as a factorial effect with order 0.*

By Definition 2.2, design $\boldsymbol{D}_3$ in Example 2.1 is a nonregular design since the main effects are not orthogonal to nor fully aliased with the grand mean. Deleting the first row from $\boldsymbol{D}_3$, we obtain another design $\boldsymbol{D}_4$ of 6 runs for three factors. Design $\boldsymbol{D}_4$ is also a nonregular design because the main effects are partially aliased with one another, although they are all orthogonal to the grand mean. As another example, let us take one more look at the 12-run nonregular design in Chapter 1. This design is an orthogonal array of strength two and has orthogonal main effects. Because $(\boldsymbol{d}_1\boldsymbol{d}_2)^T\boldsymbol{d}_3/n = -1/3$, the main effect $\boldsymbol{d}_3$ is partially aliased with the interaction effect $\boldsymbol{d}_1\boldsymbol{d}_2$.

Of most interest are nonregular designs that have orthogonal main effects. They are given by orthogonal arrays of strength two. The bulk of the book is devoted to the studies of such nonregular designs, with major exceptions being Chapters 8 and 9 where, for example, our focus in Chapter 8 is on supersaturated designs, a class of designs that do not have orthogonal main effects.

Remark 2.3. *Our understanding from standard textbooks is that a regular design consists of one or more copies of a design generated by adding interaction columns to independent columns given by a set of basic factors. For Definition 2.2 to be in full agreement with this understanding, it needs to be shown that a design can be generated in way of adding interaction columns to independent columns if any two factorial effects under the given design are either orthogonal or fully aliased with each other. This can indeed be established, and we refer to Fontana, Pistone, and Rogantin (2000) and Ye (2004) for details.*

2.8 Factorial Effects are Orthogonal Contrasts

Factorial effects are informally introduced in Section 2.2 as the coefficients in a postulated linear model under a given design. Aliasing between factorial effects under a given design is discussed in Section 2.7. As unknown parameters, factorial effects are in fact mutually orthogonal contrasts of treatment means, independent of the postulated linear model and the design used for data collection. This section formally defines factorial effects.

Suppose that m factors of two levels are considered in an experiment to examine their effects on a response variable Y. In total, there are 2^m level combinations, which are often called treatment combinations, as each level combination gives rise to a treatment. The factorial structure of the 2^m treatments allows one to study and compare them using factorial effects that have meaningful interpretations in attributing the treatment differences to certain specific factors.

Let $\mu(x_1, \ldots, x_m)$ denote the mean response at the treatment combination $(x_1, \ldots, x_m)$ where $x_j = \pm 1$. Then the factorial effects are defined by the dependence of $\mu(x_1, \ldots, x_m)$ on $(x_1, \ldots, x_m)$ expressed as follows:

$$\mu(x_1, \ldots, x_m) = \beta_0 + \sum_{j=1} \beta_j x_j + \sum_{j_1 < j_2} \beta_{j_1 j_2} x_{j_1} x_{j_2} + \cdots + \beta_{12\cdots m} x_1 x_2 \cdots x_m, \tag{2.12}$$

where β_0 is the grand mean, and all the other βs are factorial effects. In particular, β_js are the main effects and $\beta_{j_1 j_2}$s are the two-factor interactions, and so on.

In the expression for $\mu(x_1, \ldots, x_m)$ given in (2.12), there is one β corresponding to every subset of the set of indices $Z_m = \{1, \ldots, m\}$. We then obtain a compact equation by introducing β_v for every $v \subseteq Z_m$, given by

$$\mu(x_1, \ldots, x_m) = \sum_{v \subseteq Z_m} \beta_v \prod_{j \in v} x_j.$$

For each $u \subseteq Z_m$, let $\boldsymbol{x}_u = (x_{u1}, \ldots, x_{um})$ be such that $x_{uj} = -1$ if $j \in u$ and $x_{uj} = +1$ otherwise. By allowing u to run through all the subsets of Z_m, we obtain all 2^m treatment combinations. Evaluating $\prod_{j \in v} x_j$ under $\boldsymbol{x}_u = (x_{u1}, \ldots, x_{um})$, we obtain $\prod_{j \in v} x_{uj} = (-1)^{|u \cap v|}$ where $|u \cap v|$ denotes the number of elements in $u \cap v$. Letting $\mu_u = \mu(\boldsymbol{x}_u)$, we obtain

$$\mu_u = \sum_{v \subseteq Z_m} (-1)^{|u \cap v|} \beta_v. \tag{2.13}$$

We need a lemma before proceeding further.

Lemma 2.1. *The $2^m \times 2^m$ matrix $\boldsymbol{H} = \left((-1)^{|u \cap v|}\right)$ is a Hadamard matrix, where the rows of $\boldsymbol{H}$ are indexed by u and the columns indexed by v, both running through all subsets of Z_m, and for definiteness, both the rows and columns of $\boldsymbol{H}$ are arranged in the Yates order.*

The Yates order is also known as the lexicographical order or dictionary order.

Lemma 2.1 does not really need a proof, as $\boldsymbol{H}$ is simply the saturated regular design with a column of all ones added. To see this, we observe that the column corresponding to $v = \phi$ is a column of all ones, the m columns from those v's with $|v| = 1$ gives a full 2^m factorial, the columns from those v's with $|v| = 2$ are the pairwise Hadamard products of the m columns with $|v| = 1$, and so on.

Lemma 2.1 allows the inversion of equation (2.13), which gives

$$\beta_v = 2^{-m} \sum_{u \subseteq Z_m} (-1)^{|u \cap v|} \mu_u. \tag{2.14}$$

Equation (2.14) provides a formal definition of factorial effects. Together, these factorial effects besides β_ϕ are a set of $2^m - 1$ mutually orthogonal contrasts, with the main effects given by β_vs with $|v| = 1$ and the two-factor interactions given by those with $|v| = 2$, and so on. We may say that the β_vs with $|v| = r$ are the factorial effects of order r. The grand mean β_ϕ corresponding to $v = \phi$ is not a contrast but for convenience, we can still call it a factorial effect that has order 0.

The often-used intuitive definition of factorial effects, as popularized by Box, Hunter, and Hunter (1978), differs from our definition in (2.14) by a factor of 2. This is immaterial as a contrast when multiplied by a constant gives essentially the same contrast. Our definition in (2.14) for factorial effects is not uncommon in design literature, and goes back to as early as Plackett and Burman (1946). It has the convenience of having the coefficients in the model directly represent the factorial effects without having to worry about a factor of 2.

3

J-Characteristics and Projection Properties

In this chapter we first develop a theory of J-characteristics for factorial designs. A factorial design and its set of J-characteristics uniquely determine each other, hence the set of J-characteristics for a factorial design plays the same role as the defining relation for a regular factorial design. We then investigate the projection properties and hidden projection properties of orthogonal arrays using J-characteristics. The power of J-characteristics for studying factorial designs is further demonstrated in their applications to the enumeration of orthogonal arrays.

3.1 Defining J-Characteristics

Let $\boldsymbol{x}_1, \ldots, \boldsymbol{x}_k$ be k vectors of length n, and let $\boldsymbol{x}_j = (x_{1j}, \ldots, x_{nj})^T$ for $j = 1, \ldots, k$. Then the J-characteristic of these k vectors is defined as

$$J(\boldsymbol{x}_1, \ldots, \boldsymbol{x}_k) = \sum_{i=1}^{n} x_{i1} \cdots x_{ik}. \tag{3.1}$$

For $k = 1$, $J(\boldsymbol{x}_1)$ is simply the sum of the components in $\boldsymbol{x}_1$. For $k = 2$, $J(\boldsymbol{x}_1, \boldsymbol{x}_2) = \boldsymbol{x}_1^T \boldsymbol{x}_2$, the inner product of two vectors $\boldsymbol{x}_1$ and $\boldsymbol{x}_2$. Therefore, one may regard the J-characteristic as a generalization of the inner product. While the inner product is only defined for two vectors, the J-characteristic is defined for any number of vectors.

The following simple properties are evident.

Lemma 3.1. *We have that*

(i) $J(\boldsymbol{1}_n, \boldsymbol{x}) = J(\boldsymbol{x})$,

(ii) $J(\boldsymbol{x}_1\boldsymbol{x}_3, \boldsymbol{x}_2\boldsymbol{x}_3) = J(\boldsymbol{x}_1, \boldsymbol{x}_2)$, *and*

(iii) $J(\boldsymbol{x}_1, \ldots, \boldsymbol{x}_k) = J(\boldsymbol{x}_1 \cdots \boldsymbol{x}_k)$.

Although the definition of J-characteristics in (3.1) and the properties in Lemma 3.1 apply to real vectors $\boldsymbol{x}_1, \ldots, \boldsymbol{x}_k$, our interest in J-characteristics lies in their usefulness in characterizing factorial designs. As before, let $\boldsymbol{D} = (\boldsymbol{d}_1, \ldots, \boldsymbol{d}_m) = (d_{ij})_{n \times m}$ be a two-level factorial design of n runs for m factors. To any subset of $k \geq 1$ columns $\boldsymbol{d}_{j_1}, \ldots, \boldsymbol{d}_{j_k}$, there corresponds a J-characteristic, $J(\boldsymbol{d}_{j_1}, \ldots, \boldsymbol{d}_{j_k})$. The usefulness of J-characteristics is first seen in the next result.

Theorem 3.1. *Design* $\boldsymbol{D}$ *is regular if and only if* $J(\boldsymbol{d}_{j_1}, \ldots, \boldsymbol{d}_{j_k}) = 0$ *or* $\pm n$ *for all subsets of* $k \geq 1$ *columns* $\boldsymbol{d}_{j_1}, \ldots, \boldsymbol{d}_{j_k}$.

DOI: 10.1201/9781003371861-3

Proof. We can verify Theorem 3.1 in a straightforward manner using Definition 2.2. We give some details for the necessity part. Suppose that $\boldsymbol{D}$ is regular. Then any two factorial effects are either orthogonal or fully aliased with each other. If $\boldsymbol{d}_{j_1}$ and $\boldsymbol{d}_{j_2}\cdots\boldsymbol{d}_{j_k}$ are orthogonal, then $J(\boldsymbol{d}_{j_1}, \boldsymbol{d}_{j_2}\cdots\boldsymbol{d}_{j_k}) = 0$. Using Lemma 3.1, we have that $J(\boldsymbol{d}_{j_1}, \ldots, \boldsymbol{d}_{j_k}) = J(\boldsymbol{d}_{j_1}, \boldsymbol{d}_{j_2}\cdots\boldsymbol{d}_{j_k}) = 0$. If $\boldsymbol{d}_{j_1}$ and $\boldsymbol{d}_{j_2}\cdots\boldsymbol{d}_{j_k}$ are fully aliased with each other, meaning that $\boldsymbol{d}_{j_1} = \pm\boldsymbol{d}_{j_2}\cdots\boldsymbol{d}_{j_k}$, then we must have $J(\boldsymbol{d}_{j_1}, \ldots, \boldsymbol{d}_{j_k}) = \pm n$. The sufficiency part of Theorem 3.1 can be verified similarly. □

Theorem 3.1 provides an equivalent definition of regular and nonregular designs in terms of J-characteristics—a factorial design is said to be regular if the values of its J-characteristics are either 0 or $\pm n$, and is said to be nonregular if $0 < |J(\boldsymbol{d}_{j_1}, \ldots, \boldsymbol{d}_{j_k})| < n$ for at least one subset of columns.

The J-characteristics of design $\boldsymbol{D}$ contain all information about the aliasing properties of factorial effects. For example, the aliasing coefficient between $\boldsymbol{d}_1$ and $\boldsymbol{d}_2\boldsymbol{d}_3$ is simply $J(\boldsymbol{d}_1, \boldsymbol{d}_2, \boldsymbol{d}_3)/n$. For another example, the aliasing coefficient between $\boldsymbol{d}_1\boldsymbol{d}_2$ and $\boldsymbol{d}_2\boldsymbol{d}_3\boldsymbol{d}_4$ is $J(\boldsymbol{d}_1, \boldsymbol{d}_3, \boldsymbol{d}_4)/n$. Lemma 3.1 is helpful for understanding the calculations above.

Example 3.1. *Consider designs $\boldsymbol{D}_1$, $\boldsymbol{D}_2$, and $\boldsymbol{D}_3$ in Example 2.1. For design $\boldsymbol{D}_1$, we have*

$$J(\boldsymbol{d}_1) = J(\boldsymbol{d}_2) = J(\boldsymbol{d}_3) = J(\boldsymbol{d}_1, \boldsymbol{d}_2) = 0,$$

$$J(\boldsymbol{d}_1, \boldsymbol{d}_3) = J(\boldsymbol{d}_2, \boldsymbol{d}_3) = J(\boldsymbol{d}_1, \boldsymbol{d}_2, \boldsymbol{d}_3) = 0.$$

In short, $J(\boldsymbol{d}_{j_1}, \ldots, \boldsymbol{d}_{j_k}) = 0$ for any subset of k columns with $k = 1, 2, 3$. This happens because $\boldsymbol{D}_1$ is a full factorial design. For design $\boldsymbol{D}_2$, we have that $J(\boldsymbol{d}_1, \boldsymbol{d}_2, \boldsymbol{d}_3, \boldsymbol{d}_4) = 8$, and $J(\boldsymbol{d}_{j_1}, \ldots, \boldsymbol{d}_{j_k}) = 0$ for any subset of k columns with $k = 1, 2, 3$. Design $\boldsymbol{D}_2$ is an orthogonal array of strength 3. As will be seen later, any orthogonal array of strength 3 has the property that $J(\boldsymbol{d}_{j_1}, \ldots, \boldsymbol{d}_{j_k}) = 0$ for any subset of k columns with $k = 1, 2, 3$. For design $\boldsymbol{D}_3$, we obtain that

$$J(\boldsymbol{d}_1) = J(\boldsymbol{d}_2) = J(\boldsymbol{d}_1, \boldsymbol{d}_2) = -1, \quad J(\boldsymbol{d}_3) = J(\boldsymbol{d}_1, \boldsymbol{d}_3) = 1,$$

$$J(\boldsymbol{d}_2, \boldsymbol{d}_3) = -3, \quad J(\boldsymbol{d}_1, \boldsymbol{d}_2, \boldsymbol{d}_3) = 5.$$

Design $\boldsymbol{D}_3$ is not an orthogonal array of strength k for any $k \geq 1$. If we want to say that $\boldsymbol{D}_3$ is an orthogonal array, then it has strength 0.

The notation for J-characteristics introduced above is useful for examples and concrete applications but becomes rather clumsy in the general theoretical development. We employ a more concise notation for this purpose, which is done through a formal definition.

Definition 3.1. *Let $\boldsymbol{D} = (d_{ij})_{n\times m}$ be a factorial design. For any subset u of the set of column indices $Z_m = \{1, \ldots, m\}$, we define*

$$J_u = \sum_{i=1}^{n} \prod_{j\in u} d_{ij}.$$

If $u = \phi$, the empty set, we define $J_\phi = n$.

3.2 Characterization of Factorial Designs

The main purpose of this section is to show that, through the use of the indicator function, a factorial design $\boldsymbol{D}$ is completely determined by its J-characteristics. The indicator function of a factorial design is a convenient and useful representation of the design in terms of a complete factorial design.

Let $\boldsymbol{C}$ be a complete factorial of n runs for m factors. Then $\boldsymbol{C}$ is given by the collection of all 2^m runs

$$\boldsymbol{r}_v = (r_{v1}, \ldots, r_{vm}) \text{ where } r_{vj} = -1 \text{ if } j \in v \text{ and } r_{vj} = 1 \text{ otherwise,}$$

for all $v \subseteq Z_m = \{1, \ldots, m\}$.

Definition 3.2. *For a factorial design $\boldsymbol{D}$ of n runs for m factors, its indicator function N_v is defined as the number of times run $\boldsymbol{r}_v$ occurs in $\boldsymbol{D}$ for $v \subseteq Z_m$.*

Example 3.2. *We give an illustration of Definition 3.2 using design $\boldsymbol{D}_3$ in Example 2.1. The complete factorial $\boldsymbol{C}$ for $m = 3$ has the following eight runs.*

$\boldsymbol{r}_\phi$	1	1	1
$\boldsymbol{r}_1$	−1	1	1
$\boldsymbol{r}_2$	1	−1	1
$\boldsymbol{r}_{12}$	−1	−1	1
$\boldsymbol{r}_3$	1	1	−1
$\boldsymbol{r}_{13}$	−1	1	−1
$\boldsymbol{r}_{23}$	1	−1	−1
$\boldsymbol{r}_{123}$	−1	−1	−1

Since run $(1, 1, 1)$ occurs in $\boldsymbol{D}_3$ exactly once and run $(-1, 1, 1)$ is missing from $\boldsymbol{D}_3$, we have that $N_\phi = 1$ and $N_1 = 0$. Similarly, we can obtain the other N_v values for $\boldsymbol{D}_3$:

$$(N_\phi, N_1, N_2, N_{12}, N_3, N_{13}, N_{23}, N_{123})^T = (1, 0, 1, 2, 0, 2, 1, 0)^T.$$

In the above, we have used shorthand notations for $\boldsymbol{r}_v$ and N_v. For example, $\boldsymbol{r}_{12}$ means $\boldsymbol{r}_{\{1,2\}}$ and N_{12} means $N_{\{1,2\}}$.

In Example 3.2, the N_v values are listed in the Yates order, which is also known as the lexicographical order or dictionary order.

As a set of points with possible repeats, $\boldsymbol{D}$ is completely determined by its indicator function. As a matrix, $\boldsymbol{D}$ is determined by its indicator function up to a row permutation. We are now ready to present the main result of this section.

Theorem 3.2. *The following relationships between J_us and N_vs hold:*

$$J_u = \sum_{v \subseteq Z_m} (-1)^{|u \cap v|} N_v, \tag{3.2}$$

$$N_v = 2^{-m} \sum_{u \subseteq Z_m} (-1)^{|u \cap v|} J_u. \tag{3.3}$$

The above can be equivalently written as

$$\boldsymbol{J} = \boldsymbol{H}\boldsymbol{N} \quad \text{and} \quad \boldsymbol{N} = 2^{-m}\boldsymbol{H}^T\boldsymbol{J},$$

where $\boldsymbol{J}$ and $\boldsymbol{N}$ are the column vectors that, respectively, collect J_us and N_vs in the Yates order, and $\boldsymbol{H} = \left((-1)^{|u \cap v|}\right)$ is a $2^m \times 2^m$ Hadamard matrix with its rows indexed by u and columns indexed by v, both arranged in the Yates order.

Proof. By the definition of N_v, we obtain that

$$J_u = \sum_{i=1}^{n} \prod_{j \in u} d_{ij} = \sum_{v \subseteq Z_m} N_v \prod_{j \in u} r_{vj}.$$

Since $\prod_{j \in u} r_{vj} = (-1)^{|u \cap v|}$, we have that $J_u = \sum_{v \subseteq Z_m} (-1)^{|u \cap v|} N_v$, which can be written in matrix form as $\boldsymbol{J} = \boldsymbol{H}\boldsymbol{N}$. As $\boldsymbol{H}$ is a Hadamard matrix of order 2^m by Lemma 2.1, we obtain $\boldsymbol{N} = 2^{-m}\boldsymbol{H}^T\boldsymbol{J}$, which gives $N_v = 2^{-m} \sum_{u \subseteq Z_m} (-1)^{|u \cap v|} J_u$. This completes the proof. □

Example 3.3. *We use design $\boldsymbol{D}_3$ in Example 2.1 to illustrate Theorem 3.2. As $m = 3$, the equation $\boldsymbol{N} = 2^{-m}\boldsymbol{H}^T\boldsymbol{J}$ becomes*

$$\begin{bmatrix} N_\phi \\ N_1 \\ N_2 \\ N_{12} \\ N_3 \\ N_{13} \\ N_{23} \\ N_{123} \end{bmatrix} = \frac{1}{8} \begin{bmatrix} 1 & 1 & 1 & 1 & 1 & 1 & 1 & 1 \\ 1 & -1 & 1 & -1 & 1 & -1 & 1 & -1 \\ 1 & 1 & -1 & -1 & 1 & 1 & -1 & -1 \\ 1 & -1 & -1 & 1 & 1 & -1 & -1 & 1 \\ 1 & 1 & 1 & 1 & -1 & -1 & -1 & -1 \\ 1 & -1 & 1 & -1 & -1 & 1 & -1 & 1 \\ 1 & 1 & -1 & -1 & -1 & -1 & 1 & 1 \\ 1 & -1 & -1 & 1 & -1 & 1 & 1 & -1 \end{bmatrix} \begin{bmatrix} J_\phi \\ J_1 \\ J_2 \\ J_{12} \\ J_3 \\ J_{13} \\ J_{23} \\ J_{123} \end{bmatrix}.$$

In Example 3.1, we have obtained that

$$(J_\phi, J_1, J_2, J_{12}, J_3, J_{13}, J_{23}, J_{123})^T = (7, -1, -1, -1, 1, 1, -3, 5)^T.$$

In Example 3.2, we have obtained that

$$(N_\phi, N_1, N_2, N_{12}, N_3, N_{13}, N_{23}, N_{123})^T = (1, 0, 1, 2, 0, 2, 1, 0)^T.$$

One can check that the equation $\boldsymbol{N} = 2^{-m}\boldsymbol{H}^T\boldsymbol{J}$ is indeed true.

Theorem 3.2 shows that the set of J-characteristics for a general factorial design plays the same role as the defining relation for a regular design. Theorem 3.2 is a powerful result, from which, as will be gradually revealed, many useful properties of factorial designs can be derived. The first of these is a characterization of a full factorial design.

Corollary 3.1. *Design $\boldsymbol{D}$ is a full factorial or several replicates thereof if and only if $J_u = 0$ for all nonempty u.*

Proof. Suppose that $J_u = 0$ for $u \neq \phi$. Then by Theorem 3.2, we have

$$N_v = 2^{-m} \sum_{u \subseteq Z_m} (-1)^{|u \cap v|} J_u = 2^{-m} n,$$

as $J_\phi = n$ and $(-1)^{|\phi \cap v|} = 1$. This shows that design $\boldsymbol{D}$ consists of $2^{-m}n$ replicates of a 2^m full factorial and establishes the sufficiency part. To prove the necessity part, let $\boldsymbol{D}$ be a full factorial or several replicates thereof. Then $N_v = 2^{-m}n$ is a constant for all $v \subseteq Z_m$. Since $\boldsymbol{H} = \left((-1)^{|u \cap v|}\right)$ is a Hadamard matrix and has an all-one row given by $u = \phi$, any other row must contain the same number of $+1$s and -1s, meaning that $\sum_{v \subseteq Z_m} (-1)^{|u \cap v|} = 0$ for any $u \neq \phi$. This shows that $J_u = \sum_{v \subseteq Z_m} (-1)^{|u \cap v|} N_v = 0$ for any $u \neq \phi$, because N_v is a constant. □

For given $u \subseteq Z_m$, if we only collect columns $\boldsymbol{d}_j$ with $j \in u$, then we obtain a projection design $\boldsymbol{D}_u$. An application of Theorem 3.2 to this projection design leads to the next result.

Corollary 3.2. *For any $u \subseteq Z_m$, the projection design $\boldsymbol{D}_u$ is completely determined by all the J-characteristics J_w with $w \subseteq u$.*

This result may be mathematically trivial but is rather insightful in terms of revealing the difference between J_us and N_vs. While both completely describe a factorial design, J-characteristics are well suited for studying projection designs. To study a projection design $\boldsymbol{D}_u$, we only need to know J_w values for $w \subseteq u$ but would have to know all N_v values for $v \subseteq Z_m$. We could consider the indicator function N_v for each given projection design $\boldsymbol{D}_u$, in which case we would have to deal with one indicator function for each projection design, thus many different indicator functions if we are interested in studying many projection designs. On the other hand, one set of J_u values for the original design $\boldsymbol{D}$ allows the study of all projection designs.

Statistical properties of a factorial design are often determined by their projection designs, as in the case of orthogonal arrays. Combining Corollaries 3.1 and 3.2, we obtain another result.

Theorem 3.3. *Design $\boldsymbol{D}$ is an orthogonal array of strength t if and only if $J_u = 0$ for all nonempty u with $|u| \leq t$.*

The indicator function was defined in Fontana, Pistone, and Rogantin (2000) and generalized in Ye (2003). The characterization results in the present section were originally obtained in Tang (2001). The presentation of Theorem 3.2 here has a more explicit form.

3.3 Foldover Designs

As shown in Theorem 3.3, J-characteristics can be used to provide a simple characterization of orthogonal arrays. In this section, we look at their application to foldover designs.

A factorial design $\boldsymbol{D}$ of n runs for m factors is called a foldover design if it can be written as

$$\boldsymbol{D} = \begin{bmatrix} \boldsymbol{D}_1 \\ -\boldsymbol{D}_1 \end{bmatrix}. \tag{3.4}$$

Clearly, $\boldsymbol{D}_1$ is a design of $n/2$ runs for m factors; we say that $\boldsymbol{D}$ is the foldover of $\boldsymbol{D}_1$. A characterization of foldover designs is in the next theorem.

Theorem 3.4. *A factorial design $\boldsymbol{D}$ is a foldover design if and only if $J_u(\boldsymbol{D}) = 0$ for all u with odd $|u|$.*

Proof. Suppose that $\boldsymbol{D}$ is the foldover of $\boldsymbol{D}_1$ as given in (3.4). Then it is obvious that $J_u(\boldsymbol{D}) = J_u(\boldsymbol{D}_1) + J_u(-\boldsymbol{D}_1)$ for any u and that $J_u(-\boldsymbol{D}_1) = -J_u(\boldsymbol{D}_1)$ for any u with odd $|u|$. This shows that $J_u(\boldsymbol{D}) = 0$ for all u with odd $|u|$.

Suppose that $J_u(\boldsymbol{D}) = 0$ for all u with odd $|u|$. To show that $\boldsymbol{D}$ is a foldover design, we need to show that $\boldsymbol{D}$ has a form as given in (3.4). Equivalently, we can show that $N_v = N_{\bar{v}}$ for design $\boldsymbol{D}$, where $\bar{v} = Z_m \setminus v$. The assertion that $N_v = N_{\bar{v}}$ simply says that run $\boldsymbol{r}_v$ and its mirror image $\boldsymbol{r}_{\bar{v}}$ occur the same number of times in design $\boldsymbol{D}$, which is exactly what a foldover design is. By (3.3) in Theorem 3.2, we have that $N_v = 2^{-m} \sum_{u \subseteq E_m} (-1)^{|u \cap v|} J_u$ and $N_{\bar{v}} = 2^{-m} \sum_{u \subseteq E_m} (-1)^{|u \cap \bar{v}|} J_u$, where E_m denotes the collection of u's with even $|u|$, since it has been assumed that $J_u = 0$ for odd $|u|$. For even $|u|$, we must have $(-1)^{|u \cap v|} = (-1)^{|u \cap \bar{v}|}$ because $(-1)^{|u \cap v|}(-1)^{|u \cap \bar{v}|} = (-1)^{|u|} = 1$. This establishes that $N_v = N_{\bar{v}}$. □

Theorem 3.4 was obtained and reported in Balakrishnan and Yang (2006) and Cheng, Mee, and Yee (2008). See also Cheng (2014a, Theorem 15.21).

In (3.4), we now suppose that $\boldsymbol{D}_1$ has an all-one column; that is, we have that

$$\boldsymbol{D} = \begin{bmatrix} \mathbf{1}_{n/2} & \boldsymbol{D}_0 \\ -\mathbf{1}_{n/2} & -\boldsymbol{D}_0 \end{bmatrix}. \tag{3.5}$$

where $\boldsymbol{D}_0$ is a design of $n/2$ runs for $m-1$ factors. Combining Theorems 3.3 and 3.4, we obtain the following well-known result (Seiden and Zemach, 1966).

Corollary 3.3. *If design $\boldsymbol{D}_0$ is an orthogonal array of even strength t, then its foldover $\boldsymbol{D}$ in (3.5) is an orthogonal array of strength $t+1$.*

Proof. First of all, by Theorem 3.3 we have that $J_u(\boldsymbol{D}_0) = 0$ for all nonempty u with $|u| \le t$, because $\boldsymbol{D}_0$ is an orthogonal array of strength t. By Theorems 3.3 and 3.4, we only need to show that $J_u(\boldsymbol{D}) = 0$ for all nonempty u with even $|u|$ and $|u| \le t$. Let $\boldsymbol{D}_1 = (\mathbf{1}_{n/2}, \boldsymbol{D}_0)$. Since $J_u(\boldsymbol{D}) = 2J_u(\boldsymbol{D}_1)$ for even $|u|$, we only need to show that $J_u(\boldsymbol{D}_1) = 0$ for all nonempty u with even $|u|$ and $|u| \le t$. This is obvious because $J_u(\boldsymbol{D}_1)$ is a J-characteristic of $\boldsymbol{D}_0$ for $|u|$ factors if u does not contain the first column and for $|u|-1$ factors if u contains the first column. □

The corollary above allows the construction of an array of odd strength $t+1$ from an array of even strength t. However, no restriction is required if we want to construct an array of strength t from an array of strength $t+1$. Given an array $\boldsymbol{D}$ of strength $t+1$ for any integer $t \ge 1$, by permuting its rows, we can always write $\boldsymbol{D}$ as

$$\boldsymbol{D} = \begin{bmatrix} \mathbf{1}_{n/2} & \boldsymbol{A} \\ -\mathbf{1}_{n/2} & \boldsymbol{B} \end{bmatrix},$$

for some designs $\boldsymbol{A}$ and $\boldsymbol{B}$. Then design $\boldsymbol{A}$ must be an array of strength t. In the literature, $\boldsymbol{A}$ is often called a half fraction of $\boldsymbol{D}$.

By Corollary 3.3, the foldover design of a strength two array is a strength three array. But not all strength three arrays are foldover designs. The question as to when a strength three array is a foldover design is of both theoretical and practical importance. Butler (2004) has an answer to this question.

Theorem 3.5. *If design $\boldsymbol{D}$ is an $OA(n, 2^m, 3)$ with $m \ge n/3$, then it must be a foldover design.*

This result was first presented in Butler (2004) and later proved in Butler (2007). The proof is rather heavy, and thus omitted here. Interested readers are referred to Butler (2007)'s original proof. In the case of regular designs, foldover designs are called even designs, based on the fact that foldover designs only have words of even lengths in their defining relations. The regular design version of Theorem 3.5 is slightly stronger, which states that a regular resolution IV design must be even if $m \ge 5n/16+1$; see Theorem 11.14 of Cheng (2014a).

3.4 Possible Values of J-Characteristics

This section is devoted to a study of possible values of J-characteristics. Being interesting in their own right, the results are useful later in this chapter and will also be used in defining criteria of generalized resolution and aberration in Chapter 4. The main sources of the results in this section are Deng and Tang (2002) and Stufken and Tang (2007).

Let design $\boldsymbol{D}$ be an OA$(n, 2^m, t)$. We have that $n = \lambda 2^t$ with λ being the index of the array. From Theorem 3.3, we know that $J_u = 0$ for $|u| = 1, \ldots, t$. From Definition 3.1, we also know that $J_u = n$ for $|u| = 0$. How about the values of J_u for $|u| \geq t+1$? The next result answers this question.

Theorem 3.6. *Let $\boldsymbol{D}$ be an OA$(n, 2^m, t)$ where $n = \lambda 2^t$. Then J_u is a multiple of 2^t for any u. More precisely, we have that $J_u = \mu_u 2^t$ for some integer μ_u. The following further results hold for μ_u.*

(i) If λ is even, then μ_u must be even.

(ii) Suppose that λ is odd. For $|u| = t+1$, we have that μ_u is odd. For $|u| = t+2$, we have that μ_u is odd if t is even and μ_u is even if t is odd.

Proof. Taking $v = \phi$ in equation (3.3) of Theorem 3.2, we obtain

$$N_\phi = 2^{-m} \sum_{w \subseteq Z_m} J_w. \tag{3.6}$$

For a given $u \subseteq Z_m$ with $|u| \geq t+1$, consider the projection design $\boldsymbol{D}_u$, which consists of the columns of $\boldsymbol{D}$ with indices in u. Applying (3.6) to $\boldsymbol{D}_u$, we obtain that $N'_\phi = 2^{-|u|} \sum_{w \subseteq u} J_w$, where $N'_\phi = N_\phi(\boldsymbol{D}_u)$ is the number of times the all-one run occurs in $\boldsymbol{D}_u$. As $J_\phi = n = \lambda 2^t$ and $J_w = 0$ for all w with $1 \leq |w| \leq t$, we have

$$J_u = 2^{|u|} N'_\phi - \lambda 2^t - \sum_{w \subset u,\ t+1 \leq |w| \leq |u|-1} J_w. \tag{3.7}$$

If $|u| = t+1$, then (3.7) gives $J_u = 2^{t+1} N'_\phi - \lambda 2^t$, which is a multiple of 2^t. For $|u| \geq t+2$, we see that J_u must also be a multiple of 2^t by mathematical induction using (3.7). This establishes that $J_u = \mu_u 2^t$ for some integer μ_u. An almost identical argument can be used to prove part (i) of Theorem 3.6. Proving part (ii) of the theorem also relies on (3.7). This proof is straightforward as well, though slightly more complicated. □

When λ is odd, we can also derive results on μ_u values for $|u| > t+3$, but we decide not to pursue them as they get more complicated and are seldom used. Rather, we will present complete results for the most useful case of strength-two arrays.

Theorem 3.7. *Let $\boldsymbol{D}$ be an OA$(n, 2^m, 2)$. Then J_u is a multiple of 4 for any u. The following further results hold for J_u.*

(i) If n is a multiple of 8, then J_u is a multiple of 8.

(ii) If n is not a multiple of 8, then J_u is not a multiple of 8 for $|u| = 4j-1$ and $4j$, and is a multiple of 8 for $|u| = 4j+1$ and $4j+2$, where $j = 1, 2, \ldots$.

Proof. Only part (ii) of Theorem 3.7 needs a proof. Consider a u with $|u| \geq 3$. Let i and j be any two elements of u. Further let $w = u \setminus \{i, j\}$, and we define $\boldsymbol{d}_w$ to be the Hadamard product of the columns $\boldsymbol{d}_k$s for all $k \in w$. Consider the three-column design $\boldsymbol{D}^* = (\boldsymbol{d}_i, \boldsymbol{d}_j, \boldsymbol{d}_w)$. Applying equation (3.3) of Theorem 3.2 to design $\boldsymbol{D}^*$ and taking $v = \phi$, we obtain that

$$N_\phi = 2^{-3} \left(n + J(\boldsymbol{d}_w) + J(\boldsymbol{d}_i, \boldsymbol{d}_w) + J(\boldsymbol{d}_j, \boldsymbol{d}_w) + J(\boldsymbol{d}_i, \boldsymbol{d}_j, \boldsymbol{d}_w) \right),$$

which has been simplified by using $J(\boldsymbol{d}_i) = J(\boldsymbol{d}_j) = J(\boldsymbol{d}_i, \boldsymbol{d}_j) = 0$. It is also obvious that $J(\boldsymbol{d}_w) = J_{u\setminus\{i,j\}}$, $J(\boldsymbol{d}_i, \boldsymbol{d}_w) = J_{u\setminus\{j\}}$, $J(\boldsymbol{d}_j, \boldsymbol{d}_w) = J_{u\setminus\{i\}}$, and $J(\boldsymbol{d}_i, \boldsymbol{d}_j, \boldsymbol{d}_w) = J_u$. We therefore obtain that

$$8N_\phi = n + J_{u\setminus\{i,j\}} + J_{u\setminus\{j\}} + J_{u\setminus\{i\}} + J_u.$$

Now we assume that part (ii) of the theorem holds for all subsets u' with $|u'| \leq |u| - 1$. Then it follows that $J_{u\setminus\{j\}} + J_{u\setminus\{i\}}$ must be a multiple of 8. As n is not a multiple of 8, we see that $J_{u\setminus\{i,j\}} + J_u$ must not be a multiple of 8. Therefore, whether or not being a multiple of 8, $J_{u\setminus\{i,j\}}$ and J_u must be the opposite of each other. Part (ii) of Theorem 3.7 is now established. We note that J_u being a multiple of 4 has been implicitly used in the arguments above. □

If we are only interested in the main effects and two-factor interactions, then the relevant aliasing information is all contained in the J_u values for $|u| = 3$ and 4, assuming an orthogonal array of strength two is under consideration. In this practically important case, Theorem 3.7 can be expressed in a very simple way.

Corollary 3.4. *Let $\boldsymbol{D}$ be an $OA(n, 2^m, 2)$. Then $n - |J_u|$ is always a multiple of 8 for any u with $|u| = 3$ and 4.*

Theorems 3.6 and 3.7 deal with possible values of J_u for one u. The next result says that if subsets u_1 and u_2 are "close", then their J values are constrained by each other.

Theorem 3.8. *Let $\boldsymbol{D}$ be an $OA(n, 2^m, t)$. We must have*

$$|J_{u_1}| + |J_{u_2}| \leq n$$

provided that $1 \leq |u_1 \triangle u_2| \leq t$, where $u_1 \triangle u_2 = (u_1 \setminus u_2) \cup (u_2 \setminus u_1)$.

Proof. Let $\boldsymbol{d}_j$ be the jth column of $\boldsymbol{D}$ and $\boldsymbol{d}_{u_1}$ be the Hadamard product of the columns $\boldsymbol{d}_j$ for $j \in u_1$. Define $\boldsymbol{d}_{u_2}$ similarly. Consider the design $\boldsymbol{D}^*$ given by the two column vectors $\boldsymbol{d}_{u_1}$ and $\boldsymbol{d}_{u_2}$; that is, $\boldsymbol{D}^* = (\boldsymbol{d}_{u_1}, \boldsymbol{d}_{u_2})$. Obviously, $J(\boldsymbol{d}_{u_1}) = J_{u_1}$ and $J(\boldsymbol{d}_{u_2}) = J_{u_2}$. In addition, we have $J(\boldsymbol{d}_{u_1}, \boldsymbol{d}_{u_2}) = J_{u_1 \triangle u_2} = 0$, because $1 \leq |u_1 \triangle u_2| \leq t$ and $\boldsymbol{D}$ is an orthogonal array of strength t. Now, applying (3.3) to design $\boldsymbol{D}^*$ and noting that $N_v \geq 0$ for all v, we obtain

$$n + J_{u_1} + J_{u_2} \geq 0, \quad n - J_{u_1} + J_{u_2} \geq 0, \quad n + J_{u_1} - J_{u_2} \geq 0, \quad n - J_{u_1} - J_{u_2} \geq 0.$$

This shows that $|J_{u_1}| + |J_{u_2}| \leq n$. □

As an immediate application, we note that a key result in Cheng (1995a, Lemma 2.2) is a direct consequence of Theorem 3.8. Provided that the condition in Theorem 3.8 is satisfied, we must have $J_{u_2} = 0$ if $|J_{u_1}| = n$, and this is essentially what Lemma 2.2 of Cheng (1995a) means. The next result is an application of Theorem 3.8 to orthogonal arrays of strength two.

Corollary 3.5. *Let $\boldsymbol{D}$ be an $OA(n, 2^m, 2)$, where n is not a multiple of 8.*

(i) We have that $|J_u| \leq n - 8$ for all u with $1 \leq |u| \leq m - 1$.

(ii) If $|J_u| = n$ for $u = Z_m$, then we must have that $m \equiv 3 \pmod 4$ and that $\boldsymbol{D}$ cannot be embedded into an $OA(n, 2^{m+1}, 2)$.

Proof. For a given u, we want to find a subset u_2 by deleting one or two columns from or adding one or two columns to u such that J_{u_2} is not a multiple of 8; hence $|J_{u_2}| \neq 0$. Once this u_2 is found, applying Theorem 3.8, we obtain that $|J_u| + |J_{u_2}| \leq n$, showing that $|J_u| < n$. Part (i) of Corollary 3.5 follows because, by Corollary 3.4, $n - |J_u|$ is a multiple of 8.

With reference to part (ii) of Theorem 3.7, we see that this desired u_2 can always be found for $1 \leq |u| \leq m - 1$ by one of three possible ways: deleting one or two columns from u or adding one column to u. Now suppose that $\boldsymbol{D}$ cannot be embedded into an $OA(n, 2^{m+1}, 2)$ and we consider J_u for $u = Z_m$. If $m \equiv 3 \pmod 4$ is not true, then we can still find a desired u_2 by considering the deletion of one or two columns from u. This shows that if $|J_u| = n$ for $u = Z_m$, then it is necessary that $m \equiv 3 \pmod 4$ and that $\boldsymbol{D}$ cannot be embedded into an $OA(n, 2^{m+1}, 2)$. □

If $|J_u| = n$, then the set of columns $\boldsymbol{d}_j$'s with $j \in u$ is said to form a defining word, which results in full effect aliasing. Corollary 3.5 says that when the run size n is not a multiple of 8, except in rare cases, an orthogonal array of strength two does not have a defining word.

We give two examples to illustrate the results in this section.

Example 3.4. *Let $\boldsymbol{D}$ be an $OA(16, 2^{15}, 2)$ obtained by deleting the all-one column from a normalized Hadamard matrix of order 16. Because $n = 16$ is a multiple of 8, by part (i) of Theorem 3.7 we have that $J_u = 0, \pm 8, \pm 16$ for all u. Now suppose that $|J_u| = 16$ for some u, say $|J(\boldsymbol{d}_1, \boldsymbol{d}_2, \boldsymbol{d}_3)| = 16$. Then by Theorem 3.8, we must have that*

$$J(\boldsymbol{d}_1, \boldsymbol{d}_2, \boldsymbol{d}_4) = J(\boldsymbol{d}_1, \boldsymbol{d}_2, \boldsymbol{d}_3, \boldsymbol{d}_4) = J(\boldsymbol{d}_1, \boldsymbol{d}_2, \boldsymbol{d}_3, \boldsymbol{d}_4, \boldsymbol{d}_5) = 0.$$

Example 3.5. *Let $\boldsymbol{D}$ be an $OA(20, 2^{19}, 2)$ obtained by deleting the all-one column from a normalized Hadamard matrix of order 20. Since $n = 20$ is not a multiple of 8, by part (ii) of Theorem 3.7 we have that*

$$|J_u| = \begin{cases} 4, 12, 20 & \text{for all } u \text{ with } |u| = 3, 4, 7, 8, 11, 12, 15, 16, 19, \text{ and} \\ 0, 8, 16 & \text{for all } u \text{ with } |u| = 5, 6, 9, 10, 13, 14, 17, 18. \end{cases}$$

Suppose that $|J(\boldsymbol{d}_1, \boldsymbol{d}_2, \boldsymbol{d}_3, \boldsymbol{d}_4)| = 12$. Then we have that $|J_u| \leq 8$ for many u's according to Theorem 3.8. For example, we must have that

$$|J(\boldsymbol{d}_1, \boldsymbol{d}_2, \boldsymbol{d}_3)| = |J(\boldsymbol{d}_1, \boldsymbol{d}_2, \boldsymbol{d}_3, \boldsymbol{d}_5)| = 4.$$

By part (i) of Corollary 3.5, $|J_u|$ cannot be equal to 20 for all nonempty u's with $|u| < 19$. By negating some columns, one can convert the array into one that has all ones in its first row. Then all other rows must have 9 plus ones and 10 minus ones because the rows of the corresponding Hadamard matrix are orthogonal. This shows that $|J_u| = 20$ for $u = \{1, 2, \ldots, 19\}$; that is, $|J(\boldsymbol{d}_1, \boldsymbol{d}_2, \ldots, \boldsymbol{d}_{19})| = 20$. By part (ii) of Corollary 3.5, we have that $19 \equiv 3 \pmod 4$, which is obviously true, and that $\boldsymbol{D}$ cannot be embedded into an $OA(20, 2^{20}, 2)$, which is also obviously true because any $OA(20, 2^{19}, 2)$ is saturated.

3.5 Projection Properties

Corollary 3.2 in Section 3.2 shows that the projection design $\boldsymbol{D}_u$, which consists of columns $\boldsymbol{d}_j$ of $\boldsymbol{D}$ with $j \in u$, is completely determined by the collection of J-characteristics J_w with

$w \subseteq u$. When $\boldsymbol{D}$ is an OA$(n, 2^m, t)$ and $|u| = t+1$, the projection design $\boldsymbol{D}_u$ has a simple and appealing structure, fully determined by the single characteristic J_u. This is detailed in the next lemma.

Lemma 3.2. *Let $\boldsymbol{D}$ be an OA$(n, 2^m, t)$. Consider a projection design $\boldsymbol{D}_u$ with $|u| = t+1$. Then, up to a row permutation, $\boldsymbol{D}_u$ can be written as*

$$\boldsymbol{D}_u = \begin{bmatrix} \boldsymbol{D}_1 \\ \boldsymbol{D}_2 \end{bmatrix},$$

where $\boldsymbol{D}_1$ is a design of n_1 runs, consisting of α copies of a 2^{t+1} complete factorial, and $\boldsymbol{D}_2$ is a design of n_2 runs, having β copies of a half fraction of 2^{t+1} factorial with resolution $t+1$, where

$$n_1 = n - |J_u|, \qquad n_2 = |J_u|, \qquad \alpha = 2^{-(t+1)}\left(n - |J_u|\right), \qquad \beta = 2^{-t}|J_u|.$$

Proof. Applying Theorem 3.2 to $\boldsymbol{D}_u$ and noting that $\boldsymbol{D}_u$ is an array of strength t, we obtain that $N'_v = 2^{-(t+1)}\left(n + (-1)^{|u\cap v|}J_u\right)$, where $N'_v = N_v(\boldsymbol{D}_u)$ is the indicator function of $\boldsymbol{D}_u$. We see that N'_v can only take on one of the two possible values, either $2^{-(t+1)}(n + |J_u|)$ or $2^{-(t+1)}(n - |J_u|)$. This means that any run from a 2^{t+1} complete factorial occurs in $\boldsymbol{D}_u$ either $2^{-(t+1)}(n + |J_u|)$ or $2^{-(t+1)}(n - |J_u|)$ times. Let $\alpha = 2^{-(t+1)}(n - |J_u|)$ and $\boldsymbol{D}_1$ be a design with $n_1 = n - |J_u|$ runs, given by α copies of a 2^{t+1} complete factorial. Then $\boldsymbol{D}_u$ can be written as $\boldsymbol{D}_u = (\boldsymbol{D}_1^T, \boldsymbol{D}_2^T)^T$ where $\boldsymbol{D}_2$ collects the remaining runs of $\boldsymbol{D}_u$ that are not in $\boldsymbol{D}_1$. Since $J_w(\boldsymbol{D}_u) = J_w(\boldsymbol{D}_1) + J_w(\boldsymbol{D}_2)$, we have that $J_w(\boldsymbol{D}_2) = 0$ for all $|w| \le t$ and $J_u(\boldsymbol{D}_2) = J_u(\boldsymbol{D}_u) = J_u$. Thus, $\boldsymbol{D}_2$ is a design of $n_2 = n - n_1 = |J_u|$ runs, having $\beta = |J_u|/2^t$ copies of a half fraction of 2^{t+1} complete factorial with resolution $t+1$. □

Example 3.6. *Let $\boldsymbol{D} = (\boldsymbol{d}_1, \boldsymbol{d}_2, \ldots, \boldsymbol{d}_9)$ be the 20-run design in Table 1.4. This design has that $J(\boldsymbol{d}_1, \boldsymbol{d}_2, \boldsymbol{d}_3) = 12$ and $J(\boldsymbol{d}_1, \boldsymbol{d}_2, \boldsymbol{d}_4) = 4$. According to Lemma 3.2, the projection design $(\boldsymbol{d}_1, \boldsymbol{d}_2, \boldsymbol{d}_3)$ contains one copy of a 2^3 complete factorial and the projection design $(\boldsymbol{d}_1, \boldsymbol{d}_2, \boldsymbol{d}_4)$ contains two copies of a 2^3 complete factorial.*

Lemma 3.2 includes two special cases, $\boldsymbol{D}_u = \boldsymbol{D}_1$ corresponding to $J_u = 0$ and $\boldsymbol{D}_u = \boldsymbol{D}_2$ corresponding to $J_u = \pm n$. But so long as $J_u \neq \pm n$, $\boldsymbol{D}_u$ contains at least one copy of a complete 2^{t+1} factorial. When $J_u = \pm n$, we say that the columns $\boldsymbol{d}_j$ with $j \in u$ form a defining word of length $|u| = t+1$. With all the above preparation, we obtain the next result.

Theorem 3.9. *Let $\boldsymbol{D}$ be an OA$(n, 2^m, t)$. If $\boldsymbol{D}$ does not have a defining word of length $t+1$, then every projection design onto $t+1$ factors contains at least one copy of a complete 2^{t+1} factorial.*

With the help of Theorems 3.6 and 3.8, Theorem 3.9 leads to another important result.

Theorem 3.10. *Let $\boldsymbol{D}$ be an OA$(n, 2^m, t)$ with $m \ge t+2$. If n is not a multiple of 2^{t+1}, then every projection design onto $t+1$ factors has no defining word of length $t+1$, and thus contains at least one copy of a complete 2^{t+1} factorial.*

Proof. Let $n = \lambda 2^t$. Then λ is odd because n is not a multiple of 2^{t+1}. According to Theorem 3.6, for any u with $|u| = t+1$, we must have $J_u = \mu_u 2^t$ for some odd μ_u. This means that no u with $|u| = t+1$ can have $J_u = 0$. If we choose $u_1 = \{1, \ldots, t+1\}$ and $u_2 = \{2, \ldots, t+2\}$ to apply Theorem 3.8, we obtain that $|J_{u_1}| + |J_{u_2}| \le n$, which implies that $|J_{u_1}| < n$ because $|J_{u_2}|$ cannot be zero. The same argument allows us to establish that $|J_u| < n$ for any u with $|u| = t+1$, as one can let $u_1 = u$ and choose a u_2 by deleting an element from u and adding an element from outside u, which is possible because $m \ge t+2$. This means that design $\boldsymbol{D}$ does not have any defining word of length $t+1$. Theorem 3.10 then follows from Theorem 3.9. □

Orthogonal arrays of strength two and three are the most important, and thus deserve special mention.

Corollary 3.6. *(i) If n is not a multiple of 8, then every projection design of an $OA(n, 2^m, 2)$ with $m \geq 4$ onto three factors contains at least one copy of a complete 2^3 factorial.*
(ii) If n is not a multiple of 16, then every projection design of an $OA(n, 2^m, 3)$ with $m \geq 5$ onto four factors contains at least one copy of a complete 2^4 factorial.

As shown in Section 2.6, one can construct two arrays $\boldsymbol{D}_1$, an $\text{OA}(n, 2^{n-1}, 2)$, and $\boldsymbol{D}_2$, an $\text{OA}(2n, 2^n, 3)$, from a Hadamard matrix of order n. If n is not a multiple of 8, then by Corollary 3.6, we conclude that every 3-factor projection design of $\boldsymbol{D}_1$ contains at least one copy of a 2^3 complete factorial and every 4-factor projection design of $\boldsymbol{D}_2$ contains at least one copy of a 2^4 complete factorial. Examples of $\boldsymbol{D}_1$ are an $\text{OA}(12, 2^{11}, 2)$, an $\text{OA}(20, 2^{19}, 2)$, and an $\text{OA}(28, 2^{27}, 2)$, and the corresponding $\boldsymbol{D}_2$ are an $\text{OA}(24, 2^{12}, 3)$, an $\text{OA}(40, 2^{20}, 3)$, and an $\text{OA}(56, 2^{28}, 3)$.

The requirements for the run size n in Corollary 3.6 are sufficient but not necessary. If the Hadamard matrix in the preceding paragraph is obtained from Paley's first construction, then the corresponding arrays $\boldsymbol{D}_1$ and $\boldsymbol{D}_2$ still enjoy the same projection properties even if $n \geq 12$ is a multiple of 8. This topic will be dealt with in the next chapter.

One last point that needs attention is the condition $m \geq t + 2$ in Theorem 3.10 and Corollary 3.6. This is necessary, for otherwise, the projection property may not hold. An illustration of this point is an $\text{OA}(12, 2^3, 2)$ given by three copies of

$$\begin{bmatrix} 1 & 1 & 1 \\ -1 & 1 & -1 \\ 1 & -1 & -1 \\ -1 & -1 & 1 \end{bmatrix},$$

which does not contain a complete 2^3 factorial.

We make a final note that the results in this section were originally obtained by Cheng (1995a).

3.6 Hidden Projection Properties

We have seen in Section 3.5 that if n is not a multiple of 8, then an $\text{OA}(n, 2^m, 2)$ with $m \geq 4$ has the appealing projection property that every projection design onto three factors contains at least one copy of a complete 2^3 factorial. This array has another surprising property that every projection design onto four factors allows the estimation of all main effects and all two-factor interactions. This was named as a hidden projection property by Wang and Wu (1995) when studying nonregular designs of 12 and 20 runs from the estimability perspective of projection designs. We present some general results from Cheng (1995a) and Cheng (1998). The next result is due to Cheng (1995a).

Theorem 3.11. *Let $\boldsymbol{D}$ be an $OA(n, 2^m, 2)$ where $m \geq 4$. Then every projection design onto four factors allows the estimation of all main effects and all two-factor interactions if and only if $\boldsymbol{D}$ has no defining word of length three or four.*

Proof. It is evident that we only need to prove the theorem for $m = 4$. The necessity part of the theorem is obvious; only the sufficiency part needs a proof. Since $\boldsymbol{D}$ has no defining word of length four, the foldover design

$$\tilde{\boldsymbol{D}} = \begin{bmatrix} \boldsymbol{D} \\ -\boldsymbol{D} \end{bmatrix}$$

is an orthogonal array of strength three that has no defining word of length four. Applying Theorem 3.9 to $\boldsymbol{D}$ and $\tilde{\boldsymbol{D}}$, we obtain that

(i) every projection design of $\boldsymbol{D}$ onto three factors contains a complete 2^3 factorial;

(ii) for any $\boldsymbol{x} = (x_1, x_2, x_3, x_4)$ with $x_i = \pm 1$, $\boldsymbol{D}$ must contain at least one of $\boldsymbol{x}$ and $-\boldsymbol{x}$.

For the purpose of establishing estimability, we can treat the rows of $\boldsymbol{D}$ as a set of points and ignore the repeats. Theorem 3.11 follows if we can show that any $\boldsymbol{D}$ that satisfies both (i) and (ii) allows the estimation of all main effects and all two-factor interactions. This is easily done on a computer. We first generate all possible designs that satisfy condition (ii). The 16 possible vectors (x_1, x_2, x_3, x_4) where $x_i = \pm 1$ are in 8 pairs, within each of which, the two points are mirror images of the other. To obey condition (ii), from each pair, we either select one of the two points or select both points. This means that the total number of designs that can be generated while satisfying condition (ii) is $3^8 = 6561$. We then check if they satisfy condition (i). It turns out that 393 of the 6561 designs satisfy condition (i). So, there are 393 designs that satisfy both conditions (i) and (ii). It can be verified by a computer that all these 393 designs allow the estimation of all main effects and all two-factor interactions. This completes the proof. □

If n is not a multiple of 8, then it follows from Corollary 3.5 that an OA$(n, 2^m, 2)$ with $m \geq 4$ cannot have a defining word of length three or four. This gives rise to the next result.

Corollary 3.7. *Let $\boldsymbol{D}$ be an OA$(n, 2^m, 2)$ where $m \geq 4$ and n is not a multiple of 8. Then every projection design onto four factors allows the estimation of all main effects and all two-factor interactions.*

Examples of orthogonal arrays having the hidden projection property in Corollary 3.7 include an OA$(12, 2^{11}, 2)$, an OA$(20, 2^{19}, 2)$, and an OA$(28, 2^{27}, 2)$, which are all readily available from Hadamard matrices of order 12, 20, and 28, respectively.

The hidden projection property enjoyed by nonregular designs as given in Theorem 3.11 and Corollary 3.7 is nothing less than remarkable, if one looks at regular designs from this perspective. Regular designs do not have this property unless they are of resolution V or higher, which requires exceedingly large run sizes even for moderate numbers of factors. To illustrate, the run size n would have to be 128 for $m = 11$, 512 for $m = 19$, and 1024 for $m = 27$, whereas the corresponding run sizes for nonregular designs are 12, 20, and 28. The differentials in the required run sizes are astounding.

The condition of n being not a multiple of 8 is sufficient but not necessary. What is really required as given in Theorem 3.11 is the absence of defining words of length three and four. When $n \geq 12$ is a multiple of 8, an OA$(n, 2^{n-1}, 2)$ given by the Hadamard matrix from Paley's first construction does not have any defining word of length three or four. Many other Hadamard matrices can also give rise to orthogonal arrays with this property. We will revisit this issue in Section 4.4.

Theorem 3.11 was generalized by Cheng (1998) to orthogonal arrays of strength three.

Theorem 3.12. *Let $\boldsymbol{D}$ be an OA$(n, 2^m, 3)$ where $m \geq 5$. Then every projection design onto five factors allows the estimation of all main effects and all two-factor interactions if and only if $\boldsymbol{D}$ has no defining word of length four.*

Proof. We only need to prove the theorem for $m = 5$, and only need to prove the sufficiency part. Let $\boldsymbol{D} = (\boldsymbol{d}_1, \ldots, \boldsymbol{d}_5)$. Suppose that $\boldsymbol{D}$ has no defining word of length four. We need to prove that the following 16 column vectors

$$\mathbf{1}_n, \boldsymbol{d}_1, \ldots, \boldsymbol{d}_5, \boldsymbol{d}_1\boldsymbol{d}_2, \ldots, \boldsymbol{d}_4\boldsymbol{d}_5$$

are linearly independent. Since $\boldsymbol{D}$ is a strength-three array, the two-factor interaction columns $\boldsymbol{d}_i\boldsymbol{d}_j$s are orthogonal to the grand mean column $\mathbf{1}_n$ and the main effect columns $\boldsymbol{d}_j$s. This shows that we only need to prove that the interaction columns $\boldsymbol{d}_i\boldsymbol{d}_j$s are linearly independent. Consider a design $\boldsymbol{D}'$ defined by $\boldsymbol{D}' = (\boldsymbol{d}_1\boldsymbol{d}_2, \boldsymbol{d}_1\boldsymbol{d}_3, \boldsymbol{d}_1\boldsymbol{d}_4, \boldsymbol{d}_1\boldsymbol{d}_5)$. We then see that design $\boldsymbol{D}'$ is an $\mathrm{OA}(n, 2^4, 2)$ that has no defining word of length three or four. By Theorem 3.11, the main effect columns and two-factor interaction columns of design $\boldsymbol{D}'$ are linearly independent. We observe that these columns are precisely all the $\boldsymbol{d}_i\boldsymbol{d}_j$s for $1 \le i < j \le 5$. The proof is now complete. □

An application of Theorem 3.10 leads to the next corollary.

Corollary 3.8. *Let $\boldsymbol{D}$ be an $OA(n, 2^m, 3)$ where $m \ge 5$ and n is not a multiple of 16. Then any projection design onto five factors allows the estimation of all main effects and all two-factor interactions.*

Examples of orthogonal arrays having the hidden projection property in Corollary 3.8 include an $\mathrm{OA}(24, 2^{12}, 3)$, an $\mathrm{OA}(40, 2^{20}, 3)$, and an $\mathrm{OA}(56, 2^{28}, 3)$, which can readily be constructed from folding over Hadamard matrices of orders 12, 20, and 28, respectively.

3.7 Enumeration of Orthogonal Arrays

This long section is divided into two subsections. In the first subsection, we introduce the enumeration problem for orthogonal arrays, present some computational findings, investigate the relationships between the numbers of nonisomorphic Hadamard matrices and saturated orthogonal arrays of strength two and three, and discuss the embeddability of orthogonal arrays. In the second subsection, we provide some theoretical results on enumerating orthogonal arrays by using J-characteristics.

3.7.1 Design classification and enumeration

Two factorial designs are said to be isomorphic if one can be obtained from the other by a sequence of isomorphism operations that include permutations of rows, permutations of columns, and negation of columns, and are said to be nonisomorphic otherwise. The enumeration problem is to classify all designs into isomorphism classes, thereby finding a complete collection of nonisomorphic designs. It is important to note that negating rows is not permissible for design isomorphism but is allowed for Hadamard matrix isomorphism.

This enumeration problem is of theoretical interest, not just because of the simplicity rendered by focusing on nonisomorphic designs but also because all possible designs can be generated through isomorphism operations from a complete collection of nonisomorphic designs. The problem is practically important as well. In most applications of factorial designs, isomorphic designs have the same statistical properties and can thus be considered as the same design, and we therefore only need to examine all nonisomorphic designs. These situations occur when the two levels are arbitrarily designated as high and low levels, the

TABLE 3.1
The numbers of nonisomorphic $\mathrm{OA}(n, 2^m, 2)$s.

m	n=8	n=12	n=16	n=20	n=24
3	2	2	3	3	4
4	2	1	5	3	10
5	1	2	11	11	63
6	1	2	27	75	1350
7	1	1	55	474	57389
8		1	80	1603	1470157
9		1	87	2477	12952435
10		1	78	2389	38592861
11		1	58	1914	52912678
12			36	1300	51154497
13			18	730	43092737
14			10	328	31833387
15			5	124	19960039
16				40	10351396
17				11	4385567
18				6	1502242
19				3	409478
20					86725
21					13833
22					1604
23					130

factors are of equal importance and the runs are to be completely randomized. Indeed, there are applications where isomorphic designs cannot always be treated as the same; examples include but are not limited to when a baseline parametrization is appropriate, when some two-factor interactions are known a priori to be important, or when the experiments are to be run in some systematic order for cost consideration. But even in such situations, knowing a complete collection of nonisomorphic designs is still useful, since one can generate all suitable designs by certain isomorphism operations from this complete collection. For example, in experiments that call for a baseline parametrization where one level is the default level and the other is the test level, one can obtain all possible designs by conducting level permutations from the complete collection of nonisomorphic designs.

Enumerating orthogonal arrays is a very challenging problem. General theoretical results are hard to come by, and to date, there are only a few such results available in the literature, some of which will be presented in the next subsection.

Alternatively, one can resort to computer search but even with modern computers, complete enumeration of orthogonal arrays can only be done for small run sizes, for the obvious reason that a very large number of isomorphism operations have to be considered in order to determine if two designs are isomorphic. To be computationally feasible, theoretical insights are often necessary for developing efficient algorithms. Over the years, many researchers have endeavored to enumerate orthogonal arrays. Most comprehensive results were obtained by Sun, Li, and Ye (2008) and Schoen, Eendebak, and Nguyen (2010), who, among other things, completely enumerated strength-two arrays for up to 24 runs and strength-three arrays for up to 48 runs. We present a summary of these results. Table 3.1 provides the numbers of nonisomorphic orthogonal arrays of strength two and Table 3.2 does the same for strength-three arrays of up to 48 runs.

As discussed in Section 2.6, an $\mathrm{OA}(n, 2^{n-1}, 2)$ and an $\mathrm{OA}(2n, 2^n, 3)$ can be constructed from a Hadamard matrix of order n and vice versa. One may wonder if and how the

TABLE 3.2
The numbers of nonisomorphic OA$(n, 2^m, 3)$s.

m	n=16	n=24	n=32	n=40	n=48
4	2	2	3	3	4
5	2	1	5	3	10
6	1	2	10	9	45
7	1	1	17	25	397
8	1	1	33	105	8383
9		1	34	213	166081
10		1	32	353	1310006
11		1	22	260	3528089
12		1	23	235	4460865
13			12	132	3980095
14			10	96	3139653
15			5	36	2165144
16			5	26	1288460
17				7	629705
18				6	259346
19				3	84495
20				3	24012
21					4919
22					1129
23					130
24					60

numbers of nonisomorphic OA$(n, 2^{n-1}, 2)$s and OA$(2n, 2^n, 3)$s are related to the number of nonisomorphic Hadamard matrices of order n. Let $a(n)$ and $b(n)$, respectively, denote the numbers of nonisomorphic OA$(n, 2^{n-1}, 2)$s and OA$(2n, 2^n, 3)$s, and $c(n)$ denote the number of nonisomorphic Hadamard matrices of order n.

Theorem 3.13. *We have that* $b(n) = c(n) \leq a(n)$.

Proof. We first prove that $c(n) \leq a(n)$. Note that every Hadamard matrix is isomorphic to one with the first column consisting of all ones. Let $\boldsymbol{H}_1 = (\boldsymbol{1}_n, \boldsymbol{D}_1)$ and $\boldsymbol{H}_2 = (\boldsymbol{1}_n, \boldsymbol{D}_2)$ be two nonisomorphic Hadamard matrices of order n. If we show that $\boldsymbol{D}_1$ and $\boldsymbol{D}_2$ are also nonisomorphic, then we establish that $c(n) \leq a(n)$. We prove this by contradiction. Assume that $\boldsymbol{D}_1$ and $\boldsymbol{D}_2$ are isomorphic. Then there is a sequence of isomorphism operations involving permutations of rows, permutations of columns and negation of columns that can convert $\boldsymbol{D}_1$ to $\boldsymbol{D}_2$. Obviously, this same sequence of isomorphism operations also converts $\boldsymbol{H}_1$ to $\boldsymbol{H}_2$, thereby showing that $\boldsymbol{H}_1$ and $\boldsymbol{H}_2$ are isomorphic. This is a contradiction.

We need some preparation to prove $b(n) = c(n)$. According to Theorem 3.5, an OA$(2n, 2^n, 3)$ must be a foldover design. Therefore, we may assume that all OA$(2n, 2^n, 3)$s have the form of $\boldsymbol{D} = (\boldsymbol{X}^T, -\boldsymbol{X}^T)^T$ for some $\boldsymbol{X}$. Then one can show that $\boldsymbol{X}$ must be a Hadamard matrix. To see this, let $\boldsymbol{x}_i$ and $\boldsymbol{x}_j$, where $i \neq j$, be two columns of $\boldsymbol{X}$, and $\boldsymbol{d}_i$ and $\boldsymbol{d}_j$ be the corresponding columns of $\boldsymbol{D}$. Then $J(\boldsymbol{x}_i, \boldsymbol{x}_j) = J(\boldsymbol{d}_i, \boldsymbol{d}_j)/2 = 0$ because $\boldsymbol{D}$ is an OA$(2n, 2^n, 3)$.

To prove that $b(n) \leq c(n)$, we only need to show that Hadamard matrices $\boldsymbol{X}_1$ and $\boldsymbol{X}_2$ are nonisomorphic if the two OA$(2n, 2^n, 3)$s $\boldsymbol{D}_1 = (\boldsymbol{X}_1^T, -\boldsymbol{X}_1^T)^T$ and $\boldsymbol{D}_2 = (\boldsymbol{X}_2^T, -\boldsymbol{X}_2^T)^T$ are nonisomorphic. Assume the opposite that $\boldsymbol{X}_1$ and $\boldsymbol{X}_2$ are isomorphic. Then there is a sequence of isomorphism operations involving permutations of rows or columns and negation of rows or columns that can convert $\boldsymbol{X}_1$ to $\boldsymbol{X}_2$. A column or row permutation on $\boldsymbol{X}_1$ naturally corresponds to a column or row permutation on $\boldsymbol{D}_1$, and negating a column of $\boldsymbol{X}_1$ naturally corresponds to negating a column of $\boldsymbol{D}_1$. Negating a row of $\boldsymbol{X}_1$ corresponds to switching one row with its mirror image in $\boldsymbol{D}_1$. This shows that there is a sequence of isomorphism operations involving permutations of rows or columns and negation of columns that can convert $\boldsymbol{D}_1$ to $\boldsymbol{D}_2$, showing that $\boldsymbol{D}_1$ and $\boldsymbol{D}_2$ are isomorphic, which is a contradiction.

To prove that $c(n) \leq b(n)$, we only need to show that two OA$(2n, 2^n, 3)$s $\boldsymbol{D}_1 = (\boldsymbol{X}_1^T, -\boldsymbol{X}_1^T)^T$ and $\boldsymbol{D}_2 = (\boldsymbol{X}_2^T, -\boldsymbol{X}_2^T)^T$ are nonisomorphic if Hadamard matrices $\boldsymbol{X}_1$ and $\boldsymbol{X}_2$ are nonisomorphic. Assume that $\boldsymbol{D}_1$ and $\boldsymbol{D}_2$ are isomorphic. Then there is a sequence of isomorphism operations involving the permutation of rows or columns and the negation of columns that can convert $\boldsymbol{D}_1$ to $\boldsymbol{D}_2$. A column permutation on $\boldsymbol{D}_1$ naturally corresponds to a column permutation on $\boldsymbol{X}_1$, and negating a column of $\boldsymbol{D}_1$ naturally negates a column of $\boldsymbol{X}_1$. To maintain the foldover structure, only certain row permutations on $\boldsymbol{D}_1$ are permissible. These include simultaneously and identically permuting rows within the group of rows 1 to n and within the group of rows $n+1$ to $2n$, which naturally leads to a row permutation on $\boldsymbol{X}_1$. The only cross-group row permutation without damaging the foldover structure is switching a row in the first group with its mirror image in the second group. This amounts to negating a row of $\boldsymbol{X}_1$. The above arguments show that $\boldsymbol{X}_1$ and $\boldsymbol{X}_2$ are isomorphic, a contradiction. □

TABLE 3.3
The numbers of nonisomorphic OA$(n, 2^{n-1}, 2)$s, OA$(2n, 2^n, 3)$s and Hadamard matrices of order n, denoted by $a(n)$, $b(n)$, and $c(n)$, respectively.

n	12	16	20	24	28	32
$a(n)$	1	5	3	130	7570	$\geq 13,710,027$
$b(n)$	1	5	3	60	487	13,710,027
$c(n)$	1	5	3	60	487	13,710,027

Table 3.3 presents $a(n), b(n)$, and $c(n)$ for $n \leq 32$, based on Theorem 3.13 and the computational results in Schoen, Eendebak, and Nguyen (2010) and Kharaghani and Tayfeh-Rezaie (2013). The fact $a(n) > c(n)$ for $n = 24$ implies that there must exist a Hadamard matrix of order 24 that, when normalizing different columns, can produce nonisomorphic saturated orthogonal arrays of strength two.

Another important question is whether or not all OA$(n, 2^m, 2)$s can be obtained from Hadamard matrices of order n. This question can be answered by considering a more general problem for what m^* values maximal OA$(n, 2^{m^*}, 2)$s exist. An OA$(n, 2^{m^*}, 2)$ is said to be maximal if it cannot be embedded into an OA$(n, 2^{m^*+1}, 2)$. Clearly, all OA$(n, 2^m, 2)$s can be obtained from Hadamard matrices of order n if and only if $m^* = n - 1$ is the only possible value for m^*. Knowing all possible m^* values for given n is useful because all OA$(n, 2^m, 2)$s can be obtained by taking projections from all maximal OA$(n, 2^{m^*}, 2)$s. Bulutoglu and Kaziska (2009) investigated this problem, and found that the set of m^* values is precisely $\{3, 11\}$ for $n = 12$, $\{15\}$ for $n = 16$, and $\{3, 11, 19\}$ for $n = 20$. They also found that the set of m^* values includes $\{7, 11, 15, 23\}$ as a subset for $n = 24$ and $\{3, 7, 9, 10, 11, 12, 13, 15, 19, 27\}$ as a subset for $n = 28$. General theoretical results on maximal designs are difficult to obtain. One such result is that of Verheiden (1978), who showed that a set of m orthogonal columns of ± 1 can be embedded into a Hadamard matrix if $m \geq n-7$. This means that if a maximal OA$(n, 2^{m^*}, 2)$ with $m^* < n - 1$ exists, we must have that $m^* \leq n - 9$. Another theoretical result is part (ii) of Corollary 3.5, which provides a sufficient condition for checking if an OA$(n, 2^m, 2)$ is maximal. More precisely, suppose that $\boldsymbol{D}$ is an OA$(n, 2^m, 2)$, where n is not a multiple of 8. If $|J_u| = n$ for $u = Z_m$, then $\boldsymbol{D}$ must be maximal and in addition we must have that $m \equiv 3 \pmod 4$. We can verify that all maximal orthogonal arrays of 12 and 20 runs satisfy this sufficient condition. On the other hand, the existence of maximal OA$(28, 2^{m^*}, 2)$s for $m^* = 9, 10, 12, 13$ shows that the sufficient condition is not necessary. For the material on maximal designs in the present paragraph, we benefited greatly from our email discussions with Bobby Mee, to whom we are grateful.

3.7.2 Enumerating OA$(n, 2^m, t)$s with m=t+1 and t+2

To start, we make some general comments on the enumeration of regular 2^{m-p} designs, which should shed some light on the enumeration of orthogonal arrays. Whether or not two regular designs are isomorphic can be judged by comparing their defining relations. As the defining relation is invariant with respect to row permutations, one only needs to examine level permutations and column permutations for checking isomorphism. The problem is further simplified by concentrating on the principal fractions, those having no negative signs in the defining relation, since this eliminates the consideration of level permutations. Thus, two regular designs are isomorphic if the defining relation of one design can be obtained from that of the other by a column permutation, and are nonisomorphic otherwise. More reduction in computation can be achieved by examining all possible choices of independent columns instead of all possible column permutations. These are indeed the main ideas in the approaches employed by Chen, Sun, and Wu (1993) and Xu (2009).

Corresponding to the defining relation of a regular design is the set of J-characteristics of an orthogonal array. The use of J-characteristics for enumeration also removes the consideration of row permutations. Unlike regular designs, there are no such things as principal fractions for orthogonal arrays. Thus, one has to consider both level switching and column permuting when examining if two arrays are isomorphic. Starting with a given orthogonal array, all possible ways of level switching and column permuting generate a collection of isomorphic arrays. These isomorphic arrays lead to different sets of J-characteristics. If we can find a unique set of J-characteristics out of all the possibilities by using suitable constraints, then this set of J-characteristics uniquely identifies the whole class of isomorphic arrays. The above summarizes the main ideas behind the results to be presented below.

We first look at enumerating OA$(n, 2^m, t)$s for $m = t + 1$. This problem was solved by Seiden and Zemach (1966). A simple derivation of their result can be obtained using J-characteristics, which we do next. An OA$(n, 2^m, t)$ with $m = t+1$ is uniquely determined by the single J-characteristic J_u where $u = Z_m$. Permuting columns does not change J_{Z_m}. The arrays obtained from level switching the columns in all possible ways all have $\pm J_{Z_m}$. This implies that for the enumeration problem, we only need to consider those arrays with $J_{Z_m} \leq 0$. Clearly, two arrays with different $|J_{Z_m}|$ values are nonisomorphic. Any value of $J_{Z_m} \leq 0$ gives an orthogonal array so long as the corresponding N_v is a nonnegative integer for all v, thus giving a legitimate indicator function. We are ready to present the next result.

Theorem 3.14. *We have that*

(i) each solution (μ, k) to the equation $\lambda + \mu = 2k$, where $-\lambda \leq \mu \leq 0$ and $k \geq 0$ is an integer, uniquely determines an OA$(\lambda 2^t, 2^{t+1}, t)$ with $J_{Z_m} = \mu 2^t$;

(ii) a complete set of nonisomorphic OA$(\lambda 2^t, 2^{t+1}, t)$s is given by collecting the arrays obtained in (i) over all possible solutions (μ, k)s.

Proof. To prove (i), consider equation (3.3) in Theorem 3.2, which now becomes $N_v = 2^{-m}(n + (-1)^{|v|} J_{Z_m}) = 2^{-(t+1)}(\lambda 2^t + (-1)^{|v|}\mu 2^t) = 2^{-1}(\lambda + (-1)^{|v|}\mu)$. The proof for (i) is completed if we can show that N_v is a nonnegative integer for all v, which is obvious as $2^{-1}(\lambda + \mu) = k$ is a nonnegative integer and $\mu \leq 0$ is also an integer. To prove (ii), consider an OA$(\lambda 2^t, 2^{t+1}, t)$ with $J_{Z_m} \leq 0$. By Theorem 3.6, there exists an integer μ such that

$J_{Z_m} = \mu 2^t$. Take $k = N_\phi = 2^{-(t+1)}(n + J_{Z_m}) = 2^{-1}(\lambda + \mu)$, which is a nonnegative integer. Then (μ, k) is a solution to $\lambda + \mu = 2k$. □

The solutions to the equation $\lambda + \mu = 2k$, where $-\lambda \leq \mu \leq 0$ and $k \geq 0$ is an integer, can easily be obtained. In terms of μ, they are all odd numbers in the interval $[-\lambda, 0]$ if λ is odd, and all even numbers in $[-\lambda, 0]$ if λ is even. Thus, the total number of nonisomorphic OA($\lambda 2^t, 2^{t+1}, t$)s is precisely $\lfloor \lambda/2 \rfloor + 1$.

The problem of enumerating OA($\lambda 2^t, 2^{t+2}, t$)s is significantly harder than that for OA($\lambda 2^t, 2^{t+1}, t$)s. Notwithstanding, Stufken and Tang (2007) provided a complete solution to the problem, where the four cases: (a) even t and odd λ, (b) even t and even λ, (c) odd t and odd λ, and (d) odd t and even λ, have to be dealt with separately. We now present their results for case (c), and refer the reader to the original paper for a full discussion.

An OA($\lambda 2^t, 2^{t+2}, t$) has $J_\phi = n$ and $J_u = 0$ for all nonempty u with $|u| \leq t$, and therefore it is completely determined by $J_{u_1}, \ldots, J_{u_{m+1}}$ where $u_j = Z_m \setminus \{m+1-j\}$ for $j = 1, \ldots, m$ and $u_{m+1} = Z_m$. As commented at the beginning of this subsection, we want to choose one array to represent the whole class of isomorphic arrays generated by all possible ways of sign-switching and column-permuting a given array. This is done in the following lemma, which we state without a proof.

Lemma 3.3. *When t is odd, every class of isomorphic arrays OA($\lambda 2^t, 2^{t+2}, t$)s contains a unique array that satisfies*

$$J_{u_1} \leq \cdots \leq J_{u_{m-1}} \leq -|J_{u_m}|, \quad J_{u_{m+1}} \leq 0. \tag{3.8}$$

Lemma 3.3 is a key to solving the enumeration problem. It allows the main theorem to be derived when coupled with the results on possible values of J-characteristics as presented in Theorem 3.6. Note that Lemma 3.3 only requires odd t and is valid for both odd and even λ.

Consider the equation

$$\lambda + \mu_1 + \cdots + \mu_{m+1} = 4k, \tag{3.9}$$

where k and μ_j's are unknown integers, $k \geq 0$, $\mu_1, \ldots, \mu_m$ are odd with $|\mu_j| \leq \lambda - 2$, and μ_{m+1} is even with $|\mu_{m+1}| \leq \lambda - 1$. Further, the μ_j's must satisfy

$$\mu_1 \leq \cdots \leq \mu_{m-1} \leq -|\mu_m|, \quad \mu_{m+1} \leq 0. \tag{3.10}$$

Theorem 3.15. *Consider OA($\lambda 2^t, 2^{t+2}, t$)s for odd t and odd λ. Then we have that*

(i) each solution $(\mu_1, \ldots, \mu_{m+1}, k)$ to equation (3.9) under (3.10) uniquely determines an OA($\lambda 2^t, 2^{t+2}, t$) with $J_{u_j} = 2^t \mu_j$ for $j = 1, \ldots, m+1$;

(ii) a complete set of nonisomorphic OA($\lambda 2^t, 2^{t+2}, t$)s is given by collecting the arrays obtained in (i) over all the solutions to equation (3.9) under (3.10).

Though the proof for Theorem 3.15 is more involved than that for Theorem 3.14, the ideas are actually similar. Basically, to prove part (i) of Theorem 3.15, we need to show that, for given values of $J_{u_j} = 2^t \mu_j$ determined by a solution to equation (3.9) under (3.10), N_v in equation (3.3) of Theorem 3.2 is a nonnegative integer for all v, and to prove part (ii), we need to show that there exists a solution to equation (3.9) under (3.10) for every OA($\lambda 2^t, 2^{t+2}, t$) that satisfies (3.8). We omit the details.

Example 3.7. *Consider OA($8\lambda, 2^5, 3$)s. Here $t = 3$. For $\lambda = 1$, equation (3.9) has no solution under (3.10). For $\lambda = 3$, equation (3.9) has a unique solution under (3.10)* *given by* $(\mu_1, \ldots, \mu_6, k) = (-1,-1,-1,-1,1,0,0)$. *When* $\lambda = 5$, *there are three solutions* $(-3,-1,-1,-1,1,0,0)$, $(-1,-1,-1,-1,-1,0,0)$, *and* $(-1,-1,-1,-1,1,-2,0)$. *When* $\lambda = 7$, *there are seven solutions*

$$(-1,-1,-1,-1,1,0,1), (-5,-1,-1,-1,1,0,0), (-1,-1,-1,-1,1,-4,0),$$

$$(-3,-3,-1,-1,1,0,0), (-3,-1,-1,-1,-1,0,0), (-3,-1,-1,-1,1,-2,0),$$

$$(-1,-1,-1,-1,-1,-2,0).$$

The results in the example above can be generalized to OA$(\lambda 2^t, 2^{t+2}, t)$s with $\lambda = t-2, t, t+2, t+4$ for any odd $t \geq 3$. Let $g(\lambda, t)$ be the number of nonisomorphic OA$(\lambda 2^t, 2^{t+2}, t)$s with odd t and odd λ.

Corollary 3.9. *We have that* $g(\lambda, t) = 0$ *for* $\lambda \leq t-2$, *and that* $g(\lambda, t) = 1, 3, 7$ *for* $\lambda = t, t+2, t+4$, *respectively.*

Corollary 3.9 can be verified directly using Theorem 3.15. If we can find all the solutions to equation (3.9) under (3.10), then a complete set of nonisomorphic OA$(\lambda 2^t, 2^{t+2}, t)$s is obtained for odd t and odd λ. The next result addresses this issue. We need some notation first. Let $Z[a,b]$, $O[a,b]$ and $E[a,b]$, denote, respectively, the set of integers, the set of odd integers and the set of even integers within the interval $[a,b]$.

TABLE 3.4
The number $q(n)$ of nonisomorphic OA$(n, 2^5, 3)$s when $n = 8\lambda$ is not a multiple of 16. Note that $q(n) = g(\lambda, 3)$.

n	24	40	56	72	88	104	120	136	152
$q(n)$	1	3	7	15	28	49	82	130	199
n	168	184	200	216	232	248	264	280	296
$q(n)$	296	428	605	839	1142	1530	2022	2637	3399
n	312	328	344	360	376	392	408		
$q(n)$	4336	5476	6854	8509	10481	12818	15573		

Theorem 3.16. *Consider odd* t *and odd* λ. *If* $\lambda \leq t-2$, *then equation (3.9) has no solution under (3.10). If* $\lambda \geq t$, *then equation (3.9) has at least one solution under (3.10), and the complete set of solutions is given by*

$$k \in Z[\,0,\ (\lambda - t)/4\,],$$

$$\mu_{m+1} \in E[\,-(\lambda - t - 4k),\ 0\,],$$

$$\mu_m \in O[\,-(\lambda - 4k + \mu_{m+1})/m,\ (\lambda - 4k + \mu_{m+1})/(m-2)\,],$$

$$\mu_{m-1} \in O[\,-(\lambda - 4k + \mu_m + \mu_{m+1})/(m-1),\ -|\mu_m|\,],$$

$$\mu_j \in O[\,-(\lambda - 4k + \mu_{j+1} + \cdots + \mu_{m+1})/j,\ \mu_{j+1}\,] \qquad \text{for } j = m-2, m-3, \ldots, 2,$$

$$\mu_1 = -(\lambda - 4k + \mu_2 + \cdots + \mu_{m+1}).$$

The proof of Theorem 3.16 involves some tedious but rather straightforward algebra. Based on Theorems 3.15 and 3.16, one can easily find all nonisomorphic OA$(\lambda 2^t, 2^{t+2}, t)$s for given odd t and odd λ. In Table 3.4, we present the number of nonisomorphic OA$(8\lambda, 2^5, 3)$s for $\lambda = 3, 5, \ldots, 51$.

4

Generalized Resolution and Minimum G-Aberration

This chapter centers around the generalized resolution, a concept capable of offering a simple yet informative characterization for nonregular designs in much the same way as the resolution capable of for regular designs. Besides the definition and examples, we provide two statistical justifications for the usefulness of the concept, and present construction results for designs with large generalized resolutions. Lower bounds on relevant J-characteristics provide a powerful tool for proving that many constructed designs have maximum generalized resolutions. Along the way, we introduce good Hadamard matrices and quaternary code designs. Good Hadamard matrices are useful not just for constructing nonregular designs with large generalized resolutions but also for constructing supersaturated designs with low correlations. Quaternary code designs are an attractive class of nonregular designs due to their partial linearity. We further introduce the criterion of minimum G-aberration, a refined version of the generalized resolution as a design selection criterion. Computational results are surveyed and discussed as well.

4.1 Generalized Resolution and Its Justifications

The resolution is a very useful concept as it provides a simple yet informative characterization for regular factorial designs. Knowing the resolution of a regular design amounts to knowing the basic properties of the design in terms of aliasing, estimability, and projection designs (Box, Hunter, and Hunter, 1978). As the resolution is defined through the defining relation of regular factorial designs, it is not applicable to nonregular designs. The generalized resolution to be defined serves the same purpose for all factorial designs, regular or nonregular, and it gives rise to the resolution for regular factorial designs.

Definition 4.1. *Let $\boldsymbol{D}$ be a two-level factorial design of n runs for m factors. Suppose that r is the smallest positive integer such that $\max_{|u|=r}|J_u| > 0$. Then the generalized resolution of design $\boldsymbol{D}$ is defined as*

$$R(\boldsymbol{D}) = r + 1 - \max_{|u|=r}|J_u|/n.$$

Example 4.1. *Consider the 12-run factorial design in Chapter 1. It is an orthogonal array of strength two, implying that $J_u = 0$ for all u with $|u| = 1, 2$. One can check that $|J_u| = 4$ for all u with $|u| = 3$, which gives $\max_{|u|=3}|J_u| = 4$. The generalized resolution of this design is therefore $R = 3 + 1 - 4/12 = 3\frac{2}{3}$.*

Example 4.2. *Consider the two 20-run factorial designs in Chapter 1, and let $\boldsymbol{D}_1$ and $\boldsymbol{D}_2$ denote the designs in Tables 1.3 and 1.4, respectively. Both $\boldsymbol{D}_1$ and $\boldsymbol{D}_2$ are orthogonal arrays of strength two. Easy calculations show that $\max_{|u|=3}|J_u| = 4$ and 12 for $\boldsymbol{D}_1$ and $\boldsymbol{D}_2$,*

DOI: 10.1201/9781003371861-4

respectively. Thus $\boldsymbol{D}_1$ *has a generalized resolution* 3.8 *and* $\boldsymbol{D}_2$ *has a generalized resolution* 3.4.

If $\boldsymbol{D}$ is regular, then J_u equals 0 or $\pm n$, begetting that $\max_{|u|=r} |J_u| > 0$ is equivalent to $\max_{|u|=r} |J_u| = n$. Thus, if a regular design $\boldsymbol{D}$ has resolution r, then its generalized resolution R must be equal to r. We know that a regular design of resolution r is an orthogonal array of strength $r-1$. In general, it follows immediately from Theorem 3.3 that if a factorial design has generalized resolution R, then it is an orthogonal array of strength $\lfloor R \rfloor - 1$, where $\lfloor x \rfloor$ is the floor function. Given an orthogonal array of strength t, but not of strength $t+1$, again by Theorem 3.3 we must have that $t+1 \leq R < t+2$, because $J_u = 0$ for all nonempty u with $|u| \leq t$ and $J_u \neq 0$ for at least one u with $|u| = t+1$. If this array of strength t does not have a defining word of length $t+1$, then its generalized resolution R cannot be equal to $t+1$ because $\max_{|u|=t+1} |J_u| < n$. These relationships of the generalized resolution with the resolution and strength are summarized in the next result.

Lemma 4.1. *(i) The generalized resolution of a regular design is the same as its resolution. (ii) A design of generalized resolution* R *is an orthogonal array of strength* $\lfloor R \rfloor - 1$. *(iii) If a design is an orthogonal array of strength* t*, but not of strength* $t+1$*, then its generalized resolution* R *has that* $t+1 \leq R < t+2$. *If this array of strength* t *does not have a defining word of length* $t+1$*, then* $t+1 < R < t+2$.

We next present a geometrical interpretation for the generalized resolution through projection designs. Suppose that a design $\boldsymbol{D}$ has generalized resolution

$$R = r + \delta, \quad \text{where } \delta = 1 - \max_{|u|=r} |J_u|/n. \tag{4.1}$$

Then design $\boldsymbol{D}$ is an orthogonal array of strength $t = r-1$. Consider a projection design $\boldsymbol{D}_u$ onto r factors, where $|u| = r$. By Lemma 3.2, design $\boldsymbol{D}_u$ is composed of α copies of a 2^r complete factorial and β copies of a 2_r^{r-1} fractional factorial, where $\alpha = (n - |J_u|)/2^r$ and $\beta = |J_u|/2^{r-1}$. The n runs of design $\boldsymbol{D}_u$ have two portions, with $n_1 = n - |J_u|$ runs from the full factorial portion and the other $n_2 = |J_u|$ runs from the fractional factorial portion. The proportion of the runs from the full factorial portion is $\delta_u = (n - |J_u|)/n$. We therefore obtain that

$$\min_{|u|=r} \delta_u = \min_{|u|=r} (n - |J_u|)/n = 1 - \max_{|u|=r} |J_u|/n = \delta,$$

where δ is defined in (4.1). The following result is established.

Theorem 4.1. *Suppose that a design* $\boldsymbol{D}$ *has generalized resolution* $R = r + \delta$ *as in* (4.1). *Then the projection design onto any* r *factors contains at least* $\delta n/2^r$ *copies of a* 2^r *full factorial. If* $\delta > 0$*, then the projection design onto any* r *factors contains at least one copy of a* 2^r *full factorial. For regular designs, we always have* $\delta = 0$.

The geometrical interpretation given in Theorem 4.1 formally justifies the use of generalized resolution as a design selection criterion. The larger the generalized resolution is, the better is the design. One should therefore select a design that has the maximum generalized resolution among all competing designs.

The generalized resolution criterion also makes intuitive sense. Since the J values measure the amount of aliasing among factorial effects, designs with small $|J|$ values should be preferred. Furthermore, aliasing among lower order effects is more serious; therefore one should concentrate on minimizing $|J_u|$ values for small $|u|$. This is actually how the concept of generalized resolution was first conceived (Deng and Tang, 1999).

The generalized resolution has another justification, which is from the angle of minimizing the biases of estimated main effects. We focus on orthogonal arrays of strength two and consider a specialized version of what has been done in Section 2.7. Suppose that the model $\boldsymbol{Y} = \beta_0 \mathbf{1}_n + \boldsymbol{X}_1\boldsymbol{\beta}_1 + \boldsymbol{X}_2\boldsymbol{\beta}_2 + \boldsymbol{\epsilon}$ is true but $\boldsymbol{Y} = \beta_0 \mathbf{1}_n + \boldsymbol{X}_1\boldsymbol{\beta}_1 + \boldsymbol{\epsilon}$ is fitted, where β_0 is the grand mean, $\boldsymbol{\beta}_1$ is the vector of main effects, $\boldsymbol{\beta}_2$ is the vector of two-factor interactions, and $\boldsymbol{X}_1$ and $\boldsymbol{X}_2$ are the corresponding model matrices. Because $\boldsymbol{X}_1$ is an orthogonal array of strength two, the columns of $\boldsymbol{X}_1$ are mutually orthogonal and are all orthogonal to $\mathbf{1}_n$. Therefore, the least square estimator of $\boldsymbol{\beta}_1$, the main effects, from the fitted model is $\hat{\boldsymbol{\beta}}_1 = (\boldsymbol{X}_1^T\boldsymbol{X}_1)^{-1}\boldsymbol{X}_1^T\boldsymbol{Y} = n^{-1}\boldsymbol{X}_1^T\boldsymbol{Y}$, which is biased under the true model with the bias given by

$$\text{Bias}(\hat{\boldsymbol{\beta}}_1, \boldsymbol{\beta}_1) = n^{-1}\boldsymbol{X}_1^T\boldsymbol{X}_2\boldsymbol{\beta}_2 = \boldsymbol{C}\boldsymbol{\beta}_2,$$

where $\boldsymbol{C} = (c_{ij}) = n^{-1}\boldsymbol{X}_1^T\boldsymbol{X}_2$. One can easily see that $\max_{i,j}|c_{ij}| = n^{-1}\max_{|u|=3}|J_u|$. One way to minimize $\text{Bias}(\hat{\boldsymbol{\beta}}_1, \boldsymbol{\beta}_1)$ is to minimize $\max_{i,j}|c_{ij}|$, which is equivalent to maximizing the generalized resolution.

4.2 Designs from Paley's Constructions

One major focus of this chapter is the construction of nonregular designs with maximum or large generalized resolutions. In Sections 4.2, 4.3, and 4.4, we present some construction results from the three papers Shi and Tang (2018), Shi and Tang (2023), and Chen, Shi, and Tang (2023). This section provides lower bounds on the relevant J-characteristics and, with the assistance of these lower bounds, we show that many designs from Paley's two constructions of Hadamard matrices have maximum generalized resolutions. Section 4.3 considers designs from tensor product constructions. Further results are afforded by good Hadamard matrices, which is the subject of Section 4.4. The present section is divided into two subsections for a structured presentation.

4.2.1 Results on saturated OA$(n, 2^{n-1}, 2)$s

Our first set of results concerns saturated orthogonal arrays OA$(n, 2^{n-1}, 2)$s. The next theorem provides a lower bound on the value of $\max_{|u|=3}|J_u|$ for such designs.

Theorem 4.2. *Let $\boldsymbol{D}$ be an OA$(n, 2^{n-1}, 2)$. We then have that*

$$\max_{|u|=3}|J_u| \geq L(n) = n - 8\left\lfloor (n/8)\left(1 - 1/(n-3)^{1/2}\right)\right\rfloor.$$

Proof. Let $\boldsymbol{d}_j$ be the jth column of $\boldsymbol{D}$. As $\boldsymbol{D}$ is an OA$(n, 2^{n-1}, 2)$, the vectors

$$\mathbf{1}_n/n^{1/2}, \boldsymbol{d}_1/n^{1/2}, \ldots, \boldsymbol{d}_{n-1}/n^{1/2}$$

form an orthonormal basis in the n-dimensional Euclidean space. This establishes that

$$\left[J\left(\boldsymbol{d}_1\boldsymbol{d}_2, \mathbf{1}_n/n^{1/2}\right)\right]^2 + \sum_{j=1}^{n-1}\left[J\left(\boldsymbol{d}_1\boldsymbol{d}_2, \boldsymbol{d}_j/n^{1/2}\right)\right]^2 = n,$$

which simplifies to $\sum_{j=3}^{n-1}|J_{\{1,2,j\}}|^2 = n^2$ by noting the orthogonality of $\boldsymbol{d}_1\boldsymbol{d}_2$ to $\mathbf{1}_n, \boldsymbol{d}_1$ and $\boldsymbol{d}_2$. We therefore obtain that $(n-3)\max_{|u|=3}|J_u|^2 \geq n^2$, which leads to

$$(1/8)(n - \max_{|u|=3}|J_u|) \leq (n/8)(1 - 1/(n-3)^{1/2}).$$

The left-hand side of the above inequality must be an integer, since, according to Corollary 3.4, the value of $n - |J_u|$ is always a multiple of 8 for $|u| = 3$ whether or not n is a multiple of 8. We thus obtain

$$(1/8)(n - \max_{|u|=3} |J_u|) \leq \left\lfloor (n/8)(1 - 1/(n-3)^{1/2}) \right\rfloor,$$

solving which gives Theorem 4.2. □

A lower bound on $\max_{|u|=3} |J_u|$ immediately produces an upper bound on the generalized resolution R. We therefore have that

$$R \leq 3 + (8/n) \left\lfloor (n/8) \left(1 - 1/(n-3)^{1/2}\right) \right\rfloor$$

for any $\mathrm{OA}(n, 2^{n-1}, 2)$.

In Section 2.5, we have discussed Paley's two constructions of Hadamard matrices of order n. The first construction requires that $n - 1$ be a prime power, and the resulting Hadamard matrix is called a *Paley matrix* in the literature. If we delete the first column, consisting of all ones, from a Paley matrix, then we obtain a *Paley design*, which is a saturated orthogonal array $\mathrm{OA}(n, 2^{n-1}, 2)$. Based on a result of Bulutoglu and Cheng (2003), we present an important property of a Paley design.

Theorem 4.3. *A Paley design of n runs has the property that*

$$|J_u| \leq 4 + 2(n-1)^{1/2}$$

for all u with $|u| = 3, 4$.

Proof. Theorem 2.1 of Bulutoglu and Cheng (2003) states that

$$|J_u| \leq k + 1 + (k-1)(n-1)^{1/2} \text{ if } |u| = k \text{ is odd.}$$

Taking $k = 3$ leads to $|J_u| \leq 4 + 2(n-1)^{1/2}$ for $|u| = 3$. The same theorem says that $|J_u + 1| \leq k + 1 + (k-2)(n-1)^{1/2}$ if $|u| = k$ is even. With a simple modification to their proof, one can actually show that

$$|J_u| \leq k + (k-2)(n-1)^{1/2} \text{ if } |u| = k \text{ is even.}$$

Taking $k = 4$, we obtain that $|J_u| \leq 4 + 2(n-1)^{1/2}$ for $|u| = 4$. The proof is completed. □

Theorem 4.4. *If an $OA(n, 2^{n-1}, 2)$ is a Paley design, then we have that*

$$\max_{|u|=3} |J_u| \leq U(n) = n - 8 \left\lceil (1/8) \left(n - 4 - 2(n-1)^{1/2}\right) \right\rceil,$$

where $\lceil x \rceil$ is the ceiling function.

Proof. By Theorem 4.3, for a Paley design we have that $\max_{|u|=3} |J_u| \leq 4 + 2(n-1)^{1/2}$. This gives $(1/8)(n - \max_{|u|=3} |J_u|) \geq (1/8)\left(n - 4 - 2(n-1)^{1/2}\right)$. As the left hand side is an integer, we obtain

$$(1/8)(n - \max_{|u|=3} |J_u|) \geq \left\lceil (1/8) \left(n - 4 - 2(n-1)^{1/2}\right) \right\rceil,$$

from which Theorem 4.4 follows. □

TABLE 4.1
The values of $L(n)$ and $U(n)$ for $12 \le n \le 100$

n	$L(n)$	$U(n)$	Paley design exists?
12	4	4	Yes
16	8	8	No
20	12	12	Yes
24	8	8	Yes
28	12	12	Yes
32	8	8	Yes
36	12	12	No
40	8	16	No
44	12	12	Yes
48	8	16	Yes
52	12	12	No
56	8	16	No
60	12	12	Yes
64	16	16	No
68	12	20	Yes
72	16	16	Yes
76	12	20	No
80	16	16	Yes
84	12	20	Yes
88	16	16	No
92	12	20	No
96	16	16	No
100	12	20	No

If the lower bound $L(n)$ in Theorem 4.2 matches the upper bound $U(n)$ in Theorem 4.4, then a Paley design minimizes $\max_{|u|=3} |J_u|$ and thus maximizes the generalized resolution.

We calculate $L(n)$ and $U(n)$ for all $12 \le n \le 100$ that are multiples of 4 and give the results in Table 4.1. For some of these n values, Paley designs do not exist, as the corresponding values of $n-1$ are not prime powers. From Table 4.1, we see that $L(n) = U(n)$ and Paley designs exist for $n = 12, 20, 24, 28, 32, 44, 60, 72$, and 80. Therefore, Paley designs for these run sizes have maximum generalized resolutions.

Whether or not Paley designs have maximum generalized resolutions for other run sizes remains to be settled. The answer may well be affirmative as we have not found a counterexample. A sharper lower bound on the $\max_{|u|=3} |J_u|$ value would help. Nevertheless, Paley designs provide an attractive class of nonregular designs as they have a generalized resolution bounded below by

$$3 + (8/n) \left\lceil (1/8) \left(n - 4 - 2(n-1)^{1/2} \right) \right\rceil,$$

which is close to 4 for large n.

In addition to having maximum generalized resolutions, Paley designs also have small $\max_{|u|=4} |J_u|$ values due to Theorem 4.3. This makes these designs even more attractive. An immediate conclusion is that Paley designs for $n \ge 12$ enjoy the hidden projection property that every projection design onto four factors allows the estimation of all main effects and all two-factor interactions, because they cannot have defining words of length 3 or 4. Compared with other saturated OA$(n, 2^{n-1}, 2)$s that have maximum generalized resolutions, Paley designs are preferable because of their small $\max_{|u|=4} |J_u|$ values. For example, among all 7570 nonisomorphic OA$(28, 2^{27}, 2)$s found by Schoen, Eendebak, and Nguyen (2010), twelve arrays have $\max_{|u|=3} |J_u| = 12$, and all twelve have the maximum generalized resolution. Only one of these twelve arrays has $\max_{|u|=4}|J_u| = 12$, and the other eleven all have $\max_{|u|=4}|J_u| = 20$. Not surprisingly, the best one is precisely the Paley design.

4.2.2 Results on OA$(n, 2^m, 2)$s for $m \leq n-1$

We now examine the general case and consider OA$(n, 2^m, 2)$s for $m \leq n-1$. The next result generalizes Theorem 4.2.

Theorem 4.5. *Let $\boldsymbol{D}$ be an OA$(n, 2^m, 2)$ with $n/2 \leq m \leq n-1$. We then have that*

$$\max_{|u|=3}|J_u| \geq L(n,m) = n - 8\left\lfloor (n/8)(1-q^{1/2})\right\rfloor,$$

where $q = (2m-n)/((m-1)(m-2))$.

Proof. Let $\boldsymbol{d}_j$ be the jth column of $\boldsymbol{D}$ for $j = 1, \ldots, m$. Then there exist real vectors $\boldsymbol{e}_1, \ldots, \boldsymbol{e}_p$, with $p = n-m-1$, such that

$$\boldsymbol{1}_n/n^{1/2}, \boldsymbol{d}_1/n^{1/2}, \ldots, \boldsymbol{d}_m/n^{1/2}, \boldsymbol{e}_1, \ldots, \boldsymbol{e}_p$$

form an orthonormal basis in the n-dimensional Euclidean space. It follows from the mutual orthogonality of $\boldsymbol{d}_1\boldsymbol{d}_2, \ldots, \boldsymbol{d}_1\boldsymbol{d}_m$ that

$$\sum_{j=2}^{m} |\langle \boldsymbol{d}_1\boldsymbol{d}_j/n^{1/2}, \boldsymbol{e}_k\rangle|^2 \leq 1,$$

for any given $k = 1, \ldots, p$, where $\langle \boldsymbol{x}, \boldsymbol{y}\rangle$ denotes the inner product of $\boldsymbol{x}$ and $\boldsymbol{y}$. Since $\boldsymbol{d}_1\boldsymbol{d}_1 = \boldsymbol{1}_n$ and $\langle \boldsymbol{1}_n, \boldsymbol{e}_k\rangle = 0$, we have that $\sum_{j=1}^{m}|\langle \boldsymbol{d}_1\boldsymbol{d}_j, \boldsymbol{e}_k\rangle|^2 \leq n$. Similarly, $\sum_{j=1}^{m}|\langle \boldsymbol{d}_i\boldsymbol{d}_j, \boldsymbol{e}_k\rangle|^2 \leq n$ for every $i = 1, \ldots, m$. Summing over i and removing redundancy, we obtain that

$$\sum_{1\leq i<j\leq m} |\langle \boldsymbol{d}_i\boldsymbol{d}_j, \boldsymbol{e}_k\rangle|^2 \leq nm/2.$$

This establishes that

$$\sum_{1\leq i<j\leq m}\sum_{k=1}^{p} |\langle \boldsymbol{d}_i\boldsymbol{d}_j, \boldsymbol{e}_k\rangle|^2 \leq pnm/2. \tag{4.2}$$

Considering vectors $\boldsymbol{d}_i\boldsymbol{d}_j$ with $i < j$ under the orthonormal basis $\boldsymbol{1}_n/n^{1/2}$, $\boldsymbol{d}_1/n^{1/2}$, ..., $\boldsymbol{d}_m/n^{1/2}$, $\boldsymbol{e}_1, \ldots, \boldsymbol{e}_p$, we obtain that

$$n = \sum_{l\neq i,j} |\langle \boldsymbol{d}_i\boldsymbol{d}_j, \boldsymbol{d}_l/n^{1/2}\rangle|^2 + \sum_{k=1}^{p} |\langle \boldsymbol{d}_i\boldsymbol{d}_j, \boldsymbol{e}_k\rangle|^2,$$

owing to the orthogonality of $\boldsymbol{d}_i\boldsymbol{d}_j$ to $\boldsymbol{1}_n, \boldsymbol{d}_i, \boldsymbol{d}_j$. Since $|\langle \boldsymbol{d}_i\boldsymbol{d}_j, \boldsymbol{d}_l\rangle| \leq \max_{|u|=3}|J_u|$,

$$n^2 \leq (m-2)\left(\max_{|u|=3}|J_u|\right)^2 + n\sum_{k=1}^{p} |\langle \boldsymbol{d}_i\boldsymbol{d}_j, \boldsymbol{e}_k\rangle|^2.$$

Summing over all $i<j$ in the above inequality, using (4.2), and then simplifying, we obtain that

$$n^2(m-1) \leq (m-2)(m-1)\left(\max_{|u|=3}|J_u|\right)^2 + n^2 p.$$

Solving for $\max_{|u|=3}|J_u|$ with $p = n-1-m$, we obtain that $\max_{|u|=3}|J_u| \geq nq^{1/2}$, where $q = (2m-n)/((m-1)(m-2))$. The proof is completed by noting that $n - \max_{|u|=3}|J_u|$ is a multiple of eight. □

TABLE 4.2
Nonregular designs with the maximum generalized resolutions

n	$\max_{\lvert u\rvert=3} \lvert J_u\rvert$	m	source
12	4	$3, 4, \ldots, 11$	$\boldsymbol{P}_{n,m}$
20	12	$13, 14, \ldots, 19$	$\boldsymbol{P}_{n,m}$
24	8	$13, 14, \ldots, 23$	$\boldsymbol{P}_{n,m}$
28	12	$17, 18, \ldots, 27$	$\boldsymbol{P}_{n,m}$
32	8	$17, 18, \ldots, 31$	$\boldsymbol{P}_{n,m}$
44	12	$25, 26, \ldots, 43$	$\boldsymbol{P}_{n,m}$
60	12	$33, 34, \ldots, 59$	$\boldsymbol{P}_{n,m}$
72	16	$52, 53, \ldots, 71$	$\boldsymbol{P}_{n,m}$
80	16	$54, 55, \ldots, 79$	$\boldsymbol{P}_{n,m}$
36	12	$21, 22, \ldots, 35$	Shi and Tang (2018)
144	16	$83, 84, \ldots, 121$	Example 4.3
48	8	$25, 26, \ldots, 44$	Example 4.5
192	16	$106, 107, \ldots, 176$	Example 4.5
768	32	$511, 512, \ldots, 704$	Example 4.5
128	16	$75, 76, \ldots, 124$	Example 4.6

When $m = n - 1$, $q = 1/(n-3)$. This shows that the lower bound $L(n, m)$ in Theorem 4.5 reduces to that in Theorem 4.2. For $m = n/2$, Theorem 4.5 also gives sensible results as it is easily seen that $L(n, n/2) = 0$ if n is a multiple of 8 and $L(n, n/2) = 4$ otherwise.

Let $\boldsymbol{P}_n$ denote a Paley design of order n, which is a saturated design of n runs for $n-1$ factors. Further, let $\boldsymbol{P}_{n,m}$ be any design that selects m columns from $\boldsymbol{P}_n$ and, for convenience, we still call $\boldsymbol{P}_{n,m}$ a Paley design. Early in the section, we have proved that $\boldsymbol{P}_n$ has the maximum generalized resolution for $n = 12, 20, 24, 28, 32, 44, 60, 72$, and 80. Using the bound in Theorem 4.5, we can establish that for each of these orders, design $\boldsymbol{P}_{n,m}$ also has the maximum generalized resolution for a wide range of m values. Table 4.2 presents the details. Note that $\boldsymbol{P}_{12,m}$ has the maximum generalized resolution for any $m \geq 3$ because $\boldsymbol{P}_{12,m}$ has $\max_{|u|=3} |J_u| = 4$ and any other $\mathrm{OA}(12, 2^m, 2)$ must have $\max_{|u|=3} |J_u| \geq 4$. Table 4.2 also includes other designs with maximum generalized resolutions, obtained in the next section. Entries for $n = 36$ can be obtained using either of the two Hadamard matrices of order 36 in Shi and Tang (2018).

The rest of the section focuses on $\mathrm{OA}(n, 2^m, 2)$s for n's that are not multiples of 8. From Theorem 3.7, we know that J_u is also not a multiple of 8 for all u with $|u| = 3$. This means that $\max_{|u|=3} |J_u| \geq 4$. The best situation is $\max_{|u|=3} |J_u| = 4$, which happens if and only if $|J_u| = 4$ for all $|u| = 3$. An $\mathrm{OA}(n, 2^m, 2)$ with $|J_u| = 4$ for all $|u| = 3$ has a maximum generalized resolution of $4 - 4/n$.

For $n = 12$, design $\boldsymbol{P}_{12}$, an $\mathrm{OA}(12, 2^{11}, 2)$, has $|J_u| = 4$ for $|u| = 3$. For $n = 20, 28$, and 36, computational results from Deng and Tang (2002), Sun, Li, and Ye (2008) and Schoen, Vo-Thanh, and Goos (2017) showed that $\mathrm{OA}(n, 2^m, 2)$s with $\max_{|u|=3} |J_u| = 4$ can be found for $m \leq n/2$. This is not a coincidence, as the following general result can be established.

Let $n = 8k + 4$. Suppose that $s = n/2 - 1 = 4k + 1$ is a prime power. Consider the following two matrices

$$\boldsymbol{C} = \begin{bmatrix} -1 & \boldsymbol{1}_s^T \\ \boldsymbol{1}_s & \boldsymbol{Q} - \boldsymbol{I}_s \\ -1 & -\boldsymbol{1}_s^T \\ -\boldsymbol{1}_s & -\boldsymbol{Q} - \boldsymbol{I}_s \end{bmatrix}, \quad \boldsymbol{D} = \begin{bmatrix} -1 & \boldsymbol{1}_s^T \\ \boldsymbol{1}_s & \boldsymbol{Q} - \boldsymbol{I}_s \\ 1 & \boldsymbol{1}_s^T \\ -\boldsymbol{1}_s & -\boldsymbol{Q} - \boldsymbol{I}_s \end{bmatrix}, \qquad (4.3)$$

where $\boldsymbol{Q}$ is defined in equation (2.7) in Section 2.5. We see that matrix $\boldsymbol{C}$ is an $n \times (n/2)$ submatrix of the Hadamard matrix from Paley's second construction given in Section 2.5, and thus has orthogonal columns. However, $\boldsymbol{C}$ is not an orthogonal array because all of its columns have more -1s than $+1$s. Matrix $\boldsymbol{D}$ is an OA$(n, 2^{n/2}, 2)$, which is obtained from $\boldsymbol{C}$ by multiplying -1 to the $(n/2+1)$th row of $\boldsymbol{C}$ while leaving all other rows untouched.

Theorem 4.6. *Let $n = 8k+4$ and suppose that $n/2-1 = 4k+1$ is a prime power. Then the design $\boldsymbol{D}$ in (4.3) is an $OA(n, 2^{n/2}, 2)$ with $\max_{|u|=3} |J_u| = 4$, and therefore has a maximum generalized resolution $R = 4 - 4/n$.*

Proof. We need to show that $|J_u| = 4$ for all $|u| = 3$. We note that a projection design of $\boldsymbol{D}$ onto three factors has many mirror image runs. When calculating the J value of a projection design onto three factors, these mirror image runs do not contribute; so we only need to focus on those runs that have no mirror images. Looking at the projection design onto the first three factors, we see that only the following six runs need to be considered: runs 1, 2, 3, $(n/2+1)$, $(n/2+2)$, and $(n/2+3)$. Using the symmetry of matrix $\boldsymbol{Q}$, one can easily see that the J value for the three columns of the matrix consisting of these six runs must be ± 4. For other projection designs onto three factors, without loss of generality, we consider the projection design onto columns 2, 3, and 4. For this projection design, we see that we only need to consider runs 1, 2, 3, 4 and runs $(n/2+1)$, $(n/2+2)$, $(n/2+3)$, $(n/2+4)$. Due to the symmetry of $\boldsymbol{Q}$, matrix $\boldsymbol{A}_1$ collecting runs 1, 2, 3, 4 as rows and matrix $\boldsymbol{A}_2$ collecting runs $(n/2+1)$, $(n/2+2)$, $(n/2+3)$, $(n/2+4)$ as rows must have forms

$$\boldsymbol{A}_1 = \begin{bmatrix} 1 & 1 & 1 \\ -1 & q_1 & q_2 \\ q_1 & -1 & q_3 \\ q_2 & q_3 & -1 \end{bmatrix}, \qquad \boldsymbol{A}_2 = \begin{bmatrix} 1 & 1 & 1 \\ -1 & -q_1 & -q_2 \\ -q_1 & -1 & -q_3 \\ -q_2 & -q_3 & -1 \end{bmatrix}.$$

The J value for the three columns of the 8×3 matrix obtained by combining the rows of $\boldsymbol{A}_1$ and $\boldsymbol{A}_2$ is therefore $2 - 2(q_1q_2 + q_1q_3 + q_2q_3)$, which is equal to ± 4 for all possible scenarios of $q_1 = \pm 1$, $q_2 = \pm 1$, and $q_3 = \pm 1$. We have therefore shown that $J_u = \pm 4$ for all $|u| = 3$. □

Theorem 4.6 allows the construction of an OA$(n, 2^{n/2}, 2)$ with a generalized resolution $R = 4 - 4/n$ for $n = 20, 28, 36, 52, 60, 76, 84, 100$, and so on.

An inescapable question arising from Theorem 4.6 is how large m can be if an OA$(n, 2^m, 2)$ is to have $R = 4 - 4/n$ when $n = 8k+4$. Theorem 4.6 says that $m \geq n/2$ if $n/2 - 1 = 4k+1$ is a prime power. The next result gives an upper bound on m.

Theorem 4.7. *If an $OA(n, 2^m, 2)$ is to have $\max_{|u|=3} |J_u| = 4$ for $n = 8k+4 \geq 20$, then it is necessary that $m \leq n/2 + 2$.*

Proof. We only need to prove that $L(n,m) \geq 12$ for $m = n/2 + 3$, where $L(n,m)$ is the lower bound on $\max_{|u|=3} |J_u|$ in Theorem 4.5. As $n = 8k+4$, we have that

$$L(n,m) = (8k+4) - 8\left\lfloor k + 0.5 - (n/8)q^{1/2} \right\rfloor.$$

Thus, to prove $L(n,m) \geq 12$, we only need to prove that $0.5 - (n/8)q^{1/2} < 0$. This is because if $0.5 - (n/8)q^{1/2} < 0$, then $\lfloor k + 0.5 - (n/8)q^{1/2} \rfloor \leq k - 1$, showing that $L(n,m) \geq (8k+4) - 8(k-1) = 12$. Note that $0.5 - (n/8)q^{1/2} < 0$ is equivalent to $q > (4/n)^2$. Plugging-in $n = 8k+4$ and $m = n/2 + 3 = 4k+5$ into $q > (4/n)^2$ and then doing some elementary algebra, we obtain an equivalent inequality $2k(2k-1) > 3$. This last inequality holds true whenever $k \geq 2$ and correspondingly $n = 8k+4 \geq 20$. □

Theorem 4.7 cannot be improved by just using Theorem 4.5, as we can check that $L(n, n/2+2) = 4$. So, the existence of an OA$(n, 2^{n/2+1}, 2)$ or even an OA$(n, 2^{n/2+2}, 2)$ with $R = 4-4/n$ for $n \geq 44$ is still theoretically possible, even though the impossibility has been established for $n = 20, 28$, and 36 by Sun, Li, and Ye (2008) and Schoen, Vo-Thanh, and Goos (2017).

When n is a multiple of 8 and $m \leq n/2$, orthogonal arrays of strength three are available. Construction of such designs with maximum generalized resolutions will be treated in Section 4.4.

4.3 Designs from Tensor Products

We investigate in the present section how to build large designs with large generalized resolutions using small designs with large generalized resolutions. This requires the use of tensor product and the associated method of calculating J-characteristics. We give some preparation first.

Let $\boldsymbol{A} = (a_{ij})_{n_1 \times m_1}$ and $\boldsymbol{B} = (b_{ij})_{n_2 \times m_2}$ be two matrices. Then their tensor product is defined as $\boldsymbol{A} \otimes \boldsymbol{B} = (a_{ij}\boldsymbol{B})$, which is an $(n_1 n_2) \times (m_1 m_2)$ matrix. The next lemma from Tang (2006) grants an easy calculation of J-characteristics for vectors of tensor product form.

Lemma 4.2. *Let $\boldsymbol{a}_1, \ldots, \boldsymbol{a}_k$ be vectors of length n_1 and $\boldsymbol{b}_1, \ldots, \boldsymbol{b}_k$ be vectors of length n_2. We have that*

$$J(\boldsymbol{a}_1 \otimes \boldsymbol{b}_1, \ldots, \boldsymbol{a}_k \otimes \boldsymbol{b}_k) = J(\boldsymbol{a}_1, \ldots, \boldsymbol{a}_k) J(\boldsymbol{b}_1, \ldots, \boldsymbol{b}_k).$$

Proof. Let $a_{i_1 j}$ be the i_1th element of $\boldsymbol{a}_j$ and $b_{i_2 j}$ be the i_2th element of $\boldsymbol{b}_j$. Then we have

$$J(\boldsymbol{a}_1 \otimes \boldsymbol{b}_1, \ldots, \boldsymbol{a}_k \otimes \boldsymbol{b}_k) = \sum_{i_1=1}^{n_1} \sum_{i_2=1}^{n_2} a_{i_1 1} b_{i_2 1} \cdots a_{i_1 k} b_{i_2 k},$$

which is equal to

$$\left(\sum_{i_1-1}^{n_1} a_{i_1 1} \cdots a_{i_1 k} \right) \left(\sum_{i_2-1}^{n_2} b_{i_2 1} \cdots b_{i_2 k} \right) = J(\boldsymbol{a}_1, \ldots, \boldsymbol{a}_k) J(\boldsymbol{b}_1, \ldots, \boldsymbol{b}_k).$$

This finishes the proof. □

Theorem 4.8. *Let $\boldsymbol{A}$ be an OA$(n_1, 2^{m_1}, 2)$ with $R(\boldsymbol{A}) = 4 - \epsilon_1$ and $\boldsymbol{B}$ be an OA$(n_2, 2^{m_2}, 2)$ with $R(\boldsymbol{B}) = 4 - \epsilon_2$. Then $\boldsymbol{D} = \boldsymbol{A} \otimes \boldsymbol{B}$ is an OA$(n_1 n_2, 2^{m_1 m_2}, 2)$ with $R(\boldsymbol{D}) = 4 - \epsilon_1 \epsilon_2$.*

Proof. Let $\boldsymbol{a}_i$ be the ith column of $\boldsymbol{A}$ and $\boldsymbol{b}_j$ be the jth column of $\boldsymbol{B}$. By Lemma 4.2, we have

$$J(\boldsymbol{a}_{i_1} \otimes \boldsymbol{b}_{j_1}, \boldsymbol{a}_{i_2} \otimes \boldsymbol{b}_{j_2}, \boldsymbol{a}_{i_3} \otimes \boldsymbol{b}_{j_3}) = J(\boldsymbol{a}_{i_1}, \boldsymbol{a}_{i_2}, \boldsymbol{a}_{i_3}) J(\boldsymbol{b}_{j_1}, \boldsymbol{b}_{j_2}, \boldsymbol{b}_{j_3}).$$

Observing that $J(\boldsymbol{a}_{i_1}, \boldsymbol{a}_{i_2}, \boldsymbol{a}_{i_3}) = J(\boldsymbol{a}_{i_3}) = 0$ when $i_1 = i_2$, we obtain that

$$\max_{|u|=3} |J_u(\boldsymbol{D})| = \max_{|u|=3} |J_u(\boldsymbol{A})| \max_{|u|=3} |J_u(\boldsymbol{B})|,$$

from which Theorem 4.8 follows. □

Example 4.3. *Let both $\boldsymbol{A}$ and $\boldsymbol{B}$ be $\boldsymbol{P}_{12}$, an $OA(12, 2^{11}, 2)$, which has a generalized resolution $R = 4 - 1/3$. Then Theorem 4.8 asserts that design $\boldsymbol{D} = \boldsymbol{A} \otimes \boldsymbol{B}$, an $OA(144, 2^{121}, 2)$ has a generalized resolution $R(\boldsymbol{D}) = 4 - 1/9$. In terms of J-characteristics, we have that* $\max_{|u|=3} |J_u(\boldsymbol{D})| = 16$. *As $L(144, 121) = 16$, design $\boldsymbol{D}$ has the maximum generalized resolution. Deleting some columns from $\boldsymbol{D}$ gives a design with fewer factors. Since $L(144, m) = 16$ for each m value in the range of $83 \le m \le 121$, we therefore obtain an $OA(144, 2^m, 2)$ with the maximum generalized resolution.*

Example 4.4. *Let $\boldsymbol{A}$ be $\boldsymbol{P}_{12}$ and $\boldsymbol{B}$ be $\boldsymbol{P}_{32}$; so $R(\boldsymbol{A}) = 4 - 1/3$ and $R(\boldsymbol{B}) = 4 - 1/4$. Using Theorem 4.8, we obtain a design $\boldsymbol{D} = \boldsymbol{A} \otimes \boldsymbol{B}$, an $OA(384, 2^{341}, 2)$, which has a generalized resolution $R = 4 - 1/12$. This design does not achieve the lower bound in Theorem 4.5. Whether or not it has the maximum generalized resolution remains to be settled.*

The next result has a flavor similar to that of Theorem 4.8, but offers a different perspective on the tensor product construction. Given a Hadamard matrix $\boldsymbol{H}$, we define

$$\gamma(\boldsymbol{H}) = \max_{|u|=1,3} |J_u(\boldsymbol{H})|. \tag{4.4}$$

Theorem 4.9. *Let $\boldsymbol{H}$ be a Hadamard matrix of order n_1 with $\gamma = \gamma(\boldsymbol{H})$ and $\boldsymbol{A}$ be an $OA(n_2, 2^{m_2}, 2)$ with $R(\boldsymbol{A}) = 4 - \epsilon$. Then the design $\boldsymbol{D} = \boldsymbol{H} \otimes \boldsymbol{A}$ is an $OA(n_1 n_2, 2^{n_1 m_2}, 2)$ with $R(\boldsymbol{D}) = 4 - \epsilon(\gamma/n_1)$.*

Proof. It is obvious that $\boldsymbol{D} = \boldsymbol{H} \otimes \boldsymbol{A}$ is an OA$(n_1 n_2, 2^{n_1 m_2}, 2)$. Using Lemma 4.2, we obtain that

$$\max_{|u|=1,3} |J_u(\boldsymbol{D})| = \max_{|u|=1,3} |J_u(\boldsymbol{H})| \max_{|u|=1,3} |J_u(\boldsymbol{A})|.$$

Because both $\boldsymbol{A}$ and $\boldsymbol{D}$ are orthogonal arrays of strength two, we have that $J_u(\boldsymbol{D}) = J_u(\boldsymbol{A}) = 0$ for all u with $|u| = 1$. This shows that

$$\max_{|u|=3} |J_u(\boldsymbol{D})| = \max_{|u|=1,3} |J_u(\boldsymbol{H})| \max_{|u|=3} |J_u(\boldsymbol{A})| = \gamma \max_{|u|=3} |J_u(\boldsymbol{A})|,$$

entailing that $R(\boldsymbol{D}) = 4 - \epsilon(\gamma/n_1)$. □

One can easily check that the following Hadamard matrix

$$\boldsymbol{H} = \begin{bmatrix} -1 & 1 & 1 & 1 \\ 1 & -1 & 1 & 1 \\ 1 & 1 & -1 & 1 \\ 1 & 1 & 1 & -1 \end{bmatrix}$$

has that $\gamma(\boldsymbol{H}) = 2$. The next result is immediate from Theorem 4.9.

Theorem 4.10. *Consider the above Hadamard matrix $\boldsymbol{H}$ of order 4 and let $\boldsymbol{A}$ be an $OA(n, 2^m, 2)$ with $R(\boldsymbol{A}) = 4 - \epsilon$. Define*

$$\boldsymbol{D}_0 = \boldsymbol{A}, \text{ and } \boldsymbol{D}_k = \boldsymbol{H} \otimes \boldsymbol{D}_{k-1}$$

for $k \ge 1$. Then $\boldsymbol{D}_k$, an $OA(n4^k, 2^{m4^k}, 2)$, has that $R(\boldsymbol{D}_k) = 4 - \epsilon/2^k$ for any integer $k \ge 0$.

Example 4.5. *If we take $\boldsymbol{P}_{12}$ as $\boldsymbol{A}$ in Theorem 4.10, then $\boldsymbol{D}_k$ is an $OA(12{\cdot}4^k, 2^{11\cdot 4^k}, 2)$ that has $R(\boldsymbol{D}_k) = 4 - (1/3)/2^k$ for $k \ge 0$. Also obvious is that* $\max_{|u|=3} |J_u(\boldsymbol{D}_k)| = 4 \cdot 2^k = 2^{k+2}$. *Now consider $\boldsymbol{D}_1$, $\boldsymbol{D}_2$, and $\boldsymbol{D}_3$. Design $\boldsymbol{D}_1$ is an $OA(48, 2^{44}, 2)$ with* $\max_{|u|=3} |J_u(\boldsymbol{D}_1)| = 8$. *As $L(48, m) = 8$ for $25 \le m \le 44$, any design obtained by selecting m columns from $\boldsymbol{D}_1$ has the maximum generalized resolution of $4 - 1/6$. Similarly, we obtain from $\boldsymbol{D}_2$ an $OA(192, 2^m, 2)$ with the maximum generalized resolution for any m with $106 \le m \le 176$, and from $\boldsymbol{D}_3$ an $OA(768, 2^m, 2)$ with the maximum generalized resolution for any m with $511 \le m \le 704$.*

Example 4.6. *If we take $\boldsymbol{P}_{32}$ as $\boldsymbol{A}$ in Theorem 4.10, then $\boldsymbol{D}_k$ for $k \geq 0$ is an $OA(n, 2^m, 2)$ with $n = 32 \cdot 4^k = 2^{2k+5}$, and $m = 31 \cdot 4^k = 31 \cdot 2^{2k}$. We have that $R(\boldsymbol{D}_k) = 4 - (1/4)/2^k = 4 - 1/2^{k+2}$ and that $\max_{|u|=3} |J_u(\boldsymbol{D}_k)| = 8 \cdot 2^k = 2^{k+3}$. Again using the lower bound in Theorem 4.5, we obtain from $\boldsymbol{D}_1$ an $OA(128, 2^m, 2)$ that has the maximum generalized resolution for any m with $75 \leq m \leq 124$.*

Now let

$$\boldsymbol{D}'_k = \begin{bmatrix} \boldsymbol{D}_k & \boldsymbol{D}_k \\ \boldsymbol{D}_k & -\boldsymbol{D}_k \end{bmatrix}$$

for $k \geq 0$. Then $\boldsymbol{D}'_k$ is an $OA(n, 2^m, 2)$ with $n = 2^{2k+6}$ and $m = 31 \cdot 2^{2k+1}$, which also has $R(\boldsymbol{D}'_k) = 4 - 1/2^{k+2}$. Design $\boldsymbol{D}_k$ has run sizes $32, 128, 512, \ldots$, and $\boldsymbol{D}'_k$ has run sizes $64, 256, 1024, \ldots$. Together, $\boldsymbol{D}_k$ and $\boldsymbol{D}'_k$, $k \geq 0$, cover all run sizes $n \geq 32$ that are powers of 2. We will come back to this example in Section 4.5 when discussing quaternary code designs.

Theorem 4.10 is just one application of Theorem 4.9. Further applications of Theorem 4.9 require other Hadamard matrices $\boldsymbol{H}$ with small values of $\gamma(\boldsymbol{H})$ defined in (4.4). Let $\gamma_n = \min_{\boldsymbol{H}} \gamma(\boldsymbol{H})$, where the minimization is over all Hadamard matrices of order n. Finding the best $\boldsymbol{H}$ for Theorem 4.9 amounts to finding γ_n. We refer the reader to Shi and Tang (2023) for some results on γ_n.

4.4 Good Hadamard Matrices

We have seen in Section 2.6 that two classes of orthogonal arrays can be obtained from Hadamard matrices. The use of the vague term, "good Hadamard matrix", is intended to convey a message that some Hadamard matrices are better than others when it comes to the quality of the orthogonal arrays they produce. By good Hadamard matrices, we actually mean those that have large *types*. The material in this section is taken mainly from Shi and Tang (2018) and marginally from Chen, Shi, and Tang (2023).

Let $\boldsymbol{H}$ be a Hadamard matrix of order n. By permutation and negation of rows and columns, any four columns of $\boldsymbol{H}$ can be uniquely transformed into the following form

$$\begin{bmatrix} \mathbf{1}_a & \mathbf{1}_a & \mathbf{1}_a & \mathbf{1}_a \\ \mathbf{1}_b & \mathbf{1}_b & \mathbf{1}_b & -\mathbf{1}_b \\ \mathbf{1}_b & \mathbf{1}_b & -\mathbf{1}_b & \mathbf{1}_b \\ \mathbf{1}_a & \mathbf{1}_a & -\mathbf{1}_a & -\mathbf{1}_a \\ \mathbf{1}_b & -\mathbf{1}_b & \mathbf{1}_b & \mathbf{1}_b \\ \mathbf{1}_a & -\mathbf{1}_a & \mathbf{1}_a & -\mathbf{1}_a \\ \mathbf{1}_a & -\mathbf{1}_a & -\mathbf{1}_a & \mathbf{1}_a \\ \mathbf{1}_b & -\mathbf{1}_b & -\mathbf{1}_b & -\mathbf{1}_b \end{bmatrix} \tag{4.5}$$

where $a + b = n/4$, $0 \leq b \leq \lfloor n/8 \rfloor$. A set of four columns that can be transformed to the above form is said to be of *type* b. A Hadamard matrix is of *type* b if it has a set of four columns of type b but has no set of four columns of type less than b. The type of a Hadamard matrix was introduced by Kimura (1994). It is easily seen that the type of a Hadamard matrix is invariant with respect to isomorphism operations. Hadamard matrices of up to order 32 have been completely enumerated according to their types by Kharaghani and Tayfeh-Rezaie (2013). Their results are presented in Table 4.3.

From Table 4.3, we see that out of 60 nonisomorphic Hadamard matrices of order 24, one has type 2, another one has type 1, and the remaining 58 has type 0. According to the

TABLE 4.3
Number of nonisomorphic Hadamard matrices by type

	Order	12	16	20	24	28	32
Type	0	0	5	0	58	0	13,680,757
	1	1	0	3	1	486	26,369
	2	0	0	0	1	1	2,900
	3	0	0	0	0	0	1
	Total	1	5	3	60	487	13,710,027

theoretical results to be presented later, we know that the one of type 2 is given by Paley's first construction, the one of type 1 is given by Paley's second construction, and the one given by tensor product construction must be one of the remaining 58 Hadamard matrices of order 24.

The next result is useful for finding the type of a Hadamard matrix.

Lemma 4.3. *A Hadamard matrix of order n has type b if and only if* $\max_{|u|=4} |J_u| = n-8b$.

Proof. Suppose that $\boldsymbol{H}$ has type b. We will prove that $\max_{|u|=4} |J_u| = n - 8b$. From the proof, we see that the converse is also true. Consider the four columns in (4.5). Obviously, their J-characteristic is $4(a-b) = n - 8b \geq 0$, as $a + b = n/4$ and $b \leq \lfloor n/8 \rfloor$. For any other four columns of type b, their J-characteristic must be equal to $\pm(n-8b)$. This is because permutations of rows and columns and negation of rows have no effect on the J value, and only negation of columns can possibly change the sign of the J value. Now consider any four columns of $\boldsymbol{H}$, say $\boldsymbol{h}_1, \boldsymbol{h}_2, \boldsymbol{h}_3, \boldsymbol{h}_4$, which must be of type $b' \geq b$. Therefore $|J(\boldsymbol{h}_1, \boldsymbol{h}_2, \boldsymbol{h}_3, \boldsymbol{h}_4)| = n - 8b' \leq n - 8b$. That $\boldsymbol{H}$ has a set of four columns of type b implies that the equality must be attained. □

Lemma 4.3 provides a simple characterization for Hadamard matrices of type b, and can therefore be taken as a definition of type for anyone who finds the original definition cumbersome. The next two lemmas document the types of two familiar constructions of Hadamard matrices.

Lemma 4.4. *Let $\boldsymbol{H}$ and $\boldsymbol{G}$ be any two Hadamard matrices of orders 2 or higher. Then $\boldsymbol{H} \otimes \boldsymbol{G}$ has type 0.*

Proof. Suppose that $\boldsymbol{H}$ has order n_1 and $\boldsymbol{G}$ has order n_2. Let $\boldsymbol{h}_1, \boldsymbol{h}_2$ be two columns of $\boldsymbol{H}$ and $\boldsymbol{g}_1, \boldsymbol{g}_2$ be two columns of $\boldsymbol{G}$. Consider the four columns, $\boldsymbol{h}_1 \otimes \boldsymbol{g}_1, \boldsymbol{h}_1 \otimes \boldsymbol{g}_2, \boldsymbol{h}_2 \otimes \boldsymbol{g}_1$ and $\boldsymbol{h}_2 \otimes \boldsymbol{g}_2$, of $\boldsymbol{H} \otimes \boldsymbol{G}$. By Lemma 4.2, we have

$$J(\boldsymbol{h}_1 \otimes \boldsymbol{g}_1, \boldsymbol{h}_1 \otimes \boldsymbol{g}_2, \boldsymbol{h}_2 \otimes \boldsymbol{g}_1, \boldsymbol{h}_2 \otimes \boldsymbol{g}_2) = J(\boldsymbol{h}_1, \boldsymbol{h}_1, \boldsymbol{h}_2, \boldsymbol{h}_2)J(\boldsymbol{g}_1, \boldsymbol{g}_2, \boldsymbol{g}_1, \boldsymbol{g}_2) = n_1 n_2.$$

Lemma 4.4 then follows from Lemma 4.3. □

Lemma 4.5. *Hadamard matrices from Paley's second construction are of type 1.*

Proof. Let $\boldsymbol{H}$ be a Hadamard matrix of order n from Paley's second construction as given in Section 2.5. Since n is not a multiple of 8, $\boldsymbol{H}$ cannot have type 0, for otherwise a saturated orthogonal array obtained by normalizing $\boldsymbol{H}$ contains three columns with $|J_u| = n$, which means that the three columns form a defining word, contradicting Corollary 3.5. Therefore, $\boldsymbol{H}$ must have type 1 if we can find four columns of $\boldsymbol{H}$ whose J-characteristic is equal to $\pm(n-8)$. Write $\boldsymbol{H}$ as

$$\boldsymbol{H} = \begin{bmatrix} \boldsymbol{A} & \boldsymbol{B} \\ \boldsymbol{B} & -\boldsymbol{A} \end{bmatrix}, \text{ where } \boldsymbol{A} = \begin{bmatrix} 1 & \boldsymbol{1}_s^T \\ \boldsymbol{1}_s & \boldsymbol{Q} + \boldsymbol{I}_s \end{bmatrix} \text{and } \boldsymbol{B} = \begin{bmatrix} -1 & \boldsymbol{1}_s^T \\ \boldsymbol{1}_s & \boldsymbol{Q} - \boldsymbol{I}_s \end{bmatrix}.$$

Let $\boldsymbol{a}_1$ and $\boldsymbol{a}_2$ be the first two columns of $\boldsymbol{A}$, and $\boldsymbol{b}_1$ and $\boldsymbol{b}_2$ be the two corresponding columns of $\boldsymbol{B}$. Consider the four columns of $\boldsymbol{H}$:

$$\begin{bmatrix} \boldsymbol{a}_1 & \boldsymbol{a}_2 & \boldsymbol{b}_1 & \boldsymbol{b}_2 \\ \boldsymbol{b}_1 & \boldsymbol{b}_2 & -\boldsymbol{a}_1 & -\boldsymbol{a}_2 \end{bmatrix}.$$

The J-characteristic of these four columns is $2J(\boldsymbol{a}_1, \boldsymbol{a}_2, \boldsymbol{b}_1, \boldsymbol{b}_2) = 2J(\boldsymbol{a}_1\boldsymbol{b}_1, \boldsymbol{a}_2\boldsymbol{b}_2)$. One can easily see that $\boldsymbol{a}_1\boldsymbol{b}_1 = (-1, 1, 1, \ldots, 1)^T$ and $\boldsymbol{a}_2\boldsymbol{b}_2 = (1, -1, 1, \ldots, 1)^T$, showing that $2J(\boldsymbol{a}_1\boldsymbol{b}_1, \boldsymbol{a}_2\boldsymbol{b}_2) = 2((n/2 - 2) - 2) = n - 8$. □

One reason that the notion of type for Hadamard matrices is useful is its connection with generalized resolutions of foldover designs. The next result follows immediately from Lemma 4.3.

Theorem 4.11. *Let $\boldsymbol{H}$ be a Hadamard matrix of order n and of type b. Then the foldover design $\boldsymbol{D} = (\boldsymbol{H}^T, -\boldsymbol{H}^T)^T$, an $OA(2n, 2^n, 3)$, has a generalized resolution of $R = 4 + 8b/n$.*

Theorem 4.11 calls for Hadamard matrices of large types. Let $b_{\max}$ denote the largest b for Hadamard matrices of order n. Combining Theorem 4.11 with Theorem 3.5 leads to another result.

Corollary 4.1. *Let $\boldsymbol{H}$ be a Hadamard matrix of order n and of type $b_{\max}$. Then the foldover design $\boldsymbol{D} = (\boldsymbol{H}^T, -\boldsymbol{H}^T)^T$ has the maximum generalized resolution $R = 4 + 8b_{\max}/n$ among all $OA(2n, 2^n, 3)$s.*

We next present some results on Hadamard matrices of type $b_{\max}$. From Table 4.3, we have that $b_{\max} = 1, 0, 1, 2, 2,$ and 3 for $n = 12, 16, 20, 24, 28,$ and 32, respectively. The following upper bound on $b_{\max}$ can be established.

Theorem 4.12. *Let $\boldsymbol{H}$ be a Hadamard matrix of order $n \geq 4$ and type $b_{\max}$. We then have that*

$$b_{\max} \leq \left\lfloor (n/8)\left(1 - 1/(n-3)^{1/2}\right) \right\rfloor.$$

Proof. Let $\boldsymbol{h}_1, \ldots, \boldsymbol{h}_n$ be the columns of $\boldsymbol{H}$. Considering the vector $\boldsymbol{h}_1\boldsymbol{h}_2\boldsymbol{h}_3$ under the orthonormal basis $\boldsymbol{h}_1/n^{1/2}, \ldots, \boldsymbol{h}_n/n^{1/2}$ and noting the orthogonality of $\boldsymbol{h}_1\boldsymbol{h}_2\boldsymbol{h}_3$ to $\boldsymbol{h}_1, \boldsymbol{h}_2, \boldsymbol{h}_3$, we have $\sum_{j=4}^{n} |J(\boldsymbol{h}_1\boldsymbol{h}_2\boldsymbol{h}_3, \boldsymbol{h}_j)|^2 = n^2$. This implies that $n^2 \leq (n-3)\left(\max_{|u|=4}|J_u|\right)^2$. By Lemma 4.3, we obtain that $n^2 \leq (n-3)(n - 8b_{\max})^2$, solving which yields the desired upper bound on $b_{\max}$. □

For $n = 36$, Theorem 4.12 gives $b_{\max} \leq 3$. Shi and Tang (2018) have found two Hadamard matrices of type 3 for order 36; we therefore obtain that $b_{\max} = 3$ for $n = 36$.

We now examine the type of Paley matrices, Hadamard matrices given by Paley's first construction.

Theorem 4.13. *Let b_{Paley} denote the type of a Paley matrix of order n. Then*

$$b_{\text{Paley}} \geq \left\lceil (1/8)\left(n - 4 - 2(n-1)^{1/2}\right) \right\rceil.$$

Proof. First we recall that a Paley design is obtained from a Paley matrix by deleting the column of all ones. By Theorem 4.3, a Paley matrix must have that $|J_u| \leq 4 + 2(n-1)^{1/2}$ for all u with $|u| = 4$. Using Lemma 4.3, we obtain that

$$b_{\text{Paley}} = (1/8)(n - \max_{|u|=4}|J_u|) \geq (1/8)(n - 4 - 2(n-1)^{1/2}),$$

implying $b_{\text{Paley}} \geq \left\lceil (1/8)\left(n - 4 - 2(n-1)^{1/2}\right) \right\rceil$. □

The lower bound on b_{Paley} in Theorem 4.13 is quite sharp as Shi and Tang (2018) have verified that it is attained for all $n \leq 5000$ where Paley matrices are available.

Corollary 4.2. *For $n = 12, 20, 24, 28, 32, 44, 60, 72,$ and 80, Paley matrices have type $b_{\max}$.*

This result holds because the upper and lower bounds in Theorems 4.12 and 4.13 are equal, as can easily be checked for the listed n values. As mentioned earlier, Hadamard matrices of type $b_{\max}$ for order 36 have also been discovered. This means that Hadamard matrices of type $b_{\max}$ are currently available for order $n = 12, 20, 24, 28, 32, 36, 44, 60, 72,$ and 80. Let $\boldsymbol{G}_{2n}$ be the foldover design of the Hadamard matrix of type $b_{\max}$ for $n = 12, 20, 24, 28, 32, 36, 44, 60, 72,$ and 80. Corollary 4.1 states that $\boldsymbol{G}_{2n}$ has the maximum generalized resolution among all OA$(2n, 2^n, 3)$s. One wonders if deleting some columns from $\boldsymbol{G}_{2n}$ still gives designs with maximum generalized resolutions. The answer is positive (Chen, Shi, and Tang, 2023).

Theorem 4.14. *Suppose $\boldsymbol{D}$ is an OA$(n, 2^m, 3)$ with $n/3 \leq m \leq n/2$. Then*

$$\max_{|u|=4} |J_u| \geq n - 16 \left\lfloor (n/16)(1 - \zeta^{1/2}) \right\rfloor,$$

where

$$\zeta = \frac{4m^3 - 3m^2 n + mn^2 - 3mn + 4m - n^3/8 + 3n^2/4 - n}{m(m-1)(m-2)(m-3)}.$$

Design $\boldsymbol{D}$ in Theorem 4.14 must be a foldover design due to Theorem 3.5. Using this property, the proof of Theorem 4.14 is very similar to that of Theorem 4.5. We omit the details. Let $\boldsymbol{G}_{2n,m}$ be a design obtained by choosing any m columns of $\boldsymbol{G}_{2n}$. We can check that $\boldsymbol{G}_{2n,m}$ attains the lower bound in Theorem 4.14, which establishes the next result.

Corollary 4.3. *Design $\boldsymbol{G}_{2n,m}$ has the maximum generalized resolution among all OA$(2n, 2^m, 3)$s for $n = 12, 20, 24, 28, 32, 36, 44, 60, 72,$ and 80, and $2n/3 \leq m \leq n$.*

Let $\boldsymbol{H} = (\boldsymbol{1}_n, \boldsymbol{D})$ be a Hadamard matrix of type $b_{\max}$. It is obvious that

$$\max_{|u|=4} |J_u(\boldsymbol{H})| = \max_{|u|=3,4} |J_u(\boldsymbol{D})| = n - 8b_{\max}.$$

This implies that design $\boldsymbol{D}$ minimizes $\max_{|u|=3,4} |J_u(\boldsymbol{D})|$ over all OA$(n, 2^{n-1}, 2)$s, and thus has a large generalized resolution owing to $\max_{|u|=3} |J_u(\boldsymbol{D})| \leq \max_{|u|=3,4} |J_u(\boldsymbol{D})|$. If $\boldsymbol{D}$ simultaneously minimizes $\max_{|u|=3} |J_u(\boldsymbol{D})|$, then it has the maximum generalized resolution. One situation for this to occur is when the value of $\max_{|u|=3,4} |J_u(\boldsymbol{D})|$ reaches the lower bound for $\max_{|u|=3} |J_u(\boldsymbol{D})|$ in Theorem 4.2. Interestingly, all Hadamard matrices of type $b_{\max}$ we have found for $n = 12, 20, 24, 28, 32, 36, 44, 60, 72,$ and 80 have this property.

Remark 4.1. *Let $\boldsymbol{H} = (\boldsymbol{1}_n, \boldsymbol{D})$ be a Hadamard matrix of type b. Because $\max_{|u|=4} |J_u(\boldsymbol{H})| = \max_{|u|=3,4} |J_u(\boldsymbol{D})| = n - 8b$, design $\boldsymbol{D}$ has no defining words of length three or four if and only if $b \geq 1$. Therefore, whenever $\boldsymbol{H} = (\boldsymbol{1}_n, \boldsymbol{D})$ has type $b \geq 1$, design $\boldsymbol{D}$ has the four-factor hidden projection property and the design $(\boldsymbol{H}^T, -\boldsymbol{H}^T)^T$ has the five-factor hidden projection property as discussed in Section 3.6.*

Hadamard matrices of type $b_{\max}$ are also useful in constructing supersaturated designs. This topic will be discussed in Chapter 8.

4.5 Quaternary Code Designs

Xu and Wong (2007) constructed quaternary code designs from quaternary codes. Let $\boldsymbol{G}$ be a $k \times m$ matrix with entries from $Z_4 = \{0, 1, 2, 3\}$. Then a quaternary code $\boldsymbol{C}$ is a matrix obtained by collecting all row vectors that are linear combinations of the rows of $\boldsymbol{G}$ over Z_4, with all calculations being modulo 4. As Z_4 is not a field, the quaternary code $\boldsymbol{C}$ is not a linear code, but the linearity over Z_4 is still an advantage as the entire code $\boldsymbol{C}$, a $4^k \times m$ matrix, can now be represented by its generator matrix $\boldsymbol{G}$, which is a much simpler $k \times m$ matrix. Section A.5 of the appendix chapter provides an introduction to coding theory.

If we apply the Gray map

$$\phi(0) = (0, 0), \quad \phi(1) = (0, 1), \quad \phi(2) = (1, 1), \quad \phi(3) = (1, 0)$$

to the quaternary code $\boldsymbol{C}$, we obtain a two-level design of 4^k runs with $2m$ factors. Denote this design by $\boldsymbol{Q}_1$. Design $\boldsymbol{Q}_1$ has levels 0 and 1, and we make it have levels ± 1 by changing 0 to -1. For simplicity, the design is still denoted by $\boldsymbol{Q}_1$ after level conversion. From $\boldsymbol{Q}_1$, we obtain another design $\boldsymbol{Q}_2$ by

$$\boldsymbol{Q}_2 = \begin{pmatrix} \boldsymbol{Q}_1 & \boldsymbol{Q}_1 \\ \boldsymbol{Q}_1 & -\boldsymbol{Q}_1 \end{pmatrix},$$

which is the double of $\boldsymbol{Q}_1$ (Chen and Cheng, 2006).

We now construct a generator matrix $\boldsymbol{G}$ of k rows and $(4^k - 2^k)/2$ columns according to the following procedure:

(a) write down all possible columns of k elements over Z_4,

(b) delete any column if it does not contain element 1, and

(c) further delete any column whose first entry that is not 0 nor 2 is 3.

Then matrix $\boldsymbol{G}$ satisfies that (i) it does not have any column containing 0 and 2 only and (ii) none of its column is a multiple of another over Z_4. The following result holds (Xu and Wong, 2007).

Theorem 4.15. *For any integer $k \geq 2$, both $\boldsymbol{Q}_1$ and $\boldsymbol{Q}_2$ derived from the just constructed generator matrix $\boldsymbol{G}$ have a generalized resolution of* 3.5.

We omit the proof. Clearly, design $\boldsymbol{Q}_1$ has $4^k = 2^{2k}$ runs and $m = 2^{2k} - 2^k$ factors and design $\boldsymbol{Q}_2$ has 2^{2k+1} runs and $m = 2^{2k+1} - 2^{k+1}$ factors. The run size of $\boldsymbol{Q}_1$ is an even power of 2 whereas that of $\boldsymbol{Q}_2$ an odd power of 2.

Example 4.7. *We give an illustration of the construction for design $\boldsymbol{Q}_1$. For $k = 2$, using the procedure above, we obtain a generator matrix $\boldsymbol{G}$ with* 2 *rows and* 6 *columns:*

$$\boldsymbol{G} = \begin{bmatrix} 1 & 0 & 2 & 1 & 1 & 1 \\ 0 & 1 & 1 & 2 & 1 & 3 \end{bmatrix}.$$

Taking all linear combinations of the two rows over $Z_4 = \{0, 1, 2, 3\}$ with all calculations

TABLE 4.4
A quaternary code design of 16 runs for 12 factors

−1	−1	−1	−1	−1	−1	−1	−1	−1	−1	−1	−1
−1	−1	−1	1	−1	1	1	1	−1	1	1	−1
−1	−1	1	1	1	1	−1	−1	1	1	1	1
−1	−1	1	−1	1	−1	1	1	1	−1	−1	1
−1	1	−1	−1	1	1	−1	1	−1	1	−1	1
−1	1	−1	1	1	−1	1	−1	1	1	−1	−1
−1	1	1	1	−1	−1	−1	1	1	−1	1	−1
−1	1	1	−1	−1	1	1	−1	−1	−1	1	1
1	1	−1	−1	−1	−1	1	1	1	1	1	1
1	1	−1	1	−1	1	−1	−1	1	−1	−1	1
1	1	1	1	1	1	1	1	−1	−1	−1	−1
1	1	1	−1	1	−1	−1	−1	−1	1	1	−1
1	−1	−1	−1	1	1	1	−1	1	−1	1	−1
1	−1	−1	1	1	−1	−1	1	−1	−1	1	1
1	−1	1	1	−1	−1	1	−1	−1	1	−1	1
1	−1	1	−1	−1	1	−1	1	1	1	−1	−1

modulo 4, we obtain a quaternary code

$$
\boldsymbol{C} = \begin{bmatrix}
0 & 0 & 0 & 0 & 0 & 0 \\
0 & 1 & 1 & 2 & 1 & 3 \\
0 & 2 & 2 & 0 & 2 & 2 \\
0 & 3 & 3 & 2 & 3 & 1 \\
1 & 0 & 2 & 1 & 1 & 1 \\
1 & 1 & 3 & 3 & 2 & 0 \\
1 & 2 & 0 & 1 & 3 & 3 \\
1 & 3 & 1 & 3 & 0 & 2 \\
2 & 0 & 0 & 2 & 2 & 2 \\
2 & 1 & 1 & 0 & 3 & 1 \\
2 & 2 & 2 & 2 & 0 & 0 \\
2 & 3 & 3 & 0 & 1 & 3 \\
3 & 0 & 2 & 3 & 3 & 3 \\
3 & 1 & 3 & 1 & 0 & 2 \\
3 & 2 & 0 & 3 & 1 & 1 \\
3 & 3 & 1 & 1 & 2 & 0
\end{bmatrix}.
$$

Applying the Gray map to the code $\boldsymbol{C}$ *and then converting level 0 to level* -1, *we obtain a quaternary code design* $\boldsymbol{Q}_1$ *with* $n = 16$ *runs for* $m = 12$ *factors, which we display in Table* 4.4.

In terms of the generalized resolution criterion, designs that are better than quaternary code designs have been constructed in Section 4.3. Recall designs $\boldsymbol{D}_k$ and $\boldsymbol{D}'_k$ for $k \geq 0$ given in Example 4.6, where $\boldsymbol{D}_k$ is an $\mathrm{OA}(2^{2k+5}, 2^{31\cdot 2^{2k}}, 2)$ with $R(\boldsymbol{D}_k) = 4 - 1/2^{k+2}$ and $\boldsymbol{D}'_k$ is $\mathrm{OA}(2^{2k+6}, 2^{31\cdot 2^{2k+1}}, 2)$ also with $R(\boldsymbol{D}'_k) = 4 - 1/2^{k+2}$. Together, designs $\boldsymbol{D}_k$ and $\boldsymbol{D}'_k$, $k \geq 0$, cover all run sizes $n \geq 32$ that are powers of 2. Taking $k = 0, 1$, we have that design $\boldsymbol{D}_0$ is an $\mathrm{OA}(32, 2^{31}, 2)$ with $R = 4 - 1/4$, design $\boldsymbol{D}_1$ is an $\mathrm{OA}(128, 2^{124}, 2)$ with $R = 4 - 1/8$, design $\boldsymbol{D}'_0$ is an $\mathrm{OA}(64, 2^{62}, 2)$ with $R = 4 - 1/4$, and design $\boldsymbol{D}'_1$ is an $\mathrm{OA}(256, 2^{248}, 2)$ with $R = 4 - 1/8$.

Xu and Wong (2007) further examined the selection of quaternary code designs using two aberration criteria, which are to be introduced later in this chapter and the next chapter. Phoa and Xu (2009) investigated quarter fractions of quaternary code designs. A trignometric approach was considered by Zhang, Phoa, Mukerjee, and Xu (2011). Mukerjee and Tang (2013) developed a complementary design theory for quaternary code designs.

The famous Nordstrom-Robinson code (Nordstrom and Robinson, 1967) can be derived from a quaternary code, and is design $\boldsymbol{Q}_1$ in our notation based on the generator matrix

$$\boldsymbol{G} = \begin{bmatrix} 1 & 3 & 1 & 2 & 1 & 0 & 0 & 0 \\ 1 & 0 & 3 & 1 & 2 & 1 & 0 & 0 \\ 1 & 0 & 0 & 3 & 1 & 2 & 1 & 0 \\ 1 & 0 & 0 & 0 & 3 & 1 & 2 & 1 \end{bmatrix}.$$

This design of 256 runs for 16 factors has a generalized resolution of 6.5. A half fraction of $\boldsymbol{Q}_1$, a design of 128 runs for 15 factors, has a generalized resolution 5.5. An extensive study on these two and related designs can be found in Xu (2005). We remark that any 256-run regular design for 16 factors has a resolution at most 5 and any 128-run regular design for 15 factors has resolution at most 4.

4.6 Minimum G-Aberration

The generalized resolution provides a simple characterization of the aliasing properties of a nonregular design. But as a design selection criterion, it is not very discriminating since many designs can have the same generalized resolution. The minimum G-aberration was introduced in Deng and Tang (1999) as a refinement of the generalized resolution criterion and can also be viewed as an extension of minimum aberration from regular designs to nonregular designs.

To get the basic idea, consider all $|J_u|$ values for $|u| = 3$. In total there are $\binom{m}{3}$ of these $|J_u|$ values when designs with m factors are under consideration. The generalized resolution criterion attempts to minimize all these $|J_u|$ values by minimizing $\max_{|u|=3} |J_u|$. If two designs minimize $\max_{|u|=3} |J_u|$, the one with the smaller number of subsets u attaining the same $\max_{|u|=3} |J_u|$ is preferred. If two designs minimize $\max_{|u|=3} |J_u|$ and have the same frequency of attaining this $\max_{|u|=3} |J_u|$ value, we then compare them by looking at the second largest $|J_u|$ values and select the one that corresponds to the smaller of the two second largest $|J_u|$ values. If we carry this idea to the fullest extent, first for all $|J_u|$ values with $|u| = 3$, then for all $|J_u|$ values with $|u| = 4$ and so on, then we obtain the minimum G-aberration criterion.

We need the *confounding frequency vector* (CFV) to give a precise definition for minimum G-aberration. Let $\boldsymbol{D}$ be an OA($n = 4\lambda, 2^m, 2$), an orthogonal array of strength two. Recall that by Theorem 3.7, J_u must be a multiple of 4. Now consider all $|J_u|$ values with $|u| = k$ for a fixed $k = 3, 4, \ldots, m$. Let f_{kj} be the frequency of $|J_u|$ values that are equal to $n - 4j$ for $j = 0, 1, \ldots, \lambda$, and define $F_k = (f_{k0}, f_{k1}, \ldots, f_{k\lambda})$. Then the CFV of design $\boldsymbol{D}$ is defined as CFV $= (F_3, F_4, \ldots, F_m)$, which is a vector of $(m-2)(\lambda+1)$ components. The minimum G-aberration criterion is defined as one that selects designs by sequentially minimizing the components of the CFV in the order listed. It is evident that this criterion selects designs with maximum generalized resolutions.

In the original definition of Deng and Tang (1999), the last entry $f_{k\lambda}$ is omitted from each F_k as it is determined by the other entries due to $\sum_{j=0}^{\lambda} f_{kj} = \binom{m}{k}$. We now include it in F_k as we think that having this redundancy is helpful since it provides a checking on the correctness of f_{kj} values. In retrospect, it seems more appropriate to use the terminology *aliasing frequency vector* instead of *confounding frequency vector*. But the latter has been in use in the literature for two decades; so we make no change and stick with the confounding frequency vector, the CFV.

In actual implementation, one can work with a simpler CFV. Again by Theorem 3.7, we know that J_u is a multiple of 8 if n is a multiple of 8. So, we can use a simpler F_k that collects the frequencies of $|J_u|$ values equaling $n-8j$ for $j=0,1,\ldots,n/8$. When n is not a multiple of 8, we have that J_u is not a multiple of 8 for $|u|=3,4$, and J_u is a multiple of 8 for $|u|=5,6$. So, for $k=3,4$, we let F_k collect the frequencies of $|J_u|$ values that equal $n-8j$ for $j=0,1,\ldots,\lfloor n/8 \rfloor$, and for $k=5,6$, let F_k collect the frequencies of $|J_u|$ values that equal $n-4-8j$ for $j=0,1,\ldots,\lfloor n/8 \rfloor$. As we are often satisfied with designs that have small $|J_u|$ values for $|u|=3,4$, further simplification is achieved by considering the shortened CFV given by (F_3, F_4).

Theoretical construction of minimum G-aberration designs is a challenging problem. Tang and Deng (2003) obtained minimum G-aberration designs of 3, 4, and 5 factors for all run sizes that are multiples of 4. The same problem for 6 factors was solved by Evangelaras (2015). As $\mathrm{OA}(n, 2^m, t)$s have been completely enumerated for $m=t+1$ and $t+2$ (Seiden and Zemach, 1966; Stufken and Tang, 2007), effectively, construction of minimum G-aberration $\mathrm{OA}(n, 2^m, t)$s is completely solved for $m=t+1$ and $t+2$. Without complete enumeration, Evangelaras (2015) obtained all minimum G-aberration $\mathrm{OA}(n, 2^m, t)$s for $m=t+3$.

Isomorphic designs have the same CFV. If two designs have different CFVs, then they must be nonisomorphic. As a result, the CFV can be used as a fast screening tool for detecting nonisomorphism. A survey of this and other related topics was conducted by Katsaounis and Dean (2008).

4.7 Computational Results

Deng and Tang (2002) conducted a computer search of minimum G-aberration designs of $n=12,16,20$, and 24 runs. However, their search is incomplete for $n=20$ and 24. Over the years, many authors have devoted their efforts to computer search of minimum G-aberration designs. The most comprehensive results are those given in Sun, Li, and Ye (2008), Schoen, Eendebak, and Nguyen (2010) and Schoen, Vo-Thanh, and Goos (2017). While many good designs are found using the minimum G-aberration criterion, we focus in this section on surveying the results that give minimum G-aberration designs among all competing designs. It is worth repeating that minimum G-aberration designs have maximum generalized resolutions.

Orthogonal arrays of 12, 16, and 20 runs were completely enumerated by Sun, Li, and Ye (2008). Therefore, all minimum G-aberration designs of 12, 16, and 20 runs have been obtained.

Orthogonal arrays of 24 runs were completely enumerated by Schoen, Eendebak, and Nguyen (2010), which means that all minimum G-aberration designs of 24 runs have also been found.

Strength-3 orthogonal arrays of 32, 40, and 48 runs were completely enumerated by Schoen, Eendebak, and Nguyen (2010). This shows that all minimum G-aberration designs of $n=32,40$, and 48 runs and m factors have been identified for $m \leq n/2$.

$\mathrm{OA}(28, 2^m, 2)$s were completely enumerated for $m \leq 7$ and $m=27$ by Schoen, Eendebak, and Nguyen (2010). $\mathrm{OA}(28, 2^m, 2)$s with $\max_{|u|=3} |J_u| = 4$ for $m \leq 14$ were completely enumerated by Bulutoglu and Ryan (2015) and Schoen, Vo-Thanh, and Goos (2017). Therefore, all minimum G-aberration designs of 28 runs for $m \leq 14$ and $m=27$ factors have been obtained.

OA$(36, 2^m, 2)$s were completely enumerated for $m \leq 5$ by Evangelaras, Koukouvinos, and Lappas (2007). OA$(36, 2^m, 2)$s with $\max_{|u|=3} |J_u| = 4$ and $\max_{|u|=4} |J_u| \leq 12$ for $m \leq 18$ were completely enumerated by Schoen, Vo-Thanh, and Goos (2017). Therefore, all minimum G-aberration designs of 36 runs for $m \leq 18$ have been found.

To sum up, all minimum G-aberration designs with up to 24 runs are available, and all minimum G-aberration designs with $n = 28, 32, 36, 40$, and 48 runs are available for $m \leq n/2$ factors.

4.8 Further Remarks

The resolution introduced for regular designs by Box and Hunter (1961) describes the worst-case scenario of the aliasing relationships among factorial effects, and characterizes whether or not projection designs with up to a certain number of factors contain a full factorial. The generalized resolution introduced in this chapter accomplishes the same for orthogonal arrays. As a result, the generalized resolution not only provides a simple characterization for an orthogonal array in terms of both effect aliasing and projection designs, but also serves as a useful selection criterion for comparing different orthogonal arrays.

The resolution of a regular design has another interpretation in terms of the estimability of factorial effects. This can be taken as a definition of resolution for any factorial design as was done previously by Webb (1968). Specifically, a factorial design has resolution r if it permits the estimation of factorial effects involving up to $\lfloor (r-1)/2 \rfloor$ factors assuming that factorial effects involving $\lceil (r-1)/2 \rceil + 1$ or more factors are negligible. Although not suitable as a criterion for comparing different designs, this concept of resolution is useful for classifying designs in terms of the estimability of factorial effects, and it is especially convenient to use when studying non-orthogonal designs, a topic we will pursue in Chapter 9.

Box and Tyssedal (1996) introduced projectivity as a weak version of the strength for orthogonal arrays. An orthogonal array has projectivity t if its projection design onto any t factors contains a complete 2^t factorial. Suppose that an orthogonal array has a generalized resolution of $R = r+\delta$ with $0 < \delta < 1$. Then it must have projectivity r as shown in Theorem 4.1. Very often, such an array has projectivity larger than r. For example, the OA$(128, 2^{15}, 4)$ mentioned in Section 4.5 has generalized resolution 5.5 but has projectivity 6 (Xu, 2005). For another example, Paley design P_n, which has generalized resolution $3 < R < 4$ for $n \geq 12$, has projectivity 4 for $n \geq 68$ (Chen, Shi, and Tang, 2023).

Covering arrays are a class of combinatorial designs useful in software testing and more generally in testing complex engineered systems. A binary covering array has two symbols, and it is said to have strength t if any subarray of t columns contains all symbol combinations. We see that the strength t of a covering array is precisely the projectivity introduced for orthogonal arrays. Because there are no requirements for covering arrays other than their strengths, one can find a covering array with a much smaller run size than what is needed by an orthogonal array for a given strength and a number of factors. For example, for strength two and 126 factors, a covering array of 10 runs can be found while an orthogonal array needs 128 runs. Readers who are interested in covering arrays are referred to Colbourn (2004) and Lawrence, Kacker, Lei, Kuhn, and Forbes (2011) for general readings and Lekivetz and Morgan (2020) for recent research.

5

Minimum G_2-Aberration

In this chapter, a relaxed variant of minimum G-aberration, called minimum G_2-aberration, is introduced and studied. It is much easier to use than minimum G-aberration and, like minimum G-aberration, also reduces to minimum aberration when it is applied to regular designs. We provide several statistical justifications for this criterion, including minimum contamination and model robustness justifications under effect hierarchy, as well as justifications from projection and Bayesian considerations. We show that minimum G_2-aberration is equivalent to another criterion called minimum moment aberration. This equivalence facilitates both theoretical characterization and algorithmic construction of minimum G_2-aberration designs. We also present a complementary design theory, which is a powerful tool for determining minimum G_2-aberration designs when the number of factors is large.

5.1 Minimum G_2-Aberration

Consider strength-two orthogonal arrays of size $n = 4\lambda$ for m two-level factors. Recall that the minimum G-aberration criterion is to select a design by sequentially minimizing the components of the confounding frequency vector CFV $= (F_3, F_4, \ldots, F_m)$, where $F_k = (f_{k0}, f_{k1}, \ldots, f_{k\lambda})$ for $k = 3, 4, \ldots, m$, and f_{kj} is the frequency of $|J_u|$ values with $|u| = k$ that are equal to $n - 4j$ for $j = 0, 1, \ldots, \lambda$. Note that CFV is a vector of $(m-2)(\lambda+1)$ components. Sequentially comparing these components can be tedious especially when n and m are large. Tang and Deng (1999) proposed a more relaxed variant by collapsing CFV into a vector with $m-2$ components only. Specifically, the $\lambda+1$ components $f_{k0}, f_{k1}, \ldots, f_{k\lambda}$ of each F_k are replaced by a single quantity $A_k = (1/n^2)\sum_{j=0}^{\lambda}(n-4j)^2 f_{kj}$.

More directly, we define

$$A_k(\boldsymbol{D}) = \frac{1}{n^2} \sum_{|u|=k} J_u^2(\boldsymbol{D}), \tag{5.1}$$

for any two-level factorial design $\boldsymbol{D}$ of n runs for m factors. Thus $A_k(\boldsymbol{D})$ can be considered as an overall measure of the largeness of the $|J_u|$ values for subsets u of $\{1, 2, \ldots, m\}$ with $|u| = k$. For two designs $\boldsymbol{D}_1$ and $\boldsymbol{D}_2$, let k be the smallest integer such that $A_k(\boldsymbol{D}_1) \neq A_k(\boldsymbol{D}_2)$. Then we say that $\boldsymbol{D}_1$ has less G_2-aberration than $\boldsymbol{D}_2$ if $A_k(\boldsymbol{D}_1) < A_k(\boldsymbol{D}_2)$. In other words, a design has minimum G_2-aberration if it sequentially minimizes the components of $(A_1(\boldsymbol{D}), \ldots, A_m(\boldsymbol{D}))$. If orthogonal arrays of strength two are under consideration, we have that $A_1(\boldsymbol{D}) = A_2(\boldsymbol{D}) = 0$ by Theorem 3.3. In this case, a design has minimum G_2-aberration if it sequentially minimizes $A_3(\boldsymbol{D}), \ldots, A_m(\boldsymbol{D})$.

Under a regular design, $|J_u| = n$ if the factors in u form a defining word, and $|J_u| = 0$ otherwise. Thus A_k is equal to the number of defining words of length k. It follows that for regular designs, minimum G_2-aberration is the same as minimum aberration. Earlier, we saw that in this case, minimum G-aberration is also the same as minimum aberration; thus there is no difference among the three criteria.

DOI: 10.1201/9781003371861-5

For nonregular designs, minimum aberration is not defined, and since minimum G-aberration is a more refined criterion than minimum G_2-aberration, we may have a design $\boldsymbol{D}_1$ that has less G-aberration than another design $\boldsymbol{D}_2$, but they have the same performance under the minimum G_2-aberration criterion. Sometimes the two criteria can even produce inconsistent rankings of designs.

The minimum G_2-aberration criterion was extended by Xu and Wu (2001) to what they called the generalized minimum aberration criterion for factorial designs with more than two levels or even mixed levels. Following their terminology, we shall call $(A_1(\boldsymbol{D}), \ldots, A_m(\boldsymbol{D}))$ the generalized wordlength pattern of $\boldsymbol{D}$. The minimum G_2-aberration theory for general factorial designs will be studied in Chapter 10.

A more general criterion than minimum G_2-aberration is minimum G_e-aberration, which still requires the sequential minimization of $A_1(\boldsymbol{D}), \ldots, A_m(\boldsymbol{D})$, but with $A_k(\boldsymbol{D})$ defined by

$$A_k(\boldsymbol{D}) = \frac{1}{n^e} \sum_{|u|=k} |J_u(\boldsymbol{D})|^e$$

for any $e > 0$. As e increases, the minimum G_e-aberration criterion behaves more and more like the minimum G-aberration, and for given n and m, we can make the minimum G_e-aberration equivalent to the minimum G-aberration by choosing a sufficiently large e. Therefore, the minimum G_e-aberration with large e can be used as a computationally efficient implementation of the minimum G-aberration. We refer to Ingram and Tang (2005) for a detailed discussion on this topic.

5.2 Statistical Justifications

In this section, we provide justifications for the minimum G_2-aberration criterion from several different perspectives. We start with minimum contamination of the estimates of main effects in the presence of nonnegligible interactions.

5.2.1 Minimum contamination

Recall that in Section 4.1, the criterion of generalized resolution was justified via minimizing the bias of estimates of main effects in the presence of nonnegligible two-factor interactions. Continuing with this theme, suppose that main effects are of primary interest and that we are not interested in estimating interactions although some of them may not be negligible. Suppose the main effects are estimated via the main-effect model $\boldsymbol{Y} = \beta_0 \mathbf{1}_n + \boldsymbol{X}_1\boldsymbol{\beta}_1 + \boldsymbol{\epsilon}$, where $\boldsymbol{Y}$ is the vector of n observations, β_0 is the grand mean, $\mathbf{1}_n$ is the $n \times 1$ vector of all 1's, $\boldsymbol{X}_1 = \boldsymbol{D}$ is the design matrix, $\boldsymbol{\beta}_1$ is the vector of all main effects, and $\boldsymbol{\epsilon}$ is the vector of random errors, which are assumed to be uncorrelated with zero mean and constant variance. On the other hand, suppose the true model is

$$\boldsymbol{Y} = \beta_0 \mathbf{1}_n + \boldsymbol{X}_1\boldsymbol{\beta}_1 + \boldsymbol{X}_2\boldsymbol{\beta}_2 + \cdots + \boldsymbol{X}_m\boldsymbol{\beta}_m + \boldsymbol{\epsilon}, \tag{5.2}$$

where, for $2 \le j \le m$, $\boldsymbol{\beta}_j$ is the vector of all $\binom{m}{j}$ j-factor interactions, and $\boldsymbol{X}_j$ is the matrix obtained by taking the componentwise products for all possible subsets of j columns of $\boldsymbol{X}_1$. One question of interest is how much the estimates of main effects are biased by nonnegligible interactions.

Suppose we have an orthogonal array of strength two. As in Section 4.1, the least squares estimator of main effects under the main-effect model is $\hat{\boldsymbol{\beta}}_1 = n^{-1}\boldsymbol{X}_1^T\boldsymbol{Y}$, which is biased under the true model in (5.2) with the bias given by

$$\boldsymbol{C}_2\boldsymbol{\beta}_2 + \cdots + \boldsymbol{C}_m\boldsymbol{\beta}_m, \text{ where } \boldsymbol{C}_j = n^{-1}\boldsymbol{X}_1^T\boldsymbol{X}_j.$$

We try to minimize the bias, but since the $\boldsymbol{\beta}_j$'s are unknown, instead we minimize each $\|\boldsymbol{C}_j\|^2 = \text{trace}(\boldsymbol{C}_j^T\boldsymbol{C}_j)$, which is the sum of squares of the entries of $\boldsymbol{C}_j$. Under the hierarchical assumption that lower-order effects are more important than higher-order ones, we should sequentially minimize $\|\boldsymbol{C}_2\|^2, \ldots, \|\boldsymbol{C}_m\|^2$.

The following can be established (Tang and Deng, 1999).

Theorem 5.1. *The sequential minimization of $\|\boldsymbol{C}_2\|^2, \ldots, \|\boldsymbol{C}_m\|^2$ is equivalent to that of $A_3, \ldots, A_m$, that is, minimum G_2-aberration. In other words, a minimum G_2-aberration orthogonal array of strength two sequentially minimizes the contamination due to the j-factor interactions on the estimation of main effects in the order of $j = 2, \ldots, m$.*

Proof. We first establish that

$$\|\boldsymbol{C}_j\|^2 = (j+1)A_{j+1} + (m-j+1)A_{j-1} \tag{5.3}$$

for $2 \le j \le m-1$, and $\|\boldsymbol{C}_m\|^2 = A_{m-1}$. To verify (5.3), we notice that a typical entry of $n\boldsymbol{C}_j = \boldsymbol{X}_1^T\boldsymbol{X}_j$ is the inner product of two vectors, a main effect column and a j-factor interaction column. This inner product is the J-characteristic for a set of $j+1$ factors of the design if the main effect column and the j-factor interaction column do not have a factor in common, and is the J-characteristic for a set of $j-1$ factors of the design if they share a common factor. The precise expression in (5.3) is then obtained by some combinatorial counting. Since $A_1 = A_2 = 0$ for orthogonal arrays of strength two, we have

$$\|\boldsymbol{C}_2\|^2 = 3A_3 \text{ and } \|\boldsymbol{C}_3\|^2 = 4A_4. \tag{5.4}$$

It then follows from (5.4) and repeated applications of (5.3) that sequential minimization of $\|\boldsymbol{C}_2\|^2, \ldots, \|\boldsymbol{C}_m\|^2$ is equivalent to that of $A_3, \ldots, A_m$. □

Theorem 5.1 provides a justification of the minimum G_2-aberration criterion in terms of minimizing the contamination of nonnegligible interactions on the estimation of main effects according to the order of importance dictated by the hierarchical assumption.

5.2.2 Projection justification

Projection properties of factorial designs were discussed in Section 3.5. In the initial stage of experimentation, often there are a large number of potential factors, but only a small number of them are actually important. This is the principle of factor sparsity. Once the important factors have been identified, one can concentrate on the study of these factors. However, since it is unknown beforehand which factors are important, one needs to have a design that performs well no matter which factors are important. This brings us to the issue of designs with good projection properties.

As has been discussed in Section 4.8, a design with m factors is said to have *projectivity* t if in its projection onto any t factors, each treatment combination appears at least once. In this case, if there are t active factors, then one can estimate all their factorial effects no matter which t factors are active. By definition, an orthogonal array of strength t has projectivity t. Not only it has projectivity t, but also all the treatment combinations appear the same number of times in every t-factor projection. Such *uniformity* guarantees the optimality of all projection designs.

Recall that each treatment combination can be represented by $\boldsymbol{r}_v = (r_{v1}, \ldots, r_{vm})$, with $r_{vj} = -1$ if $j \in v$ and $r_{vj} = 1$ otherwise, where $v \subseteq Z_m = \{1, \ldots, m\}$; see the second paragraph of Section 3.2. Let N_v be the number of occurrences of $\boldsymbol{r}_v$ in a given two-level design $\boldsymbol{D}$ of n runs for m factors. In general, the nonuniformity of design $\boldsymbol{D}$ can be measured by

$$V(\boldsymbol{D}) = \frac{1}{2^m} \sum_{v \subseteq Z_m} (N_v - \bar{N})^2, \tag{5.5}$$

where $\bar{N} = n/2^m$ is the average of all N_v's. Obviously, $V(\boldsymbol{D}) = 0$ if all the treatment combinations appear the same number of times in $\boldsymbol{D}$.

Let $\boldsymbol{J}$ and $\boldsymbol{N}$ be column vectors that collect the J-characteristics J_u and N_v values, respectively, in the Yates order. It was shown in Theorem 3.2 that

$$\boldsymbol{N} = \frac{1}{2^m} \boldsymbol{H}^T \boldsymbol{J}, \tag{5.6}$$

where $\boldsymbol{H}^T$ is a Hadamard matrix with the first column consisting of all 1's. Projecting both sides of (5.6) onto the orthogonal complement of the vector of all 1's, we obtain

$$\sum_{v \subseteq Z_m} \left(N_v - \bar{N}\right)^2 = \frac{1}{2^m} \Big(\sum_{w \subseteq Z_m} J_w^2 - n^2 \Big).$$

Therefore, $V(\boldsymbol{D})$ in (5.5) can be expressed in terms of the J-characteristics as

$$V(\boldsymbol{D}) = \frac{1}{2^{2m}} \Big(\sum_{w \subseteq Z_m} J_w^2 - n^2 \Big). \tag{5.7}$$

Since a factorial design is completely determined by its J-characteristics, its projection onto the factors in a given subset u of Z_m is completely determined by the J_w values with $w \subseteq u$. Applying (5.7) to the projection design $\boldsymbol{D}_u$ of $\boldsymbol{D}$ onto the factors in $u \subseteq Z_m$, we obtain that

$$V(\boldsymbol{D}_u) = \frac{1}{2^{2|u|}} \Big(\sum_{w \subseteq u} J_w^2 - n^2 \Big).$$

Thus the overall nonuniformity of k-dimensional projections of $\boldsymbol{D}$ can be measured by

$$V_k(\boldsymbol{D}) = \sum_{|u|=k} V(\boldsymbol{D}_u) = \frac{1}{2^{2k}} \sum_{|u|=k} \Big(\sum_{w \subseteq u} J_w^2 - n^2 \Big).$$

Via some simple algebra, one obtains

$$V_k(\boldsymbol{D}) = \frac{1}{2^{2k}} \sum_{j=1}^{k} \binom{m-j}{k-j} \sum_{|w|=j} J_w^2. \tag{5.8}$$

This gives rise to a vector $(V_1(\boldsymbol{D}), \ldots, V_m(\boldsymbol{D}))$ that provides nonuniformity measures of the projections of $\boldsymbol{D}$ onto different dimensions. If lower-dimensional projection properties are regarded as more important than higher-dimensional ones, then we ought to sequentially minimize $V_1(\boldsymbol{D}), \ldots, V_m(\boldsymbol{D})$ when seeking good designs. The next result provides a projection justification for the minimum G_2-aberration criterion (Tang, 2001).

Theorem 5.2. *The two sequences $A_1(\boldsymbol{D}), \ldots, A_m(\boldsymbol{D})$ and $V_1(\boldsymbol{D}), \ldots, V_m(\boldsymbol{D})$ are linearly related, and minimum G_2-aberration is equivalent to sequentially minimizing $V_1(\boldsymbol{D}), \cdots, V_m(\boldsymbol{D})$.*

Proof. Since

$$A_k(\boldsymbol{D}) = \frac{1}{n^2} \sum_{|w|=k} J_w^2, \tag{5.9}$$

it follows from (5.8) and (5.9) that

$$V_k(\boldsymbol{D}) = \frac{n^2}{2^{2k}} \sum_{j=1}^{k} \binom{m-j}{k-j} A_j(\boldsymbol{D}).$$

The theorem is now obvious. □

While Theorem 5.1 applies to orthogonal arrays of strength two, Theorem 5.2 applies to all two-level factorial designs. For orthogonal arrays of strength two, Theorem 5.2 states that sequentially minimizing $A_3(\boldsymbol{D}), \ldots, A_m(\boldsymbol{D})$ is equivalent to sequentially minimizing $V_3(\boldsymbol{D}), \ldots, V_m(\boldsymbol{D})$ because $A_1(\boldsymbol{D}) = A_2(\boldsymbol{D}) = V_1(\boldsymbol{D}) = V_2(\boldsymbol{D}) = 0$.

Earlier, Chen (1998) showed that for regular designs of resolution R, the number of k-factor projections with $R + 1 \leq k \leq R + \lfloor (R-1)/2 \rfloor$ that are of resolution j, where $R \leq j \leq k$, is proportional to A_j, the number of defining words of length j. Therefore through sequential minimization of the A_j's, minimum aberration designs achieve the good projection property of having fewer projections of low resolutions.

5.2.3 Model robustness

Cheng, Steinberg, and Sun (1999) investigated a model robustness property of minimum aberration regular designs via the concept of estimation capacity introduced by Sun (1993). Given a regular two-level design $\boldsymbol{D}$, consider the models containing all the main effects and k two-factor interactions. Let $E_k(\boldsymbol{D})$ be the number of such models that can be estimated under $\boldsymbol{D}$, and call $E_1(\boldsymbol{D}), E_2(\boldsymbol{D}), \ldots$, the estimation capacity sequence of $\boldsymbol{D}$. Cheng, Steinberg, and Sun (1999) showed that a regular minimum aberration design of resolution III or higher maximizes the number of two-factor interactions that are not aliased with main effects, and tends to distribute these two-factor interactions very uniformly over the alias sets. These two properties imply that a regular minimum aberration design tends to have large, if not maximum, E_k values. In other words, minimum aberration is a good surrogate for maximum estimation capacity. Cheng, Steinberg, and Sun (1999) also provided some conditions under which a design maximizes all E_k values. Here model robustness refers to that a design $\boldsymbol{D}$ can be used to entertain many possible models.

Note that under regular designs, all the estimable effects are estimated with full efficiency. Therefore it is enough just to compare the number of estimable models under different designs. For nonregular designs, one would also have to consider efficiencies. Cheng, Deng, and Tang (2002) showed that minimum G_2-aberration is a good surrogate for model robustness under some classical design efficiency criteria.

We consider balanced n-run designs with m two-level factors. Balanced designs are orthogonal arrays of strength one, in which the two levels are equireplicated for each factor. Let $\mathcal{P}$ be the set of all the $\binom{m}{2}$ subsets of size 2 of $Z_m = \{1, \cdots, m\}$, each defining a pair of factors. Suppose there are f active two-factor interactions and all the other interactions are negligible. Let $\mathcal{F}$ be the subset of $\mathcal{P}$ that corresponds to the f active two-factor interactions. Let $\boldsymbol{D}$ be an OA$(n, 2^m, 1)$. Then under design $\boldsymbol{D}$, we have the following linear model:

$$\boldsymbol{Y} = \beta_0 \mathbf{1}_n + \boldsymbol{X}_1 \boldsymbol{\beta}_1 + \boldsymbol{X}_{\mathcal{F}} \boldsymbol{\beta}_{\mathcal{F}} + \epsilon,$$

where $\boldsymbol{Y}$ is the $n \times 1$ vector of observations, β_0 is an unknown parameter representing the grand mean, $\boldsymbol{X}_1$ is the design matrix $\boldsymbol{D}$, $\boldsymbol{\beta}_1$ is the $m \times 1$ vector of main effects, $\boldsymbol{\beta}_{\mathcal{F}}$ is an $f \times 1$

vector consisting of the active two-factor interactions, $\boldsymbol{X}_{\mathcal{F}}$ is an $n \times f$ matrix, each of whose columns is the componentwise product of the main-effect columns of the factors in an active two-factor interaction, and $\boldsymbol{\epsilon}$ is an $n \times 1$ vector of uncorrelated random errors, assumed to have zero mean and a constant variance. Write the model matrix as $\boldsymbol{W}_{\mathcal{F}} = [\mathbf{1}_n, \boldsymbol{X}_1, \boldsymbol{X}_{\mathcal{F}}]$. Then the information matrix per observation is

$$\boldsymbol{M}_{\mathcal{F}} = \frac{1}{n}\boldsymbol{W}_{\mathcal{F}}^T\boldsymbol{W}_{\mathcal{F}} = \frac{1}{n}\begin{bmatrix} n & 0 & \mathbf{1}_n^T\boldsymbol{X}_{\mathcal{F}} \\ 0 & \boldsymbol{X}_1^T\boldsymbol{X}_1 & \boldsymbol{X}_1^T\boldsymbol{X}_{\mathcal{F}} \\ \boldsymbol{X}_{\mathcal{F}}^T\mathbf{1}_n & \boldsymbol{X}_{\mathcal{F}}^T\boldsymbol{X}_1 & \boldsymbol{X}_{\mathcal{F}}^T\boldsymbol{X}_{\mathcal{F}} \end{bmatrix}. \tag{5.10}$$

Optimality criteria for choosing factorial designs were discussed in Section 2.3. It was shown there that a design is universally optimal if its information matrix is a constant multiple of the identity matrix and it maximizes the trace of the information matrix. Under a main-effect model, the information matrix under an orthogonal array of strength two is a multiple of the identity matrix. It may not be the case if there are nonnegligible interactions, but a design whose information matrix has the maximum trace and is close to a multiple of the identity matrix is expected to be highly efficient. The trace of a matrix is equal to the sum of its eigenvalues; so it is desirable to maximize the sum of the eigenvalues. Since all the eigenvalues of a multiple of the identity matrix are equal, the closeness of a matrix to one that is a multiple of an identity matrix can be measured by the dispersion of the eigenvalues; it is desirable to have them as equal as possible. With a given trace, the sum of the eigenvalues is fixed; then one simple thing to do is to minimize their sum of squares. This leads to a simple two-stage strategy: first maximize the sum, and then minimize the sum of squares of the eigenvalues. This, called the (M.S)-criterion, was proposed by Eccleston and Hedayat (1974). In our setting, since all the diagonal entries of an information matrix are equal to n, the trace of an information matrix is a constant. Therefore it reduces to the minimization of the sum of squares of eigenvalues, i.e., the sum of squares of all the entries of the information matrix. Once again using the fact that the diagonal entries of the information matrix are equal to n, it is further reduced to the minimization of the sum of squares of the off-diagonal entries of the information matrix.

Since it is not known a priori which f two-factor interactions are active, we should evaluate the average performance of a design with respect to all possible choices of f two-factor interactions, e.g., average of the D-criterion values over all possible choices of $\mathcal{F}$. A good surrogate would be to minimize the average of the sums of squares of the off-diagonal entries of the information matrix $\boldsymbol{M}_{\mathcal{F}}$ in (5.10) over all possible choices of f-subsets $\mathcal{F}$ of $\mathcal{P}$. Denote such an average by S_f^2. The following theorem provides an explicit expression of S_f^2.

Theorem 5.3. *Consider an n-run design for m two-level factors such that the two levels are equireplicated for each factor. Let $F = \binom{m}{2}$ and f be any positive integer such that $1 \leq f \leq F$. Then we have that*

$$S_f^2 = \alpha_2 A_2 + \alpha_3 A_3 + \alpha_4 A_4,$$

where the coefficients α_2, α_3, and α_4 are give by

$$\alpha_2 = 2\Big[1 + \frac{f}{F} + \frac{f(f-1)}{F(F-1)}(m-2)\Big], \quad \alpha_3 = 6\frac{f}{F}, \quad \alpha_4 = 6\frac{f(f-1)}{F(F-1)}.$$

We refer the readers to Cheng, Deng, and Tang (2002) for a detailed proof of Theorem 5.3. The proof is not difficult, basically requiring the use of some standard results on simple random sampling without replacement from a finite population.

Theorem 5.3 shows how S_f^2 depends on m, f, A_2, A_3, and A_4. In practice we expect f to be small relative to $F = \binom{m}{2}$; that is, only a small number of two-factor interactions are expected to be active. This is what the principle of effect sparsity says. Then in S_f^2, A_2 has a heavier weight than A_3, which in turn has more influence on S_f^2 than A_4. Thus for small f, sequentially minimizing A_2, A_3, A_4 is a good surrogate for minimizing S_f^2, and is thus also a good surrogate for minimizing the average of D-criterion values over all possible sets of f active two-factor interactions. This provides a justification of minimum G_2-aberration in terms of model robustness and efficiency.

Empirical studies showed strong consistency between the ranking based on minimum G_2-aberration and that based on the averages of D-criterion values (Cheng, Deng, and Tang, 2002).

5.2.4 Bayesian justification

The study in the previous subsection is limited to the case where there are no active interactions involving more than two factors, and the results only apply to the sequential minimization of A_2, A_3, and A_4. In this subsection, we revisit the issue from a Bayesian point of view, showing how such restrictions can be avoided.

In a pioneering work on computer experiments, Sacks, Welch, Mitchell, and Wynn (1989) initiated the use of Gaussian processes to model unknown deterministic functions. Mitchell, Morris, and Ylvisaker (1995) used such Gaussian processes to set up a framework for Bayesian fractional factorial designs, and the approach was further developed in Kerr (2001); also see Joseph (2006), Joseph and Delaney (2007), Ai, Kang, and Joseph (2009), Joseph, Ai, and Wu (2009), and Kang and Joseph (2009).

Let $\boldsymbol{\mu}$ be a $2^m \times 1$ vector whose components are the effects $\mu(\boldsymbol{x}) = \mu(x_1, \ldots, x_m)$ of the 2^m treatment combinations $\boldsymbol{x} = (x_1, \ldots, x_m)^T$, where $x_i = 1$ or -1, and $\boldsymbol{\beta}$ be the $2^m \times 1$ vector consisting of the grand mean and all the factorial effects (main effects and interactions) $\boldsymbol{\beta}_u$ with $u \subseteq Z_m$. When the components of $\boldsymbol{\mu}$ and $\boldsymbol{\beta}$ are arranged in the Yates order, we have

$$\boldsymbol{\mu} = \boldsymbol{P}\boldsymbol{\beta}, \tag{5.11}$$

where

$$\boldsymbol{P} = \begin{bmatrix} 1 & -1 \\ 1 & 1 \end{bmatrix} \otimes \cdots \otimes \begin{bmatrix} 1 & -1 \\ 1 & 1 \end{bmatrix}.$$

There are 2^m unknown parameters in $\boldsymbol{\beta}$. Specifying a prior joint distribution for such a large number of unknown parameters is challenging, since it requires the assignment of a large number of hyperparameters. By (5.11), we have

$$\boldsymbol{\beta} = \boldsymbol{P}^{-1}\boldsymbol{\mu}. \tag{5.12}$$

Following the approach in computer experiments, consider $\boldsymbol{\mu}$ as a realization of a Gaussian process. Then a prior distribution of the factorial effects $\boldsymbol{\beta}$ can be obtained from that of $\boldsymbol{\mu}$ via (5.12). This is Mitchell, Morris, and Ylvisaker (1995)'s approach of using a functional prior of the treatment effects to induce a prior distribution of the many parameters in a factorial treatment model.

Suppose the prior distribution of $\boldsymbol{\mu}$ is multivariate normal with zero mean and covariance matrix $\sigma_0^2\boldsymbol{R}$. Then it follows from (5.12) that

$$\boldsymbol{\beta} \sim \mathrm{N}\left(\boldsymbol{0}, \sigma_0^2 \boldsymbol{P}^{-1}\boldsymbol{R}\left(\boldsymbol{P}^{-1}\right)^T\right). \tag{5.13}$$

We have $\boldsymbol{P}^T = 2^m\boldsymbol{P}^{-1}$; thus the prior covariance matrix of $\boldsymbol{\beta}$ is

$$\operatorname{cov}(\boldsymbol{\beta}) = 2^{-2m}\sigma_0^2\boldsymbol{P}^T\boldsymbol{R}\boldsymbol{P}. \tag{5.14}$$

An important class of stochastic processes is stationary processes. In our context, $\boldsymbol{\mu}$ is stationary if and only if the correlation between $\mu(\boldsymbol{x}_1)$ and $\mu(\boldsymbol{x}_2)$ depends only on the factors where $\boldsymbol{x}_1$ and $\boldsymbol{x}_2$ differ. Mitchell, Morris, and Ylvisaker (1995) established the following result on a necessary and sufficient condition for $\boldsymbol{\mu}$ to be stationary.

Theorem 5.4. *We have that $\boldsymbol{\mu}$ is stationary if and only if the β_u's, $u \subseteq \{1, \ldots, m\}$, are independent.*

It follows from Theorem 5.4 that if $\boldsymbol{\mu}$ is stationary, then the covariance matrix in (5.14) is diagonal.

Consider, for example, the following correlation function of a stationary prior:

$$\operatorname{corr}(\mu(\boldsymbol{x}_1), \mu(\boldsymbol{x}_2)) = \prod_{i:1 \le i \le m,\, x_{1i} \ne x_{2i}} \rho_i,$$

where $\boldsymbol{x}_1 = (x_{11}, \ldots, x_{1m})^T$, $\boldsymbol{x}_2 = (x_{21}, \ldots, x_{2m})^T$, and $0 < \rho_1, \ldots, \rho_m < 1$. Let

$$r_i = (1-\rho_i)/(1+\rho_i) \text{ and } \tau^2 = \sigma_0^2 \left[\prod_{i=1}^{m}(1+r_i)\right]^{-1}.$$

Then

$$\operatorname{var}(\beta_\phi) = \tau^2, \text{ and for nonempty } u, \ \operatorname{var}(\beta_u) = \tau^2 \prod_{i \in u} r_i. \tag{5.15}$$

It follows from (5.15) that

$$\text{if } u \subseteq u', \text{ then } \operatorname{var}(\beta_u) \ge \operatorname{var}(\beta_{u'}). \tag{5.16}$$

In the case of an *isotropic* stationary prior where $\rho_1 = \cdots = \rho_m = \rho$, we have that $\operatorname{var}(\beta_u) = \sigma_0^2 2^{-m}(1-\rho)^{|u|}(1+\rho)^{m-|u|}$, where $|u|$ is the number of elements in u. Under the alternative parametrization $r = (1-\rho)/(1+\rho)$ and $\tau^2 = \sigma_0^2(1+r)^{-m}$, we have that

$$\operatorname{var}(\beta_u) = \tau^2 r^{|u|}. \tag{5.17}$$

In the rest of this section, we assume that the prior is isotropic stationary. Then, by (5.17), factorial effects of the same order have the same variance, and lower-order effects have larger variances. This is consistent with the usual hierarchical principle that factorial effects of the same order are equally important and that lower-order effects are more important than higher-order effects, a situation where minimum G_2-aberration designs are expected to perform well.

Now suppose the observations $\boldsymbol{Y}$ on the treatment combinations given $\boldsymbol{\beta}$ are independent and normally distributed with variance σ^2. That is,

$$\boldsymbol{Y}|\boldsymbol{\beta} \sim \mathrm{N}(\boldsymbol{U}\boldsymbol{\beta}, \sigma^2\boldsymbol{I}), \tag{5.18}$$

where $\boldsymbol{U}$ is the $n \times 2^m$ submatrix of $\boldsymbol{P}$ consisting of the rows that correspond to the n observed treatment combinations given by a design $\boldsymbol{D}$ of n runs for m factors. From (5.13), (5.14), and (5.18), we have that the posterior distribution of $\boldsymbol{\beta}$ given $\boldsymbol{Y}$ is normal with

$$\operatorname{cov}(\boldsymbol{\beta}|\boldsymbol{Y}) = \left[\sigma^{-2}\boldsymbol{U}^T\boldsymbol{U} + (\operatorname{cov}(\boldsymbol{\beta}))^{-1}\right]^{-1}.$$

Suppose we adopt the *D*-criterion to minimize the determinant of the posterior covariance matrix. Then since

$$\det\left[\operatorname{cov}(\boldsymbol{\beta}|\boldsymbol{Y})^{-1}\right] = \det\left[\operatorname{cov}(\boldsymbol{\beta})^{-1}\right]\det\left(\sigma^{-2}\operatorname{cov}(\boldsymbol{\beta})\boldsymbol{U}^T\boldsymbol{U} + \boldsymbol{I}_{2^m}\right),$$

and $\det\left[\mathrm{cov}(\boldsymbol{\beta})^{-1}\right]$ is a constant, minimizing $\det\left[\mathrm{cov}(\boldsymbol{\beta}|\boldsymbol{Y})\right]$ is equivalent to maximizing $\det\left(\sigma^{-2}\mathrm{cov}(\boldsymbol{\beta})\boldsymbol{U}^T\boldsymbol{U}+\boldsymbol{I}_{2^m}\right)$. A good surrogate for this, as we have discussed in Section 5.2.3, is to first maximize $\mathrm{tr}(\sigma^{-2}\mathrm{cov}(\boldsymbol{\beta})\boldsymbol{U}^T\boldsymbol{U}+\boldsymbol{I}_{2^m})$ and then minimize $\mathrm{tr}\left[(\sigma^{-2}\mathrm{cov}(\boldsymbol{\beta})\boldsymbol{U}^T\boldsymbol{U}+\boldsymbol{I}_{2^m})^2\right]$ among those that maximize $\mathrm{tr}\left(\sigma^{-2}\mathrm{cov}(\boldsymbol{\beta})\boldsymbol{U}^T\boldsymbol{U}+\boldsymbol{I}_{2^m}\right)$. Such designs are said to be Bayesian (M.S)-optimal.

By (5.14), we have that

$$\sigma^{-2}\mathrm{cov}(\boldsymbol{\beta})\boldsymbol{U}^T\boldsymbol{U}+\boldsymbol{I}_{2^m}=\sigma^{-2}2^{-2m}\sigma_0^2\boldsymbol{P}^T\boldsymbol{R}\boldsymbol{P}\boldsymbol{U}^T\boldsymbol{U}+\boldsymbol{I}_{2^m}.$$

Thus maximizing $\mathrm{tr}(\sigma^{-2}\mathrm{cov}(\boldsymbol{\beta})\boldsymbol{U}^T\boldsymbol{U}+\boldsymbol{I}_{2^m})$ is equivalent to maximizing $\mathrm{tr}(\boldsymbol{P}^T\boldsymbol{R}\boldsymbol{P}\boldsymbol{U}^T\boldsymbol{U})=\mathrm{tr}(\boldsymbol{U}\boldsymbol{P}^T\boldsymbol{R}\boldsymbol{P}\boldsymbol{U}^T)$. Using the fact that $\boldsymbol{P}^T\boldsymbol{R}\boldsymbol{P}$ is a diagonal matrix, one can readily see that $\mathrm{tr}\left(\sigma^{-2}\mathrm{cov}(\boldsymbol{\beta})\boldsymbol{U}^T\boldsymbol{U}\right)$ is a constant not depending on the design. Thus one only needs to perform the second step of the (M.S)-criterion to minimize $\mathrm{tr}\left[\left(\sigma^{-2}\mathrm{cov}(\boldsymbol{\beta})\boldsymbol{U}^T\boldsymbol{U}+\boldsymbol{I}_{2^m}\right)^2\right]$. By some matrix algebra and the fact that $\mathrm{tr}\left(\sigma^{-2}\mathrm{cov}(\boldsymbol{\beta})\boldsymbol{U}^T\boldsymbol{U}+\boldsymbol{I}_{2^m}\right)$ is a constant, we only need to minimize $\mathrm{tr}\left[\left(\mathrm{cov}(\boldsymbol{\beta})\boldsymbol{U}^T\boldsymbol{U}\right)^2\right]$. Note that

$$\begin{aligned}
&\mathrm{tr}\left[\left(\mathrm{cov}(\boldsymbol{\beta})\boldsymbol{U}^T\boldsymbol{U}\right)^2\right]\\
&=\mathrm{tr}\left[\mathrm{cov}(\boldsymbol{\beta})\boldsymbol{U}^T\boldsymbol{U}\mathrm{cov}(\boldsymbol{\beta})\boldsymbol{U}^T\boldsymbol{U}\right]\\
&=\mathrm{tr}\left[\mathrm{cov}(\boldsymbol{\beta})^{1/2}\boldsymbol{U}^T\boldsymbol{U}\mathrm{cov}(\boldsymbol{\beta})^{1/2}\mathrm{cov}(\boldsymbol{\beta})^{1/2}\boldsymbol{U}^T\boldsymbol{U}\mathrm{cov}(\boldsymbol{\beta})^{1/2}\right]\\
&=\mathrm{tr}\left[\left\{\left(\boldsymbol{U}\mathrm{cov}(\boldsymbol{\beta})^{1/2}\right)^T\left(\boldsymbol{U}\mathrm{cov}(\boldsymbol{\beta})^{1/2}\right)\right\}^2\right].
\end{aligned}$$

Let $\boldsymbol{x}_u$ be the column vector of $\boldsymbol{U}$ associated with u and let $v_u=\mathrm{var}(\beta_u)$. Recall that under an isotropic stationary prior, we have $v_u>v_{u'}$ for $|u|<|u'|$ and $v_u=v_{u'}$ for $|u|=|u'|$. Then

$$\begin{aligned}
&\mathrm{tr}\left[\left\{\left(\boldsymbol{U}\mathrm{cov}(\boldsymbol{\beta})^{1/2}\right)^T\left(\boldsymbol{U}\mathrm{cov}(\boldsymbol{\beta})^{1/2}\right)\right\}^2\right]\\
=&\sum_{u,u'\subseteq\{1,\ldots,m\}}\left(\boldsymbol{x}_u^T\boldsymbol{x}_{u'}\right)^2v_uv_{u'}\\
=&\sum_{u,u'\subseteq\{1,\ldots,m\}}\left(\boldsymbol{1}^T\boldsymbol{x}_{u\Delta u'}\right)^2v_uv_{u'}\\
=&\sum_{u\subseteq\{1,\ldots,m\}}b_u\left(\boldsymbol{1}^T\boldsymbol{x}_u\right)^2,
\end{aligned}$$

where $u\Delta u'=(u\cup u')-(u\cap u')$ and

$$b_u=\sum_{u',u''\subseteq\{1,\ldots,m\}:u'\Delta u''=u}v_{u'}v_{u''}.$$

We therefore have that

$$\mathrm{tr}\left[\left(\mathrm{cov}(\boldsymbol{\beta})\boldsymbol{U}^T\boldsymbol{U}\right)^2\right]=\sum_{u\subseteq\{1,\ldots,m\}}b_u\left(\boldsymbol{1}^T\boldsymbol{x}_u\right)^2. \tag{5.19}$$

Theorem 2.3 in Chang (2015) shows that if $v_u>v_{u'}$ for $|u|<|u'|$ and $v_u=v_{u'}$ for $|u|=|u'|$, then the b_u's also possess the same property. We therefore have that $b_u>b_{u'}$ for $|u|<|u'|$

and $b_u = b_{u'}$ for $|u| = |u'|$. Using this fact and (5.19), and by noting that $\mathbf{1}^T \boldsymbol{x}_u = J_u$, we obtain that

$$\operatorname{tr}\left[\left(\operatorname{cov}(\boldsymbol{\beta})\boldsymbol{U}^T\boldsymbol{U}\right)^2\right] = \sum_{k=0}^{m} c_k \left(\sum_{|u|=k} (\mathbf{1}^T \boldsymbol{x}_u)^2\right) = n^2 \sum_{k=0}^{m} c_k A_k,$$

where $c_k = b_u$ for the u's with $|u| = k$, and $c_0 > c_1 > \cdots > c_m$. Thus minimum G_2-aberration designs are expected to perform well with respect to minimizing $\operatorname{tr}\left[\left(\operatorname{cov}(\boldsymbol{\beta})\boldsymbol{U}^T\boldsymbol{U}\right)^2\right]$. This shows that minimum G_2-aberration is a good surrogate for the Bayesian (M.S)-criterion, especially when $c_0 >> c_1 >> \cdots >> c_m$.

We acknowledge that the proof of minimizing $\operatorname{tr}\left[\left(\operatorname{cov}(\boldsymbol{\beta})\boldsymbol{U}^T\boldsymbol{U}\right)^2\right]$ in the above was provided by Ming-Chung Chang.

The discussion of the Bayesian (M.S)-criterion in this subsection assumes no structure in experimental units. We shall apply this Bayesian approach to study factorial designs for experiments where the experimental units have certain structures. We first study the simpler cases of block designs, split-plot designs, and row-column designs in Chapter 7, and eventually present in Chapter 11 a general theory covering mixed-level factorial treatments as well as experimental units with more complicated structures.

We end this subsection with a proof of the property of the b_u's mentioned above. Let $u \subset u'$ and $u, u' \subseteq \{1, \ldots, m\}$. This implies that $v_u \geq v_{u'}$. Our goal is to demonstrate that $b_u \geq b_{u'}$. It is sufficient to establish this for $u' = u \cup \{q\}$, where $q \notin u$. If $m = 1$, the assertion is trivial. For $m = 2$, consider the four subsets: $\emptyset$, $\{1\}$, $\{2\}$, and $\{1,2\}$. The values are given by

$$b_{\{1\}} = v_\emptyset v_{\{1\}} + v_{\{2\}} v_{\{1,2\}} \text{ and } b_{\{1,2\}} = v_\emptyset v_{\{1,2\}} + v_{\{1\}} v_{\{2\}}.$$

Therefore, $b_{\{1\}} - b_{\{1,2\}} = (v_\emptyset - v_{\{2\}})(v_{\{1\}} - v_{\{1,2\}}) \geq 0$.

For $m > 2$, we partition the 2^m subsets of $\{1, \ldots, m\}$ into 2^{m-2} groups based on $\{\emptyset, u, u', \{q\}\}$, each group containing four subsets in the form: $w, w\Delta u, w\Delta u', w\Delta\{q\}$, where $w \subseteq \{1, \ldots, m\}$ and $w \notin \{\emptyset, u, u', \{q\}\}$. Consequently, we calculate

$$\begin{aligned} b_u - b_{u'} &= \sum_w \left\{ (v_w v_{w\Delta u} + v_{w\Delta u'} v_{w\Delta\{q\}}) - (v_w v_{w\Delta u'} + v_{w\Delta u} v_{w\Delta\{q\}}) \right\} \\ &= \sum_w \left(v_{w\Delta u} - v_{w\Delta u'}\right)\left(v_w - v_{w\Delta\{q\}}\right). \end{aligned}$$

If $q \notin w$, then $w\Delta u \subset w\Delta u'$ and $w \subset w\Delta\{q\}$, leading to $v_{w\Delta u} \geq v_{w\Delta u'}$ and $v_w \geq v_{w\Delta\{q\}}$. Conversely, if $q \in w$, then $w\Delta u' \subset w\Delta u$ and $w\Delta\{q\} \subset w$, resulting in $v_{w\Delta u} \leq v_{w\Delta u'}$ and $v_w \leq v_{w\Delta\{q\}}$. In both scenarios $b_u - b_{u'} \geq 0$, and hence the proof is complete.

5.3 A Theory of Complementary Designs

Each regular design can be constructed by choosing factors (columns) from a regular saturated design. The columns that are not chosen form another design called the complementary design. Tang and Wu (1996) derived identities that relate the wordlength pattern of a two-level regular design to that of its complementary design. Independently, Chen and Hedayat (1996) derived such identities for defining words of lengths three and four. These identities can be used to construct minimum aberration designs through their complementary designs. This provides a particularly useful tool for determining nearly saturated minimum aberration designs. Using coding theoretical tools, Suen, Chen, and Wu (1997) extended the results to regular designs with more than two levels.

Each nonregular two-level saturated design of strength two and size n can be obtained by deleting a column of 1's from an $n \times n$ Hadamard matrix. Any subset of m columns of the $n-1$ columns of the saturated design forms an orthogonal array of strength two, and the other columns also constitute an orthogonal array of strength two. Extending the result of Tang and Wu (1996) mentioned above, Tang and Deng (1999) derived identities relating the generalized wordlength patterns of the two complementary designs.

Theorem 5.5. *Let $\boldsymbol{D}$ and $\bar{\boldsymbol{D}}$ be complementary subdesigns of an n-run saturated orthogonal array of strength two. Suppose $\boldsymbol{D}$ has m factors. Then $\bar{\boldsymbol{D}}$ has $n-1-m$ factors. We then have that*

$$A_k(\boldsymbol{D}) = a_0 + \sum_{j=3}^{k} a_j A_j(\bar{\boldsymbol{D}}),$$

where $a_0, a_3, \ldots, a_k$ are constants, depending only on n, m, and k. Furthermore, we have that $a_k = (-1)^k$.

Theorem 5.5 says that each $A_k(\boldsymbol{D})$ is a linear combination of the $A_j(\bar{\boldsymbol{D}})$'s with $j \leq k$ and the coefficient of the leading term $A_k(\bar{\boldsymbol{D}})$ is $(-1)^k$. Thus a design $\boldsymbol{D}$ has minimum G_2-aberration among all the m-factor subdesigns of a given saturated orthogonal array of strength two and size n if and only if its complementary design sequentially minimizes $-A_3(\bar{\boldsymbol{D}}), A_4(\bar{\boldsymbol{D}}), -A_5(\bar{\boldsymbol{D}}), \ldots$ among all $(n-m-1)$-factor subdesigns of the saturated design.

From Theorem 5.5, it is immediate that if $\bar{\boldsymbol{D}}$ has 0, 1, or 2 factors, then all $A_k(\boldsymbol{D})$'s are constants and therefore $\boldsymbol{D}$ must have minimum G_2-aberration. If $\bar{\boldsymbol{D}}$ has three factors, minimum G_2-aberration designs can also be obtained. To do this, we just need to identify three columns, from all saturated orthogonal arrays, whose J-characteristic is maximized. If n is a multiple of 8, then the saturated orthogonal array obtained from the Hadamard matrix given by the tensor product of a 2×2 and an $n/2 \times n/2$ Hadamard matrices must have a defining word of length three. Deleting the three columns that form a defining word of length three then gives a minimum G_2-aberration design. If n is not a multiple of 8 and $n/2-1$ is a prime power, then the saturated orthogonal array obtained from the Hadamard matrix given by Paley's second construction must have three columns whose J-characteristic is equal to $n-8$, the largest possible value when n is not a multiple of 8. Deleting these three columns, we obtain a minimum G_2-aberration design. The reader may find Lemmas 4.4 and 4.5 helpful for understanding the above arguments.

However, unlike its counterpart for regular designs, for a general m, Theorem 5.5 only allows us to draw conclusion about having minimum G_2-aberration among m-factor subdesigns of a given saturated design, instead of over all m-factor designs. One reason is that not all nonregular designs are subdesigns of a saturated design; see the last paragraph of Section 3.7.1 for a discussion on maximal orthogonal arrays. The same paragraph also says that all OA$(n, 2^m, 2)$s are obtainable from saturated orthogonal arrays of n runs if $m \geq n-8$. So it is at least theoretically possible to find minimum G_2-aberration designs for $m \geq n-8$ by the complementary design theory, which we have done for $m = n-1, n-2, n-3$, and $n-4$ in the preceding paragraph. Computationally, the complication still remains, because, unlike regular designs, saturated 2-level designs are not unique. Although all saturated orthogonal arrays of n runs are isomorphic for $n = 4, 8$, and 12, there are 5, 3, 130, 7570, and $\geq 13,710,027$ nonisomorphic saturated orthogonal arrays of n runs for $n = 16, 20, 24, 28$, and 32, respectively. See Table 3.3 in Chapter 3.

Xu and Wu (2001) observed that the coding theoretic argument used by Suen, Chen, and Wu (1997) to establish a complementary design theory for general regular designs also works for nonregular orthogonal arrays with any number of levels. We shall present their result when designs with more than two levels are discussed in Chapter 10. The proof of Theorem 5.5 is thus skipped.

5.4 Minimum Moment Aberration

The minimum G_2-aberration criterion is based on J-characteristics. Each J-characteristic is defined in terms of the componentwise product of some columns of the design matrix, and can be viewed as a measure of the extent of aliasing (or similarity) of some factorial effects. As such, minimum G_2-aberration is a column-based criterion. Butler (2003b) and Xu (2003) showed that minimum G_2-aberration is equivalent to a row-based criterion, which was termed minimum moment aberration by Xu (2003). The minimum moment aberration criterion is based on distances (or similarities) of the rows of the design matrix (i.e. experimental runs). While Butler (2003b) only considered two-level designs, Xu (2003)'s minimum moment aberration criterion can be applied to mixed-level designs, and he established the equivalence of minimum G_2-aberration and minimum moment aberration for designs with all the factors having the same number of levels, which does not have to be two.

We provide some motivation before formally defining the minimum moment aberration criterion. As we have seen, given a linear model with model matrix $\boldsymbol{X}$, the usual optimality criteria such as A-, D-, and E-criteria are functionals of the information matrix $\boldsymbol{X}^T\boldsymbol{X}$; all these three criteria are column-based. We have also seen that a good surrogate for the A-, D-, and E-criteria is the (M.S)-criterion, which is to first maximize $\text{tr}(\boldsymbol{X}^T\boldsymbol{X})$, and then minimize $\text{tr}[(\boldsymbol{X}^T\boldsymbol{X})^2]$ $=\text{tr}(\boldsymbol{X}^T\boldsymbol{X}\boldsymbol{X}^T\boldsymbol{X})$. Now $\text{tr}(\boldsymbol{X}^T\boldsymbol{X})$ $=\text{tr}(\boldsymbol{X}\boldsymbol{X}^T)$ and $\text{tr}(\boldsymbol{X}^T\boldsymbol{X}\boldsymbol{X}^T\boldsymbol{X})$ $=\text{tr}(\boldsymbol{X}\boldsymbol{X}^T\boldsymbol{X}\boldsymbol{X}^T)$. Thus the (M.S)-criterion is equivalent to (actually, the same as) a row-based criterion. This has been used by Nguyen (1996) and Cheng (1997) in their studies of supersaturated designs.

Let $\boldsymbol{D}$ be the design matrix of an n-run factorial design for m two-level factors, with each row corresponding to one run and each column corresponding to a factor. Then we have that

$$A_1 = n^{-2}\text{tr}(\boldsymbol{D}^T\mathbf{1}_n\mathbf{1}_n^T\boldsymbol{D}).$$

On the other hand, $\text{tr}(\boldsymbol{D}^T\mathbf{1}_n\mathbf{1}_n^T\boldsymbol{D}) = \mathbf{1}_n^T\boldsymbol{D}\boldsymbol{D}^T\mathbf{1}_n$. This shows that

$$A_1 = n^{-2}\mathbf{1}_n^T\boldsymbol{T}\mathbf{1}_n, \tag{5.20}$$

where $\boldsymbol{T} = \boldsymbol{D}\boldsymbol{D}^T$. Let t_{ij} be the (i,j)th entry of $\boldsymbol{T}$, which is the inner product of the ith and jth rows of $\boldsymbol{D}$. From (5.20), we obtain that

$$A_1 = n^{-2}\sum_{i=1}^{n}\sum_{j=1}^{n} t_{ij}. \tag{5.21}$$

Next, let $\boldsymbol{d}_i$ be the ith column of $\boldsymbol{D}$. Then

$$\begin{aligned} A_2 &= \frac{1}{n^2}\sum_{1\le i<j\le m}(\boldsymbol{d}_i^T\boldsymbol{d}_j)^2 \\ &= \frac{1}{2n^2}\left[\sum_{i=1}^{m}\sum_{j=1}^{m}(\boldsymbol{d}_i^T\boldsymbol{d}_j)^2 - \sum_{i=1}^{m}(\boldsymbol{d}_i^T\boldsymbol{d}_i)^2\right]. \end{aligned}$$

Noting that $\sum_{i=1}^{m}(\boldsymbol{d}_i^T\boldsymbol{d}_i)^2 = mn^2$ and that

$$\sum_{i=1}^{m}\sum_{j=1}^{m}(\boldsymbol{d}_i^T\boldsymbol{d}_j)^2 = \mathrm{tr}(\boldsymbol{D}^T\boldsymbol{D}\boldsymbol{D}^T\boldsymbol{D}) = \mathrm{tr}(\boldsymbol{D}\boldsymbol{D}^T\boldsymbol{D}\boldsymbol{D}^T) = \mathrm{tr}(\boldsymbol{T}^2) = \sum_{i=1}^{n}\sum_{j=1}^{n} t_{ij}^2,$$

we obtain that

$$A_2 = \frac{1}{2n^2}\left[\sum_{i=1}^{n}\sum_{j=1}^{n} t_{ij}^2 - mn^2\right]. \tag{5.22}$$

By (5.21) and (5.22), minimizing A_1 and A_2 are equivalent to minimizing $\sum_{i=1}^{n}\sum_{j=1}^{n} t_{ij}$ and $\sum_{i=1}^{n}\sum_{j=1}^{n} t_{ij}^2$, respectively. We will show in the following that this generalizes to higher moments, and leads to an equivalent criterion of *minimum moment aberration*.

Definition 5.1. *A design is said to have minimum moment aberration if it sequentially minimizes $M_1, M_2, \ldots, M_m$, where*

$$M_k = n^{-2}\sum_{i=1}^{n}\sum_{j=1}^{n} t_{ij}^k.$$

The following theorem is the main result of this section.

Theorem 5.6. *A design has minimum G_2-aberration if and only if it has minimum moment aberration.*

Proof. Let $J(\boldsymbol{d}_{j_1}, \ldots, \boldsymbol{d}_{j_k}) = \sum_{i=1}^{n} d_{ij_1}\cdots d_{ij_k}$ be the J-characteristic of k column vectors $\boldsymbol{d}_{j_1}, \ldots, \boldsymbol{d}_{j_k}$ of design $\boldsymbol{D}$, where d_{ij} is the (i,j)th entry of $\boldsymbol{D}$. Then (5.21) and (5.22) can be expressed, respectively, as

$$\sum_{j=1}^{m} J^2(\boldsymbol{d}_j) = \sum_{i=1}^{n}\sum_{j=1}^{n} t_{ij}$$

$$\sum_{j_1=1}^{m}\sum_{j_2=1}^{m} J^2(\boldsymbol{d}_{j_1}, \boldsymbol{d}_{j_2}) = \sum_{i=1}^{n}\sum_{j=1}^{n} t_{ij}^2.$$

We claim that these can be extended to

$$\sum_{j_1=1}^{m}\cdots\sum_{j_k=1}^{m} J^2(\boldsymbol{d}_{j_1}, \ldots, \boldsymbol{d}_{j_k}) = \sum_{i=1}^{n}\sum_{j=1}^{n} t_{ij}^k \tag{5.23}$$

for all $k = 1, \ldots, m$. Equation (5.23) can be proved as follows. We first have that

$$\sum_{j_1=1}^{m}\cdots\sum_{j_k=1}^{m} J^2(\boldsymbol{d}_{j_1}, \ldots, \boldsymbol{d}_{j_k}) = \sum_{j_1=1}^{m}\cdots\sum_{j_k=1}^{m}\left[\sum_{i=1}^{n} d_{ij_1}\cdots d_{ij_k}\right]^2,$$

which is equal to

$$\sum_{j_1=1}^{m}\cdots\sum_{j_k=1}^{m}\sum_{i_1=1}^{n}\sum_{i_2=1}^{n} d_{i_1j_1}\cdots d_{i_1j_k}d_{i_2j_1}\cdots d_{i_2j_k} = \sum_{i_1=1}^{n}\sum_{i_2=1}^{n}\sum_{j_1=1}^{m}\cdots\sum_{j_k=1}^{m} d_{i_1j_1}\cdots d_{i_1j_k}d_{i_2j_1}\cdots d_{i_2j_k}.$$

Equation (5.23) follows if we notice that

$$\sum_{j_1=1}^{m}\cdots\sum_{j_k=1}^{m} d_{i_1j_1}\cdots d_{i_1j_k}d_{i_2j_1}\cdots d_{i_2j_k} = \left(\sum_{j_1=1}^{m} d_{i_1j_1}d_{i_2j_1}\right)\cdots\left(\sum_{j_k=1}^{m} d_{i_1j_k}d_{i_2j_k}\right)$$

and that

$$\sum_{j_1=1}^{m} d_{i_1j_1}d_{i_2j_1} = \cdots = \sum_{j_k=1}^{m} d_{i_1j_k}d_{i_2j_k} = t_{i_1i_2}.$$

To finish the proof of Theorem 5.6, we note that $J(\boldsymbol{d}_{j_1}, \ldots, \boldsymbol{d}_{j_k})$ is the J-characteristic for a set of k factors if $j_1, \ldots, j_k$ are distinct and becomes the J-characteristic for a set of $k-2$ or fewer factors otherwise. With this in mind, we see from (5.23) that M_k must be a linear combination of $A_1, \ldots, A_k$ with a positive coefficient for A_k. It is now obvious that sequentially minimizing $A_1, \ldots, A_k$ is equivalent to sequentially minimizing $M_1, \ldots, M_k$. This completes the proof. □

For orthogonal arrays of strength two, the next lemma makes it explicit the linear dependence of M_k on $A_1, \ldots, A_k$ for $k \leq 5$.

Lemma 5.1. *For orthogonal arrays of strength two, we have that* $M_1 = 0$, $M_2 = m$, $M_3 = 6A_3$, $M_4 = 24A_4 + m(3m-2)$, *and* $M_5 = 120A_5 + (60m - 120)A_3$.

We recall that a monic polynomial of degree k is a polynomial $g(t) = a_0 + a_1t + \cdots + a_kt^k$ such that $a_k = 1$.

Lemma 5.2. *Suppose that* $g_k(t)$ *is a sequence of monic polynomials of degree* k *for* $k \geq 3$. *Then an orthogonal array of strength two has minimum* G_2*-aberration if it minimizes*

$$\sum_{i=1}^{n} \sum_{j=1, i\neq j}^{n} g_k(t_{ij})$$

for each $k \geq 3$ *under the constraints that* $M_1 = 0$ *and* $M_2 = m$.

Like Theorem 5.6, Lemma 5.2 also follows from equation (5.23). Whereas Theorem 5.6 establishes the equivalence between minimum G_2-aberration and minimum moment aberration, Lemma 5.2 is much easier to use in proving a design to have minimum G_2-aberration. To apply Lemma 5.2, we also need a technical result on a lower bound of $\min_{ij}(t_{ij})$ when $m > n/2$. In this case, it is known that the strength of a design cannot be more than 2.

Lemma 5.3. *For* $m > n/2$, *orthogonal arrays of strength 2 satisfy that*

$$\min_{ij}(t_{ij}) \geq -(n-m).$$

Proof. Let $\min_{ij}(t_{ij}) = t_{i'j'}$. Suppose the stated result does not hold. Then $t_{i'j'} = -(n-m) - k$ for some $k > 0$. This implies that the componentwise product of the rows corresponding to runs i' and j' has $n/2 + k/2$ negative entries and $m - n/2 - k/2$ positive entries. In other words, $n/2 + k/2$ factors have different levels in runs i' and j', and the other factors have the same level in these two runs. Now consider a subdesign consisting of a subset of $n/2+1$ factors that have different levels in runs i' and j'. If needed, change the signs of all the entries in the same column so that all these $n/2+1$ factors have the same level in run i'. Finally, delete runs i' and j'. Then the resulting design matrix $\boldsymbol{Z}$, with $n-2$ rows and $n/2+1$ columns, is such that all the diagonal entries of $\boldsymbol{Z}^T\boldsymbol{Z}$ are equal to $n-2$ and all the off-diagonal entries are equal to -2. It follows that $\boldsymbol{Z}^T\boldsymbol{Z}\boldsymbol{1}_{n/2+1} = -2\boldsymbol{1}_{n/2+1}$. Thus $\boldsymbol{1}_{n/2+1}^T\boldsymbol{Z}^T\boldsymbol{Z}\boldsymbol{1}_{n/2+1} = -n-2 < 0$. This contradicts the fact that $\boldsymbol{Z}^T\boldsymbol{Z}$ is a positive semidefinite matrix. □

Strongly related to Lemma 5.3 is a result of Bingham, Sitter, and Tang (2009), which claims that an orthogonal array of strength 2 with $m > n/2$ cannot have a repeated run nor can it have mirror image runs.

Butler (2003a) established Theorem 5.6 for two-level regular designs. The result was extended to nonregular designs in Butler (2003b). Lemmas 5.1, 5.2, and 5.3 are also from Butler (2003b). Xu (2003) defined minimum moment aberration more generally for designs with mixed levels. He also established the equivalence of minimum moment aberration and generalized minimum aberration, a generalization of minimum G_2-aberration to mixed-level designs (Xu and Wu, 2001). Butler's version of minimum moment aberration, as presented here, was based on the inner products of rows of the design matrix (the experimental runs). This version works for two-level designs, but not for designs with more than two levels. Xu (2003)'s more general definition based on Hamming distances of experimental runs can be applied to any setting, and, for two-level designs, is equivalent to Butler's definition.

The minimum moment aberration criterion switches the roles played by rows and columns (experimental runs and factorial effects). This feature offers both theoretical and computational advantages. In the next section, we shall present several results on the theoretical characterization of minimum G_2-aberration designs via the minimum moment aberration criterion. It is also very useful in developing algorithms for computer generation and search of minimum G_2-aberration designs. The complexity of finding minimum G_2-aberration designs directly is much higher than that of algorithmic construction of minimum moment aberration designs.

5.5 Construction Results for Minimum G_2-Aberration

In this section, we present some construction results for minimum G_2-aberration designs from Butler (2003b) and Butler (2004). While all the construction results from Butler (2003b) are presented here, we only include some major results from Butler (2004).

Let $\boldsymbol{H}_1$ and $\boldsymbol{H}_2$ be two Hadamard matrices of orders n_1 and n_2, respectively. We assume both of them are normalized, meaning that their first columns consist of all plus ones. We know from Theorem 2.8 that the tensor product $\boldsymbol{H}_1 \otimes \boldsymbol{H}_2$ is a Hadamard matrix of order $n_1 n_2$, which is also normalized. The basic idea of Butler's constructions (Butler, 2003b) is that many designs obtained by carefully selecting columns from $\boldsymbol{H}_1 \otimes \boldsymbol{H}_2$ have minimum G_2-aberration among orthogonal arrays of strength two.

Theorem 5.7. *Let $\boldsymbol{H}_1$ and $\boldsymbol{H}_2$ be Hadamard matrices of orders n_1 and n_2, respectively, both with the first columns consisting of all ones, where n_1 is equal to* 2 *or a multiple of* 4 *and n_2 is equal to* 1, 2, *or a multiple of* 4.

Let $\boldsymbol{F}$ be an $n_1 \times (n_1 - r)$ matrix obtained by deleting the all-one column and $r-1$ other columns from $\boldsymbol{H}_1$, where $r = 1$ or 2. *Then the design $\boldsymbol{F} \otimes \boldsymbol{H}_2$ has minimum G_2-aberration among all orthogonal arrays of strength two with $n = n_1 n_2$ runs and $m = (n_1 - r)n_2$ factors.*

Some special cases are of interest. If we take $n_2 = 1$ in Theorem 5.7, we see that, for $r = 1$, $\boldsymbol{F} \otimes \boldsymbol{H}_2$ is a saturated orthogonal array obtained by deleting the all-one column from $\boldsymbol{H}_1$ and, for $r = 2$, is a design obtained by deleting one additional column from $\boldsymbol{H}_1$. This shows that saturated and nearly saturated orthogonal arrays have minimum G_2-aberration. If we take $n_2 = 2$ and $r = 2$, then the minimum G_2-aberration design $\boldsymbol{F} \otimes \boldsymbol{H}_2$ is the one obtained by deleting three columns that form a defining word of length 3 from a saturated orthogonal array when n is a multiple of 8, which has previously been discussed in Section 5.3. If we take $n_1 = 2$ and $r = 1$ in Theorem 5.7, we see that the minimum G_2-aberration design $\boldsymbol{F} \otimes \boldsymbol{H}_2$ is the saturated strength-3 orthogonal array obtained by folding over $\boldsymbol{H}_2$. The next example provides further illustrations of Theorem 5.7.

Example 5.1. *Take $n_1 = 12$ and $n_2 = 4$. Theorem 5.7 gives a minimum G_2-aberration design $\boldsymbol{F} \otimes \boldsymbol{H}_2$, which is an orthogonal array of 48 runs and $48-4r$ factors. Design $\boldsymbol{F} \otimes \boldsymbol{H}_2$ has 44 factors for $r = 1$ and 40 factors for $r = 2$. On the other hand, if we take $n_1 = 4$ and $n_2 = 12$, we obtain a minimum G_2-aberration design of 48 runs for 36 factors when $r = 1$ and for 24 factors when $r = 2$.*

The next theorem says that the designs obtained by adding one or two columns to those in Theorem 5.7 continue to have minimum G_2-aberration.

Theorem 5.8. *Let $\boldsymbol{H}_1$, $\boldsymbol{H}_2$, and $\boldsymbol{F} \otimes \boldsymbol{H}_2$ be as given in Theorem 5.7. Let $n = n_1 n_2$ and $u = 1$ or 2. Let $\boldsymbol{V}$ be an $n \times u$ matrix, with columns, not equal to the all-one column, that are from $\boldsymbol{H}_1 \otimes \boldsymbol{H}_2$ but not from $\boldsymbol{F} \otimes \boldsymbol{H}_2$. Then design $(\boldsymbol{F} \otimes \boldsymbol{H}_2, \boldsymbol{V})$ has minimum G_2-aberration.*

If we take $n_1 = 2$ and $r = 1$ in Theorem 5.8, we see that adding one or two columns to the saturated strength-3 orthogonal array still gives minimum G_2-aberration designs.

Example 5.2. *Take $n_1 = 12$ and $n_2 = 4$. Using Theorem 5.8 and the results in Example 5.1, we obtain minimum G_2-aberration designs of 48 runs for $41, 42, 45$, and 46 factors. Now take $n_1 = 4$ and $n_2 = 12$. Again using Theorem 5.8 and the results in Example 5.1, we obtain minimum G_2-aberration designs of 48 runs for $25, 26, 37$, and 38 factors.*

Theorems 5.7 and 5.8 are both from Butler (2003b). In view of Theorem 5.8, one wonders if we can delete some columns from $\boldsymbol{F} \otimes \boldsymbol{H}_2$ in Theorem 5.7 to obtain minimum G_2-aberration designs. Unfortunately, a general result similar to Theorem 5.8 cannot be established. But some restricted results are available.

Theorem 5.9. *Let $\boldsymbol{H}_1$, $\boldsymbol{H}_2$ and $\boldsymbol{F} \otimes \boldsymbol{H}_2$ be as given in Theorem 5.7.*
(i) If $n_1 = 2$ and $r = 1$, then the design obtained from deleting any u columns, where $u = 1, 2$, and 3, from $\boldsymbol{F} \otimes \boldsymbol{H}_2$ has minimum G_2-aberration.
(ii) If $n_1 = 4$ and $r = 1$, then the design obtained by deleting any one column from $\boldsymbol{F} \otimes \boldsymbol{H}_2$ has minimum G_2-aberration.

Theorem 5.9 is also from Butler (2003b) except for the case for $u = 3$ in part (i), which was obtained in Butler (2004). Part (i) of Theorem 5.9 asserts that deleting up to any three columns from a saturated strength-3 orthogonal array still gives a minimum G_2-aberration design. Previously, an application of Theorem 5.7 says that a saturated strength-3 array has minimum G_2-aberration. Therefore, minimum G_2-aberration OA$(n, 2^m, 3)$s have been obtained for $n/2 - 3 \leq m \leq n/2$.

Example 5.3. *Take $n_1 = 2$ and $n_2 = 24$. Part (i) of Theorem 5.9 allows the construction of minimum G_2-aberration designs of 48 runs for $m = 21, 22$, and 23 factors. Take $n_1 = 4$ and $n_2 = 12$. Part (ii) of Theorem 5.9 allows the construction of a minimum G_2-aberration design of 48 runs for 35 factors.*

Theorem 3.5 (Butler, 2004) states that an OA$(n, 2^m, 3)$ must be a foldover design if $m \geq n/3$. Butler (2004) studied the construction of minimum G_2-aberration foldover designs. Apart from part (i) of Theorem 5.9, Butler (2004) obtained another result.

Theorem 5.10. *If m is a multiple of 4, then a foldover OA$(n, 2^m, 3)$ has*

$$A_4 \geq LB = m^2(m^2 - 16)/(12n) - (m-2)(m-4)/8 + 64e(n)/n^2,$$

where $e(n) = 0$ if n is a multiple of 16 and $e(n) = 1$ if n is a multiple of 8 but not of 16. If LB is attained, then the design has minimum G_2-aberration among all foldover designs.

So far, no infinite class of minimum G_2-aberration designs has been constructed using Theorem 5.10. Despite this, the result is still very useful. Butler (2004) used the lower bound LB in Theorem 5.10 to obtain minimum G_2-aberration designs of n runs and m factors for the following (n, m) combinations:

$$(24,4),(24,8),(32,12),(48,16),(48,20),(64,24),(64,28).$$

Proving the above construction results, except for Theorem 5.10, all relies on Lemmas 5.2 and 5.3. To give the reader some ideas about how it is done without undue technical burden, we only present a proof for Theorem 5.7.

Proof. Every row of $\boldsymbol{F}\otimes\boldsymbol{H}_2$ has the form of $\boldsymbol{f}\otimes\boldsymbol{h}$, where $\boldsymbol{f}$ is a row of $\boldsymbol{F}$ and $\boldsymbol{h}$ is a row of $\boldsymbol{H}_2$. Consider rows i and j of $\boldsymbol{F}\otimes\boldsymbol{H}_2$ for $i,j=1,\ldots,n=n_1n_2$. Let the two rows be $\boldsymbol{f}_{i_1}\otimes\boldsymbol{h}_{j_1}$ and $\boldsymbol{f}_{i_2}\otimes\boldsymbol{h}_{j_2}$. Then their inner product is given by

$$t_{ij}=(\boldsymbol{f}_{i_1}\otimes\boldsymbol{h}_{j_1})(\boldsymbol{f}_{i_2}\otimes\boldsymbol{h}_{j_2})^T=(\boldsymbol{f}_{i_1}\boldsymbol{f}_{i_2}^T)(\boldsymbol{h}_{j_1}\boldsymbol{h}_{j_2}^T).$$

Obviously, $\boldsymbol{h}_{j_1}\boldsymbol{h}_{j_2}^T$ equals n_2 if $j_1=j_2$ and equals 0 otherwise.

We first prove Theorem 5.7 for $r=1$. In this case, we have that $\boldsymbol{f}_{i_1}\boldsymbol{f}_{i_2}^T=n_1-1$ if $i_1=i_2$ and $\boldsymbol{f}_{i_1}\boldsymbol{f}_{i_2}^T=-1$ otherwise. For given i, consider all possible t_{ij} values by varying $j=1,\ldots,n$. Equivalently, for given i_1 and j_1, consider all possible values of $(\boldsymbol{f}_{i_1}\otimes\boldsymbol{h}_{j_1})(\boldsymbol{f}_{i_2}\otimes\boldsymbol{h}_{j_2})^T$ by varying $i_2=1,\ldots,n_1$ and $j_2=1,\ldots,n_2$. Out of these n values of t_{ij}, one of them, corresponding to $i=j$, is equal to $(n_1-1)n_2$, n_1-1 of them are equal to $-n_2$, and all the others are equal to 0. Therefore, for $i\neq j$, t_{ij} is either 0 or $-n_2$. We now apply Lemma 5.2 by choosing $g_k(t)=t^2(t+n_2)^{k-2}$ for $k\geq 3$. First, the constraints $M_1=0$ and $M_2=m$ are satisfied because $\boldsymbol{F}\otimes\boldsymbol{H}_2$ is an orthogonal array of strength two. Now consider any strength-2 orthogonal array with $n=n_1n_2$ runs and $m=(n_1-1)n_2$ factors. By Lemma 5.3, this design has that $\min_{ij}(t_{ij})\geq-(n-m)=-n_2$, showing that $g_k(t_{ij})=t_{ij}^2(t_{ij}+n_2)^{k-2}\geq 0$ for all $i\neq j$. On the other hand, our design $\boldsymbol{F}\otimes\boldsymbol{H}_2$ has that $g_k(t_{ij})=t_{ij}^2(t_{ij}+n_2)^{k-2}=0$ for all $i\neq j$, thereby minimizing $\sum_{i=1}^n\sum_{j=1,i\neq j}^n g_k(t_{ij})$. By Lemma 5.2, we have therefore proved that design $\boldsymbol{F}\otimes\boldsymbol{H}_2$ has minimum G_2-aberration among all orthogonal arrays of strength two with $n=n_1n_2$ runs and $m=(n_1-1)n_2$ factors.

We now prove Theorem 5.7 for $r=2$. In the present situation, we have that for each given i_1, $\boldsymbol{f}_{i_1}\boldsymbol{f}_{i_2}^T=n_1-2$ if $i_1=i_2$, $\boldsymbol{f}_{i_1}\boldsymbol{f}_{i_2}^T=-2$ for $(n_1/2-1)$ rows $\boldsymbol{f}_{i_2}$, and $\boldsymbol{f}_{i_1}\boldsymbol{f}_{i_2}^T=0$ for $n_1/2$ rows $\boldsymbol{f}_{i_2}$. For given i, let us consider all possible t_{ij} values by varying $j=1,\ldots,n$. Out of these n values of t_{ij}, one of them, corresponding to $i=j$, is equal to $(n_1-2)n_2$, $n_1/2-1$ of them are equal to $-2n_2$, and all the others are equal to 0. Therefore, for $i\neq j$, t_{ij} is either 0 or $-2n_2$. To apply Lemma 5.2, we take $g_k(t)=t^2(t+2n_2)^{k-2}$ for $k\geq 3$. Again, the constraints $M_1=0$ and $M_2=m$ are satisfied because $\boldsymbol{F}\otimes\boldsymbol{H}_2$ is an orthogonal array of strength two. Now consider any strength-2 orthogonal array with $n=n_1n_2$ runs and $m=(n_1-2)n_2$ factors. By Lemma 5.3, it has that $\min_{ij}(t_{ij})\geq-n+m=-n_1n_2+(n_1-2)n_2=-2n_2$, showing that $g_k(t_{ij})=t_{ij}^2(t_{ij}+2n_2)^{k-2}\geq 0$ for all $i\neq j$. However, design $\boldsymbol{F}\otimes\boldsymbol{H}_2$ has that $g_k(t_{ij})=t_{ij}^2(t_{ij}+2n_2)^{k-2}=0$ for all $i\neq j$, and therefore minimizes $\sum_{i=1}^n\sum_{j=1,i\neq j}^n g_k(t_{ij})$. According to Lemma 5.2, $\boldsymbol{F}\otimes\boldsymbol{H}_2$ has minimum G_2-aberration among all orthogonal arrays of strength two with $n=n_1n_2$ runs and $m=(n_1-2)n_2$ factors.

□

6

Clear Effects, Requirement Sets, and SOS Designs

We begin this chapter with a study on designs with clear effects, discussing their usefulness and examining their existence and constructions. Special attention is paid to clear two-factor interactions for orthogonal arrays of strength three. A contamination-based G_2-aberration criterion is then introduced to provide a general framework for selecting designs with specified requirement sets. We next present some results on finding orthogonal arrays that allow the estimation of all main effects and certain presumably important two-factor interactions. We end this chapter with an investigation into second-order saturated (SOS) designs. As will be seen, minimal SOS designs play a key role in the study of SOS designs. The topics treated in this chapter appear more diverse than in the earlier chapters. However, our interest in estimating some two-factor interactions is a persisting theme that connects these topics together.

6.1 Clear Effects and Preliminary Results

To discuss clear effects, we assume that all interactions involving three or more factors are negligible. So, main effects and two-factor interactions are all the factorial effects the discussion of clear effects is concerned with. Throughout this chapter, whenever we need Hadamard matrices or their associated orthogonal arrays, we assume their existence. This results in a neater presentation, as we need not mention the existence of Hadamard matrices or orthogonal arrays in every individual result involving them.

We recall that in Section 2.7, under a given factorial design, two factorial effects are said to be orthogonal if their corresponding columns in the model matrix are orthogonal.

Definition 6.1. *Given an orthogonal array of strength two, a factorial effect in the collection of main effects and two-factor interactions is said to be clear if it is orthogonal to all the other effects in the collection.*

Clear effects are introduced for orthogonal arrays of strength two. Recall that orthogonal arrays of strength t also have strength t' for any $t' < t$. It follows that a strength-four orthogonal array is also a strength-two orthogonal array. Under an orthogonal array of strength four, main effects and two-factor interactions are mutually orthogonal, and therefore all of them are clear effects. For this reason, the problem of studying clear effects is interesting only for orthogonal arrays of strength two or three that do not have strength four. The next lemma is obvious.

Lemma 6.1. *Under an orthogonal array of strength three, all main effects are clear.*

DOI: 10.1201/9781003371861-6

This result may be simple, but it carries an important message that one can estimate all main effects with an orthogonal array of strength three without worrying about two-factor interactions, negligible or not. In other words, even if two-factor interactions are nonnegligible, they do not contaminate the estimation of main effects. The next lemma provides the necessary conditions for an orthogonal array of strength two to have clear main effects.

Lemma 6.2. *If an* $OA(n, 2^m, 2)$ *has a clear main effect, then it must be true that* (i) n *is a multiple of* 8 *and* (ii) $m \leq n/2$.

Proof. Let $\boldsymbol{D}$ be an $\mathrm{OA}(n, 2^m, 2)$ and $\boldsymbol{d}_j$ be its jth column. Suppose that $\boldsymbol{d}_1$ is clear. Then we have that

$$\mathbf{1}_n, \boldsymbol{d}_1, \boldsymbol{d}_2, \ldots, \boldsymbol{d}_m, \boldsymbol{d}_1\boldsymbol{d}_2, \ldots, \boldsymbol{d}_1\boldsymbol{d}_m$$

are mutually orthogonal, which can be verified as follows. Because $\boldsymbol{D}$ is an $\mathrm{OA}(n, 2^m, 2)$, it is obvious that main effects are mutually orthogonal and are all orthogonal to the grand mean $\mathbf{1}_n$. It also follows from $\boldsymbol{D}$ being an $\mathrm{OA}(n, 2^m, 2)$ that $\boldsymbol{d}_1\boldsymbol{d}_2, \ldots, \boldsymbol{d}_1\boldsymbol{d}_m$ are mutually orthogonal and all orthogonal to $\mathbf{1}_n$. What does not follow from $\boldsymbol{D}$ being an $\mathrm{OA}(n, 2^m, 2)$ is the orthogonality between $\boldsymbol{d}_i$ and $\boldsymbol{d}_1\boldsymbol{d}_j$. For this, we need the fact that $\boldsymbol{d}_1$ is clear. For example, to see why $\boldsymbol{d}_1\boldsymbol{d}_2$ is orthogonal to $\boldsymbol{d}_3$, we calculate that $J(\boldsymbol{d}_1\boldsymbol{d}_2, \boldsymbol{d}_3) = J(\boldsymbol{d}_1, \boldsymbol{d}_2\boldsymbol{d}_3) = 0$, where the last equality holds because $\boldsymbol{d}_1$ is clear. The fact that $\mathbf{1}_n, \boldsymbol{d}_1, \boldsymbol{d}_2, \ldots, \boldsymbol{d}_m, \boldsymbol{d}_1\boldsymbol{d}_2, \ldots, \boldsymbol{d}_1\boldsymbol{d}_m$ are mutually orthogonal implies that $1 + m + m - 1 \leq n$, and hence that $m \leq n/2$. The calculation above also shows that $J(\boldsymbol{d}_1, \boldsymbol{d}_2, \boldsymbol{d}_3) = 0$. This implies that $(\boldsymbol{d}_1, \boldsymbol{d}_2, \boldsymbol{d}_3)$ is an array of strength three according to Theorem 3.3. Therefore, n is a multiple of 8. □

If n is a multiple of 8 and $m \leq n/2$, then by Theorem 2.12 we can construct an $\mathrm{OA}(n, 2^m, 3)$ using a Hadamard matrix of order $n/2$. Lemmas 6.1 and 6.2 together show that if there exists an $\mathrm{OA}(n, 2^m, 2)$ that has a clear main effect, one can then find an $\mathrm{OA}(n, 2^m, 3)$ in which all main effects are clear. Therefore, it is appropriate to say that the problem of finding designs with clear main effects is completely solved.

We next study clear two-factor interactions for strength-two arrays. The same topic for strength-three arrays will be treated in the next section.

Theorem 6.1. *There exists an* $OA(n, 2^m, 2)$ *that has a clear two-factor interaction if and only if* (i) n *is a multiple of* 8 *and* (ii) $m \leq n/2$.

Proof. Let $\boldsymbol{D}$ be an $\mathrm{OA}(n, 2^m, 2)$ and $\boldsymbol{d}_j$ be its jth column. Suppose that $\boldsymbol{d}_1\boldsymbol{d}_2$ is clear. Then one can check in a way similar to what was done in proving Lemma 6.2 that

$$\mathbf{1}_n, \boldsymbol{d}_1, \boldsymbol{d}_2, \ldots, \boldsymbol{d}_m, \boldsymbol{d}_1\boldsymbol{d}_2, \boldsymbol{d}_1\boldsymbol{d}_2\boldsymbol{d}_3, \ldots, \boldsymbol{d}_1\boldsymbol{d}_2\boldsymbol{d}_m$$

are mutually orthogonal. It follows that $1+m+1+m-2 \leq n$, implying that $m \leq n/2$. Also because $\boldsymbol{d}_1\boldsymbol{d}_2$ is clear, we have that $J(\boldsymbol{d}_1, \boldsymbol{d}_2, \boldsymbol{d}_3) = 0$, showing that $(\boldsymbol{d}_1, \boldsymbol{d}_2, \boldsymbol{d}_3)$ is an array of strength three. Therefore n must be a multiple of 8. The sufficiency part of Theorem 6.1 follows from the construction we describe next. □

We construct an orthogonal array with clear two-factor interactions from two orthogonal arrays. Let $\boldsymbol{A} = (\boldsymbol{a}_1, \ldots, \boldsymbol{a}_{m_1})$ be an $\mathrm{OA}(n_1, 2^{m_1}, 2)$ and $\boldsymbol{B} = (\boldsymbol{b}_1, \ldots, \boldsymbol{b}_{m_2})$ be an $\mathrm{OA}(n_2, 2^{m_2}, 2)$. Consider the design

$$\boldsymbol{D} = (\boldsymbol{c}_1, \ldots, \boldsymbol{c}_{m_1}, \boldsymbol{d}_1, \ldots, \boldsymbol{d}_{m_2}) \text{ where } \boldsymbol{c}_i = \boldsymbol{a}_i \otimes \mathbf{1}_{n_2} \text{ and } \boldsymbol{d}_j = \mathbf{1}_{n_1} \otimes \boldsymbol{b}_j. \tag{6.1}$$

Theorem 6.2. *Design* $\boldsymbol{D}$ *constructed in (6.1) is an* $OA(n_1 n_2, 2^{m_1+m_2}, 2)$ *such that all interactions* $\boldsymbol{c}_i\boldsymbol{d}_j$*s are clear for* $i = 1, \ldots, m_1$ *and* $j = 1, \ldots, m_2$.

Proof. First, we note that $J(\boldsymbol{a}_i) = J(\boldsymbol{a}_i, \boldsymbol{a}_j) = 0$ for $i \neq j$ because $\boldsymbol{A}$ is an orthogonal array of strength two. Similarly, we have that $J(\boldsymbol{b}_i) = J(\boldsymbol{b}_i, \boldsymbol{b}_j) = 0$ for $i \neq j$.

To show that $\boldsymbol{D}$ is an orthogonal array of strength two, we need to show that $J(\boldsymbol{c}_i) = J(\boldsymbol{d}_j) = J(\boldsymbol{c}_i, \boldsymbol{d}_j) = 0$ for all i, j. All these can easily be verified using Lemma 4.2. For example, $J(\boldsymbol{c}_i, \boldsymbol{d}_j) = J(\boldsymbol{a}_i, \boldsymbol{1}_{n_1})J(\boldsymbol{1}_{n_2}, \boldsymbol{b}_j) = 0$. Since $J(\boldsymbol{c}_i\boldsymbol{d}_j, \boldsymbol{c}_k) = J(\boldsymbol{a}_i, \boldsymbol{1}_{n_1}, \boldsymbol{a}_k)J(\boldsymbol{1}_{n_2}, \boldsymbol{b}_j, \boldsymbol{1}_{n_2}) = J(\boldsymbol{a}_i, \boldsymbol{a}_k)J(\boldsymbol{b}_j) = 0$ for any i, j, k, we have that $\boldsymbol{c}_i\boldsymbol{d}_j$ is orthogonal to main effects $\boldsymbol{c}_k$s. Similarly, $\boldsymbol{c}_i\boldsymbol{d}_j$ is orthogonal to $\boldsymbol{d}_k$s. For $(i_1, j_1) \neq (i_2, j_2)$, we have that $J(\boldsymbol{c}_{i_1}\boldsymbol{d}_{j_1}, \boldsymbol{c}_{i_2}\boldsymbol{d}_{j_2}) = J(\boldsymbol{a}_{i_1}, \boldsymbol{a}_{i_2})J(\boldsymbol{b}_{j_1}, \boldsymbol{b}_{j_2}) = 0$, meaning that $\boldsymbol{c}_i\boldsymbol{d}_j$s are mutually orthogonal. This completes the proof. □

In Theorem 6.2, if we take $\boldsymbol{A}$ to be an $\mathrm{OA}(2, 2^1, 2)$ and $\boldsymbol{B}$ to be an $\mathrm{OA}(n_2, 2^{n_2-1}, 2)$, then $\boldsymbol{D}$ is an $\mathrm{OA}(2n_2, 2^{n_2}, 2)$ under which $\boldsymbol{c}_1\boldsymbol{d}_1, \ldots, \boldsymbol{c}_1\boldsymbol{d}_{n_2-1}$ are all clear. This establishes the sufficiency part of Theorem 6.1.

Clear two-factor interactions are mutually orthogonal and also orthogonal to all main effects. Therefore, the number of clear two-factor interactions in an $\mathrm{OA}(n, 2^m, 2)$ is bounded above by $n - m - 1$. In Theorem 6.2, if we take an $\mathrm{OA}(n_1, 2^{n_1-1}, 2)$ as $\boldsymbol{A}$ and an $\mathrm{OA}(n_2, 2^{n_2-1}, 2)$ as $\boldsymbol{B}$, then $\boldsymbol{D}$ has $(n_1 - 1)(n_2 - 1)$ clear two-factor interactions. Since

$$(n_1 - 1)(n_2 - 1) = n_1n_2 - (n_1 + n_2) + 1 = n - m - 1,$$

where $n = n_1n_2$ and $m = n_1 + n_2 - 2$, design $\boldsymbol{D}$ has the maximum number of clear two-factor interactions among all $\mathrm{OA}(n_1n_2, 2^{n_1+n_2-2}, 2)$s.

Orthogonal arrays of strength two with clear two-factor interactions are not very interesting because they generally do not have clear main effects. Nonetheless, $\mathrm{OA}(n, 2^m, 2)$s with clear two-factor interactions have applications in the robust parameter design problem, where interactions between control factors and noise factors are of paramount importance (Shoemaker, Tsui, and Wu, 1991). Design $\boldsymbol{D}$ in (6.1) is in fact a product array, advocated by Taguchi (1986), with the control array being $\boldsymbol{A} = (\boldsymbol{a}_1, \ldots, \boldsymbol{a}_{m_1})$ and the noise array being $\boldsymbol{B} = (\boldsymbol{b}_1, \ldots, \boldsymbol{b}_{m_2})$.

6.2 Strength-Three Arrays with Clear Interactions

All main effects are clear under an $\mathrm{OA}(n, 2^m, 3)$, a strength-three array. If an $\mathrm{OA}(n, 2^m, 3)$ also has some clear two-factor interactions, then we can estimate all main effects and these clear two-factor interactions without assuming other two-factor interactions being negligible. This renders $\mathrm{OA}(n, 2^m, 3)$s with clear two-factor interactions the most attractive designs that the notion of clear effects spawns. In some sense, they are designs intermediate between $\mathrm{OA}(n, 2^m, 3)$s and $\mathrm{OA}(n, 2^m, 4)$s.

Theorem 6.3. *There exists an* $\mathrm{OA}(n, 2^m, 3)$ *that has a clear two-factor interaction if and only if* (i) n *is a multiple of* 16 *and* (ii) $m \leq n/4 + 1$.

Proof. Let $\boldsymbol{D}$ be an $\mathrm{OA}(n, 2^m, 3)$ and $\boldsymbol{d}_j$ be its jth column. Suppose that $\boldsymbol{d}_1\boldsymbol{d}_2$ is clear. Using this in combination with the fact that $\boldsymbol{D}$ is an $\mathrm{OA}(n, 2^m, 3)$, one can easily check that

$$\boldsymbol{1}_n, \boldsymbol{d}_1\boldsymbol{d}_2;\ \boldsymbol{d}_1, \boldsymbol{d}_2, \ldots, \boldsymbol{d}_m;\ \boldsymbol{d}_1\boldsymbol{d}_3, \ldots, \boldsymbol{d}_1\boldsymbol{d}_m;\ \boldsymbol{d}_2\boldsymbol{d}_3, \ldots, \boldsymbol{d}_2\boldsymbol{d}_m;\ \boldsymbol{d}_1\boldsymbol{d}_2\boldsymbol{d}_3, \ldots, \boldsymbol{d}_1\boldsymbol{d}_2\boldsymbol{d}_m$$

are mutually orthogonal. This establishes that

$$2 + m + 3(m - 2) \leq n,$$

showing that $m \le n/4+1$. Also, because $\boldsymbol{d_1 d_2}$ is clear, we have that $J(\boldsymbol{d}_1,\boldsymbol{d}_2,\boldsymbol{d}_3,\boldsymbol{d}_4)=0$. This shows that $(\boldsymbol{d}_1,\boldsymbol{d}_2,\boldsymbol{d}_3,\boldsymbol{d}_4)$ is an $\mathrm{OA}(n,2^4,4)$. Therefore, n is a multiple of 16. The sufficiency part of Theorem 6.3 follows from the first construction method we discuss next. □

We now present two construction methods of strength-three orthogonal arrays that have clear two-factor interactions. The basic idea is to create a new design using tensor products involving the columns of two base arrays $\boldsymbol{A}$ and $\boldsymbol{B}$.

Essentially, our first construction method is based on a strength-four array $\boldsymbol{A}$ and a strength-two array $\boldsymbol{B}$. Let $\boldsymbol{A}=(\boldsymbol{a}_1,\ldots,\boldsymbol{a}_{m_1})$ with $m_1\ge 2$ be an $\mathrm{OA}(n_1,2^{m_1},\min(4,m_1))$, which is an array of strength 4 if $m_1\ge 4$ and of strength m_1 if $m_1=2,3$. For $m_1=2,3$, $\boldsymbol{A}$ is not of strength four, but it has properties similar to strength-four arrays in that it has no three or four columns that have a non-zero J-characteristic. Let $\boldsymbol{B}=(\boldsymbol{b}_1,\ldots,\boldsymbol{b}_{m_2})$ be an $\mathrm{OA}(n_2,2^{m_2},2)$. Consider the design

$$\boldsymbol{D}=(\boldsymbol{c}_1,\ldots,\boldsymbol{c}_{m_1},\boldsymbol{d}_1,\ldots,\boldsymbol{d}_{m_2}) \text{ where } \boldsymbol{c}_i=\boldsymbol{a}_i\otimes\boldsymbol{1}_{n_2} \text{ and } \boldsymbol{d}_j=(\boldsymbol{a}_1\boldsymbol{a}_2)\otimes\boldsymbol{b}_j. \tag{6.2}$$

Theorem 6.4. *Design $\boldsymbol{D}$ given in (6.2) is an $OA(n_1n_2,2^{m_1+m_2},3)$ that has the following clear interactions: $\boldsymbol{c}_i\boldsymbol{c}_j$s for all $1\le i<j\le m_1$ and $\boldsymbol{c}_i\boldsymbol{d}_j$s for all $i=1,\ldots,m_1$ and $j=1,\ldots,m_2$.*

Proof. This theorem can be verified routinely using Lemma 4.2. To show that $\boldsymbol{D}$ in (6.2) is an $\mathrm{OA}(n_1n_2,2^{m_1+m_2},3)$, one needs to check that $J(\boldsymbol{c}_i,\boldsymbol{c}_j,\boldsymbol{c}_k)=J(\boldsymbol{c}_i,\boldsymbol{c}_j,\boldsymbol{d}_k)=J(\boldsymbol{c}_i,\boldsymbol{d}_j,\boldsymbol{d}_k)=J(\boldsymbol{d}_i,\boldsymbol{d}_j,\boldsymbol{d}_k)=0$ for all possible i,j,k. For example, we have that

$$J(\boldsymbol{c}_i,\boldsymbol{d}_j,\boldsymbol{d}_k)=J(\boldsymbol{a}_i,\boldsymbol{a}_1\boldsymbol{a}_2,\boldsymbol{a}_1\boldsymbol{a}_2)J(\boldsymbol{1}_{n_2},\boldsymbol{b}_j,\boldsymbol{b}_k)=J(\boldsymbol{a}_i)J(\boldsymbol{b}_j,\boldsymbol{b}_k)=0.$$

To show that $\boldsymbol{c}_i\boldsymbol{d}_j$s are clear, one needs to check that $J(\boldsymbol{c}_i,\boldsymbol{c}_j,\boldsymbol{c}_k,\boldsymbol{d}_l)=J(\boldsymbol{c}_i,\boldsymbol{d}_j,\boldsymbol{d}_k,\boldsymbol{d}_l)=0$ for all possible i,j,k,l, and $J(\boldsymbol{c}_i,\boldsymbol{c}_j,\boldsymbol{d}_k,\boldsymbol{d}_l)=0$ for all possible i,j,k,l such that $(i,k)\neq(j,l)$. Similarly, one can check that $\boldsymbol{c}_i\boldsymbol{c}_j$s are also clear. □

Example 6.1. *If we take an $OA(4,2^2,2)$ as $\boldsymbol{A}$ and an $OA(n_2,2^{n_2-1},2)$ as $\boldsymbol{B}$, then*

$$\boldsymbol{D}=(\boldsymbol{c}_1,\boldsymbol{c}_2,\boldsymbol{d}_1,\ldots,\boldsymbol{d}_{n_2-1}) \text{ is an } OA(4n_2,2^{n_2+1},3),$$

under which $\boldsymbol{c}_1\boldsymbol{c}_2$ and all $\boldsymbol{c}_i\boldsymbol{d}_j$s are clear. This establishes the sufficiency part of Theorem 6.3.

Taking $n_2=12$, we obtain

$$\boldsymbol{D}=(\boldsymbol{c}_1,\boldsymbol{c}_2,\boldsymbol{d}_1,\ldots,\boldsymbol{d}_{11}), \text{ an } OA(48,2^{13},3),$$

which has clear $\boldsymbol{c}_1\boldsymbol{c}_2$ and clear $\boldsymbol{c}_i\boldsymbol{d}_j$s for all $i=1,2$ and $j=1,\ldots,11$.

Taking $n_2=20$, we obtain

$$\boldsymbol{D}=(\boldsymbol{c}_1,\boldsymbol{c}_2,\boldsymbol{d}_1,\ldots,\boldsymbol{d}_{19}), \text{ an } OA(80,2^{21},3),$$

which has clear $\boldsymbol{c}_1\boldsymbol{c}_2$ and clear $\boldsymbol{c}_i\boldsymbol{d}_j$s for all $i=1,2$ and $j=1,\ldots,19$.

Example 6.2. *If we take an $OA(8,2^3,3)$ as $\boldsymbol{A}$ and an $OA(n_2,2^{n_2-1},2)$ as $\boldsymbol{B}$, then*

$$\boldsymbol{D}=(\boldsymbol{c}_1,\boldsymbol{c}_2,\boldsymbol{c}_3,\boldsymbol{d}_1,\ldots,\boldsymbol{d}_{n_2-1}) \text{ is an } OA(8n_2,2^{n_2+2},3),$$

which has clear $\boldsymbol{c}_1\boldsymbol{c}_2$, $\boldsymbol{c}_1\boldsymbol{c}_3$, $\boldsymbol{c}_2\boldsymbol{c}_3$, and clear $\boldsymbol{c}_i\boldsymbol{d}_j$s for all i,j. Taking $n_2=12$, we obtain $\boldsymbol{D}=(\boldsymbol{c}_1,\boldsymbol{c}_2,\boldsymbol{c}_3,\boldsymbol{d}_1,\ldots,\boldsymbol{d}_{11})$, an $OA(96,2^{14},3)$, which has clear $\boldsymbol{c}_i\boldsymbol{c}_j$s for all $i<j$ and clear $\boldsymbol{c}_i\boldsymbol{d}_j$s for all i,j.

If we take an $OA(16, 2^5, 4)$ *as* $\boldsymbol{A}$ *and an* $OA(n_2, 2^{n_2-1}, 2)$ *as* $\boldsymbol{B}$*, then*

$$\boldsymbol{D} = (\boldsymbol{c}_1, \ldots, \boldsymbol{c}_5, \boldsymbol{d}_1, \ldots, \boldsymbol{d}_{n_2-1}) \text{ is an } OA(16n_2, 2^{n_2+4}, 3),$$

which has clear $\boldsymbol{c}_i\boldsymbol{c}_j$*s for all* $i < j$ *and clear* $\boldsymbol{c}_i\boldsymbol{d}_j$*s for all* i, j*. Taking* $n_2 = 12$*, we obtain* $\boldsymbol{D} = (\boldsymbol{c}_1, \boldsymbol{c}_2, \boldsymbol{c}_3, \boldsymbol{c}_4, \boldsymbol{c}_5, \boldsymbol{d}_1, \ldots, \boldsymbol{d}_{11})$*, an* $OA(192, 2^{16}, 3)$*, which has clear* $\boldsymbol{c}_i\boldsymbol{c}_j$*s for all* $i < j$ *and clear* $\boldsymbol{c}_i\boldsymbol{d}_j$*s for all* i, j*.*

We now present the second construction method of OA$(n, 2^m, 3)$s with clear two-factor interactions. This method uses two arrays, both of strength two. Let $\boldsymbol{A} = (\boldsymbol{a}_1, \ldots, \boldsymbol{a}_{m_1})$ be an OA$(n_1, 2^{m_1}, 2)$ and $\boldsymbol{B} = (\boldsymbol{b}_1, \ldots, \boldsymbol{b}_{m_2})$ be an OA$(n_2, 2^{m_2}, 2)$. Consider design

$$\boldsymbol{D} = (\boldsymbol{c}_2, \ldots, \boldsymbol{c}_{m_1}, \boldsymbol{d}_1, \ldots, \boldsymbol{d}_{m_2}), \quad \text{where } \boldsymbol{c}_i = \boldsymbol{a}_i \otimes \boldsymbol{b}_1 \text{ and } \boldsymbol{d}_j = \boldsymbol{a}_1 \otimes \boldsymbol{b}_j. \tag{6.3}$$

Theorem 6.5. *Design* $\boldsymbol{D}$ *constructed in* (6.3) *is an* $OA(n_1n_2, 2^{m_1+m_2-1}, 3)$ *with the following clear interactions:* $\boldsymbol{c}_i\boldsymbol{d}_j$*s for all* $i = 2, \ldots, m_1$ *and* $j = 1, \ldots, m_2$*.*

The proof of Theorem 6.5 is very similar to that of Theorem 6.4, and hence omitted.

Example 6.3. *Taking an* $OA(8, 2^7, 2)$ *as* $\boldsymbol{A}$ *and an* $OA(12, 2^{11}, 2)$ *as* $\boldsymbol{B}$*, we obtain*

$$\boldsymbol{D} = (\boldsymbol{c}_2, \ldots, \boldsymbol{c}_7, \boldsymbol{d}_1, \ldots, \boldsymbol{d}_{11}), \text{ an } OA(96, 2^{17}, 3)$$

that has clear $\boldsymbol{c}_i\boldsymbol{d}_j$*s for* $i = 2, \ldots, 7$ *and* $j = 1, \ldots, 11$*.*

Taking an $OA(12, 2^{11}, 2)$ *as* $\boldsymbol{A}$ *and an* $OA(8, 2^7, 2)$ *as* $\boldsymbol{B}$*, we obtain*

$$\boldsymbol{D} = (\boldsymbol{c}_2, \ldots, \boldsymbol{c}_{11}, \boldsymbol{d}_1, \ldots, \boldsymbol{d}_7), \text{ an } OA(96, 2^{17}, 3)$$

that has clear $\boldsymbol{c}_i\boldsymbol{d}_j$*s for* $i = 2, \ldots, 11$ *and* $j = 1, \ldots, 7$*.*

Both of the above two $OA(96, 2^{17}, 3)$*s have clear two-factor interactions, but the patterns of clear two-factor interactions are different. They are both useful depending on practical situations. The first array has* 66 *clear two-factor interactions and the second array has* 70*. So, the second array is preferred if our interest lies solely in maximizing the number of clear two-factor interactions.*

One interesting problem in the studies of clear effects is the determination of the maximum number of clear two-factor interactions. This problem is a simple one for strength-two orthogonal arrays, as we have briefly discussed in the second to last paragraph of Section 6.1. The same problem is harder for strength-three orthogonal arrays. The next result provides an upper bound.

Theorem 6.6. *The maximum number of clear two-factor interactions an* $OA(n, 2^m, 3)$ *can have is bounded above by*

$$U(n, m) = \lfloor m(n - 2m)/(m - 2) \rfloor .$$

Interested readers are referred to Lekivetz and Tang (2011) for a proof of Theorem 6.6. As careful readers may find, this proof bears some technical resemblance to that for Theorem 4.5.

Applying Theorem 6.6 to $n = 96$ and $m = 17$, we obtain $U(96, 17) = 70$. This shows that the second OA$(96, 2^{17}, 3)$ in Example 6.3 has the maximum number of clear two-factor interactions over all OA$(96, 2^{17}, 3)$s.

Clear effects for regular designs were studied by many researchers. The following is a brief survey of the relevant literature. Clear effects were first introduced by Wu and Chen (1992) and further discussed in Chen, Sun, and Wu (1993). Chen and Hedayat (1998) obtained the first set of theoretical results on the existence of clear two-factor interactions, and their results are essentially regular design versions of Theorems 6.1 and 6.3. Tang, Ma, Ingram, and Wang (2002) presented some construction results for designs with clear two-factor interactions, and studied the maximum number of clear two-factor interactions. Their results are related to Theorems 6.4, 6.5, and 6.6. Wu and Wu (2002) made an improvement over one theoretical bound in Tang, Ma, Ingram, and Wang (2002) for an important special case among other useful discussion on clear two-factor interactions. Finding designs with structured patterns of clear interactions is the subject of Ke, Tang, and Wu (2005), Wu, Mee, and Tang (2012), and Grömping (2014). For example, Ke, Tang, and Wu (2005) considered the construction of four classes of compromise plans with clear two-factor interactions.

Clear two-factor interactions for nonregular designs were studied in Tang (2006), which is the main source for our presentation of Sections 6.1 and 6.2. Useful designs when certain two-factor interactions are negligible were considered in Lekivetz and Tang (2011) through the use of *partially clear* two-factor interactions.

Clear effects are studied in Sections 6.1 and 6.2 with respect to the collection of all main effects and all two-factor interactions. Generalizations can easily be made by considering a different collection of factorial effects. One possibility is to include in the collection all factorial effects involving up to three factors. Another possibility is to consider all main effects and a special subset of two-factor interactions, which is essentially the approach taken by Lekivetz and Tang (2011).

6.3 Experiments with Requirement Sets

In many applications, researchers know a priori the importance of certain two-factor interactions. The experimental goal is then to find a design that allows estimation of these presumably important interactions as well as the main effects. The set of effects that require estimation is known as a requirement set in design literature. How to find a regular design for a given requirement set was considered by Greenfield (1976), Franklin and Bailey (1977), Franklin (1985), and Dey and Suen (2002). When the effects outside the requirement set cannot be safely assumed to be negligible, they will bias the estimation of the effects in the requirement set. Ke and Tang (2003) examined this design selection problem when there are up to four important two-factor interactions. Cheng and Tang (2005) proposed a general contamination-based criterion useful for the requirement set problem.

We adapt the work of Cheng and Tang (2005) to nonregular designs in this section, hence offering a general approach to the design selection for the requirement set problem. The framework helps us better understand designs with clear effects discussed in Sections 6.1 and 6.2, and perhaps more importantly suggests ideas for extension.

Suppose that design $\boldsymbol{D}$ under consideration is an OA$(n, 2^m, 2)$. Let β_0 be the grand mean, $\boldsymbol{\beta}_1$ the main effects, and $\boldsymbol{\beta}_j$ the interactions involving j factors for $j = 2, \ldots, m$. Let $\boldsymbol{\beta}_{21}$ denote the two-factor interactions in the requirement set and $\boldsymbol{\beta}_{22}$ the two-factor interactions outside the requirement set. Thus, $\boldsymbol{\beta}_2 = (\boldsymbol{\beta}_{21}^T, \boldsymbol{\beta}_{22}^T)^T$. The fitted model is then given by

$$\boldsymbol{Y} = \mathbf{1}_n\beta_0 + \boldsymbol{X}_1\boldsymbol{\beta}_1 + \boldsymbol{X}_{21}\boldsymbol{\beta}_{21} + \boldsymbol{\epsilon},$$

where $\boldsymbol{Y}$ is the vector of n observations on the response variable, $\boldsymbol{X}_1 = \boldsymbol{D}$ is the design matrix, $\boldsymbol{X}_{21}$ is the model matrix corresponding to $\boldsymbol{\beta}_{21}$, $\boldsymbol{\epsilon}$ is the vector of random errors assumed to have a zero mean and a constant variance σ^2. However, if the effects outside the requirement set cannot be ignored, then the true model is

$$\boldsymbol{Y} = \boldsymbol{1}_n\beta_0 + \boldsymbol{X}_1\boldsymbol{\beta}_1 + \boldsymbol{X}_{21}\boldsymbol{\beta}_{21} + \boldsymbol{X}_{22}\boldsymbol{\beta}_{22} + \boldsymbol{X}_3\boldsymbol{\beta}_3 + \cdots + \boldsymbol{X}_m\boldsymbol{\beta}_m + \boldsymbol{\epsilon},$$

where $\boldsymbol{X}_{22}$ is the model matrix corresponding to $\boldsymbol{\beta}_{22}$ and $\boldsymbol{X}_j$ is the model matrix corresponding to $\boldsymbol{\beta}_j$ for $j = 3, \ldots, m$. For ease of presentation, let $\boldsymbol{Z}_1 = (\boldsymbol{X}_1, \boldsymbol{X}_{21})$ and $\boldsymbol{\gamma}_1 = (\boldsymbol{\beta}_1^T, \boldsymbol{\beta}_{21}^T)^T$. Then the fitted and true models become

$$\boldsymbol{Y} = \boldsymbol{1}_n\beta_0 + \boldsymbol{Z}_1\boldsymbol{\gamma}_1 + \boldsymbol{\epsilon} \tag{6.4}$$

and

$$\boldsymbol{Y} = \boldsymbol{1}_n\beta_0 + \boldsymbol{Z}_1\boldsymbol{\gamma}_1 + \boldsymbol{X}_{22}\boldsymbol{\beta}_{22} + \boldsymbol{X}_3\boldsymbol{\beta}_3 + \cdots + \boldsymbol{X}_m\boldsymbol{\beta}_m + \boldsymbol{\epsilon}, \tag{6.5}$$

respectively.

Since design $\boldsymbol{D}$ is an OA$(n, 2^m, 2)$, we have that $\boldsymbol{1}_n$ is orthogonal to the columns of $\boldsymbol{Z}_1$. So, for $\boldsymbol{\gamma}_1$ in (6.4) to be estimable, $\boldsymbol{Z}_1$ needs to have full rank. Suppose this is the case. Then the least squares estimator of $\boldsymbol{\gamma}_1$ is $\hat{\boldsymbol{\gamma}}_1 = (\boldsymbol{Z}_1^T\boldsymbol{Z}_1)^{-1}\boldsymbol{Z}_1^T\boldsymbol{Y}$, which is unbiased under (6.4) but becomes biased under (6.5) with the bias given by

$$\text{Bias}(\hat{\boldsymbol{\gamma}}_1) = (\boldsymbol{Z}_1^T\boldsymbol{Z}_1)^{-1}\boldsymbol{Z}_1^T(\boldsymbol{X}_{22}\boldsymbol{\beta}_{22} + \boldsymbol{X}_3\boldsymbol{\beta}_3 + \cdots + \boldsymbol{X}_m\boldsymbol{\beta}_m).$$

Suppose that we can find an orthogonal design for estimating $\boldsymbol{\gamma}_1$, meaning that $\boldsymbol{Z}_1$ has orthogonal columns. Then $(\boldsymbol{Z}_1^T\boldsymbol{Z}_1)^{-1}\boldsymbol{Z}_1^T = n^{-1}\boldsymbol{Z}_1^T$. We obtain that

$$\text{Bias}(\hat{\boldsymbol{\gamma}}_1) = \boldsymbol{C}_2\boldsymbol{\beta}_{22} + \boldsymbol{C}_3\boldsymbol{\beta}_3 + \cdots + \boldsymbol{C}_m\boldsymbol{\beta}_m, \tag{6.6}$$

where $\boldsymbol{C}_2 = n^{-1}\boldsymbol{Z}_1^T\boldsymbol{X}_{22}$ and $\boldsymbol{C}_j = n^{-1}\boldsymbol{Z}_1^T\boldsymbol{X}_j$ for $j \geq 3$. Suppose as usual that the hierarchical assumption holds for the effects outside the requirement set. Then it is sensible to select a design that sequentially minimizes

$$\|\boldsymbol{C}_2\|^2, \|\boldsymbol{C}_3\|^2, \ldots, \|\boldsymbol{C}_m\|^2,$$

where $\|\boldsymbol{C}_j\|^2 = \text{trace}(\boldsymbol{C}_j^T\boldsymbol{C}_j)$ is the sum of squared entries of $\boldsymbol{C}_j$. This criterion can appropriately be called a minimum G_2-aberration criterion for the requirement set problem because it is equivalent to the usual minimum G_2-aberration when only main effects are required to be estimated, as has been shown in Section 5.2.1.

Remark 6.1. *How to find OA$(n, 2^m, 2)$s that allow the estimation of $\boldsymbol{\gamma}_1$ will be addressed in Section 6.4. In situations where orthogonal designs are unavailable for estimating $\boldsymbol{\gamma}_1$, we may adopt the following strategy. We first identify a class of designs that are optimal according to some optimality criterion, such as those discussed in Section 2.3, and then select one from this class of designs by sequentially minimizing $\|\boldsymbol{C}_2'\|^2$, $\|\boldsymbol{C}_3'\|^2$, ..., $\|\boldsymbol{C}_m'\|^2$, where $\boldsymbol{C}_2' = (\boldsymbol{Z}_1^T\boldsymbol{Z}_1)^{-1}\boldsymbol{Z}_1^T\boldsymbol{X}_{22}$ and $\boldsymbol{C}_j' = (\boldsymbol{Z}_1^T\boldsymbol{Z}_1)^{-1}\boldsymbol{Z}_1^T\boldsymbol{X}_j$ for $j \geq 3$. Similar ideas have been employed in Zhang and Mukerjee (2013) and Huda and Mukerjee (2013) for a slightly different problem.*

Much insight into clear effects can be gained by perusing and dissecting this general G_2-aberration criterion. If we assume that all interactions involving three or more factors are negligible, then the best design is obtained by minimizing $\|\boldsymbol{C}_2\|^2$ alone. The best-case scenario is $\|\boldsymbol{C}_2\|^2 = 0$, in which case design $\boldsymbol{D} = \boldsymbol{X}_1$ is an OA$(n, 2^m, 3)$ and the interactions in $\boldsymbol{\beta}_{21}$ are all clear.

Now let $\boldsymbol{C}_{21} = n^{-1}\boldsymbol{X}_1^T\boldsymbol{X}_{22}$ and $\boldsymbol{C}_{22} = n^{-1}\boldsymbol{X}_{21}^T\boldsymbol{X}_{22}$. Then $\boldsymbol{C}_2 = (\boldsymbol{C}_{21}^T, \boldsymbol{C}_{22}^T)^T$ and $\|\boldsymbol{C}_2\|^2 = \|\boldsymbol{C}_{21}\|^2 + \|\boldsymbol{C}_{22}\|^2$, where $\|\boldsymbol{C}_{21}\|^2$ represents the contamination on the main effects $\boldsymbol{\beta}_1$ caused by $\boldsymbol{\beta}_{22}$ and $\|\boldsymbol{C}_{22}\|^2$ the contamination on the required interactions $\boldsymbol{\beta}_{21}$ caused by $\boldsymbol{\beta}_{22}$. If a design is such that $\|\boldsymbol{C}_{21}\|^2 = 0$, then it is an $\mathrm{OA}(n, 2^m, 3)$. If it is such that $\|\boldsymbol{C}_{22}\|^2 = 0$, then the interactions in $\boldsymbol{\beta}_{21}$ are clear. In general, minimizing $\|\boldsymbol{C}_{21}\|^2$ amounts to finding a design as close as possible to an $\mathrm{OA}(n, 2^m, 3)$ and minimizing $\|\boldsymbol{C}_{22}\|^2$ amounts to finding a design that makes $\boldsymbol{\beta}_{21}$ as close as possible to being clear. So, if an $\mathrm{OA}(n, 2^m, 3)$ with clear $\boldsymbol{\beta}_{21}$ cannot be found, then one may simply minimize $\|\boldsymbol{C}_2\|^2$, as required by the general G_2-aberration criterion. Alternatively, one could first minimize $\|\boldsymbol{C}_{21}\|^2$ and then minimize $\|\boldsymbol{C}_{22}\|^2$ or first minimize $\|\boldsymbol{C}_{22}\|^2$ and then minimize $\|\boldsymbol{C}_{21}\|^2$. Designs found by these different approaches are all worthy of consideration.

The preceding discussion suggests two variations of the general G_2-aberration criterion. Let $\boldsymbol{C}_{j1} = n^{-1}\boldsymbol{X}_1^T\boldsymbol{X}_j$ and $\boldsymbol{C}_{j2} = n^{-1}\boldsymbol{X}_{21}^T\boldsymbol{X}_j$ for $j = 3, \ldots, m$. If $\boldsymbol{\beta}_1$ needs more protection from contamination by nonnegligible interactions than $\boldsymbol{\beta}_{21}$, then one may sequentially minimize

$$\|\boldsymbol{C}_{21}\|^2, \|\boldsymbol{C}_{22}\|^2, \|\boldsymbol{C}_{31}\|^2, \|\boldsymbol{C}_{32}\|^2, \ldots, \|\boldsymbol{C}_{m1}\|^2, \|\boldsymbol{C}_{m2}\|^2.$$

If $\boldsymbol{\beta}_{21}$ needs more protection from contamination by nonnegligible interactions than $\boldsymbol{\beta}_1$, then one may sequentially minimize

$$\|\boldsymbol{C}_{22}\|^2, \|\boldsymbol{C}_{21}\|^2, \|\boldsymbol{C}_{32}\|^2, \|\boldsymbol{C}_{31}\|^2, \ldots, \|\boldsymbol{C}_{m2}\|^2, \|\boldsymbol{C}_{m1}\|^2.$$

The general minimum G_2-aberration criterion introduced in this section for the requirement set problem can also be used to select blocked factorial designs with fixed block effects (Cheng and Tang, 2005). We will not pursue it here. A more fruitful approach for dealing with blocked factorial designs is the Bayesian approach discussed in Subsection 5.2.4. In Chapter 7, factorial designs under complex unit structures including those for block designs will be treated using a unified Bayesian approach, and appropriate minimum G_2-aberration criteria will be derived.

6.4 Estimation When Some Interactions Are Important

In Section 6.3, we developed a G_2-aberration criterion for selecting designs when we are interested in estimating the effects in a requirement set consisting of all main effects and some two-factor interactions. Assuming the existence of designs for the estimation of the effects in the requirement set, this criterion aims at minimizing the contamination on the estimated effects from nonnegligible effects outside the requirement set. In the present section, we examine how to find designs that allow estimation of the required effects.

We assume that all effects outside the requirement set are negligible and consider how to find an $\mathrm{OA}(n, 2^m, 2)$ from a saturated $\mathrm{OA}(n, 2^{n-1}, 2)$ for a given requirement set. We present some characterizing results on the existence and construction of orthogonal arrays for jointly estimating all main effects and some two-factor interactions. The results are simple to use, yet powerful enough to allow useful designs to be found.

For a given requirement set S, we define its *core*, denoted by $C(S)$, to be the subset of S obtained by removing those main effects that are not involved in the two-factor interactions in S. Clearly, every requirement set has a unique core. We give two examples to illustrate this concept. If

$$S = \{\beta_1, \beta_2, \beta_3, \beta_4, \beta_5, \beta_6, \beta_{12}, \beta_{34}\},$$

then we have that

$$C(S) = \{\beta_1, \beta_2, \beta_3, \beta_4, \beta_{12}, \beta_{34}\},$$

where, for example, β_1 denotes the main effect of the first factor and β_{12} is the interaction between factors 1 and 2. This is because factors 5 and 6 are not involved in any two-factor interaction in the requirement set S. Now suppose

$$S = \{\beta_1, \beta_2, \beta_3, \beta_4, \beta_5, \beta_6, \beta_{12}, \beta_{23}\}.$$

Then its core is give by

$$C(S) = \{\beta_1, \beta_2, \beta_3, \beta_{12}, \beta_{23}\}.$$

A requirement set can be represented by a graph if we associate a main effect with a vertex and a two-factor interaction with an edge. Then the core of a requirement set is obtained by simply deleting all isolated vertices. We say that a design supports a requirement set if it allows joint estimation of the effects in the requirement set.

Theorem 6.7. *Consider the orthogonal arrays obtained by selecting columns from a saturated $OA(n, 2^{n-1}, 2)$. Then an orthogonal array that supports a requirement set S exists if and only if an orthogonal array that supports its core $C(S)$ exists.*

Theorem 6.7 reduces the problem of finding a design for a requirement set S to that of finding a design for its core $C(S)$. It is particularly powerful when the number of two-factor interactions to be estimated is small. While the necessity part of Theorem 6.7 is obvious, the sufficiency part needs some explanation. Suppose that S is composed of m main effects and e two-factor interactions. Then its core $C(S)$ consists of m_1 main effects and all the e two-factor interactions in S, where $m_1 \leq m$. If a design is to support S, its run size n must satisfy that $n \geq 1 + m + e$. Let $\boldsymbol{D}$ be an $\mathrm{OA}(n, 2^{n-1}, 2)$ and $\boldsymbol{D}_1$ be a subarray having m_1 columns of $\boldsymbol{D}$ that supports $C(S)$. Then Theorem 6.7 says that one can always find a subarray $\boldsymbol{D}_2$ with $m_2 = m - m_1$ columns from $\boldsymbol{D}$ such that $(\boldsymbol{D}_1, \boldsymbol{D}_2)$ supports S. Write $\boldsymbol{D} = (\boldsymbol{D}_1, \boldsymbol{D}_2, \boldsymbol{D}_3)$. Finding $\boldsymbol{D}_2$ is equivalent to finding $\boldsymbol{D}_3$ with $m_3 = n - 1 - m$ columns. Any $\boldsymbol{D}_3$ can be deleted from $\boldsymbol{D}$ to obtain $(\boldsymbol{D}_1, \boldsymbol{D}_2)$ as long as it satisfies

$$\det\big(\boldsymbol{X}_2^T \boldsymbol{D}_3 \boldsymbol{D}_3^T \boldsymbol{X}_2\big) > 0 \tag{6.7}$$

where $\boldsymbol{X}_2$ denotes the model matrix for the e two-factor interactions. Such a $\boldsymbol{D}_3$ always exists, provided that $\boldsymbol{D}_1$ supports $C(S)$. In all, there are $\binom{m_2+m_3}{m_3}$ possible candidates for $\boldsymbol{D}_3$ and many of these may satisfy the condition in (6.7). The next theorem tells which one to choose.

Theorem 6.8. *For a given requirement set S, suppose that $\boldsymbol{D}_1$, an $OA(n, 2^{m_1}, 2)$, supports $C(S)$. Then in terms of the D-efficiency, design $(\boldsymbol{D}_1, \boldsymbol{D}_2)$ is the best for the estimation of S if $\boldsymbol{D}_3$ maximizes* $\det\big(\boldsymbol{X}_2^T \boldsymbol{D}_3 \boldsymbol{D}_3^T \boldsymbol{X}_2\big)$.

We give a combined proof of the two theorems.

Proof. Let $\boldsymbol{X}_1 = (\boldsymbol{1}_n, \boldsymbol{D}_1)$. Then $(\boldsymbol{X}_1, \boldsymbol{X}_2)$ is the model matrix for the requirement set $C(S)$. We first have

$$\det\big[(\boldsymbol{X}_1, \boldsymbol{X}_2)^T (\boldsymbol{X}_1, \boldsymbol{X}_2)\big] = \det\big(\boldsymbol{X}_1^T \boldsymbol{X}_1\big) \det\big[\boldsymbol{X}_2^T \boldsymbol{X}_2 - \boldsymbol{X}_2^T \boldsymbol{X}_1 (\boldsymbol{X}_1^T \boldsymbol{X}_1)^{-1} \boldsymbol{X}_1^T \boldsymbol{X}_2\big].$$

Noting that $(\boldsymbol{X}_1, \boldsymbol{D}_2, \boldsymbol{D}_3)$ is a Hadamard matrix of order n, we obtain

$$\det\big[(\boldsymbol{X}_1, \boldsymbol{X}_2)^T (\boldsymbol{X}_1, \boldsymbol{X}_2)\big] = n^{m_1+1-e} \det\Big[\boldsymbol{X}_2^T \boldsymbol{D}^* \boldsymbol{D}^{*T} \boldsymbol{X}_2\Big]. \tag{6.8}$$

where $\boldsymbol{D}^* = (\boldsymbol{D}_2, \boldsymbol{D}_3)$. As $\boldsymbol{D}_1$ supports $C(S)$, we have that $\det\Big[\boldsymbol{X}_2^T \boldsymbol{D}^* \boldsymbol{D}^{*T} \boldsymbol{X}_2\Big] > 0$. The fact that matrix $\boldsymbol{X}_2^T \boldsymbol{D}^* \boldsymbol{D}^{*T} \boldsymbol{X}_2$ has a full rank implies that the $e \times (m_2 + m_3)$ matrix

$\boldsymbol{X}_2^T \boldsymbol{D}^*$ has rank e. Note that $m_2 + m_3 \geq m_3 \geq e$. Then the matrix $\boldsymbol{X}_2^T \boldsymbol{D}^*$ must contain a submatrix of m_3 columns with rank e. This shows that we can select m_3 columns from $\boldsymbol{D}^*$ to obtain a $\boldsymbol{D}_3$ such that $\det\big[\boldsymbol{X}_2^T \boldsymbol{D}_3 \boldsymbol{D}_3^T \boldsymbol{X}_2\big] > 0$. Now let $\boldsymbol{X} = (\boldsymbol{1}_n, \boldsymbol{D}_1, \boldsymbol{D}_2)$ and therefore $(\boldsymbol{X}, \boldsymbol{X}_2)$ is the model matrix for the requirement set S. Similar to equation (6.8), we have that

$$\det\big[(\boldsymbol{X}, \boldsymbol{X}_2)^T (\boldsymbol{X}, \boldsymbol{X}_2)\big] = n^{m+1-e} \det\big[\boldsymbol{X}_2^T \boldsymbol{D}_3 \boldsymbol{D}_3^T \boldsymbol{X}_2\big], \tag{6.9}$$

which is strictly positive for the choice we have just made for $\boldsymbol{D}_3$. This shows that $(\boldsymbol{D}_1, \boldsymbol{D}_2)$ supports S, proving Theorem 6.7. It is also immediate that $\det\big[(\boldsymbol{X}, \boldsymbol{X}_2)^T (\boldsymbol{X}, \boldsymbol{X}_2)\big]$ in (6.9) is maximized if $\det\big[\boldsymbol{X}_2^T \boldsymbol{D}_3 \boldsymbol{D}_3^T \boldsymbol{X}_2\big]$ is maximized, which establishes Theorem 6.8. □

Remark 6.2. *In the above proof, the fact that $\boldsymbol{X}_2$ consists of only two-factor interactions never gets used. This shows that Theorems 6.7 and 6.8 are valid as well for a general requirement set that includes higher order interactions. The core of such a general requirement set is similarly defined, and given by removing all the main effects that do not occur in any interaction.*

Example 6.4. *Suppose that we want an orthogonal array of 20 runs and 15 factors for the requirement set*

$$S = \{\beta_1, \ldots, \beta_{15}, \beta'_{12}, \beta'_{23}, \beta'_{34}, \beta'_{15}\}.$$

The core of S is given by

$$C(S) = \{\beta_1, \beta_2, \beta_3, \beta_4, \beta_5, \beta'_{12}, \beta'_{23}, \beta'_{34}, \beta'_{15}\}.$$

In this concrete example, we have used β'_{ij} to denote the interaction effects to avoid double meanings of β_{15}, which could be the main effect of factor 15 or the interaction effect between factors 1 and 5. Now consider the 20-run saturated orthogonal array $\boldsymbol{D}$ generated by using

$$(1, 1, -1, -1, 1, 1, 1, 1, -1, 1, -1, 1, -1, -1, -1, -1, 1, 1, -1)$$

as the first row, cyclically shifting this row vector one place to the left 18 times and then adding a row of all minus ones. Let $\boldsymbol{D} = (\boldsymbol{d}_1, \boldsymbol{d}_2, \ldots, \boldsymbol{d}_{19})$. We can easily check that design $\boldsymbol{D}_1 = (\boldsymbol{d}_1, \boldsymbol{d}_2, \boldsymbol{d}_3, \boldsymbol{d}_4, \boldsymbol{d}_5)$ supports $C(S)$. We have that $\boldsymbol{X}_2 = (\boldsymbol{d}_1\boldsymbol{d}_2, \boldsymbol{d}_2\boldsymbol{d}_3, \boldsymbol{d}_3\boldsymbol{d}_4, \boldsymbol{d}_1\boldsymbol{d}_5)$, $\boldsymbol{D}^* = (\boldsymbol{d}_6, \ldots, \boldsymbol{d}_{19})$, and

$$\boldsymbol{X}_2^T \boldsymbol{D}^* = \begin{bmatrix} 4 & -4 & 4 & -12 & -4 & -4 & 4 & 4 & -4 & 4 & -4 & -4 & -4 & -4 \\ 4 & 4 & -4 & 4 & -12 & -4 & -4 & 4 & 4 & -4 & 4 & -4 & -4 & -4 \\ 4 & 4 & 4 & -4 & 4 & 12 & -4 & -4 & 4 & 4 & -4 & 4 & -4 & -4 \\ -4 & 4 & -4 & -4 & 4 & 4 & -4 & 4 & -12 & -4 & -4 & 4 & -4 & 4 \end{bmatrix}. \tag{6.10}$$

As $\boldsymbol{D}_1$ supports $C(S)$, the matrix in (6.10) must have rank 4. Let $\boldsymbol{A}$ be a 4×4 submatrix of the matrix in (6.10). We now evaluate $\det(\boldsymbol{A}\boldsymbol{A}^T)$ for all the $\binom{14}{4} = 1001$ submatrices $\boldsymbol{A}$ and present the results for $4^{-8}\det(\boldsymbol{A}\boldsymbol{A}^T)$ in the following table.

$4^{-8}\det(\boldsymbol{A}\boldsymbol{A}^T)$	0	64	256	576	1024	2304	4096
frequency	364	305	242	22	55	6	7

From the above table, we can see that out of the 1001 submatrices $\boldsymbol{A}$, seven of them take the largest value of $4^{-8}\det(\boldsymbol{A}\boldsymbol{A}^T) = 4096$, one of which is given by columns 2, 4, 5, 6 of the matrix in (6.10). These columns of the matrix in (6.10) correspond to columns $\boldsymbol{d}_7, \boldsymbol{d}_9, \boldsymbol{d}_{10}, \boldsymbol{d}_{11}$ in $\boldsymbol{D}^*$. Therefore, one best design for the requirement set S given by Theorems 6.7 and 6.8 is

$$(\boldsymbol{d}_1, \boldsymbol{d}_2, \ldots, \boldsymbol{d}_6, \boldsymbol{d}_8, \boldsymbol{d}_{12}, \boldsymbol{d}_{13}, \ldots, \boldsymbol{d}_{19}),$$

which is the design obtained from the 20-run saturated orthogonal array $\boldsymbol{D} = (\boldsymbol{d}_1, \boldsymbol{d}_2, \ldots, \boldsymbol{d}_{19})$ by deleting columns $\boldsymbol{d}_7, \boldsymbol{d}_9, \boldsymbol{d}_{10}, \boldsymbol{d}_{11}$.

Theorems 6.7 and 6.8 were obtained by Tang and Zhou (2009), who also considered more general applications of the theorems than that given in Example 6.4. It is worth noting that Theorem 6.8 is an optimality result conditional on a given orthogonal array that supports $C(S)$. A more ambitious research problem is to find a D-optimal orthogonal array for a given requirement set from all orthogonal arrays. Tang and Zhou (2013) obtained some results along this line. In particular they obtained all D-optimal orthogonal arrays of 12 and 20 runs for requirement sets containing up to three two-factor interactions.

6.5 Second Order Saturated Designs

6.5.1 Definition and basic results

Second order saturated (SOS) designs were first introduced by Block and Mee (2003) in the context of regular designs. This concept was generalized to orthogonal arrays by Cheng, Mee, and Yee (2008).

Definition 6.2. *An $OA(n, 2^m, 2)$ is said to be SOS if it allows the estimation of all main effects and a set of $n-1-m$ two-factor interactions, provided all other factorial effects are negligible.*

Let $\boldsymbol{D}$ be an $\text{OA}(n, 2^m, 2)$. Further let $\boldsymbol{X}_2$ denote the model matrix of all two-factor interactions, i.e., $\boldsymbol{X}_2$ consists of all the $\binom{m}{2}$ two-factor interaction columns. Then $\boldsymbol{D}$ is SOS if and only if $\text{rank}(\boldsymbol{D}, \boldsymbol{X}_2) = n - 1$.

Remark 6.3. *For any $OA(n, 2^m, 2)$ $\boldsymbol{D}$, its ability of estimating two-factor interactions can be measured by $df(\boldsymbol{D}) = \text{rank}(\boldsymbol{D}, \boldsymbol{X}_2)$, called the number of degrees of freedom of $\boldsymbol{D}$. Obviously, we have that $df(\boldsymbol{D}) \leq n - 1$. Therefore, SOS designs are those that have $df = n - 1$, the maximum number of degrees of freedom.*

SOS designs are interesting in their own right, as one can estimate a maximum number of two-factor interactions besides main effects, thus wasting no degree of freedom in terms of the design capacity for estimating effects of the first two orders. The notion of SOS designs is useful also because of the equivalence between regular resolution IV designs' being SOS and their being maximal. More importantly, SOS designs are useful for constructing strong orthogonal arrays, a class of space-filling designs for computer experiments. Space-filling designs will be discussed in Chapter 12.

For an $\text{OA}(n, 2^m, 2)$ to be SOS, it is necessary that $m + \binom{m}{2} \geq n - 1$. Thus, an $\text{OA}(n, 2^m, 2)$ with a very small m relative to n cannot be SOS. At the other extreme, we can show that an $\text{OA}(n, 2^m, 2)$ with a large m must be SOS.

Theorem 6.9. *An $OA(n, 2^m, 2)$ must be SOS if $m \geq n/2$.*

Proof. Let $\boldsymbol{D} = (\boldsymbol{d}_1, \ldots, \boldsymbol{d}_m)$ be an $\text{OA}(n, 2^m, 2)$ with $m \geq n/2$. Suppose that $\boldsymbol{D}$ is not SOS. Then there must exist a real vector $\boldsymbol{x}$ that is orthogonal to all the following vectors:

$$\boldsymbol{1}_n, \boldsymbol{d}_1, \ldots, \boldsymbol{d}_m, \boldsymbol{d}_i\boldsymbol{d}_j, \text{ for all } i < j,$$

as the linear subspace generated by them cannot have a rank of n. Now consider

$$\boldsymbol{1}_n, \boldsymbol{d}_1, \ldots, \boldsymbol{d}_m, \boldsymbol{x}, \boldsymbol{x}\boldsymbol{d}_1, \ldots, \boldsymbol{x}\boldsymbol{d}_m.$$

One can easily verify that these $2(m+1)$ vectors are mutually orthogonal. For example, $J(\boldsymbol{1}_n, \boldsymbol{x}\boldsymbol{d}_1) = J(\boldsymbol{x}, \boldsymbol{d}_1) = 0$, $J(\boldsymbol{d}_1, \boldsymbol{x}\boldsymbol{d}_2) = J(\boldsymbol{x}, \boldsymbol{d}_1\boldsymbol{d}_2) = 0$, and $J(\boldsymbol{x}\boldsymbol{d}_1, \boldsymbol{x}\boldsymbol{d}_2) = J(\boldsymbol{d}_1, \boldsymbol{d}_2) = 0$. We thus have $2(m+1) \leq n$, which contradicts $m \geq n/2$. □

Theorem 6.9 was obtained in an email discussion by the three authors of Cheng, He, and Tang (2021) after the paper was accepted. The regular version was established by Chen and Cheng (2004), and documented as Theorem 11.3 in Cheng (2014a).

An $\mathrm{OA}(n, 2^m, 3)$ is said to be maximal if it cannot be embedded into an $\mathrm{OA}(n, 2^{m+1}, 3)$. The following result gives a necessary condition for an $\mathrm{OA}(n, 2^m, 3)$ to be SOS.

Theorem 6.10. *For an $OA(n, 2^m, 3)$ to be SOS, it is necessary that it is maximal.*

Proof. Let $\boldsymbol{D} = (\boldsymbol{d}_1, \ldots, \boldsymbol{d}_m)$ be an $\mathrm{OA}(n, 2^m, 3)$, which is SOS. Suppose that $\boldsymbol{D}$ is not maximal. Then there exists an $\mathrm{OA}(n, 2^{m+1}, 3)$ of the form $\boldsymbol{D}' = (\boldsymbol{d}_1, \ldots, \boldsymbol{d}_m, \boldsymbol{d}_{m+1})$. This means that $\boldsymbol{d}_{m+1}$ is orthogonal to $\boldsymbol{d}_1, \ldots, \boldsymbol{d}_m$ and $\boldsymbol{d}_i\boldsymbol{d}_j$ for all $1 \leq i < j \leq m$. This is a contradiction because $\boldsymbol{D} = (\boldsymbol{d}_1, \ldots, \boldsymbol{d}_m)$ is SOS. □

As given in Theorem 11.4 of Cheng (2014a), the counterpart of Theorem 6.10 for regular designs is stronger, which states that a regular resolution IV design is SOS if and only if it is maximal. A systematic study of resolution IV designs is available in Chen and Cheng (2006), who provided a complete characterization of maximal designs of resolution IV for $m \geq n/4 + 1$ using the concept of doubling.

Theorem 6.10 implies that an $\mathrm{OA}(n, 2^m, 3)$ cannot be SOS if it is not maximal. Consider an $\mathrm{OA}(n, 2^{n/2}, 3)$, obtained by folding over a Hadamard matrix of order $n/2$, as given in Theorem 2.12. Then any subarray $\mathrm{OA}(n, 2^m, 3)$ of this saturated strength-three array with $m < n/2$ columns is not SOS.

Theorem 6.10 is a new result, obtained while this book is being written. It is yet to be settled whether or not an $\mathrm{OA}(n, 2^m, 3)$ must be SOS if it is maximal.

6.5.2 Minimal SOS designs and their characterization

Cheng, He, and Tang (2021) conducted an investigation into minimal SOS designs and their applications to the construction of space-filling designs. An SOS design is said to be minimal if whenever any factor is removed, the resulting design is no longer SOS. Starting with an SOS design, if it is not minimal, we can obtain another SOS design by deleting a factor. Continuing this process, one can always obtain a minimal SOS design from an SOS design by deleting some factors. This shows that any SOS design is either minimal or can be obtained by adding some factors to a minimal SOS design. Though simple, this is worth documenting.

Lemma 6.3. *Any SOS design is either minimal or can be obtained by adding factors to a minimal SOS design.*

Lemma 6.3 states that all SOS designs can be obtained if we have all minimal SOS designs. We can therefore focus our study on minimal SOS designs. The next result provides a characterization for an SOS $\mathrm{OA}(n, 2^m, 2)$ to be minimal.

Theorem 6.11. *Let $\boldsymbol{D} = (\boldsymbol{d}_1, \ldots, \boldsymbol{d}_m)$ be an SOS $OA(n, 2^m, 2)$. Then $\boldsymbol{D}$ is minimal if for any j, at least one of the following effects*

$$\boldsymbol{d}_j, \boldsymbol{d}_j\boldsymbol{d}_1, \ldots, \boldsymbol{d}_j\boldsymbol{d}_{j-1}, \boldsymbol{d}_j\boldsymbol{d}_{j+1}, \ldots, \boldsymbol{d}_j\boldsymbol{d}_m$$

is clear.

Proof. Obviously, if at least one of the effects listed above is clear, then the resulting design after removing column $\boldsymbol{d}_j$ from $\boldsymbol{D}$ is no longer SOS. Since no column can be removed from $\boldsymbol{D}$ without changing the property of being SOS, design $\boldsymbol{D}$ is minimal. □

Corollary 6.1. *If an $OA(n, 2^m, 3)$ is SOS, then it must be a minimal SOS design.*

Corollary 6.1 follows immediately from Theorem 6.11 since all main effects are clear under an orthogonal array of strength three.

We can easily obtain other SOS orthogonal arrays of strength two from a given SOS $\text{OA}(n, 2^m, 2)$. The design obtained by adding a factor to a given SOS $\text{OA}(n, 2^m, 2)$ is still SOS, unless the given $\text{OA}(n, 2^m, 2)$ is maximal. Deleting a particular factor from a given SOS $\text{OA}(n, 2^m, 2)$ also produces an SOS design, unless the given $\text{OA}(n, 2^m, 2)$ is minimal SOS. On the other hand, Theorem 6.10 and Corollary 6.1 reveal that it is impossible to obtain another SOS orthogonal array of strength three from a given SOS $\text{OA}(n, 2^m, 3)$ by deleting or adding a factor. Deleting a factor from an SOS $\text{OA}(n, 2^m, 3)$ still gives a strength-three array, but it cannot be SOS because of Corollary 6.1. Adding a factor to an SOS $\text{OA}(n, 2^m, 3)$ results in an SOS design that cannot be an array of strength three because of Theorem 6.10.

For regular designs, the condition in Theorem 6.11 for an SOS $\text{OA}(n, 2^m, 2)$ to be minimal is also necessary (Cheng, He, and Tang, 2021). This in particular implies that a regular minimal SOS $\text{OA}(n, 2^m, 2)$ must have some clear effects.

6.5.3 Construction of minimal SOS designs

We present some constructions of minimal SOS designs in this subsection. The proofs are omitted as they are, for the most part, quite tedious and not very illuminating. Let

$$\boldsymbol{H}_{n_1} = (\boldsymbol{a}_0, \boldsymbol{a}_1, \ldots, \boldsymbol{a}_{n_1-1}) \text{ with } \boldsymbol{a}_0 = \boldsymbol{1}_{n_1}$$

and

$$\boldsymbol{H}_{n_2} = (\boldsymbol{b}_0, \boldsymbol{b}_1, \ldots, \boldsymbol{b}_{n_2-1}) \text{ with } \boldsymbol{b}_0 = \boldsymbol{1}_{n_2}$$

be two Hadamard matrices of orders n_1 and n_2, respectively. Then $\boldsymbol{H}_{n_1} \otimes \boldsymbol{H}_{n_2}$ is a Hadamard matrix of order $n_1 n_2$, which is composed of column vectors $\boldsymbol{c}_{ij} = \boldsymbol{a}_i \otimes \boldsymbol{b}_j$ where $i = 0, 1, \ldots, n_1 - 1$ and $j = 0, 1, \ldots, n_2 - 1$. By carefully selecting some columns from $\boldsymbol{H}_{n_1} \otimes \boldsymbol{H}_{n_2}$, we obtain the following constructions of minimal SOS designs:

Construction 1: $\boldsymbol{D}_1 = (\boldsymbol{c}_{10}, \ldots, \boldsymbol{c}_{(n_1-1)0}, \boldsymbol{c}_{01}, \ldots, \boldsymbol{c}_{0(n_2-1)})$;

Construction 2: $\boldsymbol{D}_2 = (\boldsymbol{c}_{20}, \ldots, \boldsymbol{c}_{(n_1-1)0}, \boldsymbol{c}_{11}, \boldsymbol{c}_{02}, \ldots, \boldsymbol{c}_{0(n_2-1)})$;

Construction 3: $\boldsymbol{D}_3 = (\boldsymbol{c}_{20}, \ldots, \boldsymbol{c}_{(n_1-1)0}, \boldsymbol{c}_{11}, \ldots, \boldsymbol{c}_{1(n_2-1)})$;

Construction 4: $\boldsymbol{D}_4 = (\boldsymbol{c}_{21}, \ldots, \boldsymbol{c}_{(n_1-1)1}, \boldsymbol{c}_{11}, \ldots, \boldsymbol{c}_{1(n_2-1)})$.

All four designs have $n_1 n_2$ runs. Design $\boldsymbol{D}_1$ has $n_1 + n_2 - 2$ factors and designs $\boldsymbol{D}_2, \boldsymbol{D}_3$, and $\boldsymbol{D}_4$ have $n_1 + n_2 - 3$ factors. While $\boldsymbol{D}_1, \boldsymbol{D}_2$, and $\boldsymbol{D}_3$ are orthogonal arrays of strength two, $\boldsymbol{D}_4$ has strength three.

Theorem 6.12. *All of $\boldsymbol{D}_1, \boldsymbol{D}_2, \boldsymbol{D}_3$, and $\boldsymbol{D}_4$ are minimal SOS designs.*

Theorem 6.12 is from Cheng, He, and Tang (2021). Regular versions of Constructions 1-4 were considered in He, Cheng, and Tang (2018) and Cheng, He, and Tang (2021).

Constructions 1 and 4 are closely related to those for designs with clear two-factor interactions in Sections 6.1 and 6.2. In fact, they can be regarded as special cases of the constructions in (6.1) and (6.3), with $\boldsymbol{A}$ and $\boldsymbol{B}$ being saturated orthogonal arrays. When $n_1 = 4$, Construction 3 is related to that in (6.2). If we take $\boldsymbol{A}$ to be an $\text{OA}(4, 2^2, 2)$ and

TABLE 6.1
Classification of all regular minimal SOS designs for $n \le 64$ and $n = 128$ with $m \le 20$.

n	m	III	IV	total
8	4	1	1	2
16	5	0	1	1
16	6	1	0	1
16	8	1	1	2
32	9	1	1	2
32	10	6	1	7
32	11	1	0	1
32	16	1	1	2
64	13	7	1	8
64	14	19	0	19
64	15	14	0	14
64	16	16	0	16
64	17	48	5	53
64	18	108	1	109
64	20	1	1	2
64	32	1	1	2
128	19	5	0	5
128	20	36	0	36

$\boldsymbol{B}$ to be a saturated OA$(n_2, 2^{n_2-1}, 2)$, then the construction in (6.2) produces design $\boldsymbol{D}_3$ as given by Construction 3.

While $\boldsymbol{D}_3$ has strength two in general, it has strength three when $n_1 = 4$. More precisely, design $\boldsymbol{D}_3$ is an OA$(4n_2, 2^{n_2+1}, 3)$. Cheng, Mee, and Yee (2008) studied SOS orthogonal arrays of strength three and constructed two types of such arrays. Their arrays of type 1, which are all OA$(4n_2, 2^{n_2+1}, 3)$s, include $\boldsymbol{D}_3$ obtained for $n_1 = 4$. Their arrays of type 2 are in the next result.

Theorem 6.13. *If $\boldsymbol{H}$ is a Hadamard matrix of order n_1 and $\boldsymbol{B}$ is an SOS OA$(n_2, 2^{m_2}, 3)$. Then $\boldsymbol{H} \otimes \boldsymbol{B}$ is an SOS OA$(n_1 n_2, 2^{n_1 m_2}, 3)$, and hence is a minimal SOS design.*

Example 6.5. *If we take $n_1 = 4$ and $n_2 = 12$ in Theorem 6.12, then design $\boldsymbol{D}_3$ is an SOS OA$(48, 2^{13}, 3)$. Using this array as $\boldsymbol{B}$ in Theorem 6.13, we can obtain an SOS OA$(96, 2^{26}, 3)$, an SOS OA$(192, 2^{52}, 3)$, and so on.*

Regular SOS designs are equivalent to 1-saturating sets in projective geometries, and also equivalent to the duals of linear codes with covering radius 2 in coding theory. If one is interested in obtaining regular SOS designs with even smaller numbers of factors than those given in Constructions 2-4, one such situation being their use for constructing space-filling designs, we point out that there exist such designs for 128 or more runs. Cheng, He, and Tang (2021) presented two constructions of these smaller SOS designs; the original constructions were obtained by Davydov, Marcugini, and Pambianco (2006) and Gabidulin, Davydov, and Tombak (1991), respectively, in projective geometry and coding theory literature.

Davydov, Marcugini, and Pambianco (2006) performed a complete enumeration of all minimal 1-saturating sets in small geometries. In design language, we provide a summary of their findings in Table 6.1. For given n and m, the last column of Table 6.1 gives the total number of regular minimal SOS designs; the third and fourth columns are the numbers of minimal SOS designs of resolution III and IV, respectively.

TABLE 6.2
Classification of all minimal SOS $OA(n, 2^m, t)$s for $t = 2$ with $n = 12, 16$, and 20, and for $t = 3$ with $n = 16, 24, 32$, and 40.

n	m	t	N_{minsos}
12	5	2	1
16	5	2	3
16	6	2	2
16	8	2	2
20	6	2	15
20	7	2	22
20	9	2	1
16	5	3	1
16	8	3	1
24	12	3	1
32	9	3	6
32	10	3	1
32	16	3	5
40	20	3	3

We give some discussion on the results in Table 6.1. For regular designs of 16 runs, there are exactly four minimal SOS designs in total. There is one minimal SOS design of resolution IV (actually V) for $m = 5$ and one minimal SOS design of resolution III for $m = 6$, and two minimal SOS designs for $m = 8$, with one being of resolution III and the other being of resolution IV. For regular designs of 32 runs, minimal SOS designs exist for $m = 9, 10, 11$, and 16, and minimal SOS designs of resolution IV exist for $m = 9, 10$, and 16. Regular minimal SOS designs of 128 runs exist for $m = 19$ and 20, and they are all of resolution III. We note that if we use the regular versions of Constructions 2-4 to construct regular minimal SOS designs of 128 runs, the smallest number of factors we can achieve is 21, obtained by taking $n_1 = 8$ and $n_2 = 16$.

Using the enumeration results of Sun, Li, and Ye (2008) and Schoen, Eendebak, and Nguyen (2010), Cheng, He, and Tang (2021) carried out a complete search of minimal SOS $OA(n, 2^m, t)$s for $t = 2$ with $n = 12, 16$, and 20, and for $t = 3$ with $n = 16, 24, 32$, and 40. The last column of Table 6.2 presents the number of minimal SOS $OA(n, 2^m, t)$ for these parameters. Some discussion on Table 6.2 follows.

There exist exactly three minimal SOS $OA(16, 2^5, 2)$s. Comparing this result with those in Table 6.1, we see that one of these three minimal SOS designs is regular, and the other two are nonregular. If we compare the results in Tables 6.1 and 6.2 for $n = 16$ and $m = 8$, we see that there are in total two minimal SOS designs and they are both regular. If we look at the results for $t = 3$ and $n = 24, 40$, we see that the saturated orthogonal arrays of strength three are the only minimal SOS designs of strength three. This implies that, for $n = 24$ and 40, saturated orthogonal arrays of strength three are also the only SOS designs of strength three. In other words, there do not exist an SOS $OA(24, 2^m, 3)$ for $m < 12$ and an SOS $OA(40, 2^m, 3)$ for $m < 20$.

7

Minimum G_2-Aberration in Complex Settings

In Chapter 5, the minimum G_2-aberration criterion was defined for the simple setting where the treatments have a factorial structure, while the experimental units are unstructured. Furthermore, a hierarchical assumption was imposed on the factorial treatment structure: factorial effects of the same order are equally important and lower-order effects are more important than higher-order ones. In practice, one may encounter other more complex situations. This chapter is devoted to the formulation of appropriate minimum G_2-aberration criteria for several complex settings that often arise in practice. While these settings are treated separately in this chapter, derivations of the criteria are guided by a consistent principle. A general theory with more applications will be presented in Chapter 11 when mixed-level designs are considered.

7.1 Introduction

The minimum G_2-aberration criterion introduced in Chapter 5 was more or less based on intuition, albeit we have shown that it can be justified from several different perspectives. When the treatments and experimental units have more complicated structures, often it is not clear how an appropriate minimum aberration type criterion can be formulated.

For example, even in the relatively simple setting of block designs where the experimental units are divided into more homogeneous blocks to improve the precision of factorial-effect estimates, several different minimum aberration type criteria have been proposed in the literature in ad hoc manners. In the context of regular designs, Sitter, Chen, and Feder (1997) was the first to consider a minimum aberration criterion for the selection of blocked fractional factorial designs. Two modifications were made later by Chen and Cheng (1999) and Cheng and Wu (2002). As with an unblocked fractional factorial, a blocked fractional factorial is also specified by a defining relation containing all the defining words. The defining words for a blocked fractional factorial are of two types: treatment defining words, those only containing treatment factors, and block defining words, those containing at least one block factor. The basic idea behind the three minimum aberration criteria given in Sitter, Chen, and Feder (1997), Chen and Cheng (1999), and Cheng and Wu (2002) is the important recognition that a treatment defining word causes more serious aliasing/confounding than a block defining word does. Cheng, Li, and Ye (2004) considered the application of these and other related criteria to nonregular designs of 12, 16, and 20 runs in b blocks with b being a power of two. Schoen, Sartono, and Goos (2013) obtained more comprehensive results by making use of mixed-level orthogonal arrays, which allowed them to consider designs of 12 runs in $b = 3$ blocks and designs of 20 runs in $b = 5$ blocks. Schoen, Sartono, and Goos (2013) also examined blocked three-level fractional factorial designs. For other related work

DOI: 10.1201/9781003371861-7

along this line of research, we refer to Sun, Wu, and Chen (1997), Zhang and Park (2000), Butler (2006), Jacroux (2009), and Grömping (2021).

While these ad hoc aberration criteria are all useful, it would be more desirable to have a general coherent principle that can be used to formulate an appropriate criterion with directly interpretable statistical meaning in a systematic way. In Section 5.2, several justifications for the minimum G_2-aberration criterion were provided. Each justification shows how the minimum G_2-aberration criterion can be derived as a simple *surrogate* for a specific concrete criterion for which optimal designs are difficult to obtain. For example, it was shown in Section 5.2.1 how the generalized minimum aberration criterion can be derived by minimizing the contamination of nonnegligible interactions on the estimation of main effects according to the order of importance dictated by the hierarchical assumption. The "contamination" alluded to here is the bias of the estimates of main effects caused by possible presence of interaction effects. However, minimizing biases may not be enough; precisions of estimates should also be taken into account. In contrast, the Bayesian approach presented in Section 5.2.4 is more useful and versatile.

In this chapter, we will show how the Bayesian approach can be used to derive suitable minimum G_2-aberration criteria for several settings where the experimental units have more complex structures. These include block designs and split-plot designs in Section 7.2, nested block designs and blocked split-plot designs in Section 7.3, and row-column designs and strip-plot designs in Section 7.4.

7.2 Block Designs and Split-Plot Designs

Blocking is an effective method for improving the precision in comparative experiments. By dividing the experimental units into blocks so that those in the same block are more alike, better precision can be obtained by comparing the treatments within blocks.

7.2.1 Two models

Suppose there are t treatments and b blocks of size k. Let y_{ij} be the observation on the jth experimental unit in the ith block, $1 \le i \le b, 1 \le j \le k$. The following *fixed-effect* model is often assumed:

$$y_{ij} = \mu(\phi(i,j)) + \gamma_i + \epsilon_{ij}, \tag{7.1}$$

where $\phi(i,j)$ is the treatment applied to the jth unit in the ith block, $\mu(\phi(i,j))$ is the effect of that treatment, $\gamma_1, \cdots, \gamma_b$ are unknown constants representing effects of the blocks, and ϵ_{ij} are mutually uncorrelated random variables with

$$\mathrm{E}(\epsilon_{ij}) = 0 \text{ and } \mathrm{var}(\epsilon_{ij}) = \sigma^2. \tag{7.2}$$

Another model assumes that

$$y_{ij} = \mu(\phi(i,j)) + \delta_{ij}, \tag{7.3}$$

where

$$\mathrm{E}(\delta_{ij}) = 0, \ \mathrm{cov}(\delta_{ij}, \delta_{i'j'}) = \begin{cases} \sigma^2, & \text{if } i = i',\ j = j'; \\ \rho_1\sigma^2, & \text{if } i = i',\ j \neq j'; \\ \rho_2\sigma^2, & \text{if } i \neq i'. \end{cases} \tag{7.4}$$

That is, any two observations in the same block have the same correlation ρ_1 and those from different blocks also have the same correlation ρ_2. Since the purpose of blocking is to make the units in the same block more alike than those in different blocks, we expect to have

$$\rho_1 > \rho_2. \tag{7.5}$$

Note that the model in (7.3) and (7.4) can be justified by randomization theory; see Section 3.6 of Cheng (2014a) for details.

We point out that (7.3) and the patterns-of-covariance form (7.4) also hold under a model with *random* block effects. This can easily be shown as follows. Suppose that in (7.1), random block effects $\{\gamma_i\}$ and random errors $\{\epsilon_{ij}\}$ are uncorrelated random variables with

$$\mathrm{E}(\gamma_i) = \mathrm{E}(\epsilon_{ij}) = 0, \text{ and } \mathrm{var}(\gamma_i) = \sigma_B^2,\ \mathrm{var}(\epsilon_{ij}) = \sigma_E^2. \tag{7.6}$$

Let $\delta_{ij} = \gamma_i + \epsilon_{ij}$. Then (7.3) and (7.4) hold with

$$\sigma^2 = \sigma_B^2 + \sigma_E^2,\ \rho_1 = \frac{\sigma_B^2}{\sigma_B^2 + \sigma_E^2}, \text{ and } \rho_2 = 0. \tag{7.7}$$

Clearly, (7.5) also holds here. On the other hand, we see that (7.4) is more general as it allows nonzero, or even negative ρ_2.

We can write (7.1) and (7.2) in matrix form:

$$\boldsymbol{Y} = \boldsymbol{X}_T\boldsymbol{\mu} + \boldsymbol{X}_B\boldsymbol{\gamma} + \boldsymbol{\epsilon},\quad \mathrm{cov}(\boldsymbol{\epsilon}) = \sigma^2\boldsymbol{I}, \tag{7.8}$$

where $\boldsymbol{Y}$ is the vector of observations, $\boldsymbol{X}_T$ is the unit-treatment incidence matrix with the (i,j)th entry equal to 1 if the jth treatment is assigned to the ith unit, and is 0 otherwise, $\boldsymbol{X}_B$ is the unit-block incidence matrix with the (i,j)th entry equal to 1 if the ith unit belongs to the jth block, and is 0 otherwise, $\boldsymbol{\mu}$ is the vector of treatment effects, $\boldsymbol{\gamma}$ is the vector of block effects, and $\boldsymbol{\epsilon}$ is the vector of random errors, assumed to be uncorrelated with a zero mean and a constant variance σ^2.

For the model in (7.3) and (7.4), we note that there are three possible relations between any pair of units: they are the same unit, different units in the same block, or units in two different blocks. Let $\boldsymbol{R}_B$ be a square matrix such that the (i,j) entry is equal to 1 if the ith and jth units are in the same block, and is zero otherwise. Let $\boldsymbol{J}$ be a matrix of all ones. Then (7.3) and (7.4) can be expressed as

$$\boldsymbol{Y} = \boldsymbol{X}_T\boldsymbol{\mu} + \boldsymbol{\delta}, \tag{7.9}$$

and

$$\mathrm{cov}(\boldsymbol{\delta}) = \sigma^2\boldsymbol{I} + \rho_1\sigma^2(\boldsymbol{R}_B - \boldsymbol{I}) + \rho_2\sigma^2(\boldsymbol{J} - \boldsymbol{R}_B), \tag{7.10}$$

where $\boldsymbol{I}$, $\boldsymbol{R}_B - \boldsymbol{I}$, and $\boldsymbol{J} - \boldsymbol{R}_B$, all being $(0,1)$ matrices, describe the three aforementioned relations between any two units: the (i,j)th entry of $\boldsymbol{I}$, $\boldsymbol{R}_B - \boldsymbol{I}$, or $\boldsymbol{J} - \boldsymbol{R}_B$ is 1 if and only if unit i and unit j are the same, different units in the same block, or in different blocks, respectively.

7.2.2 Intra- and interblock estimators

We can rearrange the terms and write the matrix in (7.10) as

$$\mathrm{cov}(\boldsymbol{\delta}) = \xi_E\left(\boldsymbol{I} - \frac{\boldsymbol{R}_B}{k}\right) + \xi_B\left(\frac{\boldsymbol{R}_B}{k} - \frac{\boldsymbol{J}}{bk}\right) + \xi_U\frac{\boldsymbol{J}}{bk}, \tag{7.11}$$

where ξ_E, ξ_B, and ξ_U are certain constants. Specifically, by making (7.10) and (7.11) equal and setting up three equations according to the coefficients of $\boldsymbol{I}$, $\boldsymbol{R}_B$, and $\boldsymbol{J}$, we obtain that

$$\xi_E = (1-\rho_1)\sigma^2, \tag{7.12}$$

$$\xi_B = (1-\rho_1)\sigma^2 + k(\rho_1-\rho_2)\sigma^2, \tag{7.13}$$

$$\xi_U = (1-\rho_1)\sigma^2 + k(\rho_1-\rho_2)\sigma^2 + bk\rho_2\sigma^2. \tag{7.14}$$

What is important here is that (7.11) gives the *spectral decomposition* of $\text{cov}(\boldsymbol{\delta})$: ξ_E, ξ_B, and ξ_U are eigenvalues of $\text{cov}(\boldsymbol{\delta})$, and

$$\boldsymbol{I} - \frac{\boldsymbol{R}_B}{k}, \quad \frac{\boldsymbol{R}_B}{k} - \frac{\boldsymbol{J}}{bk}, \quad \text{and} \quad \frac{\boldsymbol{J}}{bk}$$

are orthogonal projection matrices onto three mutually orthogonal eigenspaces with dimensions $b(k-1), k-1$, and 1, respectively. The components of these projections, respectively, give deviations of individual observations from block means, deviations of block means from the overall mean, and the overall mean. The three eigenspaces, denoted by W_E, W_B, and W_U, respectively, are called *strata*. Specifically, we call W_E the *intrablock* stratum, W_B the *interblock* stratum, and W_U the mean stratum. If we denote the orthogonal projection matrices onto the three eigenspaces by

$$\boldsymbol{P}_{W_E} = \boldsymbol{I} - \frac{\boldsymbol{R}_B}{k}, \quad \boldsymbol{P}_{W_B} = \frac{\boldsymbol{R}_B}{k} - \frac{\boldsymbol{J}}{bk}, \quad \text{and } \boldsymbol{P}_{W_U} = \frac{\boldsymbol{J}}{bk},$$

respectively, then we have that

$$\text{cov}(\boldsymbol{\delta}) = \xi_U \boldsymbol{P}_{W_U} + \xi_B \boldsymbol{P}_{W_B} + \xi_E \boldsymbol{P}_{W_E}.$$

For each normalized eigenvector $\boldsymbol{c}$ in W_B (respectively, W_E), we have $\text{var}(\boldsymbol{c}^T\boldsymbol{Y}) = \xi_B$ (respectively, ξ_E). The two eigenvalues ξ_B and ξ_E (called, respectively, interblock and intrablock variances) measure, respectively, variability between blocks and that between units within the same block. From (7.12), (7.13), and (7.14), it is clear that

$$\xi_E \le \xi_B \le \xi_U. \tag{7.15}$$

Under the fixed-effect model in (7.1) and (7.2), in order to estimate the treatment effects, we need to project the data $\boldsymbol{Y}$ onto the orthogonal complement of the column space of W_B to eliminate the block effects. Effectively $\boldsymbol{Y}$ is projected onto the intrablock stratum W_E. All the normalized vectors in W_E have constant variance ξ_E, which now plays the same role as σ^2 in (7.2). Thus estimates of the treatment effects under the fixed-effect model in (7.1) and (7.2) are the same as those under the random-effect model in (7.1) and (7.6), but only using the information in the intrablock stratum. These estimators are called *intrablock* estimators. Under the model in (7.1) and (7.6) or that in (7.3) and (7.4), there is also information about the treatment effects in the interblock stratum. Estimators of treatment effects based on the projection of $\boldsymbol{Y}$ onto W_B are called *interblock* estimators. By (7.12) and (7.13), intrablock estimators have smaller variances than interblock estimators. It can be shown that under the model in (7.1) and (7.6) or that in (7.3) and (7.4), the best linear unbiased estimator of a treatment contrast is a linear combination of interblock and intrablock estimators (but the weights require knowledge of ξ_B and ξ_E). Under the fixed-effect model in (7.1) and (7.2), treatment effects are not estimable in the interblock stratum. We can think of this as the case where the interblock variance ξ_B is ∞.

7.2.3 Minimum G_2-aberration criterion for blocking – a Bayesian approach

The extension of the minimum aberration criterion to blocked fractional factorial designs was studied by Sitter, Chen, and Feder (1997), Chen and Cheng (1999), and Cheng and Wu (2002) for regular designs under the usual fixed-effect model. Some fundamental difficulties were encountered. To construct a regular 2^{m-k} fractional factorial design, one chooses a set of k independent treatment interactions, called treatment defining effects (or words). These treatment defining words are used to divide the 2^m treatment combinations into 2^k sets of size 2^{m-k} each, and then any of them is chosen for experimentation. As we have seen, observing only one such fraction causes the aliasing of some treatment factorial effects. The treatment defining words are properly chosen to minimize the aliasing of lower-order effects which are deemed more important. To divide the chosen fraction into blocks, one further chooses another set of independent treatment factorial effects, called block defining effects (or words), to divide the 2^{m-k} chosen treatment combinations into blocks. One consequence is that these block defining effects are "confounded" with interblock contrasts, and thus are not estimable under the fixed-effect model. To extend the classical minimum aberration criterion to the current setting, one can write down two wordlength patterns, one for the treatment defining words, and the other for the block defining words. One major problem is how to combine them into one single sequence. The difficulty lies in that these two types of defining words play different roles and cause very different consequences. How can one weigh the aliasing of important effects vs. their confounding with the blocks? The three papers cited above proposed different combined wordlength patterns based on different ad hoc considerations. It is desirable to have a unified approach with proper statistical justifications.

Here we present an application of the Bayesian approach discussed in Subsection 5.2.4. As in (5.11), we have

$$\boldsymbol{\mu} = \boldsymbol{P}\boldsymbol{\beta}, \tag{7.16}$$

where

$$\boldsymbol{P} = \begin{bmatrix} 1 & -1 \\ 1 & 1 \end{bmatrix} \otimes \cdots \otimes \begin{bmatrix} 1 & -1 \\ 1 & 1 \end{bmatrix}.$$

Now suppose that given $\boldsymbol{\beta}$, the distribution of the observations $\boldsymbol{Y}$ on the selected treatment combinations is normal. We have

$$\boldsymbol{Y}|\boldsymbol{\beta} \sim \mathrm{N}(\boldsymbol{U}\boldsymbol{\beta}, \xi_U \boldsymbol{P}_{W_U} + \xi_B \boldsymbol{P}_{W_B} + \xi_E \boldsymbol{P}_{W_E}), \tag{7.17}$$

where $\boldsymbol{U}$ is the $n \times 2^m$ submatrix of $\boldsymbol{P}$ consisting of the rows that correspond to the n observed treatment combinations given by a design $\boldsymbol{D}$. For convenience in the following derivations, we let

$$\boldsymbol{V} = \xi_U \boldsymbol{P}_{W_U} + \xi_B \boldsymbol{P}_{W_B} + \xi_E \boldsymbol{P}_{W_E}.$$

Recall that the prior distribution of $\boldsymbol{\beta}$ is assumed to be normal, stationary, and isotropic. To simplify the following derivations, we use $\boldsymbol{\Sigma}_{\boldsymbol{\beta}}$ to denote the prior covariance matrix $\mathrm{cov}(\boldsymbol{\beta})$ of $\boldsymbol{\beta}$. As in Section 5.2.4, the posterior distribution of $\boldsymbol{\beta}$ given $\boldsymbol{Y}$ is normal with

$$\mathrm{cov}(\boldsymbol{\beta}|\boldsymbol{Y}) = \left(\boldsymbol{U}^T \boldsymbol{V}^{-1} \boldsymbol{U} + \boldsymbol{\Sigma}_{\boldsymbol{\beta}}^{-1}\right)^{-1}.$$

Suppose we adopt the D-optimality criterion to minimize the determinant of the posterior covariance matrix, which gives rise to the Bayesian D-optimality criterion. We first note that

$$\det[\mathrm{cov}(\boldsymbol{\beta}|\boldsymbol{Y})^{-1}] = \det(\boldsymbol{\Sigma}_{\boldsymbol{\beta}})^{-1} \det\left(\boldsymbol{\Sigma}_{\boldsymbol{\beta}} \boldsymbol{U}^T \boldsymbol{V}^{-1} \boldsymbol{U} + \boldsymbol{I}_{2^m}\right).$$

Since $\det(\boldsymbol{\Sigma}_{\boldsymbol{\beta}})$ is a constant not depending on design $\boldsymbol{D}$, we have that minimizing $\det[\text{cov}(\boldsymbol{\beta}|\boldsymbol{Y})]$ is equivalent to maximizing $\det\left(\boldsymbol{\Sigma}_{\boldsymbol{\beta}}\boldsymbol{U}^T\boldsymbol{V}^{-1}\boldsymbol{U}+\boldsymbol{I}_{2^m}\right)$. As argued in Section 5.2.4, a good surrogate for the Bayesian D-optimality criterion is the Bayesian (M,S)-optimality criterion, which requires us to first maximize

$$\text{tr}\left(\boldsymbol{\Sigma}_{\boldsymbol{\beta}}\boldsymbol{U}^T\boldsymbol{V}^{-1}\boldsymbol{U}+\boldsymbol{I}_{2^m}\right)$$

and then minimize

$$\text{tr}\left(\boldsymbol{\Sigma}_{\boldsymbol{\beta}}\boldsymbol{U}^T\boldsymbol{V}^{-1}\boldsymbol{U}+\boldsymbol{I}_{2^m}\right)^2$$

among those that maximize $\text{tr}\left(\boldsymbol{\Sigma}_{\boldsymbol{\beta}}\boldsymbol{U}^T\boldsymbol{V}^{-1}\boldsymbol{U}+\boldsymbol{I}_{2^m}\right)$. Equivalently, we can first maximize $\text{tr}\left(\boldsymbol{\Sigma}_{\boldsymbol{\beta}}\boldsymbol{U}^T\boldsymbol{V}^{-1}\boldsymbol{U}\right)$ and then minimize $\text{tr}\left(\boldsymbol{\Sigma}_{\boldsymbol{\beta}}\boldsymbol{U}^T\boldsymbol{V}^{-1}\boldsymbol{U}\right)^2$ among those that maximize $\text{tr}\left(\boldsymbol{\Sigma}_{\boldsymbol{\beta}}\boldsymbol{U}^T\boldsymbol{V}^{-1}\boldsymbol{U}\right)$. The resulting designs are said to be Bayesian (M,S)-optimal.

For notational convenience, denote the three strata W_U, W_B, W_E, and the stratum variances ξ_U, ξ_B, ξ_E by W_0, W_1, W_2, and ξ_0, ξ_1, ξ_2, respectively. The prior variance of the factorial effect involving all the factors in a subset S of $\{1,\ldots,m\}$ is denoted by v_S, and the corresponding column of $\boldsymbol{U}$ is denoted by $\boldsymbol{u}_S$. Then we have

$$\xi_0 \geq \xi_1 \geq \xi_2, \tag{7.18}$$

and by (5.17), all the factorial effects involving k factors have the same prior variance v_k such that

$$v_k < v_{k'} \quad \text{if } k > k'. \tag{7.19}$$

Both $\text{tr}\left(\boldsymbol{\Sigma}_{\boldsymbol{\beta}}\boldsymbol{U}^T\boldsymbol{V}^{-1}\boldsymbol{U}\right)$ and $\text{tr}\left(\boldsymbol{\Sigma}_{\boldsymbol{\beta}}\boldsymbol{U}^T\boldsymbol{V}^{-1}\boldsymbol{U}\right)^2$ depend on the design $\boldsymbol{D}$, stratum variances, and prior variances of the factorial effects.

From $\boldsymbol{V}=\sum_{i=0}^{2}\xi_i\boldsymbol{P}_{W_i}$, we have that $\boldsymbol{V}^{-1}=\sum_{i=0}^{2}\xi_i^{-1}\boldsymbol{P}_{W_i}$. In addition, we have that $\boldsymbol{P}_{W_i}^2=\boldsymbol{P}_{W_i}$ because $\boldsymbol{P}_{W_i}$ is a projection matrix. We now obtain that

$$\begin{aligned}
\text{tr}\left(\boldsymbol{\Sigma}_{\boldsymbol{\beta}}\boldsymbol{U}^T\boldsymbol{V}^{-1}\boldsymbol{U}\right) &= \text{tr}\left(\sum_{i=0}^{2}\frac{1}{\xi_i}\boldsymbol{\Sigma}_{\boldsymbol{\beta}}\boldsymbol{U}^T\boldsymbol{P}_{W_i}\boldsymbol{U}\right)\\
&=\sum_{i=0}^{2}\frac{1}{\xi_i}\text{tr}\left[\boldsymbol{\Sigma}_{\boldsymbol{\beta}}^{1/2}(\boldsymbol{P}_{W_i}\boldsymbol{U}\boldsymbol{\Sigma}_{\boldsymbol{\beta}}^{1/2})^T(\boldsymbol{P}_{W_i}\boldsymbol{U})\right]\\
&=\sum_{i=0}^{2}\frac{1}{\xi_i}\text{tr}\left[(\boldsymbol{P}_{W_i}\boldsymbol{U}\boldsymbol{\Sigma}_{\boldsymbol{\beta}}^{1/2})^T(\boldsymbol{P}_{W_i}\boldsymbol{U}\boldsymbol{\Sigma}_{\boldsymbol{\beta}}^{1/2})\right].
\end{aligned}$$

Under the prior distribution, the factorial effects are independent and those involving k factors have prior variance v_k, $k=0,1,\ldots,m$. This means that $\boldsymbol{\Sigma}_{\boldsymbol{\beta}}$ is a diagonal matrix with diagonal entries given by v_k, $k=0,1,\ldots,m$. With this in mind, we obtain that

$$\sum_{i=0}^{2}\frac{1}{\xi_i}\text{tr}\left[(\boldsymbol{P}_{W_i}\boldsymbol{U}\boldsymbol{\Sigma}_{\boldsymbol{\beta}}^{1/2})^T(\boldsymbol{P}_{W_i}\boldsymbol{U}\boldsymbol{\Sigma}_{\boldsymbol{\beta}}^{1/2})\right]=\sum_{i=0}^{2}\frac{1}{\xi_i}\sum_{k=0}^{m}v_k\left(\sum_{S:|S|=k}\|\boldsymbol{P}_{W_i}\boldsymbol{u}_S\|^2\right).$$

Since $\sum_{i=0}^{2}\boldsymbol{P}_{W_i}=\boldsymbol{I}_n$, one can replace $\boldsymbol{P}_{W_2}$ with $\boldsymbol{I}_n-\sum_{i=0}^{1}\boldsymbol{P}_{W_i}$. Then

$$\begin{aligned}
&\sum_{i=0}^{2}\frac{1}{\xi_i}\sum_{k=0}^{m}v_k\left(\sum_{S:|S|=k}\|\boldsymbol{P}_{W_i}\boldsymbol{u}_S\|^2\right)\\
&=c-\sum_{i=0}^{1}\left(\frac{1}{\xi_2}-\frac{1}{\xi_i}\right)\sum_{k=0}^{m}v_k\left(\sum_{S:|S|=k}\|\boldsymbol{P}_{W_i}\boldsymbol{u}_S\|^2\right)\\
&=c-n\sum_{i=0}^{1}\left(\frac{1}{\xi_2}-\frac{1}{\xi_i}\right)\sum_{k=0}^{m}v_kA_{k,i}(\boldsymbol{D}),
\end{aligned}$$

where $c = \frac{1}{\xi_2}\sum_{k=0}^{m} v_k \sum_{S:|S|=k} \|\boldsymbol{u}_S\|^2$, which is a constant, and

$$A_{k,i}(\boldsymbol{D}) = \frac{1}{n}\left(\sum_{S:|S|=k} \|\boldsymbol{P}_{W_i}\boldsymbol{u}_S\|^2\right). \tag{7.20}$$

Furthermore, it is clear that $A_{0,0}(\boldsymbol{D}) = 1$ and $A_{0,i}(\boldsymbol{D}) = 0$ for all $i > 0$, because $\boldsymbol{u}_\phi$ is a column vector of all ones. Therefore maximizing $\mathrm{tr}(\sum_{i=0}^{2}\frac{1}{\xi_i}\boldsymbol{\Sigma}_\beta \boldsymbol{U}^T\boldsymbol{P}_{W_i}\boldsymbol{U})$, the M-step of Bayesian (M,S)-optimality, is equivalent to minimizing

$$\Phi_{\mathrm{M}}(\boldsymbol{D};\boldsymbol{\xi},\boldsymbol{v}) = \sum_{i=0}^{1}\left(\frac{1}{\xi_2}-\frac{1}{\xi_i}\right)\sum_{k=1}^{m} v_k A_{k,i}(\boldsymbol{D}), \tag{7.21}$$

where $\boldsymbol{\xi}$ and $\boldsymbol{v}$ are the vectors of ξ_i's and v_S's. For designs that minimize $\Phi_{\mathrm{M}}(\boldsymbol{D};\boldsymbol{\xi},\boldsymbol{v})$, minimizing $\mathrm{tr}[(\sum_{i=0}^{2}\frac{1}{\xi_i}\boldsymbol{\Sigma}_\beta\boldsymbol{U}^T\boldsymbol{P}_{W_i}\boldsymbol{U})^2]$ reduces to minimizing

$$\begin{aligned}\Phi_{\mathrm{S}}(\boldsymbol{D};\boldsymbol{\xi},\boldsymbol{v}) = &\sum_{i=0}^{2}\frac{1}{\xi_i^2}\mathrm{tr}\left[(\boldsymbol{\Sigma}_\beta\boldsymbol{U}^T\boldsymbol{P}_{W_i}\boldsymbol{U})^2\right]\\ &+2\sum_{0\le i<i'\le 2}\frac{1}{\xi_i\xi_{i'}}\mathrm{tr}\left[(\boldsymbol{\Sigma}_\beta\boldsymbol{U}^T\boldsymbol{P}_{W_i}\boldsymbol{U})(\boldsymbol{\Sigma}_\beta\boldsymbol{U}^T\boldsymbol{P}_{W_{i'}}\boldsymbol{U})\right].\end{aligned}$$

Thus a Bayesian (M,S)-optimal design $\boldsymbol{D}^*$ minimizes $\Phi_{\mathrm{M}}(\boldsymbol{D};\boldsymbol{\xi},\boldsymbol{v})$ and minimizes $\Phi_{\mathrm{S}}(\boldsymbol{D};\boldsymbol{\xi},\boldsymbol{v})$ among those that minimize $\Phi_{\mathrm{M}}(\boldsymbol{D};\boldsymbol{\xi},\boldsymbol{v})$.

Before proceeding further, we would like to comment on the expression of $\Phi_{\mathrm{M}}(\boldsymbol{D};\boldsymbol{\xi},\boldsymbol{v})$ in (7.21) to gain some insight. Here $\Phi_{\mathrm{M}}(\boldsymbol{D};\boldsymbol{\xi},\boldsymbol{v})$ is a sum of two terms. The first term involves $A_{k,0}(\boldsymbol{D})$'s and the second term involves $A_{k,1}(\boldsymbol{D})$'s. In the first term, as defined in (7.20), each $A_{k,0}(\boldsymbol{D})$ is the sum of the squared lengths of the orthogonal projections of the k-factor interaction columns onto the mean stratum. In fact, each $n\|\boldsymbol{P}_{W_0}\boldsymbol{u}_S\|^2$ is the square of a J-characteristic. Thus, $A_{k,0}(\boldsymbol{D})$ is precisely the quantity defined in (5.1) and is a component of the generalized wordlength pattern of the unblocked design. On the other hand, in the second term, each $A_{k,1}(\boldsymbol{D})$ is the sum of the squared lengths of the orthogonal projections of the k-factor interaction columns onto the interblock stratum. For regular designs, these quantities constitute what has been called blocking wordlength pattern in the literature. Here the Bayesian approach provides a coherent way to deal with the aforementioned difficulty of properly reconciling and combining two wordlength patterns that serve very different purposes.

Rewrite the expression of $\Phi_{\mathrm{M}}(\boldsymbol{D};\boldsymbol{\xi},\boldsymbol{v})$ in (7.21) as

$$\sum_{k=1}^{m} v_k\left[\sum_{i=0}^{1}\left(\frac{1}{\xi_2}-\frac{1}{\xi_i}\right)A_{k,i}(\boldsymbol{D})\right]$$

and notice that it is a linear function of v_k. If $v_i >> v_j$ for $i < j$ as the hierarchical assumption stipulates, then a good surrogate for minimizing $\Phi_{\mathrm{M}}(\boldsymbol{D};\boldsymbol{\xi},\boldsymbol{v})$ is to sequentially minimize the terms in

$$W_{\boldsymbol{\xi}}(\boldsymbol{D}) = \left(\sum_{i=0}^{1}\left(\frac{1}{\xi_2}-\frac{1}{\xi_i}\right)A_{1,i}(\boldsymbol{D}),\ldots,\sum_{i=0}^{1}\left(\frac{1}{\xi_2}-\frac{1}{\xi_i}\right)A_{m,i}(\boldsymbol{D})\right). \tag{7.22}$$

One major issue of applying the Bayesian (M,S) procedure and the minimum aberration type surrogate derived above is that the ξ_i's are unknown. We will now derive a technical result to help alleviate this difficulty of implementing the procedure in practical applications. Any given set of ξ_i values are said to be *feasible* if (7.18) holds. We have the following result on a necessary and sufficient condition for a blocked factorial design to sequentially minimize the terms in (7.22) for *all* feasible ξ_i's. It involves checking some conditions not involving ξ_i's on generalized blocking and treatment defining wordlength patterns.

Theorem 7.1. *A necessary and sufficient condition for a design to sequentially minimize the terms of $W_{\boldsymbol{\xi}}(\boldsymbol{D})$ in (7.22) for all feasible ξ_i's is that it sequentially minimizes the terms in*

$$W_0(\boldsymbol{D}) = (A_{1,0}(\boldsymbol{D}), \ldots, A_{m,0}(\boldsymbol{D})) \tag{7.23}$$

and

$$W_1(\boldsymbol{D}) = (A_{1,0}(\boldsymbol{D}) + A_{1,1}(\boldsymbol{D}), \ldots, A_{m,0}(\boldsymbol{D}) + A_{m,1}(\boldsymbol{D}))\,. \tag{7.24}$$

The terms $A_{k,0}$ in (7.23) are the same as A_k defined in (5.1). Thus the first sequence $W_0(\boldsymbol{D})$ is precisely the generalized wordlength pattern for choosing an unblocked fractional factorial design defined in Chapter 5. The sequence in (7.24) is the sum of $W_0(\boldsymbol{D})$ and the sequence $(A_{1,1}(\boldsymbol{D}), \ldots, A_{m,1}(\boldsymbol{D}))$. The latter, for regular designs, is the so-called block wordlength pattern used for choosing a blocking scheme. As mentioned earlier, different ways of combining treatment and block wordlength patterns have been proposed in the literature without proper justifications. Here the Bayesian approach leads to a simple sum.

Note that, up to constant multiples, $W_0(\boldsymbol{D})$ and $W_1(\boldsymbol{D})$ are equivalent to $W_{\boldsymbol{\xi}}(\boldsymbol{D})$ for the cases $\xi_1 = \xi_2$ and $\xi_0 = \xi_1$, respectively. Thus in applying Theorem 7.1, we do not need to know the ξ values; a small price to pay is to consider these two extreme cases. If a design has minimum aberration with respect to both $W_0(\boldsymbol{D})$ and $W_1(\boldsymbol{D})$, then it has minimum aberration with respect to $W_{\boldsymbol{\xi}}(\boldsymbol{D})$ for all feasible ξ's. Such a design may not exist, but we can use Theorem 7.1 to eliminate *inadmissible* designs in the sense that they are dominated by at least one other design. We may then compare the remaining *admissible* designs based on prior information about $\boldsymbol{\xi}$.

Since $W_0(\boldsymbol{D})$ and $W_1(\boldsymbol{D})$ are the same as $W_{\boldsymbol{\xi}}(\boldsymbol{D})$ for the cases $\xi_1 = \xi_2$ and $\xi_0 = \xi_1$, respectively, the condition in Theorem 7.1 is necessary. To show the sufficiency, suppose a design $\boldsymbol{D}_1$ sequentially minimizes the terms in both $W_0(\boldsymbol{D})$ and $W_1(\boldsymbol{D})$. Then we have

$$\begin{aligned}
&\sum_{i=0}^{1}\left(\frac{1}{\xi_2}-\frac{1}{\xi_i}\right)A_{k,i}(\boldsymbol{D})\\
&=\left(\frac{1}{\xi_2}-\frac{1}{\xi_1}\right)(A_{k,0}(\boldsymbol{D})+A_{k,1}(\boldsymbol{D}))+\left(\frac{1}{\xi_1}-\frac{1}{\xi_0}\right)A_{k,0}(\boldsymbol{D}).
\end{aligned}$$

For feasible $\boldsymbol{\xi}$'s, $\left(\frac{1}{\xi_2}-\frac{1}{\xi_1}\right) \geq 0$ and $\left(\frac{1}{\xi_1}-\frac{1}{\xi_0}\right) \geq 0$. It follows that a design sequentially minimizes the terms in $W_{\boldsymbol{\xi}}(\boldsymbol{D})$ for all feasible ξ_i's if and only if it sequentially minimizes those in $W_0(\boldsymbol{D})$ and $W_1(\boldsymbol{D})$.

Remark 7.1. The first extreme case $\xi_1 = \xi_2$ corresponds to that of unstructured units. In the second case, we have $\xi_2 < \xi_1 = \xi_0$. Let the common value of ξ_1 and ξ_0 be ξ. Since the result does not depend on the value of ξ, by making ξ equal to ∞, we see that it is the same as to consider the setting of block designs with *fixed*, instead of random block effects. These two cases lead to, respectively, the wordlength patterns $W_0(\boldsymbol{D})$ and $W_1(\boldsymbol{D})$.

As an illustration of the results obtained in this section, we present an example reported in Chang and Cheng (2018). Earlier, Chen and Cheng (1999) and Cheng and Wu (2002) introduced two different wordlength patterns, hereafter denoted as $W_{CC}(\boldsymbol{D})$ and $W_{CW}(\boldsymbol{D})$, respectively, for choosing blocked *regular* fractional factorial designs. Cheng, Li, and Ye (2004) applied these criteria directly to nonregular designs by replacing the treatment and block defining wordlength counts, $A_{k,0}$ and $A_{k,1}$, with their general forms given in (5.1) and (7.20), respectively. Even for regular designs, these criteria were not derived with a rigorous theory. Furthermore, although the consequences of treatment and block defining words on effect aliasing and confounding are well understood under regular designs, for nonregular designs the two criteria are difficult to interpret and justify due to complex aliasing and effect nonorthogonality.

Cheng, Li, and Ye (2004) considered 16-run blocked factorial designs with five two-level treatment factors and four blocks of size four, and found minimum aberration designs $\boldsymbol{D}_1$ and $\boldsymbol{D}_2$ for $W_{CW}(\boldsymbol{D})$ and $W_{CC}(\boldsymbol{D})$, respectively. Via a computer search and using Theorem 7.1, Chang and Cheng (2018) concluded that for any feasible $\boldsymbol{\xi}$, one of $\boldsymbol{D}_1$ and $\boldsymbol{D}_2$ has minimum aberration with respect to $W_{\boldsymbol{\xi}}(\boldsymbol{D})$. We have

$$(A_{3,0}(\boldsymbol{D}_1), A_{4,0}(\boldsymbol{D}_1), A_{5,0}(\boldsymbol{D}_1)) = (0,0,1),$$

$$(A_{2,0}(\boldsymbol{D}_1) + A_{2,1}(\boldsymbol{D}_1), \ldots, A_{5,0}(\boldsymbol{D}_1) + A_{5,1}(\boldsymbol{D}_1)) = (3,3,0,1),$$

$$(A_{3,0}(\boldsymbol{D}_2), A_{4,0}(\boldsymbol{D}_2), A_{5,0}(\boldsymbol{D}_2)) = (0,1,0)$$

and

$$(A_{2,0}(\boldsymbol{D}_2) + A_{2,1}(\boldsymbol{D}_2), \ldots, A_{5,0}(\boldsymbol{D}_2) + A_{5,1}(\boldsymbol{D}_2)) = (2,4,1,0).$$

By comparing these wordlength patterns of $\boldsymbol{D}_1$ and $\boldsymbol{D}_2$, we expect $\boldsymbol{D}_1$ to be optimal for the case $\xi_1 = \xi_2$; that is, when the units are unblocked. By continuity, $\boldsymbol{D}_1$ is expected to be optimal when the interblock variance is not too larger than the intrablock variance. In contrast, $\boldsymbol{D}_2$ is optimal for the case $\xi_0 = \xi_1 = \infty$, i.e., the case of fixed block effects. By continuity, it is expected to be also optimal when the interblock variance is sufficiently larger than the intrablock variance.

Define the D-efficiency of $\boldsymbol{D}_2$ relative to $\boldsymbol{D}_1$ as d_1/d_2, where d_i is the value of $[\det(\text{cov}(\boldsymbol{\beta}|\boldsymbol{Y}))]^{1/2^5}$ under $\boldsymbol{D}_i$. Such relative efficiencies for $\xi_0 = \infty, \xi_2 = 8$, and $v_k = \tau^2 r^k$, where $\tau^2 = \sigma^2(1+r)^{-5}$ (see (5.17)) with $\sigma^2 = 20, 0 < r < 1$, are shown in Figure 7.1 (taken from Chang and Cheng (2018)) in four panels for the cases

(a) $\xi_2/\xi_1 = 0$, (b) $\xi_2/\xi_1 = 0.01$, (c) $\xi_2/\xi_1 = 0.1$, and (d) $\xi_2/\xi_1 = 1$.

The conclusion drawn in the previous paragraph can be confirmed by this figure. Panel (d) shows that $\boldsymbol{D}_1$ dominates $\boldsymbol{D}_2$ completely (for all possible values of r) when $\xi_1 = \xi_2$. For the case that ξ_1/ξ_2 is greater than 1, $\boldsymbol{D}_2$ is better for some small values of r, in which case the lower-order treatment factorial effects are much more important than the higher-order ones, and as this ratio increases, the range where $\boldsymbol{D}_2$ is better gets larger. Eventually for infinite ξ_1 (panel (a)), $\boldsymbol{D}_2$ dominates $\boldsymbol{D}_1$ completely. This suggests that $\boldsymbol{D}_2$ be used under fixed block effects.

7.2.4 Split-plot designs

Consider a factorial experiment in which only a certain number of treatment combinations can be observed on each day. As a result, the experiment may have to be carried out over a sequence of days. If there are day-to-day variations out of the experimenter's control, then the observations on the same day constitute a block, and we end up with a blocked experiment. Nominally what was developed earlier in this chapter can be applied. However, it may happen that there are certain factors whose levels are difficult to change. Then out of practical consideration, these factors must be set at the same level throughout the day. The main effects of these factors then are confounded with blocks. In other words, they must be

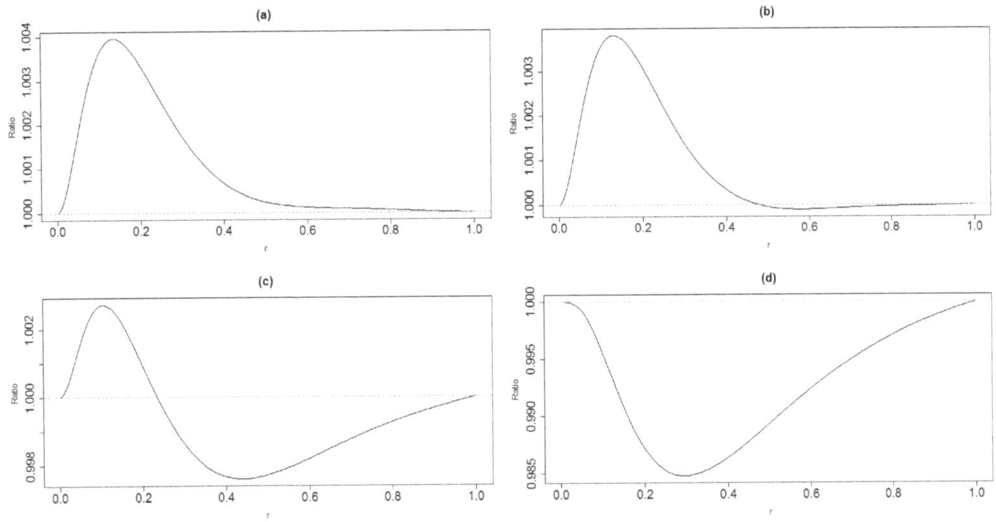

FIGURE 7.1
D-efficiencies of $\boldsymbol{D}_2$ relative to $\boldsymbol{D}_1$ for $0 < r < 1, \xi_0 = \infty, \xi_2 = 8$; (a) $\xi_2/\xi_1 = 0$; (b) $\xi_2/\xi_1 = 0.01$; (c) $\xi_2/\xi_1 = 0.1$; (d) $\xi_2/\xi_1 = 1$

estimated by less precise interblock contrasts. Under models with fixed block effects, they are even not estimable. In this context, the days are called *whole plots*, each observation is taken on a *subplot*, and such experiments are called *split-plot* experiments. As with many terms in experimental design, this terminology came from agricultural applications. The treatment factors that must take the same level on all the subplots in the same wholeplot are called wholeplot factors, and the other treatment factors are called subplot factors.

For regular designs, Huang, Chen, and Voelkel (1998), Bingham and Sitter (1999a) and Bingham and Sitter (1999b) used the usual minimum aberration criterion for selecting unblocked factorials to rank fractional factorial split-plot designs, ignoring the effects of interblock and intrablock errors on the precision of the estimators of factorial effects. This approach often cannot distinguish some different split-plot designs. Bingham and Sitter (2001) and Mukerjee and Fang (2002) proposed additional secondary criteria to break the ties. Mathematically, split-plot designs are a special kind of blocked factorial designs with a constraint on the assignment of treatments. Blocked factorial experiments and split-plot experiments have the same structures of the treatments and experimental units, and thus the same statistical model and error structure. The results developed earlier in this section, including the generalized word length patterns and Theorem 7.1, can be applied directly to split-plot designs subject to the special constraint on treatment assignment.

7.3 Nested Block Designs and Blocked Split-Plot Designs

In this section, we consider a block structure that is a bit more complicated than the one studied in the previous section by adding one layer of blocking. Simply put, we have the blocks grouped into superblocks. For example, suppose one can only perform two experimental runs on each day. There are a total of 32 experimental runs to be performed on 16 days, with four days per week and two runs per day. Then the 32 runs can be thought to have two layers of blocking (or nested blocking) with a structure which we will denote as (4 superblocks)/(4 blocks)/(2 units), or (4 blocks)/(4 whole plots)/(2 subplots) in *blocked split-plot* experiments. In the rest of this section, we shall present in the context of blocked split-plot designs. Again, nested block designs and blocked split-plot designs have the same model, error structure, and generalized G_2-aberration criterion. The only difference is that there are additional constraints on the treatment assignment under blocked split-plot designs.

Suppose there are m two-level treatment factors and bsw subplots which are split into b blocks, each block contains w whole plots, and each whole plot contains s subplots. Let y_{ijk} be the observation on the kth subplot of the jth whole plot in the ith block. Suppose

$$y_{ijk} = \mu(\phi(i,j,k)) + \gamma_i^b + \gamma_{ij}^w + \epsilon_{ijk}, \tag{7.25}$$

where $\phi(i,j,k)$ is the treatment combination applied to the kth subplot of the jth whole plot in the ith block, $\mu(\phi(i,j,k))$ is the effect of that treatment, $\{\gamma_i^b\}$, $\{\gamma_{ij}^w\}$, and $\{\epsilon_{ijk}\}$ are uncorrelated random variables with

$$\mathrm{E}(\gamma_i^b) = \mathrm{E}(\gamma_{ij}^w) = \mathrm{E}(\epsilon_{ijk}) = 0, \mathrm{var}(\gamma_i^b) = \sigma_B^2, \mathrm{var}(\gamma_{ij}^w) = \sigma_W^2, \mathrm{var}(\epsilon_{ijk}) = \sigma_E^2. \tag{7.26}$$

Another model assumes that

$$y_{ijk} = \mu(\phi(i,j,k)) + \delta_{ijk}, \tag{7.27}$$

where

$$\mathrm{E}(\delta_{ijk}) = 0, \ \mathrm{cov}(\delta_{ijk}, \delta_{i'j'k'}) = \begin{cases} \sigma^2, & \text{if } i = i', \ j = j', k = k'; \\ \rho_1\sigma^2, & \text{if } i = i', j = j', k \neq k'; \\ \rho_2\sigma^2, & \text{if } i = i', j \neq j; \\ \rho_3\sigma^2, & \text{if } i \neq i'. \end{cases} \tag{7.28}$$

As in the case of block designs, model (7.27)-(7.28) can be justified by randomization; see Section 3.6 of Cheng (2014a). For model (7.25)-(7.26), (7.27), and (7.28) also hold with $\rho_1 \geq \rho_2 \geq \rho_3$.

Following the same argument as in the derivation of (7.9), (7.10), and (7.11), we can rewrite (7.27) and (7.28) as

$$\boldsymbol{Y} = \boldsymbol{X}_T\boldsymbol{\mu} + \boldsymbol{\delta}, \tag{7.29}$$

and

$$\mathrm{cov}(\boldsymbol{\delta}) = \sigma^2\boldsymbol{I} + \rho_1\sigma^2(\boldsymbol{R}_W - \boldsymbol{I}) + \rho_2\sigma^2(\boldsymbol{R}_B - \boldsymbol{R}_W) + \rho_3\sigma^2(\boldsymbol{J} - \boldsymbol{R}_B), \tag{7.30}$$

where $\boldsymbol{I}$, $\boldsymbol{R}_W - \boldsymbol{I}$, $\boldsymbol{R}_B - \boldsymbol{R}_W$, and $\boldsymbol{J} - \boldsymbol{R}_B$ describe four relations between any two units (subplots): the (i,j)th entry of $\boldsymbol{I}$, $\boldsymbol{R}_W - \boldsymbol{I}$, $\boldsymbol{R}_B - \boldsymbol{R}_W$, and $\boldsymbol{J} - \boldsymbol{R}_B$ is 1 if subplot i and subplot j are the same, different subplots in the same whole plot, in different whole plots of the same block, or in different blocks, respectively.

Again, we can rearrange terms and write the covariance matrix in (7.30) in the following spectral form:

$$\mathrm{cov}(\boldsymbol{\delta}) = \xi_E\left(\boldsymbol{I} - \frac{\boldsymbol{R}_W}{s}\right) + \xi_W\left(\frac{\boldsymbol{R}_W}{s} - \frac{\boldsymbol{R}_B}{ws}\right) + \xi_B\left(\frac{\boldsymbol{R}_B}{ws} - \frac{\boldsymbol{J}}{bws}\right) + \xi_U\frac{\boldsymbol{J}}{bws}, \tag{7.31}$$

where ξ_E, ξ_W, ξ_B, and ξ_U are eigenvalues of $\mathrm{cov}(\boldsymbol{\delta})$. The projection matrices

$$\boldsymbol{I} - \frac{\boldsymbol{R}_W}{s}, \frac{\boldsymbol{R}_W}{s} - \frac{\boldsymbol{R}_B}{ws}, \text{ and } \frac{\boldsymbol{R}_B}{ws} - \frac{\boldsymbol{J}}{bws}$$

onto the first three strata produce, respectively, deviations from the whole plot means, deviations of whole plot means from block means, and deviations of block means from the overall mean. The corresponding eigenvalues ξ_E, ξ_W, and ξ_B measure respectively, between subplot variability within the same whole plot, between whole-plot variability within the same block, and between-block variability, respectively. Furthermore, we have

$$\xi_E \leq \xi_W \leq \xi_B \leq \xi_U.$$

As before, for simplicity, we write ξ_E, ξ_W, ξ_B, ξ_U as ξ_3, ξ_2, ξ_1, ξ_0, respectively. Then

$$\xi_3 \leq \xi_2 \leq \xi_1 \leq \xi_0. \tag{7.32}$$

Also, denote the corresponding eigenspaces as W_3, W_2, W_1, and W_0, respectively. Repeating the same argument as in Subsection 7.2.3, we can show that the M-step of the Bayesian (M,S)-optimality is to minimize

$$\Phi_{\mathrm{M}}(\boldsymbol{D};\boldsymbol{\xi},\boldsymbol{v}) = \sum_{k=1}^{m} v_k \left[\sum_{i=0}^{2}\left(\frac{1}{\xi_3}-\frac{1}{\xi_i}\right) A_{k,i}(\boldsymbol{D})\right],$$

where, as in (7.20),

$$A_{k,i}(\boldsymbol{D}) = \frac{1}{n}\left(\sum_{S:|S|=k} \|\boldsymbol{P}_{W_i}\boldsymbol{u}_S\|^2\right).$$

If $v_i >> v_j$ for $i < j$, then a good surrogate for minimizing $\Phi_{\mathrm{M}}(\boldsymbol{D};\boldsymbol{\xi},\boldsymbol{v})$ is to sequentially minimize the terms in

$$W_{\boldsymbol{\xi}}(\boldsymbol{D}) = \left(\sum_{i=0}^{2}\left(\frac{1}{\xi_3}-\frac{1}{\xi_i}\right) A_{1,i}(\boldsymbol{D}),\ldots,\sum_{i=0}^{2}\left(\frac{1}{\xi_3}-\frac{1}{\xi_i}\right) A_{m,i}(\boldsymbol{D})\right). \tag{7.33}$$

To verify this, similar to Theorem 7.1, it is enough to consider several special cases.

Theorem 7.2. *For nested block designs (and blocked split-plot designs), a necessary and sufficient condition for a design to sequentially minimize the terms in $W_{\boldsymbol{\xi}}(\boldsymbol{D})$ for all feasible ξ_i's as required by (7.32) is that it sequentially minimizes the terms in*

$$W_0(\boldsymbol{D}) = (A_{1,0}(\boldsymbol{D}),\ldots,A_{m,0}(\boldsymbol{D})),$$

$$W_1(\boldsymbol{D}) = (A_{1,0}(\boldsymbol{D})+A_{1,1}(\boldsymbol{D}),\ldots,A_{m,0}(\boldsymbol{D})+A_{m,1}(\boldsymbol{D})),$$

and

$$W_2(\boldsymbol{D}) = (A_{1,0}(\boldsymbol{D})+A_{1,1}(\boldsymbol{D})+A_{1,2}(\boldsymbol{D}),\ldots,A_{m,0}(\boldsymbol{D})+A_{m,1}(\boldsymbol{D})+A_{m,2}(\boldsymbol{D})).$$

Remark 7.2. Note that, up to constant multiples, $W_0(\boldsymbol{D}), W_1(\boldsymbol{D})$, and $W_2(\boldsymbol{D})$ are equivalent to $W_{\boldsymbol{\xi}}(\boldsymbol{D})$ with $\xi_3 = \xi_2 = \xi_1$, $\xi_3 = \xi_2 < \xi_1 = \infty$, and $\xi_2 = \xi_1 = \infty$, respectively. These correspond to, respectively, the settings of *bws* unstructured subplots, block designs with the *b* blocks of *ws* subplots as blocks, and block designs with the *bw* whole plots of *s* subplots as blocks, where, in the last two settings the blocks have fixed, rather than random effects.

7.4 Row-Column Designs and Strip-Plot Designs

Suppose 16 experimental runs are to be performed over four days with four runs per day. This can be considered as a blocked experiment with four blocks of size four to eliminate day-to-day variations. Now suppose the four experimental runs on each day are to be performed over four different time periods; then perhaps time-to-time variations should also be taken into account. Then one would consider using a *row-column design*, with the experiments conducted according to a four rows by four columns layout, where the rows correspond to days and the columns correspond to time periods. Under a block design, we randomize the treatment assignments separately on each day, while under a row-column design, we simultaneously randomize the rows and columns in the design layout.

Suppose there are m two-level treatment factors and rc experimental units that are arranged into r rows and c columns. Let y_{ij} be the observation on the unit in the ith row and jth column. Suppose

$$y_{ij} = \mu(\phi(i,j)) + \gamma_i^R + \gamma_j^C + \epsilon_{ij}, \tag{7.34}$$

where $\phi(i,j)$ is the treatment combination applied to the ith row and jth column, $\mu(\phi(i,j))$ is the effect of that treatment, $\{\gamma_i^R\}$, $\{\gamma_j^C\}$, and $\{\epsilon_{ij}\}$ are uncorrelated random variables with

$$\mathrm{E}(\gamma_i^R) = \mathrm{E}(\gamma_j^C) = \mathrm{E}(\epsilon_{ij}) = 0, \mathrm{var}(\gamma_i^R) = \sigma_R^2, \mathrm{var}(\gamma_j^C) = \sigma_C^2, \mathrm{var}(\epsilon_{ij}) = \sigma_E^2. \tag{7.35}$$

Another model assumes that

$$y_{ij} = \mu(\phi(ij)) + \delta_{ij}, \tag{7.36}$$

where

$$\mathrm{E}(\delta_{ij}) = 0,\ \mathrm{cov}(\delta_{ij}, \delta_{i'j'}) = \begin{cases} \sigma^2, & \text{if } i = i',\ j = j'; \\ \rho_1\sigma^2, & \text{if } i = i', j \neq j'; \\ \rho_2\sigma^2, & \text{if } i \neq i', j = j'; \\ \rho_3\sigma^2, & \text{if } i \neq i', j \neq j'. \end{cases} \tag{7.37}$$

Model (7.36)-(7.37)) can be justified by randomization; see Section 3.6 of Cheng (2014a). For model (7.34)-(7.35), (7.36) and (7.37) also hold with $\rho_1 > \rho_3$ and $\rho_2 > \rho_3$.

Rewrite (7.36) and (7.37) as

$$\boldsymbol{Y} = \boldsymbol{X}_T\boldsymbol{\mu} + \boldsymbol{\delta}, \tag{7.38}$$

and

$$\mathrm{cov}(\boldsymbol{\delta}) = \sigma^2\boldsymbol{I} + \rho_1\sigma^2(\boldsymbol{R}_R - \boldsymbol{I}) + \rho_2\sigma^2(\boldsymbol{R}_C - \boldsymbol{I}) + \rho_3\sigma^2(\boldsymbol{J} - \boldsymbol{R}_R - \boldsymbol{R}_C + \boldsymbol{I}), \tag{7.39}$$

where $\boldsymbol{I}$, $\boldsymbol{R}_R - \boldsymbol{I}$, $\boldsymbol{R}_C - \boldsymbol{I}$, and $\boldsymbol{J} - \boldsymbol{R}_R - \boldsymbol{R}_C + \boldsymbol{I}$ describe four relations between any two units: the (i,j)th entry of $\boldsymbol{I}$, $\boldsymbol{R}_R - \boldsymbol{I}$, $\boldsymbol{R}_C - \boldsymbol{I}$, and $\boldsymbol{J} - \boldsymbol{R}_R - \boldsymbol{R}_C + \boldsymbol{I}$ is 1 if unit i and unit j are the same, in the same row but different columns, in the same column but different rows, or in different rows and different columns, respectively.

As before, we can rearrange terms and write the covariance matrix in (7.39) in the following spectral form:

$$\mathrm{cov}(\boldsymbol{\delta}) = \xi_E\left(\boldsymbol{I} - \frac{\boldsymbol{R}_R}{c} - \frac{\boldsymbol{R}_C}{r} + \frac{\boldsymbol{J}}{rc}\right) + \xi_R\left(\frac{\boldsymbol{R}_R}{c} - \frac{\boldsymbol{J}}{rc}\right) + \xi_C\left(\frac{\boldsymbol{R}_C}{r} - \frac{\boldsymbol{J}}{rc}\right) + \xi_U\frac{\boldsymbol{J}}{rc},$$

where ξ_E, ξ_R, ξ_C, and ξ_U are eigenvalues of $\text{cov}(\boldsymbol{\delta})$ with

$$\xi_U \geq \xi_R, \xi_U \geq \xi_C, \xi_R \geq \xi_E, \xi_C \geq \xi_E.$$

We write ξ_E, ξ_R, ξ_C, ξ_U as ξ_3, ξ_2, ξ_1, ξ_0, respectively. Then

$$\xi_0 \geq \xi_1, \xi_0 \geq \xi_2, \xi_1 \geq \xi_3, \xi_2 \geq \xi_3. \tag{7.40}$$

Denote the corresponding eigenspaces as W_3, W_2, W_1, and W_0, respectively. Following the same argument as in Subsection 7.2.3, we can show that the M-step of the Bayesian (M,S)-optimality is to minimize

$$\Phi_{\text{M}}(\boldsymbol{D}; \boldsymbol{\xi}, \boldsymbol{v}) = \sum_{k=1}^{m} v_k \left[\sum_{i=0}^{2} \left(\frac{1}{\xi_3} - \frac{1}{\xi_i} \right) A_{k,i}(\boldsymbol{D}) \right],$$

where $A_{k,i}(\boldsymbol{D}) = \frac{1}{n}\left(\sum_{S:|S|=k} \|\boldsymbol{P}_{W_i}\boldsymbol{u}_S\|^2\right)$. If $v_i >> v_j$ for $i < j$, then a good surrogate for minimizing $\Phi_{\text{M}}(\boldsymbol{D}; \boldsymbol{\xi}, \boldsymbol{v})$ is to sequentially minimize the terms in

$$W_{\boldsymbol{\xi}}(\boldsymbol{D}) = \left(\sum_{i=0}^{2} \left(\frac{1}{\xi_3} - \frac{1}{\xi_i} \right) A_{1,i}(\boldsymbol{D}), \ldots, \sum_{i=0}^{2} \left(\frac{1}{\xi_3} - \frac{1}{\xi_i} \right) A_{m,i}(\boldsymbol{D}) \right). \tag{7.41}$$

To search for minimum aberration designs with respect to (7.41), we can apply the following theorem.

Theorem 7.3. *A necessary and sufficient condition for a row-column design to sequentially minimize the terms in $W_{\boldsymbol{\xi}}(\boldsymbol{D})$ for all feasible ξ_i's as required by (7.40) is that it sequentially minimizes the terms in*

$$W_0(\boldsymbol{D}) = (A_{1,0}(\boldsymbol{D}), \ldots, A_{m,0}(\boldsymbol{D})),$$

$$W_1(\boldsymbol{D}) = (A_{1,0}(\boldsymbol{D}) + A_{1,1}(\boldsymbol{D}), \ldots, A_{m,0}(\boldsymbol{D}) + A_{m,1}(\boldsymbol{D})),$$

and

$$W_2(\boldsymbol{D}) = (A_{1,0}(\boldsymbol{D}) + A_{1,2}(\boldsymbol{D}), \ldots, A_{m,0}(\boldsymbol{D}) + A_{m,2}(\boldsymbol{D})).$$

Under split-plot designs, some treatment factors must take the same level on all the subplots in a whole plot. Likewise, in the row-column design setting, it may happen that certain treatment factors must take the same level on all the units in the same row, and the other treatment factors must take the same level on all the units in the same column. Then we have the so-called *strip-plot designs*. Strip-plot designs and row-column designs have the same model and error structure. The only difference is that the treatment assignment under a strip-plot design is subject to an additional constraint. Therefore the minimum G_2-aberration criterion derived in this section and Theorem 7.3 can be applied to strip-plot designs subject to the constraint that some treatment factors must take the same level on all rows or all columns.

Remark 7.3. As in Remarks 7.1 and 7.2, the three wordlength patterns $W_0(\boldsymbol{D})$, $W_1(\boldsymbol{D})$, and $W_2(\boldsymbol{D})$ are associated with, respectively, the settings of rc unstructured units, block designs with the r rows of c units as blocks, and block designs with the c columns of r units as blocks, where all the block designs have fixed block effects.

Example 7.1. Miller (1997) considered an experiment for studying the effects of six factors of washer configurations and four factors of dryer configurations, using four washers and four dryers. At first four cloth samples are washed in each washer. Then the 16 cloth samples are divided into four groups with each group containing one sample from each washer. Each of these groups is then assigned to one dryer. If we identify the four washers with four rows and the four dryers with four columns, then we have a 4×4 row-column layout. All the observations in the same row are taken on samples that are washed in the same washer, so they must be subject to the same washer configurations. That is, each washer factor must take the same level on all the observations in the same row; thus all the main effects and interactions of washer factors must be estimated in the row stratum. Similarly, all the observations in the same column must have the same dryer configurations; so the factorial effects of the dryer factors must be estimated in the column stratum. This would result in lower precision for the estimates of some effects of interest, which is a price to pay for convenience and perhaps lower cost in conducting the experiment. Then we have a 2^{10-6} fractional factorial experiment conducted under a strip-plot design. The minimum G_2-aberration criterion derived in this section and Theorem 7.3 can be applied. This can also be considered as an experiment that is processed in two stages: washing stage and drying stage. In fact, Miller (1997) considered an experiment with 32 runs conducted in two days. Then it is an example of *blocked strip-plot designs*, with an additional block stratum at the top. Mee and Bates (1998) discussed design of experiments with multiple processing stages in the fabrication of integrated circuits.

In this chapter, we have seen how the minimum G_2-aberration criterion can be extended to several cases where the experimental units have certain structures, particularly when they are partitioned into blocks, or are arranged in rows and columns. That derivations of the results for the cases considered in this chapter follow the same line indicates that there is a more general underlying theory for experimental units with even more complicated structures. This is indeed the case and a general theory will be presented in Chapter 11.

8

Supersaturated Designs

In this chapter, we give a systematic treatment of supersaturated designs, which are a special class of nonregular designs allowing the study of n or more two-level factors using only n experimental runs. As orthogonality is an impossibility for supersaturated designs, efforts are directed, quite naturally, toward finding designs that minimize certain measures of nonorthogonality. Two most useful measures of nonorthogonality are the $E(s^2)$ and $\max_{i<j} |s_{ij}|$ criteria, where the first aims at minimizing the average of squared correlations and the second the maximum correlation. After an introductory section, $E(s^2)$-optimal supersaturated designs take the center stage in this chapter and we allocate two sections to them. We then present some results on the $\max_{i<j} |s_{ij}|$ criterion; consideration of this criterion is practically important as it can better control the correlations than the $E(s^2)$ criterion. The rest of the chapter is devoted to the discussion of other design criteria and related topics, including MDS-resolution, estimation capacity, D_f-criterion, $UE(s^2)$-optimality, robust supersaturated designs, group screening designs, etc.

8.1 Introduction

The problem considered here is an ambitious one: we want to use n experimental runs to study n or more factors. To be successful, a strong assumption of effect sparsity is indispensable. The scenario generally agreed upon is that all interactions are negligible and all main effects except for a few are also negligible. A simple example is helpful for understanding the logic behind using supersaturated designs. If 100 balls have an identical appearance but only one of them has a different weight, we are able to identify this special ball with a smaller number of weighings than 100.

This supersaturated situation rose to importance because of the paper by Satterthwaite (1959), who advocated the use of random balance designs. The approach was heavily criticized by several famed contemporaries in the accompanying discussions of the paper. The sharpest criticisms came from George Box and Oscar Kempthorne. Even though Box called Satterthwaite's paper propaganda, he commended Satterthwaite for pointing out the importance of the supersaturated situation. In his critique, Box made a constructive suggestion that systematic supersaturated designs ought to be sought by minimizing the correlations between design columns, and he even gave a small example to illustrate his idea.

Following Box's suggestion, Booth and Cox (1962) generated seven systematic supersaturated designs using an electronic computer, and did so under a specified control of the maximum allowable correlation. They showed the superior performance of their designs in comparison with random balance designs using $E(s^2)$, the average of squared correlations, which was justified by its approximate relationship with an A-efficiency criterion.

DOI: 10.1201/9781003371861-8

This fundamental paper of Booth and Cox's proposed two criteria for evaluating supersaturated designs, which we now formally introduce. Let $\boldsymbol{D} = (\boldsymbol{d}_1, \ldots, \boldsymbol{d}_m)$ be a two-level supersaturated design of n runs for m factors, where $m \geq n$ and $\boldsymbol{d}_j$ is the jth design column having entries ± 1. We assume that n is even and all design columns are balanced in that each $\boldsymbol{d}_j$ has the same number of ± 1. Because $m \geq n$, $\boldsymbol{D}$ cannot be made orthogonal. In this case a natural course of action is to find supersaturated designs that are as orthogonal as possible, which can be done by minimizing some measure of nonorthogonality. Let $s_{ij} = \boldsymbol{d}_i^T \boldsymbol{d}_j$ and therefore s_{ij}/n is the correlation of columns $\boldsymbol{d}_i$ and $\boldsymbol{d}_j$. Then the $E(s^2)$ criterion requires the minimization of

$$E(s^2) = \frac{1}{\binom{m}{2}} \sum_{i<j} s_{ij}^2,$$

the average of the squared correlations up to a constant, when seeking good supersaturated designs. A more stringent way of controlling the correlations is to minimize $\max_{i<j} |s_{ij}|$. If more than one design has the same minimum $\max_{i<j} |s_{ij}|$ value, we then go on to minimize the frequency that this minimum value is reached.

The theory of $E(s^2)$-optimal supersaturated designs has been richly developed and, as a result, we commit Sections 8.2 and 8.3 to the presentation of major results on this topic. The $\max_{i<j} |s_{ij}|$ criterion can be used alone but is often used as a secondary criterion after the $E(s^2)$; Section 8.4 gives an account of relevant results. Section 8.5 discusses other considerations of supersaturated designs, including MDS-resolution, estimation capacity, D_f-criterion, $UE(s^2)$-optimality, and robust supersaturated designs. We conclude this chapter with group screening designs and their generalizations in Section 8.6.

Remark 8.1. *It is worth noting that the $E(s^2)$ and $\max_{i<j} |s_{ij}|$ criteria are closely related to the minimum G_2-aberration and G-aberration criteria. For balanced supersaturated designs, the minimum G_2-aberration criterion requires the sequential minimization of $A_2, A_3, \ldots, A_m$, where $A_j = \sum_{|u|=j} (J_u/n)^2$. Clearly, A_2 is proportional to $E(s^2)$. Similarly, we see that minimizing $\max_{i<j} |s_{ij}|$ is equivalent to minimizing the leading term in the minimum G-aberration criterion for balanced supersaturated designs.*

8.2 $E(s^2)$-Optimal Supersaturated Designs

Nguyen (1996) and Tang and Wu (1997) independently obtained a lower bound for the $E(s^2)$ value of a supersaturated design. As will be seen, a lower bound on the $E(s^2)$ criterion greatly facilitates the study of $E(s^2)$-optimal supersaturated designs.

Theorem 8.1. *Let $\boldsymbol{D}$ be a supersaturated design of n runs for m factors, where $m \geq n$. Then its $E(s^2)$ value satisfies that*

$$E(s^2) \geq \frac{m-n+1}{(m-1)(n-1)} n^2. \tag{8.1}$$

If $\boldsymbol{D}$ attains the lower bound, then it is necessary that m is a multiple of $n-1$.

The next result (Cheng, 1997) provides a complete characterization of supersaturated designs whose $E(s^2)$ values attain the lower bound in Theorem 8.1.

Theorem 8.2. *Let $m=k(n-1)$ for some positive integer k. A design $\boldsymbol{D}$ attains the lower bound in (8.1) if and only if the matrix $\boldsymbol{X}$, obtained by adding k columns of all ones to $\boldsymbol{D}$, has mutually orthogonal rows.*

The necessary condition that m is a multiple of $n-1$ for a design to attain the lower bound in (8.1) can be strengthened using Theorem 8.2.

Corollary 8.1. *Suppose that a design attains the lower bound in* (8.1). *We then have that*
(i) *m is a multiple of $n-1$ if n is a multiple of* 4;
(ii) *m is an even multiple of $n-1$ if n is even but not a multiple of* 4.

We give a combined proof of Theorem 8.1, Theorem 8.2, and Corollary 8.1.

Proof. Consider the matrix $\boldsymbol{D}^T\boldsymbol{D}$. The diagonal entries of this matrix are equal to n, and the (i,j)th off-diagonal entry is $s_{ij} = \boldsymbol{d}_i^T\boldsymbol{d}_j$. The sum of squared entries of $\boldsymbol{D}^T\boldsymbol{D}$ is $\text{tr}(\boldsymbol{D}^T\boldsymbol{D}\boldsymbol{D}^T\boldsymbol{D})$. Therefore

$$\text{tr}(\boldsymbol{D}^T\boldsymbol{D}\boldsymbol{D}^T\boldsymbol{D}) = mn^2 + m(m-1)E(s^2). \tag{8.2}$$

Similarly, we obtain that

$$\text{tr}(\boldsymbol{D}\boldsymbol{D}^T\boldsymbol{D}\boldsymbol{D}^T) = nm^2 + \sum_{i \neq j}^{n} r_{ij}^2, \tag{8.3}$$

where r_{ij} is the inner product of row i and row j of $\boldsymbol{D}$. In obtaining (8.3), we have used $r_{ii} = m$ for $i = 1, \ldots, n$. As $\boldsymbol{D}$ is balanced, the sum of all rows of $\boldsymbol{D}$ is a row vector of zeros. This implies that $\sum_{j=1}^{n} r_{ij} = 0$ for every $i = 1, \ldots, n$. Consequently, $\sum_{i \neq j}^{n} r_{ij} = -nm$. We therefore obtain that $\sum_{i \neq j}^{n} r_{ij}^2$ is minimized if and only if all r_{ij}s for $i \neq j$ are equal to $r = -nm/(n(n-1)) = -m/(n-1)$. Thus

$$\sum_{i \neq j}^{n} r_{ij}^2 \geq n(n-1)(-m/(n-1))^2 = m^2n/(n-1). \tag{8.4}$$

Noting that $\text{tr}(\boldsymbol{D}^T\boldsymbol{D}\boldsymbol{D}^T\boldsymbol{D}) = \text{tr}(\boldsymbol{D}\boldsymbol{D}^T\boldsymbol{D}\boldsymbol{D}^T)$ and combining equations (8.2), (8.3), and (8.4), we establish that

$$mn^2 + m(m-1)E(s^2) \geq nm^2 + m^2n/(n-1),$$

which leads to the lower bound in (8.1). Since r_{ij} is the inner product of rows i and j of $\boldsymbol{D}$, it has to be an integer. We have seen that $\boldsymbol{D}$ attains the lower bound in (8.1) if and only if all r_{ij}s are equal to $r = -m/(n-1)$, in which case r is an integer. Therefore m is a multiple of $n-1$. This completes the proof of Theorem 8.1.

Let $m = k(n-1)$; so $k = -r$, where r is given above. If all r_{ij}s for $i \neq j$ are equal to $r = -k$, then the row vectors of $\boldsymbol{X}$, obtained by adding k columns of all ones to $\boldsymbol{D}$, are mutually orthogonal. The converse is obviously true. Theorem 8.2 is proved.

Matrix $\boldsymbol{X}$ has n rows and $m+k$ columns. If $\boldsymbol{X}$ has orthogonal rows, then $\boldsymbol{X}^T$ has orthogonal columns. From $\boldsymbol{X}^T$, we can construct an orthogonal array of strength two with run size $m+k$ for $n-1$ factors in the same way as we obtain a saturated orthogonal array from a Hadamard matrix. Thus $m+k$ is a multiple of 4. As $m+k = k(n-1)+k = kn$, k must be even if n is even but not a multiple of 4. This proves Corollary 8.1. □

The next two theorems show that the necessary conditions in Corollary 8.1 are also sufficient, with the following two caveats. The first is the existence of Hadamard matrices since they are used in our construction. The other caveat is that the constructed designs need to have no fully aliased columns, as it is generally accepted that a legitimate supersaturated design should not permit mutually aliased main effects.

If a supersaturated design attains the $E(s^2)$ lower bound in (8.1), then it is $E(s^2)$-optimal. We next provide constructions of such designs, separately for the two cases determined by whether or not n is a multiple of 4.

Let n be a multiple of 4 and assume the existence of a Hadamard matrix of order n. We know from Chapter 2 that a saturated orthogonal array of strength two can be constructed from a Hadamard matrix. Let $\boldsymbol{H}_1, \ldots, \boldsymbol{H}_k$ be k Hadamard matrices of order n, all normalized to have a column of ones as their first columns. Deleting the column of ones from each $\boldsymbol{H}_j$, we obtain a saturated orthogonal array $\boldsymbol{D}_j$, an OA$(n, 2^{n-1}, 2)$ to be precise. Consider design $\boldsymbol{D} = (\boldsymbol{D}_1, \ldots, \boldsymbol{D}_k)$. This construction and the $E(s^2)$-optimality as given below are from Tang and Wu (1997).

Theorem 8.3. *Let n be a multiple of 4. Suppose that design $\boldsymbol{D} = (\boldsymbol{D}_1, \ldots, \boldsymbol{D}_k)$, where each $\boldsymbol{D}_j$ is an OA$(n, 2^{n-1}, 2)$, does not contain fully aliased columns. Then it attains the $E(s^2)$ lower bound in (8.1) and is therefore $E(s^2)$-optimal.*

Proof. Adding k columns of ones to design $\boldsymbol{D} = (\boldsymbol{D}_1, \ldots, \boldsymbol{D}_k)$ gives $\boldsymbol{X} = (\boldsymbol{H}_1, \ldots, \boldsymbol{H}_k)$ up to a column permutation. Matrix $\boldsymbol{X} = (\boldsymbol{H}_1, \ldots, \boldsymbol{H}_k)$ has mutually orthogonal rows because each $\boldsymbol{H}_j$ is a Hadamard matrix and thus has mutually orthogonal rows. Theorem 8.3 then follows from Theorem 8.2. □

It is worthwhile to point out that design $\boldsymbol{D}$ in Theorem 8.3 having no fully aliased columns is not a condition for attaining the $E(s^2)$ lower bound. Rather, it is a requirement for $\boldsymbol{D}$ to be a legitimate supersaturated design.

Next consider the case that n is even but not a multiple of 4. Assume the existence of a Hadamard matrix $\boldsymbol{H}$ of order $2n$ and, without loss of generality, let it have the following form

$$\boldsymbol{H} = \begin{bmatrix} \mathbf{1}_n & \mathbf{1}_n & \boldsymbol{C} \\ \mathbf{1}_n & -\mathbf{1}_n & \boldsymbol{C}' \end{bmatrix}. \tag{8.5}$$

The method of constructing design $\boldsymbol{C}$ was proposed by Lin (1993), who called $\boldsymbol{C}$ a half fraction of a Hadamard matrix. The $E(s^2)$-optimality of design $\boldsymbol{C}$, first established by Nguyen (1996), is immediate owing to Theorem 8.2. A more general construction is given below.

Let $\boldsymbol{H}_1, \ldots, \boldsymbol{H}_p$ be p Hadamard matrices of order $2n$. By making each $\boldsymbol{H}_j$ have the form in equation (8.5), we obtain design $\boldsymbol{C}_j$. Consider design $\boldsymbol{D} = (\boldsymbol{C}_1, \ldots, \boldsymbol{C}_p)$.

Theorem 8.4. *Let n be even but not a multiple of 4. Suppose that design $\boldsymbol{D} = (\boldsymbol{C}_1, \ldots, \boldsymbol{C}_p)$, where each $\boldsymbol{C}_j$ is a half fraction of a Hadamard matrix of order $2n$, does not contain fully aliased columns. Then $\boldsymbol{D} = (\boldsymbol{C}_1, \ldots, \boldsymbol{C}_p)$, a design of n runs for $2p(n-1)$ factors, attains the $E(s^2)$ lower bound in (8.1) and is therefore $E(s^2)$-optimal.*

Theorem 8.4 follows immediately from Theorem 8.2.

There remains a problem of how to obtain $\boldsymbol{D}_j$s and $\boldsymbol{C}_j$s to ensure that the final designs $(\boldsymbol{D}_1, \ldots, \boldsymbol{D}_k)$ in Theorem 8.3 and $(\boldsymbol{C}_1, \ldots, \boldsymbol{C}_p)$ in Theorem 8.4 do not have fully aliased columns. As shown in Tang and Wu (1997), a simple device of row permutation will allow us to easily find $\boldsymbol{D}_j$s such that $(\boldsymbol{D}_1, \ldots, \boldsymbol{D}_k)$ does not have fully aliased columns, provided that n is not too small and k is not too large. More specifically, let $\boldsymbol{D}_1$ be an OA$(n, 2^{n-1}, 2)$. We can then obtain $\boldsymbol{D}_j$s for $j = 2, \ldots, k$ by considering row permutations of $\boldsymbol{D}_1$. As for the application of Theorem 8.4, since $2n$ is not a multiple of 8, a half fraction $\boldsymbol{C}$ of a Hadamard matrix $\boldsymbol{H}$ of order $2n$ does not contain fully aliased columns. Starting with $\boldsymbol{C}_1$, we can then find $\boldsymbol{C}_j$ for $j = 2, \ldots, p$ by considering row permutations of $\boldsymbol{C}_1$. We give an example to illustrate Theorem 8.3.

Example 8.1. *Starting with a design $\boldsymbol{D}_1$, an $OA(12, 2^{11}, 2)$, Tang and Wu (1997) found 11 other $OA(12, 2^{11}, 2)$s, $\boldsymbol{D}_2, \ldots, \boldsymbol{D}_{12}$, by considering row permutations of $\boldsymbol{D}_1$ such that the final design $(\boldsymbol{D}_1, \ldots, \boldsymbol{D}_{12})$ has no fully aliased columns. They did this simply by trying some random row permutations; no sophisticated algorithm is required. Also by trying some random row permutations, Tang and Wu (1997) found two other $OA(16, 2^{15}, 2)$s from a given $OA(16, 2^{15}, 2)$ to obtain a supersaturated design of 16 runs for 45 factors that has no fully aliased columns.*

Using Theorem 8.2, it can easily be shown that if a supersaturated design of n runs for m factors with no fully aliased columns attains the $E(s^2)$ lower bound in (8.1), then it is equivalent to a balanced incomplete block design (BIBD) with $n-1$ treatments and m distinct blocks of size $n/2-1$. Constructions in Theorems 8.3 and 8.4 are just two simple ways of finding such BIBDs. Some general theoretical construction results for this kind of BIBDs were presented in Bulutoglu and Cheng (2004). Considering the whole class of BIBDs with $n-1$ treatments and m distinct blocks of size $n/2-1$ is advantageous since it allows supersaturated designs to be found with better $\max_{i<j}|s_{ij}|$ properties, as clearly demonstrated in Nguyen (1996) and Liu and Zhang (2000).

There is also some interest in supersaturated designs with odd numbers of runs. Results similar to those presented in this section can also be obtained. We refer to Nguyen and Cheng (2008) for a detailed discussion on supersaturated designs with odd numbers of runs. For a general treatment of both even and odd numbers of runs, we refer to Section 15.15 of Cheng (2014a).

8.3 Further Results on $E(s^2)$-Optimality

In the last section we have obtained $E(s^2)$-optimal supersaturated designs that reach the lower bound in (8.1). These designs can be used as building blocks to construct other $E(s^2)$-optimal supersaturated designs. Cheng (1997) presented the first set of results in this direction; the next theorem summarizes his findings. Two balanced columns $\boldsymbol{c}$ and $\boldsymbol{d}$ cannot be orthogonal when n is even but not a multiple of 4, and we say that they are *nearly orthogonal* if $\boldsymbol{c}^T\boldsymbol{d} = \pm 2$.

Theorem 8.5. *Let $\boldsymbol{D}$ be an $E(s^2)$-optimal supersaturated design of n runs and $m = k(n-1)$ factors that attains the lower bound in (8.1), where $k \geq 1$ is an integer when n is a multiple of 4 and an even integer when n is even but not a multiple of 4. We then have that*
(i) a design given by deleting from or adding to $\boldsymbol{D}$ any column is $E(s^2)$-optimal;
(ii) a design given by deleting from or adding to $\boldsymbol{D}$ any two orthogonal or nearly orthogonal columns is $E(s^2)$-optimal.

It should be noted that when adding columns to $\boldsymbol{D}$, though not explicitly stated in Theorem 8.5, we only consider balanced columns that are not fully aliased with the columns that are already in $\boldsymbol{D}$.

The idea of using supersaturated designs attaining the lower bound in (8.1) to construct other $E(s^2)$-optimal supersaturated designs is systematically explored in Butler, Mead, Eskridge, and Gilmour (2001). Their approach is based on an improved lower bound on the $E(s^2)$ value, as detailed below.

Theorem 8.6. *For a supersaturated design of n runs with $m = k(n-1) + r$ factors, where $|r| < n/2$ and $k \geq 1$ is an integer when n is a multiple of 4 and an even integer when n is* *even but not a multiple of* 4, *we have that*

$$E(s^2) \geq LB = \frac{m-n+1}{(m-1)(n-1)}n^2 + \frac{n}{m(m-1)}\left\{D(n,r) - \frac{|r|^2}{n-1}\right\}, \tag{8.6}$$

where

$$D(n,r) = \begin{cases} 4|r| & \textit{for } |r| \equiv 0 \textit{ mod } (4), \\ n+2|r|-3 & \textit{for } |r| \equiv 1 \textit{ mod } (4), \\ 2n-4 & \textit{for } |r| \equiv 2 \textit{ mod } (4) \textit{ and } n \equiv 0 \textit{ mod } (4), \\ 2n-4+8/n & \textit{for } |r| \equiv 2 \textit{ mod } (4) \textit{ and } n \equiv 2 \textit{ mod } (4), \\ n+2|r|+1 & \textit{for } |r| \equiv 3 \textit{ mod } (4). \end{cases}$$

The general lower bound LB in (8.6) improves upon that in (8.1) as it can easily be checked that LB is greater than or equal to the lower bound in (8.1) with equality reached only when $r = 0$.

The lower bound LB in (8.6) can be used to prove Theorem 8.5, which can be done by simply checking that the designs in Theorem 8.5 attain the lower bound LB. Proving Theorem 8.6, however, is an onerous task. Instead of giving a full proof, we are content with an outline of the basic idea. Recalling the proof of Theorem 8.1, we see that the lower bound on the $E(s^2)$ in (8.1) is obtained by minimizing $\sum_{i \neq j} r_{ij}^2$ under the constraints that $\sum_j r_{ij} = 0$. But r_{ij}, as the inner product of two rows of a design with levels ± 1, can only take on certain possible integer values. The proof of Theorem 8.6 basically involves some very careful and elaborate analysis of further constraints on r_{ij} values and their impact on the minimization of $\sum_{i \neq j} r_{ij}^2$.

Supersaturated designs attaining the lower bound in (8.1) have restricted numbers of factors, as shown in Corollary 8.1. A theorem from Butler et al. (2001) shows how to use designs attaining the lower bound in (8.1) to construct $E(s^2)$-optimal supersaturated designs with flexible numbers of factors.

Theorem 8.7. *If* $\boldsymbol{D}_0$ *attains the lower bound LB in* (8.6) *and* $\boldsymbol{D}_1$ *attains the lower bound in* (8.1), *then* $(\boldsymbol{D}_0, \boldsymbol{D}_1)$ *attains the lower bound LB in* (8.6) *and is thus* $E(s^2)$*-optimal.*

If we go through the proof of Theorem 8.6, we will easily see the validity of Theorem 8.7. Without much effort, one can also prove Theorem 8.7 directly by a joint use of Theorems 8.2 and 8.6.

Theorem 8.7 is a powerful result. Designs attaining the lower bound in (8.1) are readily available, as shown in Section 8.2. For a given run size, Theorem 8.7 states that we only need to find designs attaining the lower bound LB in (8.6) for small numbers of factors in order to obtain $E(s^2)$-optimal supersaturated designs for large numbers of factors.

Complementing Theorem 8.7 is the next result of Butler et al. (2001) that gives a method of constructing $E(s^2)$-optimal supersaturated designs of large run sizes from those of small run sizes. To construct an $E(s^2)$-optimal supersaturated design of n runs for $m = k(n-1)+r$ factors, this method needs two designs $\boldsymbol{D}_0$ and $\boldsymbol{D}_1$, both of $n/2$ runs, where $\boldsymbol{D}_0$ has $m_0 = k(n/2-1) + r_0$ factors and $\boldsymbol{D}_1$ has $m_1 = kn/2 + r_1$ factors. The two integers r_0 and r_1 are functions of r and they have the same sign as r and the following absolute values:

$$|r_1| = 4\lfloor(|r|+4)/8\rfloor, \quad |r_0| = |r| - |r_1|.$$

Further requirements for $\boldsymbol{D}_0$ and $\boldsymbol{D}_1$ are also needed for the method, as given in the next theorem.

Theorem 8.8. *Suppose that $m = k(n-1)+r$ with $|r| < n/2$, where n, k, r satisfy either that n is a multiple of 8 and $|r| \not\equiv 3 \bmod (8)$ or that n is a multiple of 4 but not of 8, k is even, $|r| \not\equiv 2 \bmod (4)$, and $|r| \not\equiv 3 \bmod (8)$. Then we have that*

$$\begin{bmatrix} \boldsymbol{D}_0 & \boldsymbol{D}_1 \\ \boldsymbol{D}_0 & -\boldsymbol{D}_1 \end{bmatrix}$$

is an $E(s^2)$-optimal supersaturated design that attains the lower bound LB in (8.6) for n runs and $m = k(n-1)+r$ factors, if $\boldsymbol{D}_1$ has orthogonal rows and $\boldsymbol{D}_0$ is an $E(s^2)$-optimal supersaturated design attaining the lower bound LB in (8.6).

Theorem 8.8 is a very useful result. Starting with an $E(s^2)$-optimal design of n runs that attains the lower bound LB in (8.6), we can then construct a series of $E(s^2)$-optimal supersaturated designs of $2^l n$ runs for $l = 1, 2, 3, ...$ by repeatedly applying Theorem 8.8. This idea was formalized as another theorem in Butler et al. (2001).

However, we should emphasize that Theorem 8.8 has some restrictions in its applicability. Designs $\boldsymbol{D}_0$ and $\boldsymbol{D}_1$ in Theorem 8.8 have $m_0 = k(n/2-1)+r_0$ and $m_1 = kn/2+r_1$ factors, respectively. Because r_0 and r_1 are completely determined by r, we see that m_0 and m_1 are also completely determined by $m = k(n-1)+r$. Most unpleasant restrictions are those on r values, which severely limit the power of Theorem 8.8 especially when n is a multiple of 4 but not of 8. The final theorem of Butler et al. (2001) shows how to construct $E(s^2)$-optimal designs of $k(n-1)\pm 3$ factors from those of $k(n-1)\pm 4$ factors, and this partially alleviates the restrictions on r in Theorem 8.8.

We conclude this section with some remarks on lower bounds on the $E(s^2)$ value. The lower bound in Theorem 8.6 has an imperfection—it does not cover the values of $m = k(n-1)+r$ for $n/2 \leq |r| \leq n-1$ when n is even but not a multiple of 4, as k is assumed to be even in this case. Bulutoglu and Cheng (2004) derived a lower bound on the $E(s^2)$ for all cases. We choose to present the lower bound from Butler et al. (2001) because it has a simpler form and also because it is directly linked to the two construction results in Theorems 8.7 and 8.8. A modest improvement over the bound of Bulutoglu and Cheng (2004) was made for two special cases by Ryan and Bulutoglu (2007) and Das, Dey, Chan, and Chatterjee (2008), the latter of which also established the equivalence between the bound of Butler et al. (2001) and that of Bulutoglu and Cheng (2004).

8.4 Controlling Correlations

As the $E(s^2)$ criterion minimizes only the average of squared correlations, an $E(s^2)$-optimal supersaturated design may have high correlations for certain pairs of columns. To have a full control over all the correlations, we can select designs by minimizing the $\max_{i<j}|s_{ij}|$ criterion. A side benefit of using this criterion is that the requirement of no fully aliased columns is automatically accommodated. Using $\max_{i<j}|s_{ij}|$ alone as a design selection criterion is discussed in Subsection 8.4.1, while in Subsection 8.4.2 we examine the use of $\max_{i<j}|s_{ij}|$ as a criterion secondary to the $E(s^2)$.

8.4.1 Using the $\max_{i<j}|s_{ij}|$ criterion alone

The $\max_{i<j}|s_{ij}|$ criterion requires us to find a supersaturated design that minimizes $\max_{i<j}|s_{ij}|$ among all supersaturated designs. This amounts to finding

$$S(n,m) = \min_{\boldsymbol{D}} \max_{i<j} |s_{ij}|,$$

where the minimization is over all supersaturated designs $\boldsymbol{D}$ of n runs for m factors.

To study $S(n,m)$, it is more convenient to consider $M(n,\rho)$, which denotes the largest number m of factors a supersaturated design of n runs can have while satisfying $\max_{i<j}|s_{ij}| \leq \rho$ for given ρ. Requiring $\max_{i<j}|s_{ij}| \leq \rho$ effectively specifies ρ/n as the permissible maximum correlation. This is exactly the approach taken by Booth and Cox (1962) and Lin (1995) in their computer generation of supersaturated designs. The two functions $S(n,m)$ and $M(n,\rho)$ completely determine each other, as shown below.

Theorem 8.9. *We have that*

$$M(n,\rho) = \max\{m : S(n,m) \leq \rho\},$$

$$S(n,m) = \min\{\rho : M(n,\rho) \geq m\}.$$

Theorem 8.9 does not require a formal proof. A moment of thought is all that needs to be convinced of its validity.

We now focus on $M(n,\rho)$. If a supersaturated design of n runs for m factors is found to satisfy $\max_{i<j}|s_{ij}| \leq \rho$, then we have that $M(n,\rho) \geq m$, giving a lower bound on $M(n,\rho)$. We next discuss how to find an upper bound on $M(n,\rho)$ and present some results from Cheng and Tang (2001).

Only certain ρ values in $M(n,\rho)$ need to be considered. We know that for an even n, the inner product of two balanced columns is always even and is a multiple of 4 if and only if n is a multiple 4. This implies that when studying $M(n,\rho)$, we only need to consider $\rho = 4, 8, 12, \ldots$ if $n \equiv 0 \bmod (4)$ and $\rho = 2, 6, 10, \ldots$ if $n \equiv 2 \bmod (4)$.

The first method of Cheng and Tang (2001) for finding an upper bound on $M(n,\rho)$ is by connecting a supersaturated design to an error-correcting code. A binary code is a $k \times n$ matrix with entries 0 and 1, where each row is called a codeword. The Hamming weight of a codeword is the number of ones among its entries, and the Hamming distance of two codewords is the number of entries where they differ. A code has a constant weight if all codewords have the same weight. Let $A(n,d,w)$ be the maximum number of codewords of length n, constant weight w and distance at least d. See Section A.5 of the Appendix for an introduction to error-correcting codes.

Theorem 8.10. *We have that*

$$M(n,\rho) \leq \Big\lfloor (1/2)A\big(n,(n-\rho)/2, n/2\big)\Big\rfloor.$$

Proof. Let $\boldsymbol{D} = (d_{ij})_{n\times m}$ be a supersaturated design of n runs and m factors that satisfies $\max_{i<j}|s_{ij}| \leq \rho$. Consider $\boldsymbol{C} = (\boldsymbol{C}_1, \boldsymbol{C}_2)^T$ where $\boldsymbol{C}_1 = (c_{ij})_{n\times m}$ and $\boldsymbol{C}_2 = (c'_{ij})_{n\times m}$ are defined via

$$c_{ij} = (1-d_{ij})/2 \quad\text{and}\quad c'_{ij} = (1+d_{ij})/2.$$

We see that $\boldsymbol{C}$ is a binary code with $2m$ codewords of length n and of constant weight $n/2$. With just a bit of algebra, one can show that any two codewords have a distance of at least $(n-\rho)/2$. Theorem 8.10 now follows. □

Theorem 8.10 reveals that any upper bound on $A\big(n,(n-\rho)/2,n/2\big)$ can be used to obtain an upper bound on $M(n,\rho)$. The best known general upper bound on $A(n,d,w)$ is the Johnson bound, one version of which is

$$A(n,d,w) \leq \left\lfloor \frac{n}{w}\left\lfloor \frac{n-1}{w-1}\left\lfloor \frac{n-2}{w-2}\cdots\left\lfloor \frac{n-w+\delta}{\delta}\right\rfloor\cdots\right\rfloor\right\rfloor\right\rfloor, \tag{8.7}$$

where d is even and $\delta = d/2$ (MacWilliams and Sloane, 1977, Corollary 5, Ch. 17).

For $n = 12$ and $\rho = 4$, a joint use of Theorem 8.10 and the Johnson bound in (8.7) gives

$$M(12,4) \le \left\lfloor \frac{1}{2}A(12,4,6) \right\rfloor \le \left\lfloor \frac{1}{2}\left\lfloor \frac{12}{6}\left\lfloor \frac{11}{5}\left\lfloor \frac{10}{4}\left\lfloor \frac{9}{3}\left\lfloor \frac{8}{2}\right\rfloor\right\rfloor\right\rfloor\right\rfloor\right\rfloor\right\rfloor = 66.$$

Since Wu (1993) and Lin (1995) have found supersaturated designs of 12 runs and 66 factors that have $\max_{i<j}|s_{ij}| = 4$, we obtain that $M(12,4) = 66$.

The other method of Cheng and Tang (2001) to obtain an upper bound on $M(n,\rho)$ is to use a lower bound L on $E(s^2)$. Obviously, $L \le E(s^2) \le \rho^2$. Using as L the lower bound in Theorem 8.1, we obtain

$$n^2(m-n+1)/((n-1)(m-1)) \le \rho^2.$$

Solving for m gives an upper bound for $M(n,\rho)$, which we present as a theorem.

Theorem 8.11. *We have that*

$$M(n,\rho) \le \lfloor (n-1)(n^2-\rho^2)/(n^2-n\rho^2+\rho^2) \rfloor$$

provided that $\rho < n/\sqrt{n-1}$.

The condition that $\rho < n/\sqrt{n-1}$ is to ensure that $n^2 - n\rho^2 + \rho^2 > 0$. Theorem 8.11 does not apply if the condition is not satisfied.

If we use the better lower bound LB in Theorem 8.6 or that of Bulutoglu and Cheng (2004), we can obtain a better upper bound on $M(n,\rho)$ although an explicit expression like that in Theorem 8.11 is no longer available.

Consider $n \equiv 2 \pmod 4$. Taking $\rho = 2$ in Theorem 8.11, we obtain that

$$M(n,2) \le \lfloor (n+2)(n-1)/(n-2) \rfloor = \lfloor n+3+4/(n-2) \rfloor = n+3$$

provided $n \ge 10$. We will establish that $M(n,2) \le n+2$ for $n \ge 10$ if we can show that it is impossible for a design of $n \equiv 2 \pmod 4$ runs with $n+3$ factors to have $\max_{i<j}|s_{ij}| = 2$. This can be proved by using a better lower bound on the $E(s^2)$ value. The LB in Theorem 8.6 does not apply to this situation, but the lower bound of Bulutoglu and Cheng (2004) is applicable. We omit the details.

Theorem 8.12. *For* $n \equiv 2 \pmod 4$ *with* $n \ge 10$, *we have that* $M(n,2) \le n+2$.

Since Lin (1995) has found a design of 10 runs and 12 factors with $\max_{i<j}|s_{ij}| = 2$, we have $M(10,2) = 12$. Further results on $M(n,2)$ were afforded by Morales, Bulutoglu, and Arasu (2019), who established that $M(n,2) = n+1$ for $n = 14, 18, 22, 30$ and $M(n,2) \ge n+1$ for all other $n \le 270$ with $n \equiv 2 \pmod 4$ by using a combination of theoretical arguments and computational techniques. These authors also showed that $M(n,2) \ge n+1$ if $n+1$ is a prime power with $n+1 \equiv 3 \pmod 4$.

8.4.2 Using $\max_{i<j}|s_{ij}|$ as a criterion secondary to $E(s^2)$

The constructions of supersaturated designs proposed by Lin (1993) and Wu (1993) deserve special attention because they are simple to use and each produces a unique supersaturated design starting with a Hadamard matrix with a column of ones. Let $\boldsymbol{H} = (\boldsymbol{1}_n, \boldsymbol{D}_0)$ be a Hadamard matrix of order n. Write $\boldsymbol{H}$ as

$$\boldsymbol{H} = \begin{bmatrix} \boldsymbol{1}_{n/2} & \boldsymbol{1}_{n/2} & \boldsymbol{C} \\ \boldsymbol{1}_{n/2} & -\boldsymbol{1}_{n/2} & \boldsymbol{C}' \end{bmatrix}. \tag{8.8}$$

Then Lin's construction gives design $\boldsymbol{C}$, which has $n/2$ runs and $n-2$ factors. The method applies whenever a Hadamard matrix of order n exists. Earlier in Section 8.2, Lin's construction was introduced for n that is a multiple of 4 but not of 8. Wu's method constructs a design $\boldsymbol{D}$, given by adding all two-factor interaction columns to design $\boldsymbol{D}_0$. Design $\boldsymbol{D}$ has n runs and $(n-1)+(n-2)(n-1)/2 = n(n-1)/2$ factors.

Designs $\boldsymbol{C}$ and $\boldsymbol{D}$ are both $E(s^2)$-optimal. We have discussed this for $\boldsymbol{C}$ in Section 8.2. The $E(s^2)$-optimality of $\boldsymbol{D}$ was established by Bulutoglu and Cheng (2003) and can also easily be seen using Theorem 8.2.

Both $\boldsymbol{C}$ and $\boldsymbol{D}$ are uniquely determined by the underlying Hadamard matrix $\boldsymbol{H} = (\boldsymbol{1}_n, \boldsymbol{D}_0)$. The question as to which Hadamard matrix we should choose can be settled using the $\max_{i<j}|s_{ij}|$ criterion (Shi and Tang, 2018). The simplest answer is to use Paley matrices. A more precise answer is that we should use Hadamard matrices of the largest types. The type of a Hadamard matrix was previously defined in Section 4.4.

We first consider design $\boldsymbol{D}$ from Wu's construction. Let $\boldsymbol{D}_0 = (\boldsymbol{d}_1, \ldots, \boldsymbol{d}_{n-1})$, which is a saturated OA$(n, 2^{n-1}, 2)$ obtained by deleting $\boldsymbol{1}_n$ from $\boldsymbol{H} = (\boldsymbol{1}_n, \boldsymbol{D}_0)$. Then design $\boldsymbol{D}$ collects all main effect columns $\boldsymbol{d}_1, \ldots, \boldsymbol{d}_{n-1}$ and all interaction columns $\boldsymbol{d}_i\boldsymbol{d}_j$ for $1 \le i < j \le n-1$. Clearly,

$$\max_{i<j}|s_{ij}|(\boldsymbol{D}) = \max_{|u|=3,4}|J_u(\boldsymbol{D}_0)| = \max_{|u|=4}|J_u(\boldsymbol{H})| = n - 8b,$$

where the last equality is from Lemma 4.3 with b being the type of $\boldsymbol{H}$. Therefore, minimizing $\max_{i<j}|s_{ij}|(\boldsymbol{D})$ is equivalent to maximizing b.

Theorem 8.13. *In Wu's construction, the best design $\boldsymbol{D}$ under the $\max_{i<j}|s_{ij}|$ criterion is given by using a Hadamard matrix $\boldsymbol{H}$ of type $b_{\max}$.*

We know from Theorem 4.13 and Corollary 4.2 that Paley matrices have large types and they actually have type $b_{\max}$ for n = 12, 20, 24, 28, 32, 44, 60, 72, and 80. By computer search, Shi and Tang (2018) found Hadamard matrices of type $b_{\max}$ for $n = 36$. These Hadamard matrices of type $b_{\max}$ are the best to use in Wu's construction of supersaturated designs according to Theorem 8.13. We take a look at two interesting cases.

For $n = 24$, using a Hadamard matrix of type $b_{\max} = 2$ in Wu's construction produces a design $\boldsymbol{D}$ of $n = 24$ runs for $m = 276$ factors with $\max_{i<j}|s_{ij}| = 8$. By Theorem 8.11, we obtain that $M(24,4) \le 61$. This shows that any supersaturated design of 24 runs for $m \ge 62$ factors must have that $\max_{i<j}|s_{ij}| \ge 8$. We therefore conclude that design $\boldsymbol{D}$ from Wu's construction minimizes $\max_{i<j}|s_{ij}|$ among all supersaturated designs of 24 runs and 276 factors, and hence is jointly optimal under the $E(s^2)$ and $\max_{i<j}|s_{ij}|$ criteria. Similarly, for $n = 32$, if we use the Paley matrix of type $b_{\max} = 3$ in Wu's construction, we obtain a supersaturated design of 32 runs and 496 factors with $\max_{i<j}|s_{ij}| = 8$. This design is also jointly optimal under the $E(s^2)$ and $\max_{i<j}|s_{ij}|$ criteria. The optimality under the $\max_{i<j}|s_{ij}|$ criterion follows from $M(32,4) \le 59$, also by Theorem 8.11.

We now consider design $\boldsymbol{C}$ from Lin's construction. As displayed in equation (8.8), column $\boldsymbol{d}_1$ in $\boldsymbol{D}_0 = (\boldsymbol{d}_1, \ldots, \boldsymbol{d}_{n-1})$ is used as a branching column to obtain design $\boldsymbol{C}$. We can easily show that

$$\max_{i<j}|s_{ij}|(\boldsymbol{C}) = (1/2)\max_{2\le i<j\le n-1}|J(\boldsymbol{d}_1, \boldsymbol{d}_i, \boldsymbol{d}_j)|.$$

Obviously,

$$\max_{2\le i<j\le n-1}|J(\boldsymbol{d}_1, \boldsymbol{d}_i, \boldsymbol{d}_j)| \le \max_{|u|=3}|J_u(\boldsymbol{D}_0)| \le \max_{|u|=4}|J_u(\boldsymbol{H})| = n - 8b.$$

We therefore have that

$$\max_{i<j}|s_{ij}|(\boldsymbol{C}) \le (n-8b)/2. \tag{8.9}$$

TABLE 8.1
Half fraction designs from Hadamard matrices of $b_{\max}$, with n runs and $m = 2n - 2$ factors

n	6	10	12	14	16	18	22	30	36	40
m	10	18	22	26	30	34	42	58	70	78
$\max_{i<j} \lvert s_{ij} \rvert$	2	6	4	6	4	6	6	6	8	8

By equation (8.9), we see that a Hadamard matrix of type $b_{\max}$ gives a good design $\boldsymbol{C}$. If this Hadamard matrix of type $b_{\max}$ attains the bound in Theorem 4.12, then from the proof of Theorem 4.12 we see that $\max_{2\le i<j\le n-1} |J(\boldsymbol{d}_1, \boldsymbol{d}_i, \boldsymbol{d}_j)| = n - 8b_{\max}$, showing that equality in (8.9) holds. This gives the next result.

Theorem 8.14. *In Lin's construction, the best design* $\boldsymbol{C}$ *under the* $\max_{i<j} |s_{ij}|$ *criterion is given by using a Hadamard matrix* $\boldsymbol{H}$ *of type* $b_{\max}$ *that attains the bound in Theorem 4.12.*

Theorem 8.14 may seem more complicated than Theorem 8.13. The good news is that all Hadamard matrices of type $b_{\max}$ we know so far for n = 12, 20, 24, 28, 32, 36, 44, 60, 72, and 80 attain the bound in Theorem 4.12. Therefore, they are the best to use in Lin's construction. Table 8.1 displays the results. These designs are $E(s^2)$-optimal. By using Theorems 8.11 and 8.12, one can show that all the designs in Table 8.1 are also optimal under the $\max_{i<j} |s_{ij}|$ criterion.

The rest of the section discusses some computational results. The construction of Tang and Wu (1997) has an interesting attribute—it gives designs consisting of sets of mutually orthogonal columns. But in terms of the $\max_{i<j} |s_{ij}|$ criterion, the method generally does not perform well. For example, to construct a design of 16 runs and 30 factors, $\max_{i<j} |s_{ij}| = 8$ is unavoidable due to the restriction of two sets of mutually orthogonal columns. In contrast, we see that the corresponding design in Table 8.1 has $\max_{i<j} |s_{ij}| = 4$. By searching BIBDs, Nguyen (1996) and Liu and Zhang (2000) also found designs of 16 runs for 30 factors that have $\max_{i<j} |s_{ij}| = 4$.

Most comprehensive computational results on designs that are both $E(s^2)$ and $\max_{i<j} |s_{ij}|$ optimal were given by Morales and Vega (2014). For $n \le 26$, they obtained a rich collection of supersaturated designs that are optimal under both criteria, some of which have many more columns than those in Table 8.1. For example, for $n = 16$, they found a design for 60 factors with $\max_{i<j} |s_{ij}| = 4$, and for $n = 18$, they found a design for 136 factors with $\max_{i<j} |s_{ij}| = 6$.

8.5 Other Selection Criteria

The columns of a supersaturated design are necessarily linearly dependent, causing problems in model estimation and identification. A set of dependent columns is said to be a minimal dependent set (MDS) if the remaining ones become linearly independent whenever any column is removed. The MDS's of a supersaturated design capture all linear dependencies among its columns as any dependent set of columns can be obtained by adding columns to an MDS. Useful design selection criteria can be formed by utilizing MDS's. The size of an MDS is the number of columns it contains. Clearly, small MDS's are undesirable. The size of the smallest MDS is called the MDS-resolution, which, when used as a design criterion, requires the selection of designs with the maximum MDS-resolution. The MDS-aberration criterion, generalizing the MDS-resolution criterion, is to select designs by sequentially minimizing $a_1, a_2, \ldots$, where a_f is the number of MDS's of size f. The notion of MDS and related criteria were introduced and studied by Miller and Sitter (2004) and Lin, Miller, and Sitter (2008) in the context of nonorthogonal foldover designs. The use of these concepts for supersaturated designs was examined in Miller and Tang (2012) and Miller and Tang (2013).

Miller and Tang (2012) carried out an investigation into the number and structures of MDS's of a supersaturated design, and constructed a design of n runs with n factors for any even $n \geq 6$ that has the largest possible MDS-resolution of n. Miller and Tang (2013) conducted a very extensive computer search of MDS-aberration supersaturated designs of n runs with $m = n, n+1, n+2, n+3, n+4$ factors for $n = 6, 8, 10, \ldots, 36$. Their computer search is actually complete for 66 out of 80 considered cases. Of particular interest among their findings are designs with the largest possible MDS-resolution. Let $p(n)$ be the maximum number of factors for a supersaturated design of n runs to have an MDS-resolution of n. Then their computational results establish that

- $p(n) = n$ for $n = 6, 8$, $p(n) \geq n$ for $n = 10$,
- $p(n) \geq n + 2$ for $n = 12, 14, \ldots, 20$, and
- $p(n) \geq n + 4$ for $n = 22, 24, \ldots, 36$.

Let $p(n, r)$ be the maximum number of factors for a supersaturated design of n runs to have an MDS-resolution of r for given $r \leq n$. Then the above results of Miller and Tang (2013) are for $p(n, r)$ with $r = n$. The general problem of obtaining $p(n, r)$ for $r \leq n$ was considered, with computational results, by Jones, Li, Nachtsheim, and Ye (2009) and Gutman, Bulutoglu, and White (2015). Table 17 of Gutman, Bulutoglu, and White (2015) contains comprehensive results for $n =$ 6, 8, 10, and 12 with $4 \leq r \leq n$. A summary is given below:

- $p(8, 4) = 35$, $p(8, 6) = 12$, $p(8, 8) = 8$,
- $p(10, 6) \geq 23$, $p(10, 8) \geq 14$, $p(10, 10) \geq 11$, and
- $p(12, 6) \geq 41$, $p(12, 8) \geq 22$, $p(12, 10) \geq 17$, $p(12, 12) \geq 14$.

If a supersaturated design has an MDS-resolution of r, then any $r - 1$ columns are linearly independent, thus allowing the estimation of all models with up to $r - 1$ main effects. Because there is at least one MDS of size r, the *resolution rank* is precisely $r - 1$, where the resolution rank, as defined by Deng, Lin, and Wang (1999), is the largest number r' such that all r' columns are linearly independent. Therefore, the criteria of resolution rank and MDS-resolution are equivalent. The computational results of Gutman, Bulutoglu, and White (2015) were actually given under the resolution rank. We have discussed them in terms of MDS-resolution for consistency of presentation.

Related to the MDS-aberration is an aberration criterion that seeks designs by sequentially minimizing $b_1, b_2, \ldots$ where b_f is the number of dependent sets of size f. Obviously, $a_f \leq b_f$, since any MDS is a dependent set. The first non-zero entries of the two sequences $(a_1, a_2, \ldots)$ and $(b_1, b_2, \ldots)$ must be identical because a dependent set of the minimal size must be an MDS. Therefore, the two aberration criteria are very similar although they cannot be proved to be equivalent because it is impossible to determine one sequence from the other beyond the first non-zero entries, barring some simple cases.

Since the number b_f of dependent sets of size f represents the number of non-estimable models with f main effects, the sequence $(b_1, b_2, \ldots)$ is directly linked to the *estimation capacity* sequence $(E_1, E_2, \ldots)$, where E_f is the proportion of estimable models with f main effects. In fact, it holds that $b_f + \binom{m}{f}E_f = \binom{m}{f}$. The estimation capacity sequence was first proposed in Sun (1993), later studied for regular designs in Cheng, Steinberg, and Sun (1999), and its use in assessing supersaturated designs was considered by Jones, Li, Nachtsheim, and Ye (2009).

The quantity E_f only considers the estimability of models with f main effects. To take efficiency into consideration, one can use

$$D_f = \frac{1}{\binom{m}{f}} \sum_f \left\{\det\left(\boldsymbol{X}_f^T \boldsymbol{X}_f / n\right)\right\}^{1/(f+1)},$$

the average of the D-efficiencies of all models with f main effects, to evaluate the projection designs onto f factors, where $\boldsymbol{X}_f$ is a model matrix of $f+1$ columns with a column of all ones corresponding to the intercept and f columns corresponding to the main effects, and the summation is over all such model matrices. The D_f, appropriately dubbed *information capacity* by Sun (1993), was proposed by Wu (1993) for evaluating supersaturated designs. Cheng, Das, Singh, and Tsai (2018) provided a justification for the $E(s^2)$ criterion using a good surrogate of D_f based on the $(M.S)$-criterion.

A breakaway from the $E(s^2)$-criterion is the $UE(s^2)$-criterion, defined by Jones and Majumdar (2014) as

$$UE(s^2) = \frac{1}{\binom{m+1}{2}} \left\{ \sum_{i=1}^{m} (\boldsymbol{1}_n^T \boldsymbol{d}_i)^2 + \sum_{1 \le i < j \le m} (\boldsymbol{d}_i^T \boldsymbol{d}_j)^2 \right\},$$

where $\boldsymbol{1}_n$ is the all-ones column and $\boldsymbol{D} = (\boldsymbol{d}_1, \ldots, \boldsymbol{d}_m)$ is a supersaturated design of n runs with m columns, which is not required to be balanced. This is the same as $E(s^2)$ but applied to the model matrix $(\boldsymbol{1}_n, \boldsymbol{d}_1, \ldots, \boldsymbol{d}_m)$. We use the $E(s^2)$ as a design selection criterion among all supersaturated designs with balanced columns. The $UE(s^2)$ criterion drops the condition of column-balance, and thus "unconditionalizes" the $E(s^2)$—this is how "U" comes into the picture. Jones and Majumdar (2014) justified their criterion using a non-traditional D-optimality argument. Cheng, Das, Singh, and Tsai (2018) conducted an in-depth study on the $UE(s^2)$ criterion and its relationship with the $E(s^2)$ criterion. They argued that, in the process of justifying the $E(s^2)$ using a surrogate of D_f, if $UE(s^2)$-optimal designs are to be used, then one should choose *superior* $UE(s^2)$-optimal designs. A superior $UE(s^2)$-optimal design is a $UE(s^2)$-optimal design that is the most column-balanced, which is defined as the one that minimizes $\sum_{i=1}^{m}(\boldsymbol{1}_n^T \boldsymbol{d}_i)^2$ among all $UE(s^2)$-optimal designs. Cheng, Das, Singh, and Tsai (2018) also identified some supersaturated designs that are both $E(s^2)$- and $UE(s^2)$-optimal, among which the most interesting are those of Lin (1993), given by half fractions of Hadamard matrices.

Whether or not the $UE(s^2)$ is statistically more appropriate than the $E(s^2)$, the $UE(s^2)$ criterion has an appealing feature that $UE(s^2)$-optimal supersaturated designs can be constructed very easily and in fact effortlessly. Let $\boldsymbol{D}_0$ be any design with n orthogonal rows. Then $\boldsymbol{D}_0$ is $UE(s^2)$-optimal. Designs obtained by adding any column to or deleting any column from design $\boldsymbol{D}_0$ are also $UE(s^2)$-optimal. While deleting two columns from $\boldsymbol{D}_0$ may not give $UE(s^2)$-optimal designs, adding to $\boldsymbol{D}_0$ any two columns that has $n/2$ rows of either $(1,1)$ or $(-1,-1)$ and $n/2$ rows of either $(1,-1)$ or $(-1,1)$ always gives a $UE(s^2)$-optimal design. In Jones and Majumdar (2014), the rows of a Hadamard matrix are used as design $\boldsymbol{D}_0$, which puts a slight restriction on all possible $UE(s^2)$-optimal supersaturated designs since not every set of orthogonal rows can be extended to a Hadamard matrix.

The $UE(s^2)$ criterion has an application in the construction of robust supersaturated designs (Shi and Tang, 2019a). As discussed in Section 8.1, supersaturated designs are proposed under the situation of effect sparsity that all interactions are negligible and all main effects except a few are negligible. In certain applications, it is conceivable that factor sparsity may be more appropriate than effect sparsity. The scenario of factor sparsity we have in mind is that only a few factors are potentially active and active factors can potentially interact with one another. This calls for new supersaturated designs. It turns out that foldover supersaturated designs are well suited for this purpose. As main effects are orthogonal to two-factor interactions under a foldover supersaturated design, the existence of two-factor interactions will not affect the identification of active main effects. Once active main effects are identified, one can then consider their interactions in the next stage of analysis. This analysis strategy for foldover designs is well known in design literature, and Miller and Sitter (2001) may well be the first clear documentation.

A foldover supersaturated design of n runs for $m \geq n$ factors takes the form $\boldsymbol{D} = (\boldsymbol{B}^T, -\boldsymbol{B}^T)^T$, where $\boldsymbol{B}$ of $n/2$ runs is called a base design. Design $\boldsymbol{D}$ is always column-balanced whether or not $\boldsymbol{B}$ is column-balanced. Obviously,

$$E(s^2)(\boldsymbol{D}) = 2E(s^2)(\boldsymbol{B}), \quad \max_{i<j} |s_{ij}|(\boldsymbol{D}) = 2 \max_{i<j} |s_{ij}|(\boldsymbol{B}).$$

Therefore, $\boldsymbol{D}$ is $E(s^2)$-optimal if $\boldsymbol{B}$ is $E(s^2)$-optimal, and $\boldsymbol{D}$ is $\max_{i<j} |s_{ij}|$-optimal if $\boldsymbol{B}$ is $\max_{i<j} |s_{ij}|$-optimal. Without loss of generality, we may assume that $\boldsymbol{B}$ has its first column consisting of all ones. Then $\boldsymbol{B}$ is $E(s^2)$-optimal if and only if it is the model matrix of a $UE(s^2)$-optimal supersaturated design. This shows that $E(s^2)$-optimal foldover supersaturated designs are readily available since $E(s^2)$-optimal base designs $\boldsymbol{B}$ can easily be constructed. Shi and Tang (2019a) further optimized $E(s^2)$-optimal foldover supersaturated designs using $\max_{i<j} |s_{ij}|$ as a secondary criterion, and presented some computer search results by taking $\boldsymbol{B}$ to be transposes of regular designs as well as matrices given by removing rows from half fractions of Hadamard matrices.

8.6 Group Screening and Generalizations

Proposed by Watson (1961), group screening is a sequential strategy of design and analysis for experiments involving a large number of factors. The idea of group screening stems from that of group testing. The first paper on the group testing method appears to be Dorfman (1943). As an illustration, consider the problem of detecting a rare disease in the individual members of a large population. Suppose that a test is available that allows the disease to be detected with certainty from a small blood sample. The group testing method suggests that pooled blood samples be tested. If a pooled sample gives a positive result, the individual samples contributing to the pool are then tested. This strategy achieves substantial savings compared to that of testing all individual blood samples.

When a large number of factors are to be investigated in an experiment, the group screening method suggests the division of factors into groups to form group-factors. A group-factor takes the high level when all member factors within the group are set at the high level and takes the low level when all member factors are set at the low level. An orthogonal array is first employed to study group-factors. The member factors from significant group-factors are then examined in a subsequent experiment. A critical assumption for the success of the group screening method is that a group-factor is significant if and only if at least one member factor is significant. This assumption holds if the member factors within a group all have positive effects. The key is actually the knowledge about directions of the effects, as we can make a negative effect positive by switching the labels of the two levels. When no knowledge is available about directions of the effects, the possibility exists that two member factors that have appreciable effects of about the same magnitude but of opposite signs result in an insignificant group-factor.

Let $\boldsymbol{A} = (\boldsymbol{a}_1, \ldots, \boldsymbol{a}_{n-1})$ be a saturated orthogonal array $\mathrm{OA}(n, 2^{n-1}, 2)$. Then a group screening design based on $\boldsymbol{A}$ is simply

$$\boldsymbol{D} = (\boldsymbol{G}_1, \ldots, \boldsymbol{G}_{n-1}), \quad \text{where } \boldsymbol{G}_j = (\boldsymbol{a}_j, \ldots, \boldsymbol{a}_j)_{n \times k} \, . \tag{8.10}$$

Design $\boldsymbol{D}$ has $n-1$ groups of k factors, and each $\boldsymbol{G}_j$, consisting of k copies of the same $\boldsymbol{a}_j$, represents a group-factor whose member factors are mutually aliased with each other.

Generalizing group screening designs, Jones, Lekivetz, Majumdar, Nachtsheim, and Stallrich (2020) proposed a class of group-orthogonal supersaturated designs. Let $\boldsymbol{B}$ be an $n_1 \times k$ matrix of ± 1 and $\boldsymbol{A} = (\boldsymbol{a}_1, \ldots, \boldsymbol{a}_{n-1})$ be an $\mathrm{OA}(n, 2^{n-1}, 2)$. Then a group-orthogonal supersaturated design is

$$\boldsymbol{D}_1 = (\boldsymbol{G}_1, \ldots, \boldsymbol{G}_{n-1}), \quad \text{where } \boldsymbol{G}_j = \boldsymbol{a}_j \otimes \boldsymbol{B}. \tag{8.11}$$

This design of nn_1 runs has $n-1$ groups of k factors. For $\boldsymbol{D}_1$ in (8.11) to be a legitimate supersaturated design, we need to require $\boldsymbol{B}$ to have no fully aliased columns and k to satisfy $k(n-1) \geq nn_1$. Design $\boldsymbol{D}_1$ is called group-orthogonal because any two columns from different groups are orthogonal. In the initial construction of Jones et al. (2020), an extra group $\boldsymbol{G}_0 = \boldsymbol{1}_n \otimes \boldsymbol{B}$ is included and $\boldsymbol{B}$ is chosen to have orthogonal rows. With the row orthogonality of $\boldsymbol{B}$, the design given by augmenting $\boldsymbol{D}_1$ by $\boldsymbol{G}_0$ is $UE(s^2)$-optimal. As the factors in $\boldsymbol{G}_0$ may not be column-balanced, Jones et al. (2020) treated them as "fake" factors which can be used to estimate the error variance.

No column-balance is required of $\boldsymbol{B}$ to have column-balanced $\boldsymbol{D}_1$ in (8.11). But if we start with a column-balanced $\boldsymbol{B}$, then another class of group-orthogonal supersaturated designs can be constructed and this is given by $\boldsymbol{D}_2 = (\boldsymbol{G}_0, \boldsymbol{G}_1, \ldots, \boldsymbol{G}_{n-1})$, where $\boldsymbol{G}_0 = \boldsymbol{1}_n \otimes \boldsymbol{B}$ and $\boldsymbol{G}_j = \boldsymbol{a}_j \otimes \boldsymbol{B}$ for $j = 1, \ldots, n-1$. Both designs $\boldsymbol{D}_1$ and $\boldsymbol{D}_2$ enjoy some robust properties because each $\boldsymbol{G}_j$ for $j = 1, \ldots, n-1$ has a foldover structure. For example, the estimation of main effects from group $\boldsymbol{G}_1$ will not be affected by the presence of interactions of two factors from $\boldsymbol{G}_1$.

The group-orthogonal supersaturated designs just discussed may be more accurately termed between-group-orthogonal. Generalizing the group screening design $\boldsymbol{D}$ in (8.10) in a perpendicular direction, we can obtain a class of within-group-orthogonal supersaturated designs. After a column permutation, design $\boldsymbol{D}$ can be written as $\boldsymbol{D} = (\boldsymbol{A}_1, \ldots, \boldsymbol{A}_k)$ where $\boldsymbol{A}_j = \boldsymbol{A}$, the same $\mathrm{OA}(n, 2^{n-1}, 2)$. If we allow different $\mathrm{OA}(n, 2^{n-1}, 2)$'s to be used as $\boldsymbol{A}_j$'s so long as $\boldsymbol{D} = (\boldsymbol{A}_1, \ldots, \boldsymbol{A}_k)$ contains no fully aliased columns, then we obtain a supersaturated design with the property that factors within the same group are mutually orthogonal. This is precisely the construction of Tang and Wu (1997).

Group screening for detecting interactions has also been considered by many authors. We refer to Lewis and Dean (2001) and Vine, Lewis, and Dean (2005) for basic ideas and methods.

9

Other Nonregular Designs

Some selected topics on two-level nonregular designs that have not been covered in previous chapters are treated in this chapter. While the focus of Sections 9.1, 9.2, and 9.3 is on nonorthogonal designs, the subject of Sections 9.4 and 9.5 is about designs under a nonorthogonal parametrization of factorial effects, called the baseline parametrization. Section 9.1 introduces nonorthogonal main-effect plans and presents some optimality results. Estimation of main effects in the presence of two-factor interactions is considered in Section 9.2. Rechtschaffner designs and their generalizations are discussed in Section 9.3. Section 9.4 introduces the baseline parametrization and studies designs for estimating baseline main effects with and without assuming negligible interactions. Section 9.5 investigates the relationship between orthogonal and baseline parametrizations, and examines its applications to design construction. Section 9.6 presents an optimality result on spring balance weighing designs. We conclude the chapter in Section 9.7 with a brief discussion on search designs and parallel-flat designs.

9.1 Nonorthogonal Main-Effect Plans

The run sizes of nonregular orthogonal designs are multiples of 4, and hence are flexible compared with regular designs whose run sizes are powers of 2. Greatest run-size flexibility can be achieved by considering nonorthogonal designs, for which no restriction on run sizes is necessary.

From the estimability point of view, Webb (1968) gave a definition of resolution for nonorthogonal designs. A design has

- resolution III if it allows the estimation of all main effects under the assumption that all interactions are negligible,
- resolution IV if it allows the estimation of all main effects under the assumption that all interactions involving three or more factors are negligible, and
- resolution V if it allows the estimation of all main effects and all two-factor interactions under the assumption that all interactions involving three or more factors are negligible.

This definition of resolution has the same interpretation of design estimability as that of resolution by Box and Hunter (1961) for regular designs. In contrast, the generalized resolution discussed in Chapter 4 extends the resolution concept of regular designs to nonregular designs from the view point of projection designs.

Nonorthogonal designs of resolution III are the focus of the present section. Nonorthogonal designs of resolution IV and V are treated in Sections 9.2 and 9.3, respectively.

DOI: 10.1201/9781003371861-9

Let $\boldsymbol{D} = (\boldsymbol{d}_1, \ldots, \boldsymbol{d}_m)$ be a resolution III design of n runs for m factors and suppose all interactions are negligible. Then the main-effect model

$$\boldsymbol{Y} = \boldsymbol{X}\boldsymbol{\beta} + \boldsymbol{\epsilon} \tag{9.1}$$

is estimable under $\boldsymbol{D}$, where $\boldsymbol{\beta} = (\beta_0, \beta_1, \ldots, \beta_m)^T$ with β_0 being the intercept and $\beta_1, \ldots, \beta_m$ being the main effects, $\boldsymbol{X} = (\boldsymbol{1}_n, \boldsymbol{d}_1, \ldots, \boldsymbol{d}_m)$ is the model matrix, $\boldsymbol{Y}$ is the vector of n observations on the response variable, and $\boldsymbol{\epsilon}$ is the vector of random errors that have a zero mean and a constant variance σ^2.

The present design objective is to find the best design for estimating $\boldsymbol{\beta}$ in (9.1). By Theorem 2.4, we know that a design $\boldsymbol{D}$ is universally optimal if it is an OA$(n, 2^m, 2)$, an orthogonal array of strength two. However, such a design cannot be found unless $n = 4k$ for some integer k. Let $n = 4k+e$ where $e = 1, 2, 3$. Our intuition tells us that so long as n is not too small, designs obtained by adding e runs to an OA$(4k, 2^m, 2)$ or by deleting $4 - e$ runs from an OA$(4k+4, 2^m, 2)$ should perform well in terms of estimating $\boldsymbol{\beta}$. This procedure has a simple justification using the $(M.S)$-criterion. Noting that since $\text{tr}(\boldsymbol{X}^T\boldsymbol{X}) = (m+1)n$ is a constant, we only need to minimize $\text{tr}\{(\boldsymbol{X}^T\boldsymbol{X})^2\}$, as required by the $(M.S)$-criterion. If the columns of $\boldsymbol{X}$ cannot be made orthogonal, then an $\boldsymbol{X}$ that has nearly orthogonal columns should minimize $\text{tr}\{(\boldsymbol{X}^T\boldsymbol{X})^2\}$. This idea is formalized in the next result on $(M.S)$-optimality.

Theorem 9.1. *Assuming the existence of an OA$(4k, 2^m, 2)$ and an OA$(4k+4, 2^m, 2)$, we then have that*

(i) a design obtained by adding any one run to an OA$(4k, 2^m, 2)$ is (M.S)-optimal;

(ii) a design obtained by adding two runs $\boldsymbol{r}_1$ and $\boldsymbol{r}_2$ to an OA$(4k, 2^m, 2)$ is (M.S)-optimal, provided that $\boldsymbol{r}_1$ is the all-one run and $\boldsymbol{r}_2$ has $\lfloor m/2 \rfloor$ plus ones and $m - \lfloor m/2 \rfloor$ minus ones;

(iii) a design obtained by deleting any one run from an OA$(4k+4, 2^m, 2)$ is (M.S)-optimal.

The reader should be reminded that part (ii) of Theorem 9.1 will not hold if the second run $\boldsymbol{r}_2$ consists of $\lfloor m/2 \rfloor$ minus ones and $m - \lfloor m/2 \rfloor$ plus ones when m is odd. This is because we need the corresponding row $(1, \boldsymbol{r}_2)$ of the model matrix to have equal or nearly equal numbers of plus and minus ones.

The $(M.S)$-optimality is just a convenient surrogate for statistically meaningful criteria such as D, A and E optimality. Justification for intuitively appealing designs will be more compelling if they can be shown to be D-, A- or E-optimal.

Cheng (1978) introduced a class of optimality criteria, called type 1 criteria. We describe a simpler version of type 1 criteria, which was also used in Cheng (1980), Cheng (2014b), and elsewhere. Consider the general linear model in (2.1), and let $\lambda_1, \ldots, \lambda_p$ be the eigenvalues of the information matrix $\boldsymbol{M} = \boldsymbol{X}^T\boldsymbol{X}$, where $\boldsymbol{X}$ is the $n \times p$ model matrix that depends on a design. Then

$$\Phi(\boldsymbol{M}) = f(\lambda_1) + \cdots + f(\lambda_p)$$

is called a type 1 criterion if f is a real-valued function defined on $(0, \infty)$ such that $f'(x) < 0$, $f''(x) > 0$ and $f'''(x) < 0$ for $x \in (0, \infty)$, and $\lim_{x \to 0^+} f(x) = \infty$.

Though weaker than the universal optimality, the class of type 1 criteria is very rich. It includes all Φ_s criteria defined in Section 2.3, as can be seen by taking $f(x) = 1/x^s$; thus D-, A- and E-criteria are covered. With reference to the main-effect model in (9.1), we are now ready for the next result.

Theorem 9.2. *Assuming the existence of an OA$(4k, 2^m, 2)$, we then have that*

(i) a design obtained by adding any one run to an $OA(4k, 2^m, 2)$ is optimal under all type 1 criteria;

(ii) a design obtained by adding two runs $\boldsymbol{r}_1$ and $\boldsymbol{r}_2$ to an $OA(4k, 2^m, 2)$ is optimal under all type 1 criteria, where $\boldsymbol{r}_1$ is the all-one run and $\boldsymbol{r}_2$ has $\lfloor m/2 \rfloor$ plus ones and $m - \lfloor m/2 \rfloor$ minus ones; and

(iii) a design obtained by adding any three runs to an $OA(4k, 2^m, 2)$ is E-optimal.

Parts (i) and (iii) of Theorem 9.2 are due to Cheng (1980), and part (ii) of Theorem 9.2 is from Jacroux, Wong, and Masaro (1983). The proofs of these results are highly technical and thus not given here.

Apparently missing from Theorem 9.2 are designs obtained by deleting one run from an orthogonal array. Unfortunately and counter-intuitively, such designs are not always optimal, but limited results are available. Suppose an orthogonal array of strength two has an all-one run, which can always be achieved by switching the signs of certain columns. Then the design given by deleting this all-one run is D-optimal if $n \geq 2m - 3$ (Galil and Kiefer, 1980). The same design can be shown to be A-optimal under a similar but more complicated condition (Sathe and Shenoy, 1989); specifically, for each m, there is a number $n_A(m)$ such that the design given by deleting an all-one run from an orthogonal array is A-optimal as long as $n \geq n_A(m)$, and we have $n_A(m) > 2m - 3$. The Φ_s-optimality of this design was given by Corollary 4.2 of Cheng (1992). In this case, for any m and s such that $s > 0$, there is an $n_s(m)$ such that the design obtained by deleting an all-one run from an orthogonal array of strength 2 is Φ_s-optimal as long as $n \geq n_s(m)$. Here of course we have $n_0(m) = 2m - 3$ and $n_1(m) = n_A(m)$. Furthermore, if $s_1 > s_2$, then $n_{s_1}(m) > n_{s_2}(m)$.

Fractional factorial designs are intimately related to weighing designs. The earliest work on weighing designs can be found in Yates (1935) and Hotelling (1944). For a review on the subject, we refer to Shah and Sinha (1989). A weighing design can be either a chemical balance or a spring balance weighing design. Suppose that n weighings are granted to ascertain the weights of q objects. Then a chemical balance weighing design is an $n \times q$ matrix $\boldsymbol{W} = (w_{ij})$ of ± 1, which means that in the ith weighing, the jth object is put in the left pan if $w_{ij} = -1$ and in the right pan if $w_{ij} = 1$. A spring balance weighing design is an $n \times q$ matrix $\boldsymbol{W} = (w_{ij})$ of 0 and 1, with the interpretation that in the ith weighing, the jth object is present if $w_{ij} = 1$ and absent if $w_{ij} = 0$.

Under a chemical balance weighing design $\boldsymbol{W} = (w_{ij})_{n \times q}$, the linear model is given by

$$\boldsymbol{Y} = \boldsymbol{W}\boldsymbol{\theta} + \boldsymbol{\epsilon},$$

where $\boldsymbol{Y}$ is the vector of n measurements, $\boldsymbol{\theta}$ is the vector of the weights of q objects, and $\boldsymbol{\epsilon}$ is the vector of random errors that are independent and have a zero mean and a constant variance. Comparing this model with the main-effect model for factorial designs in (9.1), we see that if $\boldsymbol{W}$ has a column of all ones, then deleting this column of all ones from $\boldsymbol{W}$ leads to a factorial design of n runs for $q - 1$ factors. Every chemical balance weighing design can be assumed to have its first column consisting of all ones, which is so because if an entry in the first column is -1, we can make it $+1$ by multiplying the entire row by -1 without changing the information matrix $\boldsymbol{W}^T\boldsymbol{W}$. We therefore conclude that

Every optimal chemical balance weighing design of n runs for q objects gives rise to an optimal two-level factorial design of n runs for $m = q - 1$ factors, and vice versa.

In fact, some of the optimality results in Theorem 9.2 were originally presented in terms of weighing designs. Upon a closer look at this theorem, we see that it does not cover all (n, m) combinations. For example, the case of $n = 4k + 2$ and $m = 4k + 1$ is missing from

Theorem 9.2. Using the just mentioned connection between weighing designs and factorial designs, optimal factorial designs of $n = 4k + 2$ runs for $m = 4k + 1$ factors are obtainable from optimal chemical balance weighing designs of $n = 4k + 2$ runs for $q = 4k + 2$ objects. The latter can be found in Yang (1966) and Yang (1968).

Theorem 9.2 is about optimal designs for estimating the main-effect model in (9.1). If the intercept β_0 is of little interest, it is then sensible to find optimal designs for estimating only the main effects $\beta_1, \ldots, \beta_m$, in which case the intercept is treated as a nuisance parameter. For this slightly different design problem, results similar to parts (i) and (ii) of Theorem 9.2 are available in Cheng (2014b), presented in terms of biased chemical balance weighing designs. The design in part (i) of Theorem 9.2 continues to be optimal for estimating main effects with respect to all type 1 criteria. The optimality result for the design in part (ii) of Theorem 9.2 is weaker under the new setting, but suffices for all practical purposes as this design is still D-, A- and E-optimal.

9.2 Nonorthogonal Designs of Resolution IV

Suppose that all interactions involving three or more factors are negligible. Then a design $\boldsymbol{D} = (\boldsymbol{d}_1, \ldots, \boldsymbol{d}_m)$ of resolution IV allows unbiased estimation of all main effects even when two-factor interactions are nonnegligible. The present situation indicates the following linear model

$$\boldsymbol{Y} = \boldsymbol{1}_n\beta_0 + \boldsymbol{X}_1\boldsymbol{\beta}_1 + \boldsymbol{X}_2\boldsymbol{\beta}_2 + \boldsymbol{\epsilon}, \tag{9.2}$$

where β_0 is the intercept, $\boldsymbol{\beta}_1$ is the vector of main effects and $\boldsymbol{X}_1 = \boldsymbol{D}$ is the design matrix, $\boldsymbol{\beta}_2$ represents the vector of all two-factor interactions with $\boldsymbol{X}_2$ being the corresponding model matrix, and $\boldsymbol{\epsilon}$ is the vector of independent random errors that have a zero mean and a constant variance σ^2.

Under a resolution IV design $\boldsymbol{D}$, the true model in (9.2) may not be estimable but the main effects $\boldsymbol{\beta}_1$ are. This indicates that there exists an $m \times n$ matrix $\boldsymbol{A}$ such that $E(\boldsymbol{AY}) = \boldsymbol{\beta}_1$ for all possible values of β_0 and $\boldsymbol{\beta}_2$. More concretely, we have that

$$E(\boldsymbol{AY}) = \boldsymbol{A}(\boldsymbol{1}_n\beta_0 + \boldsymbol{X}_1\boldsymbol{\beta}_1 + \boldsymbol{X}_2\boldsymbol{\beta}_2) = \boldsymbol{\beta}_1 \text{ for all } \beta_0 \text{ and } \boldsymbol{\beta}_2,$$

which implies that

$$\boldsymbol{AX}_1 = \boldsymbol{I}_m, \quad \boldsymbol{A1}_n = \boldsymbol{0}_m, \quad \boldsymbol{AX}_2 = \boldsymbol{0}_{m \times \binom{m}{2}}, \tag{9.3}$$

where $\boldsymbol{I}_m$ is the identity matrix of order m, $\boldsymbol{0}_m$ is a column vector of m zeros, and $\boldsymbol{0}_{m \times \binom{m}{2}}$ is an $m \times \binom{m}{2}$ matrix of all zeros. Webb (1968) established the following necessary condition for the existence of a resolution IV design.

Theorem 9.3. *Let $\boldsymbol{D}$ be a resolution IV design of n runs for m factors. Then we must have that $n \geq 2m$.*

Proof. From $\boldsymbol{AX}_1 = \boldsymbol{I}_m$ in (9.3), we conclude that $\boldsymbol{X}_1$ must have rank m, meaning that the columns $\boldsymbol{d}_1, \ldots, \boldsymbol{d}_m$ of $\boldsymbol{D} = \boldsymbol{X}_1$ are linearly independent. This further implies that $\boldsymbol{1}_n = \boldsymbol{d}_1\boldsymbol{d}_1, \boldsymbol{d}_1\boldsymbol{d}_2, \ldots, \boldsymbol{d}_1\boldsymbol{d}_m$ are linearly independent. We next show that the $2m$ columns

$$\boldsymbol{d}_1, \ldots, \boldsymbol{d}_m, \boldsymbol{1}_n, \boldsymbol{d}_1\boldsymbol{d}_2, \ldots, \boldsymbol{d}_1\boldsymbol{d}_m$$

must also be linearly independent. Let $\alpha_1, \ldots, \alpha_m, \gamma_1, \ldots, \gamma_m$ be real constants such that

$$\alpha_1\boldsymbol{d}_1 + \cdots + \alpha_m\boldsymbol{d}_m + \gamma_1\boldsymbol{1}_n + \gamma_2(\boldsymbol{d}_1\boldsymbol{d}_2) + \cdots + \gamma_m(\boldsymbol{d}_1\boldsymbol{d}_m) = \boldsymbol{0}_n. \tag{9.4}$$

From (9.3), we have $\boldsymbol{A}\mathbf{1}_n = \mathbf{0}_m$ and $\boldsymbol{A}(\boldsymbol{d}_1\boldsymbol{d}_j) = \mathbf{0}_m$ for $j = 2, \ldots, m$, as $\boldsymbol{d}_1\boldsymbol{d}_j$ is one column of $\boldsymbol{X}_2$. Left-multiplying $\boldsymbol{A}$ to both sides of equation (9.4), we obtain that

$$\boldsymbol{A}(\alpha_1\boldsymbol{d}_1 + \cdots + \alpha_m\boldsymbol{d}_m) = \mathbf{0}_m.$$

Because $\boldsymbol{A}\boldsymbol{X}_1 = \boldsymbol{I}_m$, we obtain that

$$\boldsymbol{A}(\alpha_1\boldsymbol{d}_1 + \cdots + \alpha_m\boldsymbol{d}_m) = \boldsymbol{A}\boldsymbol{X}_1(\alpha_1, \ldots, \alpha_m)^T = (\alpha_1, \ldots, \alpha_m)^T,$$

showing that $\alpha_1 = \cdots = \alpha_m = 0$. Therefore

$$\gamma_1\mathbf{1}_n + \gamma_2(\boldsymbol{d}_1\boldsymbol{d}_2) + \cdots + \gamma_m(\boldsymbol{d}_1\boldsymbol{d}_m) = \mathbf{0}_n,$$

from which we obtain that $\gamma_1 = \cdots = \gamma_m = 0$ owing to the linear independence of $\mathbf{1}_n, \boldsymbol{d}_1\boldsymbol{d}_2, \ldots, \boldsymbol{d}_1\boldsymbol{d}_m$. This shows that $\boldsymbol{d}_1, \ldots, \boldsymbol{d}_m, \mathbf{1}_n, \boldsymbol{d}_1\boldsymbol{d}_2, \ldots, \boldsymbol{d}_1\boldsymbol{d}_m$ are linearly independent in the n-dimensional Euclidean space. Therefore $n \geq 2m$. □

The proof of Theorem 9.3 actually proves that, for a resolution IV design, the grand mean, the main effects, and all two-factor interactions involving any given factor are jointly estimable under the assumption that the remaining two-factor interactions are negligible. With just a bit more effort, one can show that the converse is also true. This provides a characterization of resolution IV designs, which was also obtained by Webb (1968).

Theorem 9.4. *A design has resolution IV if and only if it allows joint estimation of the grand mean, the main effects, and all two-factor interactions involving any given factor under the assumption that the remaining two-factor interactions and all higher-order interactions are negligible.*

An OA$(n, 2^m, 3)$ is a resolution IV design. Theorems 9.3 and 9.4 under the orthogonal array setting are well-known results. For example, Theorem 9.4 under the orthogonal array setting is Theorem 2.6. The major difference here is that under an OA$(n, 2^m, 3)$, the $2m$ columns $\boldsymbol{d}_1, \ldots, \boldsymbol{d}_m, \mathbf{1}_n, \boldsymbol{d}_1\boldsymbol{d}_2, \ldots, \boldsymbol{d}_1\boldsymbol{d}_m$ are not just linearly independent; they are also mutually orthogonal. The next result, immediate from Theorem 9.4, does not have an orthogonal array counterpart.

Corollary 9.1. *Augmenting a design of resolution IV with any number of additional runs still gives a design of resolution IV.*

Resolution IV designs are obtainable using the foldover technique. Suppose that n is even and $n \geq 2m$. Let $\boldsymbol{D}_1$ be any $(n/2) \times m$ matrix of ± 1 with independent columns. Then $\boldsymbol{D} = (\boldsymbol{D}_1^T, -\boldsymbol{D}_1^T)^T$ is a resolution IV design of n runs for m factors, which is easily verifiable using Theorem 9.4. Adding one run to the above design $\boldsymbol{D}$ is still a resolution IV design according to Corollary 9.1. We have therefore shown that the run size condition $n \geq 2m$ is also sufficient for the existence of a resolution IV design.

Optimal foldover designs for estimating main effects can be obtained by taking $\boldsymbol{D}_1$ to be the model matrix of an optimal main-effect plan, which has an additional column of all ones for the intercept besides the design matrix. Optimal main-effect plans have been covered in the last section.

A minimal resolution IV design for m factors has $n = 2m$ runs. Webb (1968) conjectured that foldover designs are the only minimal resolution IV designs. The conjecture was proved by Margolin (1969a) and Margolin (1969b).

Theorem 9.5. *A resolution IV design $\boldsymbol{D}$ of $n = 2m$ runs for m factors must be a foldover design.*

Proof. As the full proof given by Margolin (1969b) is fairly tedious, we present a simpler proof from Margolin (1969a), which was done under the condition that design $\boldsymbol{D}$ has balanced columns. The simpler proof is interesting because it makes use of the theory of J-characteristics. It is amazing to see that Margolin (1969a) used J-characteristics, albeit informally. By Theorem 3.4, we need to show that $J_u(\boldsymbol{D}) = 0$ for all u with odd $|u|$. The proof is by induction. As $\boldsymbol{D}$ has balanced columns, we have that $J_u(\boldsymbol{D}) = 0$ for $|u| = 1$. Now assume that $J_u(\boldsymbol{D}) = 0$ for all u with odd $|u|$ and $|u| \le 2k-1$. We will show that $J_u(\boldsymbol{D}) = 0$ for all u with $|u| = 2k+1$.

Consider the J-characteristic $J(\boldsymbol{d}_{j_1}, \ldots, \boldsymbol{d}_{j_{2k+1}})$ for any $2k+1$ distinct columns $\boldsymbol{d}_{j_1}, \ldots, \boldsymbol{d}_{j_{2k+1}}$ of design $\boldsymbol{D}$. Since $\boldsymbol{d}_1, \ldots, \boldsymbol{d}_m, \boldsymbol{d}_{j_1}\boldsymbol{d}_1, \ldots, \boldsymbol{d}_{j_1}\boldsymbol{d}_m$ are linearly independent and thus form a basis in the $n = 2m$ dimensional Euclidean space, we must have that

$$\boldsymbol{d}_{j_2}\boldsymbol{d}_{j_3} = \sum_{j=1}^{m} \alpha_j \boldsymbol{d}_j + \sum_{j=1}^{m} \gamma_j (\boldsymbol{d}_{j_1}\boldsymbol{d}_j) \tag{9.5}$$

for some real constants $\alpha_1, \ldots, \alpha_m, \gamma_1, \ldots, \gamma_m$. Left-multiplying (9.5) by $\boldsymbol{A}$ and using the facts in (9.3), we obtain $\alpha_j = 0$ for all $j = 1, \ldots, m$. This leads to $\boldsymbol{d}_{j_2}\boldsymbol{d}_{j_3} = \sum_{j=1}^{m} \gamma_j (\boldsymbol{d}_{j_1}\boldsymbol{d}_j)$ and hence $\boldsymbol{d}_{j_1}\boldsymbol{d}_{j_2}\boldsymbol{d}_{j_3} = \sum_{j=1}^{m} \gamma_j \boldsymbol{d}_j$. We therefore obtain that

$$\boldsymbol{d}_{j_1}\boldsymbol{d}_{j_2}\boldsymbol{d}_{j_3} \cdots \boldsymbol{d}_{j_{2k+1}} = \sum_{j=1}^{m} \gamma_j (\boldsymbol{d}_j \boldsymbol{d}_{j_4} \cdots \boldsymbol{d}_{j_{2k+1}}).$$

Taking the J-characteristic on both sides gives

$$J(\boldsymbol{d}_{j_1}, \ldots, \boldsymbol{d}_{j_{2k+1}}) = \sum_{j=1}^{m} \gamma_j J(\boldsymbol{d}_j, \boldsymbol{d}_{j_4}, \ldots, \boldsymbol{d}_{j_{2k+1}}).$$

The left hand side of this equation must be zero because the J-characteristics on the right hand side involve either $2k-1$ columns or $2k-3$ columns, and are all equal to zeros by the induction hypothesis. Theorem 9.5 is now proved. □

The orthogonal array version of Theorem 9.5 is Theorem 3.5, which states that an $OA(n, 2^m, 3)$ must be a foldover design if $2m \le n \le 3m$. One then wonders if it is possible to establish Theorem 9.5 for other n values in the range of $2m < n \le 3m$. The answer is in the negative. Consider a minimal resolution IV design of $2m$ runs for m factors, which must be a foldover design according to Theorem 9.5. Adding any $n - 2m$ runs still gives a resolution IV design as per Corollary 9.1. So long as the added $n-2m$ runs do not constitute a foldover design, the augmented design cannot be a foldover design.

As can be seen from Theorem 9.4, nonorthogonal resolution IV designs allow the estimation of certain two-factor interactions besides the main effects. However, more common and sensible situations in practice are that the experimenter is willing to assume that there are a small number of active two-factor interactions but does not feel comfortable to specify which two-factor interactions are active. Under such circumstances, appropriate criteria for evaluating resolution IV designs are those of estimation capacity and information capacity, previously discussed for supersaturated designs. The estimation capacity sequence $(E_1, E_2, \ldots)$ now has the meaning that E_f is the proportion of estimable models among all models that consist of the intercept, all main effects, and f two-factor interactions. In the information capacity sequence $(D_1, D_2, \ldots)$, D_f is defined as the average of the D-efficiencies of all models that consist of the intercept, all main effects, and f two-factor interactions.

Lin, Miller, and Sitter (2008) conducted a dedicated study on nonorthogonal foldover designs. They suggested the use of MDS-resolution and MDS-aberration criteria for design selection. A minimal dependent set (MDS) under the present setting is defined as a set of two-factor interactions such that the model that contains the intercept, all main effects, and this set of two-factor interactions is not estimable, but if any one of the two-factor interactions is removed, then the resulting model is estimable. The MDS-resolution and MDS-aberration criteria are then defined using the sizes of MDS's, in the same way as that for supersaturated designs. As discussed in Section 8.5, the MDS-aberration criterion is very similar to that of estimation capacity. Lin, Miller, and Sitter (2008) also created catalogs of nonisomorphic foldover designs by computer search. Their computer search is complete for $n = 10, 12, 14, 16$, and 18, and quite extensive for $n = 20, 22$, and 24. These catalogs provide valuable resources as one can use them to easily find a suitable design under whatever criterion that is deemed appropriate.

If $n < 2m$, then resolution IV designs do not exist but resolution III designs are available so long as $n > m$. For a given resolution III design, one can still fit the main-effect model $\boldsymbol{Y} = \mathbf{1}_n\beta_0 + \boldsymbol{X}_1\boldsymbol{\beta}_1 + \boldsymbol{\epsilon}$ to obtain the least squares estimator for $\boldsymbol{\beta}_1$, where β_0 is the intercept, $\boldsymbol{\beta}_1$ is the vector of the main effects and $\boldsymbol{X}_1$ is the design matrix. This least squares estimator for the main effects is biased unless all interactions are negligible. Provided that the effect hierarchy principle applies, useful designs can be found using the idea of minimum G_2-aberration. More specifically, we select a design, among all optimal designs for the estimation of main effects, by first minimizing the bias component due to two-factor interactions, then that due to three-factor interactions, and so on. Minimum G_2-aberration nonorthogonal designs of this kind were considered by Zhang and Mukerjee (2013) for $n = 4k + 1$ runs and by Huda and Mukerjee (2013) for $n = 4k + 2$ runs.

The idea of Bayesian D-optimal designs in DuMouchel and Jones (1994) bears some resemblance to that of minimum G_2-aberration designs. In DuMouchel and Jones (1994), main effects are primary terms in the model and two-factor interactions are potential terms. By assigning differential priors to the primary terms and potential terms, the resulting Bayesian D-optimal designs mitigate the impact of two-factor interactions on the estimation of main effects.

9.3 Rechtschaffner Designs and Generalizations

A resolution V design allows estimation of all main effects and all two-factor interactions under the assumption that all interactions involving three or more factors are negligible, and thus allows the estimation of the model in (9.2). The total number of parameters in this full second-order model is $1 + m + m(m-1)/2$. A resolution V design is said to be saturated if it has the minimum number ($n = 1 + m + m(m-1)/2$) of runs.

Without a proof, Rechtschaffner (1967) provided a simple construction of saturated resolution V designs. For m factors, his design consists of three sets of runs:

(i) one run by setting all factors at level $+1$;

(ii) all the m runs by setting exactly one factor at level -1 and the remaining factors at level $+1$; and

(iii) all the $m(m-1)/2$ runs by setting exactly two factors at level -1 and the remaining factors at level $+1$.

Rechtschaffner's idea of construction is rooted in the one-factor-at-a-time design for the main-effect model, which consists of the run in (i) for the intercept and the runs in (ii) for the main effects. When the two-factor interactions are added to the main-effect model, Rechtschaffner (1967) simply added the runs in (iii) for the two-factor interactions.

We now generalize this idea to general settings. To every subset u of the set $\{1, \ldots, m\}$ of factors, there corresponds a factorial effect β_u, which involves all factors in u. The empty set naturally corresponds to the intercept. To every u, there also corresponds a run $\boldsymbol{x}_u$ whose components are -1 for all factors in u and $+1$ otherwise. Let $\mathcal{C}$ be a collection of subsets of $\{1, \ldots, m\}$. Corresponding to $\mathcal{C}$ are a model consisting of all effects β_u for $u \in \mathcal{C}$ and a design consisting of all runs $\boldsymbol{x}_u$ for $u \in \mathcal{C}$. Let $M_{\mathcal{C}}$ and $\boldsymbol{D}_{\mathcal{C}}$ denote the model and design corresponding to $\mathcal{C}$, respectively. It is hoped that $\boldsymbol{D}_{\mathcal{C}}$ allows the estimation of $M_{\mathcal{C}}$. This is indeed the case provided that $\mathcal{C}$ is an *echelon collection*, meaning that $\mathcal{C}$ has the property that if a u is in $\mathcal{C}$, then all subsets v of u are also in $\mathcal{C}$. For an echelon collection $\mathcal{C}$, we call $M_{\mathcal{C}}$ an echelon model and $\boldsymbol{D}_{\mathcal{C}}$ an echelon design.

Theorem 9.6. *Suppose that $\mathcal{C}$ is an echelon collection. Then design $\boldsymbol{D}_{\mathcal{C}}$ allows the estimation of model $M_{\mathcal{C}}$.*

The applicability of Theorem 9.6 is fairly broad as the most commonly used models are echelon models. Examples include the main-effect model, the full second-order model, and the model that contains the intercept, all main effects, and any set of two-factor interactions. A simple example of a model that is not an echelon model is the main-effect model without an intercept.

We will prove Theorem 9.6 in Section 9.5 with the assistance of the relationship between orthogonal and baseline parametrizations. A general version of Theorem 9.6 for multi-level designs is also available; see Theorem 15.25 of Cheng (2014a). This result was proved by Caboara, Pistone, Riccomagno, and Wynn (1999) using the theory of generalized confounding of Pistone and Wynn (1996). An unpublished elementary proof was also obtained by Dursun Bulutoglu and Ching-Shui Cheng around the same time.

We now return to the Rechtschaffner design for the full second-order model. Unless m is small, this design is not very efficient. The Rechtschaffner design we presented earlier in the section has seven other variants. The eight different versions of the Rechtschaffner design can be obtained by switching the two levels in each of the three sets of runs. Qu (2007) showed that the version recommended by Rechtschaffner (1967), which is the one by switching the two levels of the runs in sets (i) and (iii), is indeed D- and A-optimal among the eight versions. Tobias (1996) demonstrated that saturated designs that are better than Rechtschaffner designs in terms of D- and A-optimality can be found using a computer search for $m > 6$.

9.4 Designs for Estimating Baseline Main Effects

The present section introduces the baseline parametrization for factorial effects and presents results on designs for estimating baseline main effects. Some general results on baseline designs will be discussed in Section 9.5.

9.4.1 Baseline parametrization

The factorial effects that have been in use so far in this book can be termed as ones under an orthogonal parametrization since they provide a set of orthogonal treatment contrasts. With the two levels labeled as ± 1, a linear model as in (2.12) or its concise form as in (2.13) that expresses the treatment mean as a linear function of factorial effects automatically leads to the orthogonal factorial effects as in (2.14). Alternatively, one can define orthogonal factorial effects directly and explicitly using (2.14), which then results in the linear model in (2.12) or that in (2.13). This orthogonal parametrization of factorial effects offers mathematical elegance as well as computational advantage, and is thus adopted in almost all theory and practice of factorial designs. The appropriateness of the orthogonal parametrization hinges on the symmetry of the two levels in that no importance can be attached to one or the other level, which is indeed the case for most studies.

In certain applications, however, the two levels are asymmetrical. One level, called the control or baseline level, is the default, which is deemed more important than the other level, called the test level. Kerr (2006) gave an example of a toxicological study on several binary factors with each factor representing the presence or absence of a toxin. Then the state of absence is a natural baseline level of each factor. In industrial experiments, if one seeks to improve the current settings of factors by examining a different level for each factor, then the currently used level of each factor constitutes the baseline level and the level to be examined is the test level. The rising importance of situations with asymmetrical levels is largely due to recent work on microarray experiments; see Yang and Speed (2002), Glonek and Solomon (2004), and Banerjee and Mukerjee (2008) for related work.

Let the two levels of each factor be denoted by 0 and 1, with 0 representing the baseline level and 1 the test level. Let $\mu(z_1, \ldots, z_m)$ be the mean response at the level combination $(z_1, \ldots, z_m)$ where $z_j = 0, 1$. Then the full model

$$\mu(z_1, \ldots, z_m) = \theta_0 + \sum_{j=1}^{m} \theta_j z_j + \sum_{j_1<j_2} \theta_{j_1 j_2} z_{j_1} z_{j_2} + \cdots + \theta_{1\cdots m} z_1 \cdots z_m \tag{9.6}$$

provides an implicit definition for the baseline factorial effects θ's, where θ_0 is the intercept, θ_j's are the main effects, $\theta_{j_1 j_2}$'s are the two-factor interactions, and so on. A simple form of the model in (9.6) can be obtained with a bit of notation. To every subset u of the set $\{1, 2, \ldots, m\}$ of factor labels, there correspond a factorial effect θ_u and a mean response $\mu_u = \mu(\boldsymbol{z}_u)$, where $\boldsymbol{z}_u = (z_{u1}, \ldots, z_{um})$ with $z_{uj} = 1$ if $j \in u$ and $z_{uj} = 0$ if $j \notin u$. For example, for $u = \phi$, we have $\theta_\phi = \theta_0$ and $\mu_\phi = \mu(0, \ldots, 0)$, and for $u = \{1, 2\}$, we have θ_{12} and $\mu_{12} = \mu(1, 1, 0, \ldots, 0)$. Then the model in (9.6) has the following compact form

$$\mu_u = \sum_{v \subseteq u} \theta_v, \tag{9.7}$$

where $v \subseteq u$ means that v is a subset of u, which includes the possibility $v = u$. Equation (9.7) then leads to an explicit and recursive definition of baseline factorial effects θ's:

$$\theta_\phi = \mu_\phi, \quad \theta_u = \mu_u - \sum_{v \subsetneq u} \theta_v, \tag{9.8}$$

where $v \subsetneq u$ means that v is a proper subset of u.

Example 9.1. *Consider $m = 2$, the case of two factors. Let $\mu(0,0)$, $\mu(1,0)$, $\mu(0,1)$, $\mu(1,1)$ be denoted by $\mu_{00}, \mu_{10}, \mu_{01}, \mu_{11}$, respectively. Then by (9.8), we have that*

$$\theta_0 = \mu_{00}, \quad \theta_1 = \mu_{10} - \mu_{00},$$

$$\theta_2 = \mu_{01} - \mu_{00}, \quad \theta_{12} = \mu_{11} - \mu_{10} - \mu_{01} + \mu_{00}.$$

We see that θ_1, θ_2, and θ_{12} are still treatment contrasts, but they are not mutually orthogonal. We now label the two levels 0 and 1 as $+1$ and -1, respectively. By (2.12), the corresponding factorial effects under the orthogonal parametrization are

$$\beta_0 = (\mu_{00} + \mu_{10} + \mu_{01} + \mu_{11})/4, \quad \beta_1 = (\mu_{00} - \mu_{10} + \mu_{01} - \mu_{11})/4,$$

$$\beta_2 = (\mu_{00} + \mu_{10} - \mu_{01} - \mu_{11})/4, \quad \beta_2 = (\mu_{00} - \mu_{10} - \mu_{01} + \mu_{11})/4.$$

Clearly, β_1, β_2, and β_{12} are mutually orthogonal contrasts.

For general m, the definition of θ_0, θ_1, θ_2, and θ_{12} is almost identical to the case of $m = 2$, which still involves only four treatment means as given by

$$\theta_0 = \mu_{0\cdots0}, \quad \theta_1 = \mu_{10\cdots0} - \mu_{0\cdots0},$$

$$\theta_2 = \mu_{010\cdots0} - \mu_{0\cdots0}, \quad \theta_{12} = \mu_{110\cdots0} - \mu_{10\cdots0} - \mu_{010\cdots0} + \mu_{0\cdots0}.$$

The usefulness of the baseline factorial effects becomes crystal clear if we consider industrial experiments where one attempts to improve the current settings by testing another level for each factor. If we only change the setting for the first factor, then the change in mean response is precisely θ_1. If we want to see what happens when we change the settings for the first and second factors, then the change in mean response is simply $\theta_1 + \theta_2 + \theta_{12}$. In contrast, the corresponding expressions are much more complicated under the orthogonal parametrization.

We refer the readers to a recent work by Chen and Tang (2024) for a general framework of mixed parametrization to deal with situations where some factors have baseline levels while the two levels of the other factors are symmetrical.

9.4.2 Minimum aberration designs

Suppose that interactions are all negligible. Then the model in (9.6) becomes

$$\mu(z_1, \ldots, z_m) = \theta_0 + \sum_{j=1}^{m} \theta_j z_j. \tag{9.9}$$

Our first objective is to find designs that provide best estimation for the main effects $\theta_1, \ldots, \theta_m$. Let $\boldsymbol{Z} = (z_{ij})_{n\times m}$ be an n-run design for m factors, where $z_{ij} = 0, 1$. Then the linear model corresponding to (9.9) under the data setting can be written as

$$\boldsymbol{Y} = \boldsymbol{Z}_1\boldsymbol{\theta}_1 + \boldsymbol{\epsilon}, \tag{9.10}$$

where $\boldsymbol{Z}_1 = (\boldsymbol{1}_n, \boldsymbol{Z})$, $\boldsymbol{\theta}_1 = (\theta_0, \theta_1, \ldots, \theta_m)^T$, $\boldsymbol{Y} = (Y_1, \ldots, Y_n)^T$ is the vector of responses, and $\boldsymbol{\epsilon} = (\epsilon_1, \ldots, \epsilon_n)^T$ is the vector of random errors that have a zero mean and a constant variance σ^2.

Let $\hat{\theta}_j$ be the least squares estimator of θ_j under the model in (9.10). The next theorem summarizes the optimality results presented in Mukerjee and Tang (2012), who based their findings on the previous results of Moriguti (1954) and Cheng (1980).

Theorem 9.7. *Let $\boldsymbol{Z}$ be an $OA(n, 2^m, 2)$. We then have that*

(i) *design $\boldsymbol{Z}$ minimizes $Var(\hat{\theta}_j)$ for every $j = 1, \ldots, m$;*

(ii) *design $\boldsymbol{Z}$ is universally optimal for estimating $\theta_1, \ldots, \theta_m$.*

The design problem for estimating the main effects $\theta_1, \ldots, \theta_m$ under (9.10) is equivalent to that of biased spring balance weighing designs. Cheng (2014b) explored the relationships among four kinds of weighing designs: unbiased chemical balance, biased chemical balance, unbiased spring balance, and biased spring balance weighing designs.

However, if interactions cannot be safely ignored, then the model in (9.10) is inadequate. In this case, the true model is given by

$$Y = Z_1\theta_1 + Z_2\theta_2 + \cdots + Z_m\theta_m + \epsilon, \tag{9.11}$$

where θ_j collects all j-factor interactions and Z_j is the corresponding model matrix for $j = 2, \ldots, m$. The least squares estimator $\hat{\theta}_1 = (Z_1^T Z_1)^{-1} Z_1^T Y$ of θ_1 from the main-effect model (9.10) is biased under the true model (9.11), with the bias given by

$$(Z_1^T Z_1)^{-1} Z_1^T (Z_2\theta_2 + \cdots + Z_m\theta_m). \tag{9.12}$$

As we are concerned with the estimation of main effects, we only need to consider the bias in the main effect estimators. Let B_j be the matrix obtained by deleting the first row of $(Z_1^T Z_1)^{-1} Z_1^T Z_j$ for $j = 2, \ldots, m$. Then the size $K_j = \text{trace}(B_j B_j^T)$ of B_j provides a measure of the contribution of θ_j to the bias of the main effect estimation. Under the effect hierarchy principle, one should select designs that sequentially minimize $K_2, \ldots, K_m$. This gives a minimum G_2-aberration criterion under the baseline parametrization. An explicit expression for K_j has been found by Mukerjee and Tang (2012). To present this result, we need α_u, which denotes the number of times the all-one run $(1, \ldots, 1)$ occurs as a row in the subarray Z_u, which is the projection design of Z onto the columns in u, where u is a subset of $\{1, \ldots, m\}$.

Theorem 9.8. *Let Z be an $OA(n, 2^m, 2)$. For $j = 2, \ldots, m-1$, we have that*

$$K_j = (4/n^2)\left(j \sum_{|u|=j} \alpha_u^2 + \sum_{|u|=j+1} \left(\sum_{|v|=j, v\subset u} (2\alpha_u - \alpha_v)^2 \right)\right).$$

For $j = m$, we have that $K_m = (4m/n^2)\alpha_{\{1,\ldots,m\}}^2$.

Miller and Tang (2016) obtained the next corollary.

Corollary 9.2. *If Z is an $OA(n, 2^m, t)$ with $t \geq 2$, then we have that for $2 \leq j \leq t-1$, $K_j = j\binom{m}{j}/2^{2j-2}$, a constant not depending on Z, and that*

$$K_t = (1/2^{2t-2})\left(t\binom{m}{t} + (t+1)A_{t+1}\right),$$

where A_{t+1} is given in (5.1), which is used to define minimum G_2-aberration under the orthogonal parametrization.

Corollary 9.2 implies that minimizing the leading term K_t under the baseline parametrization is equivalent to minimizing the leading term A_{t+1} under the orthogonal parametrization. Thus, an OA$(n, 2^m, t)$ with $t \geq 2$ that has minimum G_2-aberration under the orthogonal parametrization will do well under the minimum G_2-aberration under the baseline parametrization. The two criteria cease to agree beyond the leading terms.

General results on minimum G_2-aberration baseline designs are hard to come by. Nonetheless, Mukerjee and Tang (2012) found a result for the saturated and nearly saturated situations.

Theorem 9.9. *Let Z be an $OA(n, 2^m, 2)$ for $m = n-1$ or $n-2$, with a row of zeros. Then design Z has minimum G_2-aberration under the baseline parametrization.*

Given design Z that is an OA$(n, 2^m, 2)$, interchanging levels 0 and 1 in one or more columns of Z has an effect on the K_j values for $j \geq 3$, which is obvious because K_j as given in Theorem 9.8 involves level-dependent α_u's. This phenomenon, which does not occur for minimum G_2-aberration under the orthogonal parametrization, in part explains the difficulty of finding theoretical results on minimum G_2-aberration baseline designs. The difficulty manifests itself in computer search as well. For given n and m, suppose we have a complete list of nonisomorphic orthogonal arrays $\mathrm{OA}(n, 2^m, 2)$. Then in order for the search of minimum G_2-aberration baseline designs to be complete, for each $\mathrm{OA}(n, 2^m, 2)$ on the list, one needs to consider all 2^m arrays that can be generated from this given $\mathrm{OA}(n, 2^m, 2)$ by interchanging levels 0 and 1 in all possible subsets of columns. Using the enumeration results of Sun, Li, and Ye (2008), Mukerjee and Tang (2012) found all minimum G_2-aberration baseline designs for $n = 8, 12$, and 16 with $m \leq n-1$ by complete search. For $n = 20$, Li, Miller, and Tang (2014) obtained minimum G_2-aberration baseline designs for $m \leq 13$ also by complete search. For $n = 20$ and $m \geq 14$, Li, Miller, and Tang (2014) used an algorithmic search; their findings, though not guaranteed to have minimum G_2-aberration, should come close to being optimal under the criterion, as the algorithm performs very well for the cases where complete search results are available.

No other general result than Theorem 9.9 has been found since the introduction of minimum G_2-aberration baseline designs. Good progress can be made if we restrict to regular designs; some results on this front were reported by Miller and Tang (2016) and Mukerjee and Tang (2016). Based on their theoretical results, Mukerjee and Tang (2016) provided detailed constructions of minimum G_2-aberration regular baseline designs with up to 256 runs for highly fractionated situations.

We omit the proofs for Theorems 9.8 and 9.9. The proof of Theorem 9.8 is tedious, though not very difficult, and that of Theorem 9.9 is quite difficult. Interested readers may want to read the original paper by Mukerjee and Tang (2012).

9.4.3 One-factor-at-a-time designs

The baseline parametrization has an attractive feature not shared by the orthogonal parametrization, which is the existence of designs that allow unbiased estimation of main effects without having to assume negligible interactions. These are one-factor-at-a-time (OFAT) designs. A basic OFAT design for m factors consists of the following $n = m + 1$ runs:

$$\boldsymbol{r}_0 = (0, 0, \ldots, 0), \quad \boldsymbol{r}_1 = (1, 0, \ldots, 0), \quad \ldots, \quad \boldsymbol{r}_m = (0, 0, \ldots, 1).$$

This design allows unbiased estimation of main effects in the presence of all interactions because $\boldsymbol{Z}_j$ in (9.11) and (9.12) is a matrix of all zeros for any $j = 2, \ldots, m$. For $n > m+1$, we can repeatedly use the runs in the basic OFAT design, and designs so obtained are all called OFAT designs. Any OFAT design gives zero bias in the main-effect estimation in the presence of interactions. Consider a general OFAT design that has run $\boldsymbol{r}_j$ repeated f_j times for $j = 0, 1, \ldots, m$. Then A-optimal OFAT designs for estimating the main effects $\theta_1, \ldots, \theta_m$ can be obtained by minimizing

$$\sum_{j=1}^{m} \mathrm{Var}(\hat{\theta}_j) = m/f_0 + \sum_{j=1}^{m} (1/f_j)$$

subject to $\sum_{j=0}^{m} f_j = n$ (Mukerjee and Tang, 2012).

The approach of using OFAT designs first minimizes the biases of the main-effect estimators, and then minimizes their variances. This is in sharp contrast with the minimum G_2-aberration approach taken in Subsection 9.4.2, which first seeks designs that minimize the variances of main-effect estimators, and then minimize their biases in the order of effect hierarchy. Mukerjee and Tang (2012) made a comparison between the two approaches via a Bayesian-inspired formulation. Karunanayaka and Tang (2017) proposed a class of compromise designs that attempt to balance the considerations of bias and variance. Their designs are given by adding some runs to the basic OFAT design without restricting the extra runs coming from the basic OFAT design. They showed that adding a run of all ones is D-, A- and E-optimal within the class of all designs obtained from adding one run to the basic OFAT design. A different class of compromise designs was later considered by Li, Liu, and Tang (2022a).

9.5 Some General Results on Baseline Designs

This section provides a general discussion on baseline designs. The material is largely based on a recent paper by Sun and Tang (2022).

9.5.1 Relation between baseline and orthogonal parametrizations

In the Yates order, let $\boldsymbol{\mu}$ be a vector collecting all treatment means μ_u and $\boldsymbol{\theta}$ a vector collecting all baseline effects θ_u, where μ_u is the mean response under treatment combination $\boldsymbol{z}_u = (z_{u1}, \ldots, z_{um})$ with $z_{uj} = 1$ if $j \in u$ and $z_{uj} = 0$ if $j \notin u$. If we label levels 0 and 1 as $+1$ and -1, then $\boldsymbol{z}_u$ becomes $\boldsymbol{x}_u$ and μ_u becomes the mean response at $\boldsymbol{x}_u$, where $\boldsymbol{x}_u = (x_{u1}, \ldots, x_{um})$ with $x_{uj} = -1$ if $j \in u$ and $x_{uj} = +1$ if $j \notin u$. Further let $\boldsymbol{\beta}$ be a vector collecting all β_u in the Yates order, where β_u denotes a factorial effect under the orthogonal parametrization. Equation (9.7) shows how μ_u depends on θ_v's and equation (2.13) shows how μ_u depends on β_v's. Writing equations (9.7) and (2.13) in matrix form, we obtain that

$$\boldsymbol{\mu} = \boldsymbol{A}\boldsymbol{\theta} = \boldsymbol{B}\boldsymbol{\beta} \quad \text{for some matrices } \boldsymbol{A} \text{ and } \boldsymbol{B}. \tag{9.13}$$

This implies a linear relationship between $\boldsymbol{\theta}$ and $\boldsymbol{\beta}$, which is not surprising. What is interesting is the special way in which the two are linearly related.

Theorem 9.10. *We have that*

(i) $\beta_u = \sum_{v \supseteq u} a_v \theta_v$, *where* $a_v = (-1)^{|u|} 2^{-|v|}$, *and*

(ii) $\theta_u = \sum_{v \supseteq u} b_v \beta_v$, *where* $b_v = (-2)^{|u|}$.

Proof. For convenience in the proof, we use different notations for $\boldsymbol{A}$ and $\boldsymbol{B}$ in (9.13) that describe relations among $\boldsymbol{\mu}, \boldsymbol{\theta}$, and $\boldsymbol{\beta}$. Let $\boldsymbol{H}_m$ be the m-fold tensor product of $\boldsymbol{H}$ and $\boldsymbol{L}_m$ the m-fold tensor product of $\boldsymbol{L}$, as given by

$$\boldsymbol{H}_m = \boldsymbol{H} \otimes \cdots \otimes \boldsymbol{H} \text{ and } \boldsymbol{L}_m = \boldsymbol{L} \otimes \cdots \otimes \boldsymbol{L},$$

where

$$\boldsymbol{H} = \begin{bmatrix} 1/2 & 1/2 \\ 1/2 & -1/2 \end{bmatrix} \text{ and } \boldsymbol{L} = \begin{bmatrix} 1 & 0 \\ 1 & 1 \end{bmatrix}.$$

We then have $\boldsymbol{\beta} = \boldsymbol{H}_m \boldsymbol{\mu}$ and $\boldsymbol{\mu} = \boldsymbol{L}_m \boldsymbol{\theta}$. Therefore $\boldsymbol{\beta} = \boldsymbol{H}_m \boldsymbol{L}_m \boldsymbol{\theta}$ and $\boldsymbol{\theta} = (\boldsymbol{H}_m \boldsymbol{L}_m)^{-1} \boldsymbol{\beta}$. We conclude the proof of Theorem 9.10 by noting that $\boldsymbol{H}_m \boldsymbol{L}_m$ is the m-fold tensor product of $\boldsymbol{HL}$ and $(\boldsymbol{H}_m \boldsymbol{L}_m)^{-1}$ is the m-fold tensor product of $(\boldsymbol{HL})^{-1}$, as given by

$$\boldsymbol{H}_m \boldsymbol{L}_m = (\boldsymbol{HL}) \otimes \cdots \otimes (\boldsymbol{HL}) \text{ and } (\boldsymbol{H}_m \boldsymbol{L}_m)^{-1} = (\boldsymbol{HL})^{-1} \otimes \cdots \otimes (\boldsymbol{HL})^{-1},$$

where

$$\boldsymbol{HL} = \begin{bmatrix} 1 & 1/2 \\ 0 & -1/2 \end{bmatrix}, \text{ and } (\boldsymbol{HL})^{-1} = \begin{bmatrix} 1 & 1 \\ 0 & -2 \end{bmatrix}.$$

□

By Theorem 9.10, if $\theta_u = 0$ for all $|u| \geq k$, then $\beta_u = 0$ for all $|u| \geq k$, and vice versa. This means that if interactions involving k or more factors are negligible under one parametrization, then they must be negligible under the other parametrization. Theorem 9.10 asserts that the validity of a model consisting of all effects up to order $k-1$ under one parametrization implies the same under the other parametrization. This in particular implies that if the main-effect model holds under one parametrization, then it must hold under the other parametrization, and that if the full second-order model holds under one parametrization, then it must hold under the other parametrization.

The above implication of Theorem 9.10 can be made general using echelon models, which have been introduced in Section 9.3. For an echelon collection $\mathcal{C}$ of subsets of $\{1,\ldots,m\}$, let $M_b(\mathcal{C})$ be the model consisting of θ_u's for all u's in $\mathcal{C}$ and $M_o(\mathcal{C})$ be the model consisting of β_u's for all u's in $\mathcal{C}$. Then because of Theorem 9.10, we have that $\beta_u = 0$ for all $u \notin \mathcal{C}$ implies that $\theta_u = 0$ for all $u \notin \mathcal{C}$, and vice versa. We summarize the above discussion as a corollary.

Corollary 9.3. *Let $\mathcal{C}$ be an echelon collection, and let $M_b(\mathcal{C})$ and $M_o(\mathcal{C})$ be the corresponding models under the baseline and orthogonal parametrizations. Then the validity of one model implies that of the other.*

If $M_b(\mathcal{C})$ is valid, then we have $\boldsymbol{\mu} = \boldsymbol{A}_{\mathcal{C}}\boldsymbol{\theta}_{\mathcal{C}}$, where $\boldsymbol{\theta}_{\mathcal{C}}$ is a subvector of $\boldsymbol{\theta}$ that collects θ_u for all $u \in \mathcal{C}$, and $\boldsymbol{A}_{\mathcal{C}}$ is a submatrix of $\boldsymbol{A}$ in (9.13) that consists of those columns corresponding to θ_u for all $u \in \mathcal{C}$. Corollary 9.3 then implies that $\boldsymbol{\mu} = \boldsymbol{B}_{\mathcal{C}}\boldsymbol{\beta}_{\mathcal{C}}$, where $\boldsymbol{B}_{\mathcal{C}}$ and $\boldsymbol{\beta}_{\mathcal{C}}$ are defined in the same way as $\boldsymbol{A}_{\mathcal{C}}$ and $\boldsymbol{\theta}_{\mathcal{C}}$. Conversely, if $\boldsymbol{\mu} = \boldsymbol{B}_{\mathcal{C}}\boldsymbol{\beta}_{\mathcal{C}}$, then $\boldsymbol{\mu} = \boldsymbol{A}_{\mathcal{C}}\boldsymbol{\theta}_{\mathcal{C}}$. Therefore, if one of the two models $M_b(\mathcal{C})$ and $M_o(\mathcal{C})$ is correct, then we have that $\boldsymbol{B}_{\mathcal{C}}\boldsymbol{\beta}_{\mathcal{C}} = \boldsymbol{A}_{\mathcal{C}}\boldsymbol{\theta}_{\mathcal{C}}$. This establishes that

$$\boldsymbol{\theta}_{\mathcal{C}} = (\boldsymbol{A}_{\mathcal{C}}^T\boldsymbol{A}_{\mathcal{C}})^{-1}\boldsymbol{A}_{\mathcal{C}}^T\boldsymbol{B}_{\mathcal{C}}\boldsymbol{\beta}_{\mathcal{C}}, \tag{9.14}$$

where, obviously, $(\boldsymbol{A}_{\mathcal{C}}^T\boldsymbol{A}_{\mathcal{C}})^{-1}\boldsymbol{A}_{\mathcal{C}}^T\boldsymbol{B}_{\mathcal{C}}$ is a square matrix of full rank.

Note that one cannot obtain $\boldsymbol{\theta}_{\mathcal{C}} = \boldsymbol{A}_{\mathcal{C}}^{-1}\boldsymbol{B}_{\mathcal{C}}\boldsymbol{\beta}_{\mathcal{C}}$ by inverting $\boldsymbol{A}_{\mathcal{C}}$ because $\boldsymbol{A}_{\mathcal{C}}$ is not a square matrix. In fact, both matrices $\boldsymbol{A}_{\mathcal{C}}$ and $\boldsymbol{B}_{\mathcal{C}}$ have 2^m rows and q columns, where q is the size of $\mathcal{C}$.

9.5.2 Designs for echelon models

We now consider using a baseline design $\boldsymbol{Z}$ of n runs to estimate an echelon model $M_b(\mathcal{C})$. The corresponding problem under the orthogonal parametrization is to use design $\boldsymbol{X}$, obtained from $\boldsymbol{Z}$ by changing 0 to $+1$ and 1 to -1, to estimate $M_o(\mathcal{C})$. The next result is immediate from (9.14).

Theorem 9.11. *Suppose that $\mathcal{C}$ is an echelon collection. Then we have that*

(i) *design $\boldsymbol{Z}$ allows the estimation of $M_b(\mathcal{C})$ if and only if design $\boldsymbol{X}$ allows the estimation of $M_o(\mathcal{C})$, and that*

(ii) *design $\boldsymbol{Z}$ is D-optimal for estimating $M_b(\mathcal{C})$ if and only if design $\boldsymbol{X}$ is D-optimal for estimating $M_o(\mathcal{C})$.*

Suppose a design $\boldsymbol{Z}$ is such that the model matrix of $M_o(\mathcal{C})$ has orthogonal columns. By Corollary 2.1, design $\boldsymbol{X}$ is universally optimal and in particular is D-optimal for estimating $M_o(\mathcal{C})$. Then by Theorem 9.11(ii), design $\boldsymbol{Z}$ is D-optimal for estimating $\boldsymbol{\theta}_{\mathcal{C}}$. Design $\boldsymbol{Z}$ has another optimality property, which is discussed below.

Because the model matrix of $M_o(\mathcal{C})$ has orthogonal columns, by a result of Moriguti (1954), design $\boldsymbol{X}$ minimizes $\mathrm{Var}(\hat{\beta}_u)$ for every $u \in \mathcal{C}$. We call θ_u a *cap effect* if no w in $\mathcal{C}$ contains u as a proper subset. By Theorem 9.10, θ_u is proportional to β_u if θ_u is a cap effect. This shows that $\mathrm{Var}(\hat{\theta}_u)$ is minimized for every cap effect.

Theorem 9.12. *Consider using a baseline design $\boldsymbol{Z}$ to estimate $M_b(\mathcal{C})$. If $\boldsymbol{Z}$ is such that the model matrix of $M_o(\mathcal{C})$ has orthogonal columns, then it minimizes $Var(\hat{\theta}_u)$ for every cap effect θ_u.*

The main-effect model in (9.10) is an echelon model and every main effect is a cap effect while the intercept is not. We see that Theorem 9.12 includes Theorem 9.7(i) as a special case. In a model consisting of the intercept, all main effects, and all two-factor interactions, the two-factor interactions are cap effects, whereas the intercept and the main effects are not cap effects. For a model consisting of the intercept, all main effects, and some two-factor interactions, these two-factor interactions are cap effects and so are those main effects that are not involved in any of these two-factor interactions. Theorem 9.12 is useful because, as seen in the above examples, cap effects should be the first in line to be tested if one wants to find a simpler model.

To a baseline model $M_b(\mathcal{C})$, there corresponds a baseline Rechtschaffner design $\boldsymbol{Z}_{\mathcal{C}}$ that is composed of all runs $\boldsymbol{z}_u$ with $u \in \mathcal{C}$, where $\boldsymbol{z}_u$ has its entries equal to 1 for the factors in u and equal to 0 otherwise. Rechtschaffner designs have been introduced under the orthogonal parametrization in Section 9.3. Clearly, a Rechtschaffner design under the baseline parametrization becomes a Rechtschaffner design under the orthogonal parametrization when levels 0 and 1 are labeled as $+1$ and -1, respectively.

We know from (9.13) that $\boldsymbol{\mu} = \boldsymbol{A\theta}$. With the help of (9.7), we see that $\boldsymbol{A}$ is actually a lower triangular matrix with all the diagonal entries equal to 1. Given any collection $\mathcal{C}$ of subsets of $\{1, \ldots, m\}$, suppose that model $M_b(\mathcal{C})$ is correctly specified. Then under design $\boldsymbol{Z}_{\mathcal{C}}$ and model $M_b(\mathcal{C})$, we have that $\boldsymbol{\mu}_{\mathcal{C}} = \boldsymbol{A}_{\mathcal{C}\times\mathcal{C}}\boldsymbol{\theta}_{\mathcal{C}}$, where $\boldsymbol{\mu}_{\mathcal{C}}$, a subvector of $\boldsymbol{\mu}$, consists of all μ_u with $u \in \mathcal{C}$, and $\boldsymbol{A}_{\mathcal{C}\times\mathcal{C}}$ is a square submatrix of $\boldsymbol{A}$ that has its rows and columns both indexed by $\mathcal{C}$. This shows that $\boldsymbol{A}_{\mathcal{C}\times\mathcal{C}}$ must also be a lower triangular matrix with its diagonal entries equal to 1, and is therefore invertible. The above arguments lead to the next theorem.

Theorem 9.13. *For any collection $\mathcal{C}$ of subsets of $\{1, \ldots, m\}$, the baseline Rechtschaffner design $\boldsymbol{Z}_{\mathcal{C}}$ allows the estimation of model $M_b(\mathcal{C})$.*

Theorem 9.13 does not require that $\mathcal{C}$ be an echelon collection. But if $\mathcal{C}$ is an echelon collection, then combining Theorem 9.13 with Corollary 9.3 and Theorem 9.11(i) shows that the Rechtschaffner design $\boldsymbol{D}_{\mathcal{C}}$ under the orthogonal parametrization allows the estimation of model $M_o(\mathcal{C})$. This is precisely Theorem 9.6 in Section 9.3.

If $\boldsymbol{C}$ is an echelon collection, then design $\boldsymbol{Z}_{\mathcal{C}}$ has a stronger property than that in Theorem 9.13. Suppose that $M_b(\mathcal{C})$ is a tentative model, meaning that effects outside $M_b(\mathcal{C})$ may not be negligible. For any effect θ_v outside $M_b(\mathcal{C})$, we have $v \notin \mathcal{C}$. Since $\mathcal{C}$ is an echelon collection, there is no w in $\mathcal{C}$ such that $v \subseteq w$. This implies that the column $\boldsymbol{c}_v$ corresponding to effect θ_v in the full model matrix under design $\boldsymbol{Z}_{\mathcal{C}}$, which is a Hadamard product of the columns $\boldsymbol{c}_j$ of $\boldsymbol{Z}_{\mathcal{C}}$ for all $j \in v$, must be a column vector of all zeros. Therefore, the existence of effect θ_v outside model $M_b(\mathcal{C})$ does not contribute a bias to the estimation of $\boldsymbol{\theta}_{\mathcal{C}}$.

Theorem 9.14. *If $\mathcal{C}$ is an echelon collection, then design $\boldsymbol{Z}_{\mathcal{C}}$ allows unbiased estimation of $\boldsymbol{\theta}_{\mathcal{C}}$ even though the effects θ_v for $v \notin \mathcal{C}$ are nonnegligible.*

A special case of Theorem 9.14 has been discussed in Subsection 9.4.3, which says that OFAT designs allow unbiased estimation of all baseline main effects in the presence of all baseline interaction effects.

Finally, we note that if $\mathcal{C}$ selects all u with $|u| \le k$, then the Rechtschaffner design $\boldsymbol{Z}_{\mathcal{C}}$ is a balanced array. Chakravarti (1956), Srivastava and Chopra (1971), and Rafter and Seiden (1974) are some fundamental papers on balanced arrays.

9.6 Spring Balance Weighing Designs

We have introduced weighing designs in Section 9.1. The goal of the present section is to present an optimality result on spring balance weighing designs. We begin our discussion with chemical balance weighing designs.

The model matrix $\boldsymbol{W}$ of a chemical balance weighing design has ± 1 entries. Let the n rows of $\boldsymbol{W}$ be $\boldsymbol{w}_1^T, ..., \boldsymbol{w}_n^T$. Then the information matrix per observation is

$$\frac{1}{n}\boldsymbol{W}^T\boldsymbol{W} = \frac{1}{n}\sum_{i=1}^{n} \boldsymbol{w}_i \boldsymbol{w}_i^T. \tag{9.15}$$

If we view each $\boldsymbol{w}_i$ as a vertex of the q-dimensional cube $[-1, 1]^q$, then each design can be considered as a probability distribution on $[-1, 1]^q$ (actually on $\{-1, 1\}^q$, the set of all vertices) that assigns probability $1/n$ to each row of $\boldsymbol{W}$; of course the probabilities of repeated rows are added. So here the probabilities represent the proportions of observations at different sites. Denote this probability distribution as $\xi_{\boldsymbol{W}}$. Then in view of (9.15), we can write the information matrix per observation of $\xi_{\boldsymbol{W}}$ as

$$\boldsymbol{M}(\xi_{\boldsymbol{W}}) = \int_{[-1,1]^q} \boldsymbol{w}\boldsymbol{w}^T \xi_{\boldsymbol{W}}(d\boldsymbol{w}). \tag{9.16}$$

We can expand the problem to the consideration of all the probability distributions ξ on $[-1, 1]^q$, called *approximate* designs, not just those induced by *exact* designs such as $\xi_{\boldsymbol{W}}$. For each approximate design ξ, following (9.16), we define its information matrix $\boldsymbol{M}(\xi)$ as

$$\boldsymbol{M}(\xi) = \int_{[-1,1]^q} \boldsymbol{w}\boldsymbol{w}^T \xi(d\boldsymbol{w}).$$

Then we consider the more general problem of minimizing a certain optimality criterion $\Phi(\boldsymbol{M}(\xi))$ of information matrices over the set of *all* probability distributions ξ on $[-1, 1]^q$. This approach of embedding a discrete problem into a continuous setting was pioneered by Kiefer and Wolfowitz (1960) in the investigation of optimal regression designs. It leads to the approximate theory of optimal experimental design. We refer the readers to Pukelsheim (1993) for an in-depth treatment of this subject.

A fundamental contribution in Kiefer and Wolfowitz (1960) is the *equivalence theorem* on some necessary and sufficient conditions for an approximate design ξ to be optimal. It was first established for the D-criterion and was extended to general optimality criteria in Kiefer (1974); see also Pukelsheim (1993). One can use these conditions both to come up with good candidates for optimal designs and to verify their optimality. Furthermore, they are also useful for developing algorithms for generating optimal designs.

However, the approximate theory of optimal chemical balance weighing designs is quite simple and not too interesting. Any design that is supported on the vertices of $[-1, 1]^q$ clearly maximizes the trace of the information matrix. Therefore if its information matrix is a multiple of the identity matrix, then by Theorem 2.3, the design is universally optimal. One simple example is the uniform distribution on all the vertices of the cube $[-1, 1]^q$. Such a design has a rather large support. On the other hand, the uniform distribution supported on the rows of a Hadamard matrix is a universally optimal approximate design with the smallest support. A counterpart of this result in the exact design theory is the universal optimality of Hadamard matrices established in Chapter 2.

The approximate theory for spring balance weighing designs, on the other hand, is much more interesting and complicated. In this case, the design region is $\{0, 1\}^q$, the set of vertices of the unit cube, or more generally the q-dimensional unit cube $[0,1]^q$. In either case, it is easy to see that in order for a design to maximize the trace of the information matrix, one of the conditions for the design to be universally optimal, it must be supported on the point with all the coordinates equal to 1. This gives a single-point design, which can only be used to estimate the total weight of all the q objects and provides no information on the individual weights. As a result, there is no universally optimal design in the setting of spring balance weighing designs.

Spring balance weighing designs have applications in Hadamard transform optics. For a treaties on this subject, we refer the readers to the monograph by Harwit and Sloane (1979). To determine the intensities of different frequency components of a spectrum, an optical separator is used to separate the frequency components into different slots of a mask. Each slot can be open or closed. The total intensity of the frequency components that are not blocked is measured. This is the same as using a spring balance to determine the weights of individual objects. The slots on the mask can also be partially open. This allows the entries of the design matrix to be strictly between 0 and 1, but one would expect loss of information.

There are also applications to functional magnetic resonance imaging (fMRI); see, e.g., Kao (2015), Cheng and Kao (2015), and Cheng, Kao, and Phoa (2017).

Let $D_{n,q}$ (respectively, $\overline{D}_{n,q}$) be the set of all $n \times q$ matrices with entries $0 \leq x_{ij} \leq 1$ (respectively, $x_{ij} = 0$ or 1). In the case of $n = q$, Harwit and Sloane (1979) conjectured that

$$\operatorname{tr}(\boldsymbol{X}^T\boldsymbol{X})^{-1} \geq 4n^2/(n+1)^2$$

for any $\boldsymbol{X} \in D_{n,n}$ or $\overline{D}_{n,n}$. This conjectured lower bound is attained by the so-called S-matrices. An $n \times n$ matrix is said to be an S-matrix if it can be obtained by deleting the first row and first column of an $(n+1) \times (n+1)$ Hadamard matrix with all the entries of the first row and first column equal to 1, and then changing 1's to 0's and -1's to 1's. Such matrices are also incidence matrices of balanced incomplete block designs with n treatments and n blocks of size $(n+1)/2$.

For any integer k such that $\lfloor (q+1)/2 \rfloor \leq k \leq q$, where $\lfloor x \rfloor$ is the integer part of x, let ξ_k be the uniform distribution on all the vertices of $[0,1]^q$ with k coordinates equal to 1 and $q-k$ coordinates equal to 0. We define two functions f and g on the integers $\lfloor (q+1)/2 \rfloor \leq k \leq q$ as follows:

$$g(k) = \begin{cases} \frac{\ln\{(2k+1-q)/(2k+1)\}}{\ln\{k(q-1)/(q-k)\}} + 1, \text{ if } \lfloor (q+1)/2 \rfloor \leq k \leq q-1, \\ 1, \text{ if } k = q, \end{cases}$$

and

$$f(k) = \begin{cases} \frac{\ln\{(2k-1-q)/(2k-1)\}}{\ln\{k(q-1)/(q-k)\}} + 1, \text{ if } \lfloor (q+1)/2 \rfloor + 1 \leq k \leq q-1, \\ -\infty, \text{ if } k = \lfloor (q+1)/2 \rfloor \\ 1, \text{ if } k = q. \end{cases}$$

Here $-\infty = f\big(\lfloor (q+1)/2 \rfloor\big) < g\big(\lfloor (q+1)/2 \rfloor\big) < f(k) < g(k) < f(k+1) < g(k+1) < 1 = f(q) = g(q)$ for all k such that $\lfloor (q+1)/2 \rfloor + 1 \leq k \leq q-2$. It follows that the set of closed intervals $\big[f(k), g(k)\big]$ for $k = \lfloor (q+1)/2 \rfloor, \ldots, q$ and the set of open intervals $\big(g(k), f(k+1)\big)$ for $k = \lfloor (q+1)/2 \rfloor, \ldots, q-1$ together form a partition of $[-\infty, 1]$.

By an application of the Kiefer-Wolfowitz equivalence theorem, Cheng (1987) obtained the following result on j_a-optimal designs for all $a \in [-\infty, 1]$:

Theorem 9.15. *Let k be an integer such that $\lfloor (q+1)/2 \rfloor \leq k \leq q$. Then the design ξ_k is j_a-optimal for all the a's such that $f(k) \leq a \leq g(k)$. For $k < q$ and $g(k) < a < f(k+1)$,* *the mixture* $\epsilon_k\xi_k + (1-\epsilon_k)\xi_{k+1}$, *with* ϵ_k *given by*

$$\frac{(k+1)^2(q-1)-(k+1)(q-k-1)\{(2k+1-q)/(2k+1)\}^{1/(a-1)}}{(2k+1)(q-1)+(2k+1-q)\{(2k+1-q)/(2k+1)\}^{1/(a-1)}} \tag{9.17}$$

is j_a*-optimal.*

Each j_a-optimal design is either a ξ_k or a mixture of ξ_k and ξ_{k+1} with $k \geq \lfloor (q+1)/2 \rfloor$. Specifically, for each k, let A_k be the interval $[f(k), g(k)]$ and B_k be the interval $(g(k), f(k+1))$. We have that the A_k's and B_k's together form a partition of $[-\infty, 1]$. For each $\lfloor (q+1)/2 \rfloor \leq k \leq q$, ξ_k is j_a-optimal for *all* $a \in A_k$, and for all $a \in B_k$, each j_a-optimal design is a mixture of ξ_k and ξ_{k+1}, with the weight ϵ_k of ξ_k given in (9.17). In other words, in the former case each optimal design is optimal for an interval of (infinitely many) a values, while in the latter case, each design is optimal for one single criterion of the j_a family. We can see that ϵ_k is a decreasing function of a. Therefore as a increases, the weight shifts completely from ξ_k to ξ_{k+1}. Because the A_k's and B_k's together form a partition of $[-\infty, 1]$, Theorem 9.15 gives a complete solution of j_a-optimal designs for all $a \in [-\infty, 1]$, the entire range of a where the j_a criteria are defined.

In the following we give examples to illustrate some results obtained from using the approximate theory. Among the optimal designs obtained above, the most important one is ξ_k with $k = \lfloor (q+1)/2 \rfloor$, a uniform distribution on all the vertices with $\lfloor (q+1)/2 \rfloor$ components equal to 1. This design is j_a-optimal for the widest range of a values, which covers the most important and commonly used criteria. For odd q's, $f((q+1)/2) = -\infty$, and $g((q+1)/2) = 1 - \ln(q/2+1)/\ln(q+1)$. So $\xi_{(q+1)/2}$ is j_a-optimal for all $a \in [-\infty, 1-\ln(q/2+1)/\ln(q+1)]$. We have that $1-\ln(q/2+1)/\ln(q+1) > 0$; therefore $\xi_{(q+1)/2}$ is j_a-optimal for all $a \in [-\infty, 0]$. In other words, it is Φ_s-optimal for all $s \in [0, \infty]$; in particular, it is A-, D-, and E-optimal. For the A-criterion, this proves Harwit and Sloane's conjecture mentioned earlier.

When q is even, $g(q/2) = 1 - \ln(q+1)/\ln(q-1)$, which is negative, although quite close to 0. It follows that $\xi_{q/2}$ is not D-optimal. However, since $1 - \ln(q+1)/\ln(q-1) \geq -1$ if $q \geq 3$, it is A-optimal for all even $q \geq 4$. It is also E-optimal for all even q. In this case, the D-optimal design turns out to be a mixture of $\xi_{q/2}$ and $\xi_{q/2+1}$. Since it is a mixture, according to Theorem 9.15, it is not Φ_s-optimal for any other s's. This mixture is the uniform distribution on the vertices with $q/2$ or $q/2+1$ 1's. In terms of exact designs, this establishes the D-optimality of the matrix whose rows consist of those of the incidence matrix of a BIBD with q treatments and blocks size $q/2$ and those of the incidence matrix of a BIBD with q treatments and block size $q/2+1$.

While the results presented in this section were derived and proved by using the equivalence theorem, there are alternative methods. Without using the equivalence theorem, Cheng (1995b) considered linear regression without the intercept term on general permutation-invariant design regions (such as spheres and simplexes) that can be applied to cubes.

9.7 Search Designs and Parallel-Flat Designs

We provide in this section a brief account of two other interesting developments in nonregular factorial designs, namely search designs and parallel-flat designs. All discussion is under the orthogonal parametrization.

Srivastava (1975) pioneered the research on search designs. The settings he considered are that the experimenter requires the estimates of certain factorial effects and also wants to search for nonnegligible effects. Let β_1 denote the vector of effects whose estimates are required, and let β_2 denote the vector that collects all potentially nonnegligible effects. This specifies the linear model

$$Y = X_1\beta_1 + X_2\beta_2 + \epsilon, \tag{9.18}$$

where X_1 and X_2 are model matrices for β_1 and β_2, respectively, and ϵ is the vector of uncorrelated random errors that have a zero mean and a constant variance σ^2. For example, β_1 may contain the grand mean and all main effects, and β_2 may contain all two-factor interactions.

In search design settings, β_2 generally contains a large number of effects, and consequently designs allowing estimation of the model in (9.18) are too expensive to use. On the other hand, we believe that only a few of the effects in β_2 are actually active although we have no knowledge of which effects they are. Suppose that there are q effects in β_2 that are nonnegligible. The design objective is to estimate β_1 as well as identify these q nonnegligible effects in β_2. Srivastava (1975) showed that to achieve this objective, it is necessary that the matrix (X_1, X_2') has a full column rank for every submatrix X_2' consisting of $2q$ columns of X_2, and he further showed that this necessary condition is sufficient if $\sigma^2 = 0$.

Construction of search designs was considered under a number of scenarios. For some representative work, we refer to Srivastava and Ghosh (1977), Srivastava and Gupta (1979), and Ghosh (1996). For an excellent review of search designs and other contributions by J.N. Srivastava, the readers are referred to Morgan, Ghosh, and Dean (2014).

Parallel-flat designs are nonregular designs constructed from regular designs. Simply put, a parallel-flat design is obtained by selecting some blocks from the set of all blocks created from a regular design using a blocking scheme. These designs started to emerge in design literature around 1960 with contributions from a number of authors. Most well known are three-quarter replicates of regular designs studied by John (1961) and John (1962). These designs are obtained by selecting three of four blocks from a blocked regular design. Interesting examples from these two papers are a design of 12 runs for 4 factors and a design of 48 runs for 8 factors, both being of resolution V. These two designs do not have orthogonal main effects. To study 4 factors using 12 runs, one may simply use an $OA(12, 2^4, 2)$, which has orthogonal main effects and, according to Corollary 3.7, also permits the estimation of all main effects and all two-factor interactions.

Cheng (2014a) presented two theorems on the orthogonality of parallel-flat designs. One theorem, originally obtained by Srivastava and Throop (1990), concerns the condition for a parallel-flat design to be an orthogonal array. While this result is theoretically interesting, its implementation appears difficult. As examples, Srivastava and Throop (1990) constructed two orthogonal arrays, an $OA(32, 2^6, 4)$ and an $OA(192, 2^8, 4)$, using their method. We know that a 2_V^{6-1} design and a 2_{VI}^{6-1} design can easily be constructed, both being $OA(32, 2^6, 4)$s. A 2_V^{8-2} design is also available; three replicates of this design give an $OA(192, 2^8, 4)$. It remains to be seen if the theoretical result can yield new orthogonal arrays.

The other theorem aims to find orthogonal parallel-flat designs for the estimation of a set of required effects. This result is due to Srivastava and Li (1996), who also compiled a table of orthogonal designs for estimating the main effects and certain specified two-factor interactions. A further work in this direction is Liao, Iyer, and Vecchia (1996). In the design table of Srivastava and Li (1996), many designs have one flat, which correspond to regular designs. Other designs in the table have $12, 20$, and 24 flats and their constructions appear somewhat tedious, as can be seen from the paper. Upon close examination of these designs with $12, 20$, and 24 flats, we find that better designs can be constructed using a simple tensor product method from Chapter 6. For example, one design in the table of Srivastava and Li (1996) has 48 runs for 13 factors and allows orthogonal estimation of all main effects and interactions

$$d_1d_2, d_1d_3, \ldots, d_1d_{13}, d_2d_3, \ldots, d_2d_{13},$$

where $\boldsymbol{d}_1, \ldots, \boldsymbol{d}_{13}$ are the 13 design columns. Example 6.1 in Chapter 6 contains a design of 48 runs and 13 factors. In addition to permitting orthogonal estimation of all main effects and the two-factor interactions on the above list, this design guarantees that all these effects are clear, meaning that they can all be estimated in the presence of other possibly nonnegligible two-factor interactions.

10

Multi-Level Nonregular Designs

The previous chapters have so far been on the studies of two-level factorial designs. In this long chapter, we start our discussion on multi-level and mixed-level factorial designs. After a couple of introductory sections on factorial effects and orthogonal arrays, we introduce and justify the criterion of minimum G_2-aberration through the use of J-characteristics. We next discuss an equivalent criterion, called the minimum moment aberration criterion, which provides a powerful machinery for both theoretical studies and computational searches of minimum G_2-aberration designs. Construction results for minimum G_2-aberration designs are then presented. We also examine the use of minimum G-aberration for further design selection and of generalized resolution as a concise summary of a design's properties. Lastly, fair shares of treatment are given to the topics of supersaturated designs, three-level factorial designs for quantitative factors, and definitive screening designs. All designs considered in this chapter are for unstructured experimental units. Multi-level designs under more complicated unit structures will be studied in the next chapter. Chapter 12 gives a treatment of space-filling designs.

10.1 Multi-Level Factorial Experiments

In a multi-level or mixed-level factorial experiment, the investigator studies how a response variable depends on m factors of interest, where the jth factor has s_j levels denoted by $0, 1, \ldots, s_j - 1$. We refer to such an experiment as an $s_1 \times \cdots \times s_m$ factorial experiment. In total there are $s_1 \cdots s_m$ level combinations (also called treatment combinations), which are denoted by $(x_1, \ldots, x_m)$ with $x_j = 0, 1, \ldots, s_j - 1$ for $j = 1, \ldots, m$. Let $\mu(x_1, \ldots, x_m)$ be the mean response at the treatment combination $(x_1, \ldots, x_m)$. The dependence of the mean response on the factors is best studied via various factorial effects. One commonly used definition of factorial effects is via an analysis-of-variance (ANOVA) decomposition

$$\mu(x_1, \ldots, x_m) = \sum_u \theta_u(x_1, \ldots, x_m), \tag{10.1}$$

where the summation is over all subsets u of $Z_m = \{1, \ldots, m\}$. Here $\theta_u(x_1, \ldots, x_m)$ only depends on the x_j's with $j \in u$ and is recursively defined as follows.

First of all, for each subset u of $Z_m = \{1, \ldots, m\}$, let $\mu_u(x_1, \ldots, x_m)$ be the average of the $\prod_{j \notin u} s_j$ values of $\mu(x_1', \ldots, x_m')$'s with $x_i' = x_i$ for all $i \in u$. In particular, $\mu_\phi(x_1, \ldots, x_m)$ is a constant which is the average of all the $\mu(x_1, \ldots, x_m)$'s. Now let $\theta_\phi(x_1, \ldots, x_m) = \mu_\phi(x_1, \ldots, x_m)$, and for $u \neq \phi$, let

$$\theta_u(x_1, \ldots, x_m) = \mu_u(x_1, \ldots, x_m) - \sum_{v \subsetneq u} \theta_v(x_1, \ldots, x_m).$$

From the above definition, we see that $\theta_u(x_1, \ldots, x_m)$ depends only on those x_j's with $j \in u$. In particular, $\theta_\phi(x_1, \ldots, x_m) = \theta_\phi$ is the grand mean, a constant not depending on any x_j.

DOI: 10.1201/9781003371861-10

For $u = \{j\}$, $\theta_u(x_1, \ldots, x_m) = \theta_j(x_j)$ is the main effect of the jth factor, only depending on x_j, and for $u = \{j_1, j_2\}$, $\theta_u(x_1, \ldots, x_m) = \theta_{j_1 j_2}(x_{j_1}, x_{j_2})$ is the interaction effect of the two factors j_1 and j_2, which depends only on x_{j_1} and x_{j_2}.

To illustrate, let us look at the simple case of $m = 2$. In the familiar notation of a standard textbook, the ANOVA model in (10.1) has the form

$$\mu_{ij} = \mu + \alpha_i + \beta_j + \gamma_{ij}, \qquad \text{for } i = 0, 1, \ldots, s_1 - 1 \text{ and } j = 0, 1, \ldots, s_2 - 1,$$

where μ_{ij} is the mean response given by level i of the first factor and level j of the second factor,

$$\mu = \frac{1}{s_1 s_2} \sum_{i=0}^{s_1-1} \sum_{j=0}^{s_2-1} \mu_{ij}$$

is the grand mean,

$$\alpha_i = \frac{1}{s_2} \sum_{j=0}^{s_2-1} \mu_{ij} - \mu \quad \text{and} \quad \beta_j = \frac{1}{s_1} \sum_{i=0}^{s_1-1} \mu_{ij} - \mu$$

are main effects of the first and second factors, respectively, and $\gamma_{ij} = \mu_{ij} - \mu - \alpha_i - \beta_j$ is the interaction of the ith level of the first factor and the jth level of the second factor.

Back to the general case in (10.1), one can easily verify that the terms on the right hand side are mutually orthogonal in the sense that

$$\sum_{x_1, \ldots, x_m} \theta_u(x_1, \ldots, x_m) \theta_v(x_1, \ldots, x_m) = 0$$

for any $u \neq v$. This leads to the sum of squares decomposition

$$\text{SS}_{\text{treatment}} = \sum_{k=1}^{m} \sum_{u:|u|=k} \text{SS}_u, \tag{10.2}$$

where

$$\text{SS}_{\text{treatment}} = \sum_{x_1, \ldots, x_m} [\mu(x_1, \ldots, x_m)]^2 - (s_1 \cdots s_m)\theta_\phi^2 = \sum_{x_1, \ldots, x_m} [\mu(x_1, \ldots, x_m) - \theta_\phi]^2$$

is the treatment sum of squares, and

$$\text{SS}_u = \sum_{x_1, \ldots, x_m} [\theta_u(x_1, \ldots, x_m)]^2$$

is the sum of squares of $\theta_u(x_1, \ldots, x_m)$, which is the sum of squares for the main effects of a factor when $|u| = 1$ and the sum of squares for the interaction effects of two factors when $|u| = 2$, and so on.

The ANOVA model in terms of factorial effects $\theta_u(x_1, \ldots, x_m)$ as in (10.1) is attractive because it leads to a meaningful decomposition of the treatment sum of squares into a sum of sums of squares, each representing the contribution of a subset of factors. This model is often employed in standard textbooks when dealing with two or three factors. For given u, the function $\theta_u(x_1, \ldots, x_m)$ is meaningful, but individual values of $\theta_u(x_1, \ldots, x_m)$s when evaluated at different combinations $(x_1, \ldots, x_m)$, being unknown parameters, are less useful. They are contrasts but not mutually orthogonal, and are a set of parameters with constraints. For example, in the simple case of $m = 2$, the α_is for $i = 0, 1, \ldots, s_1 - 1$ are contrasts but are not mutually orthogonal, and they satisfy that $\sum_{i=0}^{s_1-1} \alpha_i = 0$.

We next discuss a model for $\mu(x_1, \dots, x_m)$ using a set of orthogonal contrasts. For $j = 1, \dots, m$, let

$$\boldsymbol{P}_{s_j} = (\boldsymbol{p}_0^j, \boldsymbol{p}_1^j, \dots, \boldsymbol{p}_{s_j-1}^j)$$

be an $s_j \times s_j$ matrix with orthogonal columns, all of which have squared lengths equal to s_j, where $\boldsymbol{p}_0^j = \boldsymbol{1}_{s_j}$, the all-one vector. Thus, $\boldsymbol{p}_1^j, \dots, \boldsymbol{p}_{s_j-1}^j$ provide $s_j - 1$ mutually orthogonal contrasts for the s_j levels of the jth factor. For $s_j = 2$, essentially there is only one choice for $\boldsymbol{P}_2$, which is

$$\boldsymbol{P}_2 = \begin{bmatrix} 1 & -1 \\ 1 & 1 \end{bmatrix}.$$

For $s_j \geq 3$, there are many ways one can choose $\boldsymbol{P}_{s_j}$. For a quantitative factor with s_j equally spaced levels, one may use orthogonal polynomials to find a suitable $\boldsymbol{P}_{s_j}$. For $s_j = 3$, this is given by

$$\boldsymbol{P}_3 = \begin{bmatrix} 1 & -\sqrt{3/2} & \sqrt{2}/2 \\ 1 & 0 & -\sqrt{2} \\ 1 & \sqrt{3/2} & \sqrt{2}/2 \end{bmatrix},$$

where the second and third columns represent the linear and quadratic effects, respectively.

Consider

$$\boldsymbol{P} = \boldsymbol{P}_{s_1} \otimes \cdots \otimes \boldsymbol{P}_{s_m}. \tag{10.3}$$

By an application of Lemma 4.2, we can easily see that $\boldsymbol{P}$, a square matrix of order $s_1 \cdots s_m$, has mutually orthogonal columns all having squared lengths equal to $s_1 \cdots s_m$. The first column is an all-one vector and the remaining columns are mutually orthogonal contrasts. A typical column of $\boldsymbol{P}$ has the form of

$$\boldsymbol{p}_{i_1}^1 \otimes \cdots \otimes \boldsymbol{p}_{i_m}^m, \quad i_j = 0, 1, \dots, s_j - 1,$$

which corresponds to a main effect if all $i_1, \dots, i_m$ but one are zero, to a two-factor interaction if all $i_1, \dots, i_m$ but two are zero, and to a j-factor interaction if all $i_1, \dots, i_m$ but j of them are zero.

Let $\boldsymbol{\mu}$ be a vector that collects all $\mu(x_1, \dots, x_m)$ in the Yates order. If we parametrize $\boldsymbol{\mu}$ as

$$\boldsymbol{\mu} = \boldsymbol{P}\boldsymbol{\beta}, \text{ or equivalently } \boldsymbol{\beta} = \boldsymbol{P}^{-1}\boldsymbol{\mu},$$

then we obtain another definition of various factorial effects. More precisely, the first component of $\boldsymbol{\beta}$ is the grand mean, the components of $\boldsymbol{\beta}$ that correspond to $\boldsymbol{p}_{i_1}^1 \otimes \cdots \otimes \boldsymbol{p}_{i_m}^m$ with all $i_1, \dots, i_m$ but one being zero are the main effects, the components of $\boldsymbol{\beta}$ that correspond to $\boldsymbol{p}_{i_1}^1 \otimes \cdots \otimes \boldsymbol{p}_{i_m}^m$ with all $i_1, \dots, i_m$ but two being zero are two-factor interactions, and so on. For a subset u of factors, let $\boldsymbol{P}_u$ collect all the columns $\boldsymbol{p}_{i_1}^1 \otimes \cdots \otimes \boldsymbol{p}_{i_m}^m$ of $\boldsymbol{P}$ such that $i_j \neq 0$ if $j \in u$ and $i_j = 0$ if $j \notin u$, and $\boldsymbol{\beta}_u$ be the corresponding components of $\boldsymbol{\beta}$. Then we obtain that

$$\boldsymbol{\mu} = \sum_{u \subseteq Z_m} \boldsymbol{P}_u \boldsymbol{\beta}_u, \tag{10.4}$$

which has a form well in line with the model in (10.1).

We make some final remarks on the choice of orthogonal contrasts. For theoretical investigations, it is sometimes convenient to use a $\boldsymbol{P}_{s_j}$ with complex entries, which leads to complex contrasts (Bailey, 1982). Furthermore, besides the tensor product construction of interaction contrasts using the main-effect contrasts as in (10.3), one can also use finite projective geometries to construct interaction contrasts when $s_1 = \cdots = s_m = s$ is a prime power (Bose, 1947). Even more general is the use of Abelian groups to define orthogonal contrasts (Bailey, Gilchrist, and Patterson, 1977). For a general discussion on the factorial treatment structure, we refer to Chapter 6 of Cheng (2014a).

10.2 Factorial Designs and Orthogonal Arrays

Let $\boldsymbol{D} = (d_{ij})_{n\times m}$ be a factorial design for an $s_1 \times \cdots \times s_m$ factorial experiment, where $d_{ij} = 0, 1, \ldots, s_j - 1$. While the m columns of $\boldsymbol{D}$ correspond to the m factors to be studied, the n rows of $\boldsymbol{D}$ represent the n experimental runs. Let $\boldsymbol{\mu}(\boldsymbol{D})$ be the vector of the components of $\boldsymbol{\mu}$ selected by the rows of $\boldsymbol{D}$, and let $\boldsymbol{P}_u(\boldsymbol{D})$ be similarly defined. Then (10.4) reduces to

$$\boldsymbol{\mu}(\boldsymbol{D}) = \sum_u \boldsymbol{P}_u(\boldsymbol{D})\boldsymbol{\beta}_u. \tag{10.5}$$

Let $\boldsymbol{Y}$ be the vector of responses collected under design $\boldsymbol{D}$, and $\boldsymbol{\epsilon}$ be the vector of the corresponding random errors that are assumed to be uncorrelated with a zero mean and a constant variance σ^2. We then obtain a linear model under design $\boldsymbol{D}$

$$\boldsymbol{Y} = \sum_u \boldsymbol{P}_u(\boldsymbol{D})\boldsymbol{\beta}_u + \boldsymbol{\epsilon}. \tag{10.6}$$

If we want to estimate all factorial effects $\boldsymbol{\beta}_u$s, we will need a design $\boldsymbol{D}$ that is or contains a full factorial, namely, a design having as its runs all the treatment combinations $(x_1, \ldots, x_m)$s exactly once. Unless m is very small, fractional factorials have to be employed owing to cost considerations. As a result, not all factorial effects $\boldsymbol{\beta}_u$s can be estimated. In the present scenario, we need to seek designs that allow efficient estimation of lower-order factorial effects under the assumption that higher-order factorial effects are negligible. Such designs are provided by orthogonal arrays.

Definition 10.1. *A design $\boldsymbol{D} = (d_{ij})_{n\times m}$ for an $s_1 \times \cdots \times s_m$ factorial experiment is an orthogonal array of strength t, having n runs and m factors with s_j levels $0, 1, \ldots, s_j - 1$ for the jth factor, if any $n \times t$ submatrix of $\boldsymbol{D}$ contains all level combinations with the same frequency. We use $OA(n, s_1 \times \cdots \times s_m, t)$ to denote such an array. The array is said to be symmetric if $s_1 = \cdots = s_m = s$, (in which case the notation $OA(n, s^m, t)$ is used), and asymmetric otherwise. An asymmetric orthogonal array is also said to have mixed levels. We use $OA(n, s_1^{m_1} \times \cdots \times s_k^{m_k}, t)$ to denote an asymmetric orthogonal array of strength t that has m_j factors with s_j levels for $j = 1, \ldots, k$.*

Obviously, for an $\mathrm{OA}(n, s_1 \times \cdots \times s_m, t)$ to exist, the run size n has to be a multiple of the product of any t of the m integers $s_1, \ldots, s_m$. In the symmetric case, for an $\mathrm{OA}(n, s^m, t)$ to exist, we must have that $n = \lambda s^t$ for some integer λ, which is called the index of the array. In addition to this simple requirement, the run size n of an orthogonal array also needs to satisfy Rao's bound (Rao, 1947), which we will discuss after presenting optimality results on orthogonal arrays.

Theorem 10.1. *Suppose that $\boldsymbol{D}$ is an $OA(n, s_1 \times \cdots \times s_m, t)$. Then it is universally optimal for any model containing factorial effects involving at most $\lfloor t/2 \rfloor$ factors.*

Theorem 10.1 is a general version of Theorem 2.4, which is for two-level designs. The proof of Theorem 10.1 is also similar, involving a straightforward, though tedious, verification of column-orthogonality of the model matrix. More precisely, consider a reduced model of the full model in (10.6) that consists of all factorial effects of up to order $\lfloor t/2 \rfloor$:

$$\boldsymbol{Y} = \sum_{u:|u|\le\lfloor t/2\rfloor} \boldsymbol{P}_u(\boldsymbol{D})\boldsymbol{\beta}_u + \boldsymbol{\epsilon}. \tag{10.7}$$

To establish Theorem 10.1, we only need to verify that the model matrix of (10.7), given by juxtaposing all $\boldsymbol{P}_u(\boldsymbol{D})$s with $|u| \le \lfloor t/2 \rfloor$, has orthogonal columns.

The next result is a general version of Theorem 2.5.

Theorem 10.2. *Suppose that $\boldsymbol{D}$ is an $OA(n, s_1 \times \cdots \times s_m, t)$ with odd t. Then it is universally optimal for the estimation of factorial effects involving at most $\lfloor t/2 \rfloor$ factors under a model comprising all factorial effects involving up to $\lfloor t/2 \rfloor + 1$ factors.*

Besides its ability of estimating all factorial effects involving up to $\lfloor t/2 \rfloor$ factors, an orthogonal array of odd strength t has another important property: allowing estimation of all interactions of order $\lfloor t/2 \rfloor + 1$ involving any given factor.

Theorem 10.3. *Suppose that $\boldsymbol{D}$ is an $OA(n, s_1 \times \cdots \times s_m, t)$ with odd t. Then it is universally optimal for the estimation of a model that consists of all factorial effects of order up to $\lfloor t/2 \rfloor$ and those of order $\lfloor t/2 \rfloor + 1$ involving any given factor.*

Applying Theorem 10.1 to even t and Theorem 10.3 to odd t, in each case we obtain that the run size n must be equal to or greater than the number of columns in the model matrix. The resulting inequality gives rise to Rao's bound on the number m of factors in an orthogonal array (Rao, 1947). For simplicity, we only present this result for symmetric orthogonal arrays.

Theorem 10.4. *Suppose that an $OA(n, s^m, t)$ exists. Then we have that*

(i) $n \geq \sum_{i=0}^{k} \binom{m}{i}(s-1)^i$ *if* $t = 2k$ *and*

(ii) $n \geq \sum_{i=0}^{k} \binom{m}{i}(s-1)^i + \binom{m-1}{k}(s-1)^{k+1}$ *if* $t = 2k+1$.

Taking $t = 2$ and $t = 3$ in Theorem 10.4, we obtain Rao's bound for orthogonal arrays of strength two and three.

Corollary 10.1. *Consider orthogonal arrays of strength two and three.*

(i) For an $OA(n, s^m, 2)$ *to exist, it is necessary that*

$$m \leq (n-1)/(s-1).$$

(ii) For an $OA(n, s^m, 3)$ *to exist, it is necessary that*

$$m \leq (n/s - 1)/(s-1) + 1.$$

Rao's bound as given in Theorem 10.4 is intuitive and easy to obtain. A more sophisticated, but generally better, bound is the linear programming bound of Delsarte (1973). Both Rao's bound and the linear programming bound are generally applicable. Besides these two general bounds, specific results are also available. Most famous are that of Bose and Bush (1952) for orthogonal arrays of strengths two and three, and that of Bush (1952) for orthogonal arrays of index unity. We dwell no further on this topic and refer instead to Hedayat, Sloane, and Stufken (1999) for more results and details.

We next survey some fundamental construction results for orthogonal arrays. We only present final products without giving construction details. Again, interested readers are referred to Hedayat, Sloane, and Stufken (1999).

Theorem 10.5. *If s is a prime power, then an $OA(n, s^m, 2)$ with $n = s^k$ and $m = (s^k - 1)/(s-1)$ can be constructed.*

The method of producing the orthogonal arrays in Theorem 10.5 is called Rao-Hamming construction. Such arrays are actually regular fractional factorials for s levels, an introduction to which is provided in Section A.4 of the Appendix. In the language of finite projective geometries, the columns of an orthogonal array in Theorem 10.5 are simply all the points in $PG(k-1, s)$, a $(k-1)$-dimensional projective geometry based on the Galois field of order s. The next result documents Addelman and Kempthorne's construction (Addelman and Kempthorne, 1961).

Theorem 10.6. *If s is a prime power, then an $OA(n, s^m, 2)$ with $n = 2s^k$ and $m = 2(s^k - 1)/(s - 1) - 1$ can be constructed.*

Bush's construction deals with orthogonal arrays of index unity (Bush, 1952). We present his result for the strength-three case.

Theorem 10.7. *If s is a prime power, then an $OA(s^3, s^m, 3)$ can be constructed for $m = s + 1$ if s is odd, and for $m = s + 2$ if s is even.*

The next result is also on orthogonal arrays of strength three, and it uses an "ovoid" in $PG(3, s)$, a 3-dimensional projective geometry based on the Galois field of order s. See Section 5.9 of Hedayat, Sloane, and Stufken (1999) for details.

Theorem 10.8. *If s is a prime power, then an $OA(s^4, s^m, 3)$ can be constructed for $m = s^2 + 1$.*

Theorems 10.5 and 10.6 provide strength-two arrays, and Theorems 10.7 and 10.8 can be used to construct strength-three arrays. All these arrays are symmetric. We give a brief account of the construction of asymmetric (or mixed-level) orthogonal arrays to conclude the present section.

The simplest way of constructing an asymmetric orthogonal array is the expansive replacement method, which was first proposed by Addelman and Kempthorne (1961) and extensively studied by Hedayat, Pu, and Stufken (1992). As an illustration, consider an $\text{OA}(16, 4^5, 2)$ constructed in Theorem 10.5 by taking $s = 4$ and $k = 2$. If we replace a 4-level factor in the $\text{OA}(16, 4^5, 2)$ by three 2-level factors via

$$\begin{array}{ccccc} 0 & \rightarrow & 0 & 0 & 0 \\ 1 & \rightarrow & 0 & 1 & 1 \\ 2 & \rightarrow & 1 & 0 & 1 \\ 3 & \rightarrow & 1 & 1 & 0, \end{array}$$

then we obtain an $\text{OA}(16, 2^3 4^4, 2)$. Doing the same to another 4-level factor in the just obtained $\text{OA}(16, 2^3 4^4, 2)$ produces an $\text{OA}(16, 2^6 4^3, 2)$. Working in the reverse direction is the contractive replacement method, which, for example, allows one to contract three 2-level factors into a 4-level factor. Though the idea is similar, the contractive replacement method is harder to implement because one needs to identify suitable groups of lower-level factors to be contracted into higher-level factors. For details, see Wu (1989) and Wu, Zhang, and Wang (1992).

A powerful method of constructing asymmetric orthogonal arrays is to utilize difference schemes. The method has been successful in generating many useful asymmetric orthogonal arrays; see Wang and Wu (1991) for example. Using matrix images is an intriguing idea in constructing asymmetric orthogonal arrays; see Zhang, Lu, and Pang (1999) for the basic idea and Pang, Wang, Lin, and Liu (2021) for recent work. The downside of this method is that it is extremely difficult to apply.

Computational results have significantly contributed to this line of research, and are a major source of asymmetric orthogonal arrays of strength three or higher, for the obvious reason that most of the aforementioned methods are developed for the construction of strength-two asymmetric arrays. We refer to Xu (2002) and Schoen, Eendebak, and Nguyen (2010) for some significant developments.

10.3 Minimum G_2-Aberration

10.3.1 J-characteristics for multi-level designs

Consider a factorial design $\boldsymbol{D} = (d_{ij})_{n\times m}$, where $d_{ij} = 0, 1, \ldots, s_j - 1$. Let $N(x_1, \ldots, x_m)$ be its indicator function; that is, $N(x_1, \ldots, x_m)$ is the number of times treatment combination $(x_1, \ldots, x_m)$ appears in design $\boldsymbol{D}$. As previously done to $\mu(x_1, \ldots, x_m)$ in (10.1), we obtain an ANOVA decomposition of $N(x_1, \ldots, x_m)$:

$$N(x_1, \ldots, x_m) = \sum_u Q_u(x_1, \ldots, x_m), \tag{10.8}$$

where the summation is over all subsets u of $\{1, \ldots, m\}$. In (10.8), $Q_u(x_1, \ldots, x_m)$ is recursively defined as follows:

$$Q_\phi(x_1, \ldots, x_m) = \frac{1}{s_1 \cdots s_m} \sum_{x_1, \ldots, x_m} N(x_1, \ldots, x_m),$$

with the summation being over all the treatment combinations $(x_1, \ldots, x_m)$, and

$$Q_u(x_1, \ldots, x_m) = \frac{1}{\prod_{j \notin u} s_j} \sum_{x_j, j \notin u} N(x_1, \ldots, x_m) - \sum_{v \subsetneq u} Q_v(x_1, \ldots, x_m),$$

where the first summation is over all the treatment combinations $(x_1, \ldots, x_m)$ with x_j held fixed for $j \in u$, and the second summation is over all proper subsets v of u; that is, all subsets v of u excluding $v = u$.

In the ANOVA decomposition for $\mu(x_1, \ldots, x_m)$ in Section 10.1, $\theta_u(x_1, \ldots, x_m)$ in (10.1) is the *treatment effect* for the factors in u. In the present situation, the ANOVA decomposition is for $N(x_1, \ldots, x_m)$, representing the design $\boldsymbol{D}$. It is therefore appropriate to construe $Q_u(x_1, \ldots, x_m)$ as the *design effect* for the factors in u. Consequently, $Q_\phi(x_1, \ldots, x_m) = Q_\phi = n/(s_1 \cdots s_m)$ is the design grand mean; for $u = \{j\}$, $Q_u(x_1, \ldots, x_m) = Q_j(x_j)$ is the design main effect of the jth factor, and for $u = \{j_1, j_2\}$, $Q_u(x_1, \ldots, x_m) = Q_{j_1 j_2}(x_{j_1}, x_{j_2})$ is the design interaction effect of the two factors j_1 and j_2.

The J-characteristics of design $\boldsymbol{D}$ are then defined as

$$J_u(x_1, \ldots, x_m) = s_1 \cdots s_m Q_u(x_1, \ldots, x_m). \tag{10.9}$$

There are a number of reasons for preferring $J_u(x_1, \ldots, x_m)$ to $Q_u(x_1, \ldots, x_m)$ in defining J-characteristics for multi-level designs. Unlike $Q_u(x_1, \ldots, x_m)$, $J_u(x_1, \ldots, x_m)$ is consistent with the definition of J-characteristics for two-level designs. Being integer-valued is another desirable feature of $J_u(x_1, \ldots, x_m)$. Most importantly, $J_u(x_1, \ldots, x_m)$ only depends on $\boldsymbol{D}_u$, the projection design of $\boldsymbol{D}$ onto u, but not on the entire design $\boldsymbol{D}$.

Consider a two-level design with its two levels denoted by ± 1. Relabel the two levels $+1$ and -1 as 0 and 1, respectively. Then one can easily verify that the J-characteristics as defined in (10.9), after level conversion, have the properties that

- the absolute value of $J_u(x_1, \ldots, x_m)$ is a constant when varying $x_1, \ldots, x_m$, and
- $J_u(0, \ldots, 0)$ is precisely the J-characteristic for u defined in Definition 3.1 of Chapter 3.

An illustrative example is in order.

Example 10.1. *Consider a 5-run design $\boldsymbol{D}$ for a 2×3 experiment:*

$$\boldsymbol{D}^T = \begin{bmatrix} 0 & 0 & 1 & 1 & 1 \\ 0 & 1 & 0 & 1 & 2 \end{bmatrix}.$$

The following table gives the indicator function and J-characteristics of this design.

x_1	x_2	N	J_ϕ	$J_{\{1\}}$	$J_{\{2\}}$	$J_{\{1,2\}}$
0	0	1	5	−1	1	1
0	1	1	5	−1	1	1
0	2	0	5	−1	−2	−2
1	0	1	5	1	1	−1
1	1	1	5	1	1	−1
1	2	1	5	1	−2	2

Just like J-characteristics for two-level designs, J-characteristics for multi-level designs also have many useful and interesting properties. Some of the properties are obvious; others are less so, but are fairly easy to verify. We document them in a lemma.

Lemma 10.1. *Let $J_u(x_1, \dots, x_m)$ be the J-characteristics of a factorial design $\boldsymbol{D} = (d_{ij})_{n \times m}$, where $d_{ij} = 0, 1, \dots, s_j - 1$. Then the following properties of $J_u(x_1, \dots, x_m)$ hold.*

(i) *Design $\boldsymbol{D}$, in terms of the indicator function $N(x_1, \dots, x_m)$, is uniquely determined by its J-characteristics $J_u(x_1, \dots, x_m)$, and vice versa.*

(ii) *Design $\boldsymbol{D}$ is a full factorial or several replicates thereof if and only if $J_u(x_1, \dots, x_m) = 0$ for all $(x_1, \dots, x_m)$ and for all nonempty u.*

(iii) *Design $\boldsymbol{D}$ is an orthogonal array of strength t if and only if $J_u(x_1, \dots, x_m) = 0$ for all $(x_1, \dots, x_m)$ and for all nonempty u with $|u| \le t$.*

(iv) *In terms of the indicator function, the projection design $\boldsymbol{D}_u$ of $\boldsymbol{D}$ onto the factors in u is uniquely determined by $J_v(x_1, \dots, x_m)$ for all $v \subseteq u$.*

(v) *If $\boldsymbol{D}$ is an $OA(n, s^m, t)$, a symmetric array of strength t, then $J_u(x_1, \dots, x_m)$ must be a multiple of s^t for all $(x_1, \dots, x_m)$ and all u.*

10.3.2 Minimum G_2-aberration for multi-level designs

Using J-characteristics, we now define the minimum G_2-aberration criterion for multi-level designs. Let

$$S_u^2 = \sum_{x_1, \dots, x_m} [Q_u(x_1, \dots, x_m)]^2.$$

In terms of J-characteristics, we have that

$$S_u^2 = \frac{1}{(s_1 \cdots s_m)^2} \sum_{x_1, \dots, x_m} [J_u(x_1, \dots, x_m)]^2. \tag{10.10}$$

Define

$$A_j = \frac{1}{n^2}(s_1 \cdots s_m) \sum_{|u|=j} S_u^2. \tag{10.11}$$

We will call $(A_1, \ldots, A_m)$ the word length pattern of design $\boldsymbol{D}$. Note that A_0 is also defined: $A_0 = 1$ due to $J_\phi = n$. Then the minimum G_2-aberration criterion is to select designs by sequentially minimizing A_1, A_2, ..., A_m.

When applied to two-level designs, the minimum G_2-aberration criterion defined here is the same as that defined in Chapter 5. To see this, write

$$A_j = \frac{1}{n^2} \sum_{|u|=j} \frac{1}{s_1 \cdots s_m} \sum_{x_1,\ldots,x_m} \left[J_u(x_1, \ldots, x_m)\right]^2.$$

For a two-level design, we have that $|J_u(x_1, \ldots, x_m)| = |J_u(0, \ldots, 0)|$ for all possible $x_1, \ldots, x_m$, and $J_u(0, \ldots, 0)$ is equal to the J-characteristic defined in Chapter 3. This shows that

$$\frac{1}{s_1 \cdots s_m} \sum_{x_1,\ldots,x_m} \left[J_u(x_1, \ldots, x_m)\right]^2 = \left[J_u(0, \ldots, 0)\right]^2,$$

and therefore

$$A_j = \frac{1}{n^2} \sum_{|u|=j} \left[J_u(0, \ldots, 0)\right]^2,$$

which is precisely what is defined in (5.1).

The minimum G_2-aberration criterion defined above makes a lot of intuitive sense. By Lemma 10.1, all design effects are zero for a full factorial design and all design effects of up to order t are zero for an orthogonal array of strength t. Under the hierarchical principle that the treatment effects of lower orders are more important than those of higher orders, a good design should, therefore, minimize the design effects with priority given to those of lower orders. A formal implementation of this idea is to sequentially minimize $A_1, \ldots, A_m$, where $S_u^2 = \sum_{x_1,\ldots,x_m} \left[Q_u(x_1, \ldots, x_m)\right]^2$ is the sum of squares of the design effects for the set u of factors, and A_j is proportional to the total sum of squares of the design effects of order j.

Though obvious but worth documenting is the following property of the word length pattern $(A_1, \ldots, A_m)$ for orthogonal arrays.

Lemma 10.2. *A design $\boldsymbol{D}$ is an orthogonal array of strength t if and only if $A_1 = \cdots = A_t = 0$.*

The minimum G_2-aberration criterion for multi-level designs, first defined in Xu and Wu (2001), was called the generalized minimum aberration criterion. For consistency, we still call it the minimum G_2-aberration criterion. The original definition in Xu and Wu (2001), making use of orthogonal contrasts, differs from what is given above, which is based on A_j in (10.11). More specifically, consider model (10.4) under design $\boldsymbol{D}$:

$$\boldsymbol{\mu}(\boldsymbol{D}) = \sum_u \boldsymbol{P}_u(\boldsymbol{D})\boldsymbol{\beta}_u.$$

Xu and Wu (2001) defined the minimum G_2-aberration criterion as the one that selects designs by sequentially minimizing $B_1, \ldots, B_m$, where

$$B_j = \frac{1}{n^2} \sum_{|u|=j} \left\|\boldsymbol{P}_u^T(\boldsymbol{D})\mathbf{1}_n\right\|^2. \tag{10.12}$$

The two seemingly different definitions of the minimum G_2-aberration criterion are actually the same.

Theorem 10.9. *We have that $A_j = B_j$ for $j = 1, \ldots, m$, where A_j and B_j are defined in (10.11) and (10.12), respectively.*

Proof. We only need to show that

$$(s_1 \cdots s_m) S_u^2 = \left\| \boldsymbol{P}_u^T(\boldsymbol{D}) \boldsymbol{1}_n \right\|^2. \tag{10.13}$$

We recall that S_u^2 is the sum of squares of the design effects for the set u of factors. Let $\boldsymbol{N}$ be the vector that collects $N(x_1, \ldots, x_m)$ in the Yates order. Applying the orthogonal contrast model $\boldsymbol{\mu} = \sum_u \boldsymbol{P}_u \boldsymbol{\beta}_u$ as in (10.4) to $\boldsymbol{N}$, we obtain that

$$\boldsymbol{N} = \sum_u \boldsymbol{P}_u \boldsymbol{\alpha}_u,$$

where $\boldsymbol{\alpha}_u = (s_1 \cdots s_m)^{-1} \boldsymbol{P}_u^T \boldsymbol{N}$. We see that $(s_1 \cdots s_m) \|\boldsymbol{\alpha}_u\|^2$ is also the sum of squares of the design effects for the set u of factors, showing that $S_u^2 = (s_1 \cdots s_m) \|\boldsymbol{\alpha}_u\|^2$. Equation (10.13) now follows because

$$(s_1 \cdots s_m) S_u^2 = (s_1 \cdots s_m)^2 \|\boldsymbol{\alpha}_u\|^2 = \left\| \boldsymbol{P}_u^T \boldsymbol{N} \right\|^2 = \left\| \boldsymbol{P}_u^T(\boldsymbol{D}) \boldsymbol{1}_n \right\|^2.$$

The last equality is due to the simple fact that $\boldsymbol{P}_u^T(\boldsymbol{D}) \boldsymbol{1}_n = \boldsymbol{P}_u^T \boldsymbol{N}$. □

The approach of introducing the minimum G_2-aberration via multi-level J-characteristics as given in (10.10) and (10.11) is from an unpublished manuscript of Boxin Tang who presented the work in DAE 2002 in Vancouver. Some details and further exploration of the approach are available in Ryan Lekivetz's PhD thesis at Simon Fraser University (Lekivetz, 2011). The approach has the interpretation of minimizing the design effects of lower orders, and also appears easier conceptually, as it is through the ANOVA decomposition of the indicator function without the need of specifying orthogonal contrasts. The equivalence between this approach and the one in Xu and Wu (2001) as given in Theorem 10.9 also shows that B_j in (10.12) is invariant to the choice of orthogonal contrasts.

For symmetric designs with $s_1 = \cdots = s_m = s$, the word length patterns are in fact the MacWilliams transforms of their distance distributions. Some details follow. The Hamming distance of any two runs of design $\boldsymbol{D}$ is the number of components where they differ. Thus the distance distribution of $\boldsymbol{D}$ is $(\Delta_0, \Delta_1, \ldots, \Delta_m)$, where

$n\Delta_j$ is the number of ordered pairs of runs that are at distance j.

The MacWilliams transforms of $(\Delta_0, \Delta_1, \ldots, \Delta_m)$ are $(\Delta_0', \Delta_1', \ldots, \Delta_m')$ with Δ_j' defined by

$$\Delta_j' = n^{-1} \sum_{i=0}^{m} \Delta_i P_j(i; m, s) \text{ for } j = 0, 1, \ldots, m,$$

where

$$P_j(x; m, s) = \sum_{i=0}^{j} (-1)^i (s-1)^{j-i} \binom{x}{i} \binom{m-x}{j-i}$$

are the Krawtchouk polynomials.

Theorem 10.10. *We have that $A_j = \Delta_j'$ for $j = 1, \ldots, m$.*

This result, presented in Xu and Wu (2001), was based on a result of Delsarte's (Delsarte, 1973), who used Δ_j' as constraints to obtain the linear programming bound on the size of a code. The only apparent statistical connection in Delsarte (1973) is that the index of the first nonzero Δ_j' for $j = 1, \ldots, m$ determines the strength of the design as an orthogonal array. Without explanation, Ma and Fang (2001) utilized Δ_j' directly to generalize the minimum aberration criterion.

Computation of the word length pattern $(A_1, \ldots, A_m)$ via either (10.11) or (10.12) is cumbersome because one needs to deal with all subsets u of $Z_m = \{1, \ldots, m\}$. In contrast, implementation of the minimum G_2-aberration criterion through the MacWilliams transforms $(\Delta'_0, \Delta'_1, \ldots, \Delta'_m)$ of the distance distribution $(\Delta_0, \Delta_1, \ldots, \Delta_m)$ as shown in Theorem 10.10 is considerably easier. An even better approach to implementing the minimum G_2-aberration is the minimum moment aberration criterion, which we will discuss in Section 10.5.

10.4 Statistical Justifications of Minimum G_2-Aberration

For two-level designs, a number of statistical justifications for the minimum G_2-aberration criterion is provided in Section 5.2. The present section does the same for multi-level designs. We focus on presenting two statistical justifications, one from the perspective of projection designs and the other from that of minimum contamination. It is also possible to obtain a model robustness justification, but the derivation is inevitably complicated, because of which we decide not to pursue any further in that direction. Bayesian justification is left to Chapter 11, where a general Bayesian approach is taken to deal with multi-level designs under a broad class of unit structures.

Following Tang (2001), Ai and Zhang (2004) gave a projection justification of the minimum G_2-aberration for multi-level designs. We derive this result through the use of S_u^2 in (10.10), the sum of squares of design effects for the factors in u. Compared to Ai and Zhang (2004), the approach given here is much simpler as it bypasses orthogonal contrasts.

Since the ANOVA decomposition for the indicator function $N(x_1, \ldots, x_m)$ in (10.8) is an orthogonal decomposition, we obtain that

$$V(\boldsymbol{D}) = \sum_{u:\phi \subsetneq u} S_u^2, \tag{10.14}$$

where

$$V(\boldsymbol{D}) = \sum_{x_1,\ldots,x_m} [N(x_1, \ldots, x_m) - Q_\phi]^2$$

is the total design sum of squares with $Q_\phi = \bar{N}$, the average of all $N(x_1, \ldots, x_m)$s, and

$$S_u^2 = \sum_{x_1,\ldots,x_m} [Q_u(x_1, \ldots, x_m)]^2$$

is the sum of squares of the design effects for the factors in u. Clearly, $V(\boldsymbol{D}) = 0$ if and only if design $\boldsymbol{D}$ is a full factorial or several replicates thereof, or simply an orthogonal array of strength m. If no design has $V(\boldsymbol{D}) = 0$, we prefer one that has the smallest $V(\boldsymbol{D})$ value, as such a design is closest to an orthogonal array of strength m. Applying (10.14) to the projection design $\boldsymbol{D}_u$ gives

$$V(\boldsymbol{D}_u) = \sum_{v:\phi \subsetneq v \subseteq u} S_v^2.$$

Then

$$V_k(\boldsymbol{D}) = \sum_{|u|=k} V(\boldsymbol{D}_u) = \sum_{|u|=k} \sum_{v:\phi \subsetneq v \subseteq u} S_v^2$$

provides a goodness measure for how close design $\boldsymbol{D}$ is to an orthogonal array of strength k. The smaller the $V_k(\boldsymbol{D})$ value is, the closer is $\boldsymbol{D}$ to an orthogonal array of strength k. If lower dimensions are given more importance than higher dimensions, as is normally done, one should choose a design by sequentially minimizing $V_1(\boldsymbol{D}), \ldots, V_m(\boldsymbol{D})$. This may appropriately be called the projection criterion for design selection. Some simple combinatorial counting shows that

$$V_k(\boldsymbol{D}) = \sum_{j=1}^{k} \binom{m-j}{k-j} \sum_{|v|=j} S_v^2.$$

Since $A_j(\boldsymbol{D}) = n^{-2}(s_1 \cdots s_m) \sum_{|v|=j} S_v^2$, we obtain that

$$V_k(\boldsymbol{D}) = n^2(s_1 \cdots s_m)^{-1} \sum_{j=1}^{k} \binom{m-j}{k-j} A_j(\boldsymbol{D}), \tag{10.15}$$

which establishes a projection justification for the minimum G_2-aberration criterion.

Theorem 10.11. *The minimum G_2-aberration criterion is equivalent to the projection criterion. More specifically, sequentially minimizing $A_1(\boldsymbol{D}), \ldots, A_m(\boldsymbol{D})$ is equivalent to sequentially minimizing $V_1(\boldsymbol{D}), \ldots, V_m(\boldsymbol{D})$.*

We should point out that, when applied to two-level designs, the $V_k(\boldsymbol{D})$ in (10.15) differs by a constant from that in (5.8). Obviously, this causes no worry about the validity of Theorem 10.11.

The projection justification for the minimum G_2-aberration criterion given in Theorem 10.11 applies to all factorial designs. On the other hand, the minimum contamination justification to be discussed next applies only to orthogonal arrays of strength two.

Let $\boldsymbol{D}$ be an orthogonal array of strength two. Under design $\boldsymbol{D}$, we rewrite the full model in (10.6) as

$$\boldsymbol{Y} = \boldsymbol{X}_0\boldsymbol{\beta}_0 + \boldsymbol{X}_1\boldsymbol{\beta}_1 + \cdots + \boldsymbol{X}_m\boldsymbol{\beta}_m + \boldsymbol{\epsilon}, \tag{10.16}$$

where $\boldsymbol{X}_j$ collects $\boldsymbol{P}_u(\boldsymbol{D})$ for all u with $|u| = j$, and $\boldsymbol{\beta}_j$ collects $\boldsymbol{\beta}_u$ for all u with $|u| = j$ in the same order. Suppose that our interest lies in the estimation of the main effects $\boldsymbol{\beta}_1$. Fitting the main-effect model

$$\boldsymbol{Y} = \boldsymbol{X}_0\boldsymbol{\beta}_0 + \boldsymbol{X}_1\boldsymbol{\beta}_1 + \boldsymbol{\epsilon}$$

to the data, we obtain the least-squares estimator of $\boldsymbol{\beta}_1$ as $\hat{\boldsymbol{\beta}}_1 = (\boldsymbol{X}_1^T\boldsymbol{X}_1)^{-1}\boldsymbol{X}_1^T\boldsymbol{Y}$, which under the full model in (10.16) has a bias

$$E(\hat{\boldsymbol{\beta}}_1) - \boldsymbol{\beta}_1 = \boldsymbol{C}_2\boldsymbol{\beta}_2 + \cdots + \boldsymbol{C}_m\boldsymbol{\beta}_m,$$

where $\boldsymbol{C}_j = n^{-1}\boldsymbol{X}_1^T\boldsymbol{X}_j$ for $j = 2, \ldots, m$. Under the hierarchical assumption that effects of lower orders are more important than those of higher orders, designs should be selected by sequentially minimizing $\|\boldsymbol{C}_2\|^2, \ldots, \|\boldsymbol{C}_m\|^2$, where $\|\boldsymbol{C}_j\|^2 = \text{trace}(\boldsymbol{C}_j^T\boldsymbol{C}_j)$ and $\boldsymbol{C}_j$ is the contamination matrix due to j-factor interactions.

It is easily verified that $\|\boldsymbol{C}_2\|^2 = 3A_3$ for any design $\boldsymbol{D}$ that is an orthogonal array of strength two. If $\boldsymbol{D}$ is an OA$(n, s^m, 2)$, a symmetrical array, then, as shown in Xu and Wu (2001), we can obtain that

$$\|\boldsymbol{C}_j\|^2 = (j+1)A_{j+1} + j(s-2)A_j + (m-j+1)(s-1)A_{j-1} \tag{10.17}$$

for $j = 2, \ldots, m-1$, and $\|\boldsymbol{C}_m\|^2 = m(s-2)A_m + (s-1)A_{m-1}$. This provides a minimum contamination justification for the minimum G_2-aberration criterion.

Theorem 10.12. *The minimum G_2-aberration criterion is equivalent to a contamination criterion that sequentially minimizes the contamination $\|\boldsymbol{C}_j\|^2$ from the j-factor interactions on the estimation of the main effects in the order of $j = 2, \ldots, m$.*

If design $\boldsymbol{D}$ is an asymmetric orthogonal array, then a simple relationship as in (10.17) between $\|\boldsymbol{C}_j\|^2$s and A_js is not possible. But we do have that $\|\boldsymbol{C}_j\|^2 = (j+1)A_{j+1}$ for $j = 2, \ldots, t$ if design $\boldsymbol{D}$ is an asymmetric orthogonal array of strength t.

Instead of giving a proof for (10.17), we offer some pointers which should be useful to the readers for understanding the relationship between $\|\boldsymbol{C}_j\|^2$s and A_js as given in (10.17). If we compare (10.17) with the two-level version in (5.3), two differences are noticeable. A relatively minor one is the extra $s-1$ in the coefficient for A_{j-1}. The reason that A_{j-1} comes into play when calculating $\|\boldsymbol{C}_j\|^2$ is the cancellation of a factor that is involved in a j-factor interaction. An s-level factor has $s-1$ degrees of freedom for main-effect contrasts, thus contributing $s-1$ times to A_{j-1}. A more prominent difference is the dependence of $\|\boldsymbol{C}_j\|^2$ on A_j for designs of three or more levels. This is because we need to consider, when calculating $\|\boldsymbol{C}_j\|^2$ for a given factor, the contamination on one of its main-effect contrasts by a j-factor interaction involving another main-effect contrast of this given factor. The existence of such a situation for designs with $s \geq 3$ levels entails the presence of A_j in the expression for $\|\boldsymbol{C}_j\|^2$ as in (10.17).

10.5 Minimum Moment Aberration

Just as we have done for two-level designs, we now develop a minimum moment aberration criterion for s-level symmetric designs. Let $\boldsymbol{D}$ be an s-level design of n runs for m factors. Let d_{pq} be the Hamming distance between the pth and qth runs of design $\boldsymbol{D}$. Then $\delta_{pq} = m - d_{pq}$ is the number of coincidences of the two runs. Our intuition tells us that a good design should have small δ_{pq}s because two runs that are close to each other will produce overlapping information. The minimum moment aberration criterion is a formulation of this idea. Define the jth moment of the δ_{pq}s by

$$K_j = \frac{1}{\binom{n}{2}} \sum_{1 \leq p < q \leq n} \delta_{pq}^j. \tag{10.18}$$

Then the minimum moment aberration, as proposed by Xu (2003), is a criterion that selects designs by sequentially minimizing $K_1, \ldots, K_m$.

We set $K_0 = 1$, using which will help simplify our later presentation.

For designs with two levels ± 1 , the minimum moment aberration is defined in Chapter 5 as a criterion that sequentially minimizes $M_1, \ldots, M_m$ where

$$M_j = \frac{1}{n^2} \sum_{p=1}^{n} \sum_{q=1}^{n} t_{pq}^j,$$

with t_{pq} being the inner product of runs p and q. Obviously, sequentially minimizing $M_1, \ldots, M_m$ is equivalent to sequentially minimizing $M_1', \ldots, M_m'$, where

$$M_j' = \frac{1}{\binom{n}{2}} \sum_{1 \leq p < q \leq n} t_{pq}^j,$$

as $t_{pp} = m$, a constant. Since $t_{pq} = \delta_{pq} - d_{pq}$ and $m = \delta_{pq} + d_{pq}$, we have that $t_{pq} = 2\delta_{pq} - m$. This shows that sequentially minimizing $M_1', \ldots, M_m'$ is equivalent to sequentially minimizing $K_1, \ldots, K_m$. We therefore have that the minimum moment aberration criterion defined here for s-level designs indeed includes that defined for two-level designs as a special case.

Recall the distance distribution $(\Delta_0, \Delta_1, \ldots, \Delta_m)$ of design $\boldsymbol{D}$ introduced in Section 10.3, where $n\Delta_j$ is the number of ordered pairs of runs that are at distance j. Then we have that

$$\sum_{p=1}^{n}\sum_{q=1}^{n} d_{pq}^j = n\sum_{i=0}^{m} i^j \Delta_i.$$

This, in conjunction with $d_{pq} = m - \delta_{pq}$, shows that $(K_0, K_1, \ldots, K_m)$ and $(\Delta_0, \Delta_1, \ldots, \Delta_m)$ must be linearly related to each other. Theorem 10.10 reveals that $(\Delta_0, \Delta_1, \ldots, \Delta_m)$ is linearly related to $(A_0, A_1, \ldots, A_m)$. Thus, $(K_0, K_1, \ldots, K_m)$ must be linearly related to $(A_0, A_1, \ldots, A_m)$. What is special about this linear relationship is that K_j only depends on A_i for $i = 0, 1, \ldots, j$.

Theorem 10.13. *For any s-level design of n runs and m factors, we have that*

$$K_j = \sum_{i=0}^{j} c_{ji} A_i,$$

where c_{ji}'s are constants that depend only on n, m, and s with $c_{jj} > 0$. This relation reveals that sequentially minimizing $K_0, K_1, \ldots, K_m$ is equivalent to sequentially minimizing $A_0, A_1, \ldots, A_m$, showing that minimum moment aberration is equivalent to minimum G_2-aberration.

Our discussion leading to Theorem 10.13 has already given an outline of the proof. Essentially, the linear relationship between $(K_0, K_1, \ldots, K_m)$ and $(A_0, A_1, \ldots, A_m)$ can be established through their respective linear relationships with $(\Delta_0, \Delta_1, \ldots, \Delta_m)$. The rest of the proof is some painstaking algebra. We refer to Xu (2003) for details.

We provide explicit coefficients for the expressions of K_1 and K_2, as they will be needed for the discussion of multi-level supersaturated designs in Section 10.8.

Corollary 10.2. *For any s-level design of n runs and m factors, we have that*

$$K_1 = \frac{1}{(n-1)s}\Big(nA_1 + nm - ms\Big),$$

and

$$K_2 = \frac{1}{(n-1)s^2}\Big(2nA_2 + n(2m+s-2)A_1 + nm(m+s-1) - (ms)^2\Big).$$

Because of their linear relationship, $(K_0, K_1, \ldots, K_m)$ and $(\Delta_0, \Delta_1, \ldots, \Delta_m)$ bring forth the same amount of distance-related information about design $\boldsymbol{D}$. The usefulness of the distance distribution $(\Delta_0, \Delta_1, \ldots, \Delta_m)$ lies only in providing a quicker way of calculating the word length pattern $(A_0, A_1, \ldots, A_m)$. The sequence of moments $(K_0, K_1, \ldots, K_m)$, on the other hand, accomplishes more than just a computational advantage–one can use them directly as a tool to study minimum G_2-aberration designs as underscored by Theorem 10.13. A testament to the potency of this tool is the construction results to be presented in the next section.

Xu and Wu (2001) developed a complementary design theory for s-level designs, which is a general version of the theory for two-level designs in Chapter 5. As shown in Xu (2003), the minimum moment aberration criterion offers a simple derivation of the complementary design theory, which we describe next. The problem on the table is to find a minimum G_2-aberration design $\boldsymbol{D}$ from a saturated strength-two orthogonal array, say $\boldsymbol{H}$, by examining the complementary design $\bar{\boldsymbol{D}} = \boldsymbol{H} \setminus \boldsymbol{D}$, which consists of all the columns of $\boldsymbol{H}$ outside $\boldsymbol{D}$. Any two runs of a saturated strength-two orthogonal array have the same number γ of coincidences (Mukerjee and Wu, 1995), meaning that

$$\delta_{pq}(\boldsymbol{D}) = \gamma - \delta_{pq}(\bar{\boldsymbol{D}}),$$

where γ is a positive constant depending only on n and s. As a result, we obtain that

$$K_j(\boldsymbol{D}) = \sum_{i=0}^{j} \binom{j}{i} (-1)^i \gamma^{j-i} K_i(\bar{\boldsymbol{D}}).$$

Using the relationship between the K_js and A_js in Theorem 10.13 for both $\boldsymbol{D}$ and $\bar{\boldsymbol{D}}$, we establish the next result.

Theorem 10.14. *We have that*

$$A_j(\boldsymbol{D}) = \sum_{i=0}^{j} \gamma_{ji} A_i(\bar{\boldsymbol{D}}),$$

where $\gamma_{jj} = (-1)^j$ *and the other* γ_{ji}*'s are all constants depending only on* n, m, *and* s. *Therefore, finding a minimum* G_2*-aberration design* $\boldsymbol{D}$ *from a saturated strength-two orthogonal array* $\boldsymbol{H}$ *is equivalent to finding a complementary design* $\bar{\boldsymbol{D}}$ *that sequentially minimizes* $-A_3(\bar{\boldsymbol{D}}), A_4(\bar{\boldsymbol{D}}), -A_5(\bar{\boldsymbol{D}})$, *and so on.*

As we have commented on in Chapter 5, while theoretically interesting, the complementary design theory for nonregular designs is less useful compared to that for regular designs. This drawback is more pronounced for s-level designs because saturated symmetrical orthogonal arrays only exist for limited sets of n and s values. The Rao-Hamming construction presented in Section 10.3 does produce a saturated strength-two orthogonal array, but it requires that s be a prime power and n be a power of s. On the other hand, the orthogonal arrays given by the Addelman-Kempthorne construction are not saturated.

For mixed-level designs, with some suitable modifications, a minimum moment aberration criterion can also be developed, but a strong result as Theorem 10.13 is no longer possible. Interested readers may wish to read Xu (2003) for details.

10.6 Construction Results for Minimum G_2-Aberration

We present in this section Butler's construction results (Butler, 2005) for minimum G_2-aberration designs. These are some comprehensive results for symmetrical orthogonal arrays of strength two, i.e., $\mathrm{OA}(n, s^m, 2)$s, indisputably the most important case. Consider an $\mathrm{OA}(n, s^m, 2)$. As before, let d_{pq} be the Hamming distance of runs p and q of the design. It is convenient to work with

$$L_j = \frac{1}{\binom{n}{2}} \sum_{1 \le p < q \le n} d_{pq}^j$$

instead of $K_j = \sum_{1 \le p < q \le n} \delta_{pq}^j / \binom{n}{2}$. Since $d_{pq} = m - \delta_{pq}$, we see that L_j is a linear combination of $K_0, K_1, \ldots, K_j$ with the coefficient of K_j being $(-1)^j$. Consequently sequentially minimizing $K_1, \ldots, K_m$ is equivalent to sequentially minimizing

$-L_1, L_2, -L_3, L_4, \ldots, (-1)^m L_m$. Because $A_1 = A_2 = 0$ for an orthogonal array of strength two, Theorem 10.13 implies that K_1 and K_2 are constants only depending on n, m, and s. This shows that L_1 and L_2 are also constants only depending on n, m, and s. Therefore, a minimum G_2-aberration OA$(n, s^m, 2)$ maximizes

$$L_3 = \sum_{1 \le p < q \le n} d_{pq}^3 \Big/ \binom{n}{2}$$

while having constant L_1 and L_2.

Let $d_{\max}$ denote the maximum possible distance across all OA$(n, s^m, 2)$s with given n, m, and s. The key idea of Butler's constructions is contained in the next result.

Theorem 10.15. *If the distances d_{pq} of an OA$(n, s^m, 2)$ for all $p < q$ have only three possible values $d_{\max}, d_1$ and $d_1 + 1$ for some d_1, then this array has minimum G_2-aberration among all OA$(n, s^m, 2)$s.*

Proof. Because $L_1 = \sum_{1 \le p < q \le n} d_{pq} / \binom{n}{2}$ and $L_2 = \sum_{1 \le p < q \le n} d_{pq}^2 / \binom{n}{2}$ are constants for any OA$(n, s^m, 2)$, maximizing $L_3 = \sum_{1 \le p < q \le n} d_{pq}^3 / \binom{n}{2}$ is equivalent to maximizing

$$\sum_{1 \le p < q \le n} (d_{pq} - d_{\max})(d_{pq} - d_1)(d_{pq} - d_1 - 1). \tag{10.19}$$

For an arbitrary OA$(n, s^m, 2)$, we always have that

$$d_{pq} - d_{\max} \le 0 \quad \text{and} \quad (d_{pq} - d_1)(d_{pq} - d_1 - 1) \ge 0.$$

Therefore,

$$(d_{pq} - d_{\max})(d_{pq} - d_1)(d_{pq} - d_1 - 1) \le 0 \quad \text{for all } p < q.$$

This establishes that

$$\sum_{1 \le p < q \le n} (d_{pq} - d_{\max})(d_{pq} - d_1)(d_{pq} - d_1 - 1) \le 0.$$

If an OA$(n, s^m, 2)$ has $d_{pq} = d_{\max}, d_1$ or $d_1 + 1$, then it has

$$\sum_{1 \le p < q \le n} (d_{pq} - d_{\max})(d_{pq} - d_1)(d_{pq} - d_1 - 1) = 0,$$

and therefore maximizes (10.19) over all OA$(n, s^m, 2)$s for given n, m, s.

This design has minimum G_2-aberration because its distance distribution is uniquely determined if it has $d_{pq} = d_{\max}, d_1$ or $d_1 + 1$. Consequently, all L_j for $j \ge 4$ are also uniquely determined. The uniqueness of the distance distribution can be shown as follows. Let f_1, f_2, and f_3 be the numbers of pairs of runs p and q with $p < q$ that have $d_{pq} = d_{\max}, d_1$, and $d_1 + 1$, respectively. Then we have that

$$\binom{n}{2} = f_1 + f_2 + f_3,$$

$$\binom{n}{2} L_1 = \sum_{1 \le p < q \le n} d_{pq} = f_1 d_{\max} + f_2 d_1 + f_3(d_1 + 1),$$

$$\binom{n}{2} L_2 = \sum_{1 \le p < q \le n} d_{pq}^2 = f_1 d_{\max}^2 + f_2 d_1^2 + f_3(d_1 + 1)^2.$$

These equations uniquely determine f_1, f_2, and f_3. □

Theorem 10.15 does not preclude the possibilities that d_{pq} has only one or only two of the three values $d_{\max}, d_1$, and $d_1 + 1$. The next result, available in Xu (2003), applies to some of these situations.

Corollary 10.3. *If the distances d_{pq} for $p < q$ of an $OA(n, s^m, 2)$ differ by at most one, then the $OA(n, s^m, 2)$ has minimum G_2-aberration.*

For an $\mathrm{OA}(n, s^m, 2)$, it is obvious that $d_{\max} \leq m$. Butler (2005) established a further result on $d_{\max}$.

Lemma 10.3. *For any $OA(n, s^m, 2)$, we have that $d_{\max} \leq n/s$.*

The proof of Lemma 10.3 parallels that of Lemma 5.3, though it gets more complicated in the present situation.

The construction results to be presented all make use of saturated asymmetrical orthogonal arrays $\mathrm{OA}(n, s^m g^1, 2)$s. Saturation here requires that $m(s-1) + (g-1) = n-1$.

Lemma 10.4. *For a saturated $OA(n, s^m g^1, 2)$, let d_{pq} be the distance between runs p and q over the m factors with s levels. We have that*

$$d_{pq} = \begin{cases} n/s & \text{if runs } p \text{ and } q \text{ have the same level of the } g\text{-level factor,} \\ n/s - g/s & \text{otherwise.} \end{cases}$$

The above lemma is a special case of a result of Mukerjee and Wu's (Mukerjee and Wu, 1995). We are now ready to present Butler's construction results.

Theorem 10.16. *Suppose that an $OA(s^2, s^{s+1}, 2)$ exists. Then any $OA(s^2, s^m, 2)$ for $m \leq s+1$, a subarray of the $OA(s^2, s^{s+1}, 2)$, has minimum G_2-aberration.*

Proof. For the $\mathrm{OA}(s^2, s^{s+1}, 2)$, $d_{pq} = n/s = s$ for all $p < q$, which can be verified using Lemma 10.4 by taking one factor as a g-level factor, where $g = s$. This shows that any two runs of the $\mathrm{OA}(s^2, s^{s+1}, 2)$ differ in s entries and agree in one entry. The array $\mathrm{OA}(s^2, s^m, 2)$, obtained by taking m columns from the $\mathrm{OA}(s^2, s^{s+1}, 2)$, therefore has $d_{pq} = m-1$ or m. We obtain the desired result by applying Corollary 10.3. □

In Theorem 10.5, if we take $k = 2$, then we obtain an $\mathrm{OA}(s^2, s^{s+1}, 2)$, provided that s is a prime power. Combining this result with Theorem 10.16, we obtain the following minimum G_2-aberration designs:

- an $\mathrm{OA}(9, 3^m, 2)$ for $m \leq 4$,
- an $\mathrm{OA}(16, 4^m, 2)$ for $m \leq 5$,
- an $\mathrm{OA}(25, 5^m, 2)$ for $m \leq 6$,
- an $\mathrm{OA}(49, 7^m, 2)$ for $m \leq 8$,
- an $\mathrm{OA}(64, 8^m, 2)$ for $m \leq 9$, and
- an $\mathrm{OA}(81, 9^m, 2)$ for $m \leq 10$.

Theorem 10.17. *Suppose that an $OA(2s^2, s^{2s}(2s)^1, 2)$ exists. Then any $OA(2s^2, s^m, 2)$, obtained by taking m of its s-level factors, where $m \leq 2s$, has minimum G_2-aberration.*

Proof. Consider the $\mathrm{OA}(2s^2, s^{2s}, 2)$ obtained by deleting the $(2s)$-level factor from the saturated $\mathrm{OA}(2s^2, s^{2s}(2s)^1, 2)$. This array has $d_{pq} = n/s = 2s$ or $d_{pq} = n/s - g/s = 2s - 2$, by Lemma 10.4. Therefore, for any $\mathrm{OA}(2s^2, s^m, 2)$, a subarray of the $\mathrm{OA}(2s^2, s^{2s}, 2)$, has $d_{pq} = m$, $m-1$, or $m-2$. Theorem 10.17 then follows from Theorem 10.15 by taking $d_{\max} = m$ and $d_1 = m-2$. □

The Addelman-Kempthorne's construction in Section 10.2 can also be used to obtain an $\mathrm{OA}(2s^2, s^{2s}(2s)^1, 2)$ when s is a prime power. Applying Theorem 10.17 to this array, we obtain the following minimum G_2-aberration designs:

- an $\mathrm{OA}(18, 3^m, 2)$ for $m \leq 6$,
- an $\mathrm{OA}(32, 4^m, 2)$ for $m \leq 8$,
- an $\mathrm{OA}(50, 5^m, 2)$ for $m \leq 10$, and
- an $\mathrm{OA}(98, 7^m, 2)$ for $m \leq 14$.

Theorem 10.18. *Suppose that an $OA(s^3, s^{s^2}(s^2)^1, 2)$ exists. Then for $m \leq s+1$, any $OA(s^3, s^{s^2+m}, 2)$, obtained by expanding the s^2-level factor into m factors of s levels using an $OA(s^2, s^m, 2)$, has minimum G_2-aberration.*

Proof. For the $\mathrm{OA}(s^3, s^{s^2}, 2)$ obtained from deleting the s^2-level factor from the saturated $\mathrm{OA}(s^3, s^{s^2}(s^2)^1, 2)$, by Lemma 10.4 we have that $d_{pq} = n/s = s^2$ or $d_{pq} = n/s - g/s = s(s-1)$. If we add m factors by expanding the s^2-level factor into m factors of s levels using an $\mathrm{OA}(s^2, s^m, 2)$, then we have that $d_{pq} = s^2 = d_{\max}$, $d_{pq} = s(s-1) + m$, or $d_{pq} = s(s-1) + m - 1$. Theorem 15.15 can now be applied with $d_1 = s(s-1) + m - 1$. □

When s is a prime power, we can construct an $\mathrm{OA}(s^3, s^{s^2+s+1}, 2)$ by the Rao-Hamming construction. Then a saturated $\mathrm{OA}(s^3, s^{s^2}(s^2)^1, 2)$ can be obtained from the $\mathrm{OA}(s^3, s^{s^2+s+1}, 2)$ by contracting $s+1$ factors, corresponding to the points on a line in the $PG(2, s)$, into an s^2-level factor. Taking $s = 3, 4$, and 5 and applying Theorem 10.18, we obtain the following minimum G_2-aberration designs:

- an $\mathrm{OA}(27, 3^{9+m}, 2)$ for $0 \leq m \leq 4$,
- an $\mathrm{OA}(64, 4^{16+m}, 2)$ for $0 \leq m \leq 5$, and
- an $\mathrm{OA}(125, 5^{25+m}, 2)$ for $0 \leq m \leq 6$.

Theorems 10.16, 10.17, and 10.18 correspond to Theorems 5, 6, and 10 of Butler (2005). We next present Theorems 7, 8, and 9 of Butler (2005) in one combined theorem. With just a slight loss of generality, we are able to present the results in a concise fashion.

Theorem 10.19. *Suppose that an $OA(rs^2, s^{rs}(rs)^1, 2)$ exists. Then all the following three arrays have minimum G_2-aberration:*
(a) the $OA(rs^2, s^{rs}, 2)$ obtained by taking all s-level factors,
(b) any $OA(rs^2, s^{rs-1}, 2)$ obtained by taking all but one s-level factor, and
(c) the $OA(rs^2, s^{rs+1}, 2)$ obtained by collapsing the (rs)-level factor into an s-level factor.

Using Theorem 10.19, one can obtain the following minimum G_2-aberration designs:

- an $\mathrm{OA}(36, 3^m, 2)$ for $m = 11, 12, 13$ from an $\mathrm{OA}(36, 3^{12}12^1, 2)$,
- an $\mathrm{OA}(48, 4^m, 2)$ for $m = 11, 12, 13$ from an $\mathrm{OA}(48, 4^{12}12^1, 2)$, and
- an $\mathrm{OA}(54, 3^m, 2)$ for $m = 17, 18, 19$ from an $\mathrm{OA}(54, 3^{18}18^1, 2)$.

10.7 G-Aberration and Generalized Resolution

In this section, we discuss a minimum G-aberration criterion and define a concept of generalized resolution for multi-level designs. Our discussion focuses on symmetric designs. The concept of generalized resolution is useful because it offers a concise description of the properties of a design. Just like that for two-level designs, we need a minimum G-aberration criterion for multi-level designs for two reasons. The A_j value in defining the minimum G_2-aberration measures the average behavior of all j-factor projection designs. These j-factor projection designs may not be the same, and some are better than the others. It is important to eliminate the worst-case scenarios if at all possible. The second reason is that we need to be able to distinguish those designs that all have minimum G_2-aberration. For example, according to Evangelaras, Koukouvinos, and Lappas (2007), all the three nonisomorphic OA$(18, 3^7, 2)$s have minimum G_2-aberration, but are very different if we look at their projection designs onto three factors. The minimum G-aberration criterion to be introduced will be able to tell apart the three designs (Schoen, 2009).

Rewrite A_j in (10.11) as

$$A_j = \sum_{|u|=j} A_u, \quad \text{where } A_u = s^m n^{-2} S_u^2. \tag{10.20}$$

We see that up to a constant, A_u is the sum of squares for the design effects of the factors in u. The minimum G_2-aberration criterion requires that we sequentially minimize A_j, for $j = 1, \ldots, m$.

We define the minimum G-aberration criterion as the one that compares designs by carrying out the following steps.

For $j = 1, \ldots, m$,

(i) minimize the largest value of A_us for all $|u| = j$,

(ii) minimize the frequency of the u's that attain this largest value,

(iii) minimize the second largest value of A_us for all $|u| = j$,

(iv) minimize the frequency of the u's that attain this second largest value,

(v) and so on and so forth.

Obviously, the minimum G-aberration criterion just defined for multi-level designs is the same as that introduced in Chapter 4 when it is applied to two-level designs.

For orthogonal arrays of strength two, we have that $A_u = 0$ for $|u| = 1, 2$. Consequently, we only need to consider the A_u values for $|u| \geq 3$ in applying the minimum G-aberration criterion. If we consider the A_u values only for $|u| = 3$, then we obtain a weak version of the minimum G-aberration, which is termed the *projection aberration criterion* by Xu, Cheng, and Wu (2004).

Example 10.2. *According to Evangelaras, Koukouvinos, and Lappas (2007) and Schoen (2009), there are exactly three nonisomorphic $OA(18, 3^7, 2)$s, which are labeled as* 18.7.1, 18.7.2, *and* 18.7.3 *here. The A_3 values and the frequency vectors of A_u values for $|u| = 3$ of the three designs are given below.*

design	A_3	frequency of A_u values			
		1/2	2/3	1	2
18.7.1	22	16	18	0	1
18.7.2	22	20	12	2	1
18.7.3	22	28	0	6	1

The three designs have the same $A_3 = 22$, and thus all have minimum G_2-aberration. On the other hand, the minimum G-aberration criterion ranks 18.7.1 as the best and 18.7.3 as the worst, with 18.7.2 being in between. The three designs have the same $\max_{|u|=3} A_u = 2$ with the same frequency. The second largest value of A_us is 1, with frequencies 0, 2, and 6 for designs 18.7.1, 18.7.2, and 18.7.3, respectively.

All nonisomorphic $OA(18, 3^m, 2)$'s for $m \leq 7$ are enumerated by Evangelaras, Koukouvinos, and Lappas (2007), and their minimum G_2- and G-aberration properties are given in Evangelaras, Koukouvinos, and Lappas (2007) and in Schoen (2009), respectively. Some results on minimum G_2- and G-aberration orthogonal arrays of 27 runs are available in Xu, Cheng, and Wu (2004), Evangelaras, Koukouvinos, and Lappas (2011), and Lekivetz (2011). In view of the complete enumeration results for orthogonal arrays of 27 runs, minimum G_2- and G-aberration $OA(27, 3^m, 2)$s with $m \leq 13$ are all available, at least in principle.

We next discuss the concept of generalized resolution.

Definition 10.2. *Let $\boldsymbol{D}$ be a design of n runs for m factors of s levels. Suppose that r is the smallest positive integer such that $\max_{|u|=r} A_u > 0$. Then the generalized resolution of design $\boldsymbol{D}$ is defined as*

$$R(\boldsymbol{D}) = r + 1 - \sqrt{\max_{|u|=r} A_u/(s-1)}.$$

For two-level designs, we have that $S_u^2 = J_u^2/2^m$ and therefore $A_u = (J_u/n)^2$. This shows that Definition 10.2 reduces to Definition 4.1 for two-level designs.

The reasonableness of Definition 10.2 is grounded in the following result (Grömping and Xu, 2014).

Theorem 10.20. *Suppose that a design $\boldsymbol{D}$ of n runs for r s-level factors has strength $r-1$ but not r. Then we have that*

$$A_r \leq s - 1,$$

with equality being reached if and only if the design is regular, meaning that any one column is completely determined by the other $r-1$ columns.

Proof. Since design $\boldsymbol{D}$ has strength $r-1$, we have that $S_v^2 = 0$ for any v with $1 \leq |v| \leq r-1$. Applying (10.14), we obtain that

$$V(\boldsymbol{D}) = S_u^2 \quad \text{with } u = \{1, 2, \ldots, r\},$$

where

$$V(\boldsymbol{D}) = \sum_{x_1,\ldots,x_r} \left(N(x_1, \ldots, x_r) - \bar{N}\right)^2$$

is the total design sum of squares with $\bar{N}$ being the average of all $N(x_1, \ldots, x_r)$s. Now, we index $N(x_1, \ldots, x_r)$ as N_{ij} with $i = 1, \ldots, s^{r-1}$, which runs through all level combinations of the first $r-1$ factors, and $j = 1, \ldots, s$, which runs through all the levels of the last factor. By grouping the N_{ij}s according to i, we have that the total sum of squares for N_{ij}s, which is $V(\boldsymbol{D})$, can be decomposed into a sum of the between-group sum of squares and within-group sum of squares as

$$\sum_{i=1}^{s^{r-1}}\sum_{j=1}^{s}(N_{ij}-\bar{N}_{..})^2 = s\sum_{i=1}^{s^{r-1}}(\bar{N}_{i.}-\bar{N}_{..})^2+\sum_{i=1}^{s^{r-1}}\sum_{j=1}^{s}(N_{ij}-\bar{N}_{i.})^2,$$

where $\bar{N}_{..}$ is the average of all N_{ij}s and $\bar{N}_{i.} = (1/s)\sum_{j=1}^{s} N_{ij}$. Because design $\boldsymbol{D}$ is an orthogonal array of strength $r-1$, we must have that

$$\sum_{j=1}^{s} N_{ij} = n/s^{r-1},$$

leading to that

$$\bar{N}_{i.} = \bar{N}_{..} = n/s^r$$

for all i. This shows that the between-group sum of squares, $s\sum_{i=1}^{s^{r-1}}\left(\bar{N}_{i.}-\bar{N}_{..}\right)^2$, must be zero. Now consider the within-group sum of squares, $\sum_{i=1}^{s^{r-1}}\sum_{j=1}^{s}\left(N_{ij}-\bar{N}_{i.}\right)^2$. It is maximized if and only if for every given i, $N_{ij} = s\bar{N}_{i.}$ for one j and $N_{ij} = 0$ for all the other j values. This amounts to saying that the rth factor is completely determined by the first $r-1$ factors. The maximum value of the within-group sum of squares is therefore given by

$$s^{r-1}\left((s\bar{N}_{i.}-\bar{N}_{i.})^2+(s-1)(-\bar{N}_{i.})^2\right) = s^r(s-1)\bar{N}_{i.}^2 = s^r(s-1)(n/s^r)^2 = (s-1)s^{-r}n^2.$$

This shows that

$$S_u^2 = V(\boldsymbol{D}) \le (s-1)s^{-r}n^2, \quad \text{where } u = \{1, 2, \ldots, r\}.$$

We therefore obtain from (10.20) that

$$A_r = A_u = s^r n^{-2} S_u^2 \le s-1.$$

□

The appealing geometrical interpretation of generalized resolution for two-level designs, as given in Theorem 4.1, is no longer available for s-level designs when $s \ge 3$. However, as the proof of Theorem 10.20 reveals, the generalized resolution for a design of $s \ge 3$ levels does capture the worst r-factor projection design in terms of the uniformity of design points. The higher the generalized resolution is, the more uniform are the design points of the worst r-factor projection design.

In addition to providing a concise description of a design, the generalized resolution can also be used as a design criterion, which requires the selection of a design with the maximum generalized resolution. It is clear that maximizing the generalized resolution is consistent with the minimum G-aberration criterion. In other words, a minimum G-aberration design must have the maximum generalized resolution.

Lemma 4.1 continues to hold for s-level designs. We present the results in the next lemma for completeness.

Lemma 10.5. *(i) If a design has generalized resolution R, then it is an orthogonal array of strength $\lfloor R \rfloor - 1$.*
(ii) If a design is an orthogonal array of strength t, but not of strength $t+1$, then its generalized resolution R satisfies that $t+1 \le R < t+2$, and $R = t+1$ holds if and only if the design has a defining word of length $t+1$.

Part (i) of Lemma 10.5 is obvious. Part (ii) is an immediate consequence of Theorem 10.20.

Example 10.3. *There are exactly four non-isomorphic* $OA(18, 3^3, 2)$*s (Evangelaras, Koukouvinos, and Lappas, 2007), labeled here as* 18.3.1, 18.3.2, 18.3.3, *and* 18.3.4. *Their* A_3 *values are given below:*

design	18.3.1	18.3.2	18.3.3	18.3.4
A_3	1/2	2/3	1	2

Simple calculations show that designs 18.3.1, 18.3.2, 18.3.3, *and* 18.3.4 *have generalized resolutions of* 3.5, 3.42, 3.29, *and* 3, *respectively.*

Grömping and Xu (2014) conducted a comprehensive study on the concept of generalized resolution for multi-level designs, and introduced two definitions of generalized resolution, which are applicable to both symmetric and asymmetric designs. Definition 10.2 given here corresponds to their first definition. Grömping and Xu (2014) also provided interpretations through the coefficient of determination R^2 for their first definition, and through the canonical correlation for their second definition. Some success was also achieved by earlier attempts at defining generalized resolution; see Evangelaras, Koukouvinos, Dean, and Dingus (2005) and Pang and Liu (2010).

10.8 Supersaturated Designs

Two-level supersaturated designs have previously been treated in Chapter 8. A multi-level factorial design is called supersaturated if it does not have enough degrees of freedom to estimate all main effects. In the symmetric case, an s-level design of n runs for m factors is supersaturated if $n \leq m(s-1)$, since each s-level factor has $s-1$ degrees of freedom for main effects, and the grand mean is part of the main-effect model. A supersaturated design cannot be an orthogonal array of strength two, for otherwise Rao's bound in Corollary 10.1 is violated. For simplicity, we only consider balanced supersaturated designs, meaning that the s levels occur equally often in each column of the designs. A balanced design is an orthogonal array of strength one, so the run size n needs to be a multiple of s.

The minimum G_2- and G-aberration criteria can be used to select good supersaturated designs. Since we only consider balanced supersaturated designs, we have that $A_u = 0$ for all $|u| = 1$, where A_u was previously given in (10.20). So, only A_u values for $|u| \geq 2$ need to be considered in implementing the G_2- and G-aberration criteria. If we focus on the A_u values for $|u| = 2$, we should then select supersaturated designs by minimizing

$$A_2 = \sum_{|u|=2} A_u$$

according to the G_2-aberration, or by minimizing

$$\max_{|u|=2} A_u$$

according to the G-aberration.

In the literature of supersaturated designs, one generally uses A_2 as the main criterion for design selection. The $\max_{|u|=2} A_u$-criterion can be used alone but more often as a secondary criterion to select one from A_2-optimal supersaturated designs. When applied to two-level designs, the A_2-criterion is equivalent to the $E(s^2)$-criterion and the $\max_{|u|=2} A_u$-criterion is equivalent to the $\max_{i<j} |s_{ij}|$-criterion.

Works on multi-level supersaturated designs were ushered in first by Yamada and Lin (1999) and subsequently by Fang, Lin, and Ma (2000) and Lu and Sun (2001). The A_2-criterion is equivalent to the ave(χ^2)-criterion of Yamada and Lin (1999) and the $E(d^2)$-criterion of Lu and Sun (2001), and the $\max_{|u|=2} A_u$-criterion is equivalent to the max(χ^2)-criterion of Yamada and Lin (1999) and the max(d^2)-criterion of Lu and Sun (2001). To see this, we note that, for $|u| = 2$, A_u is proportional to S_u^2, which is proportional to $V(\boldsymbol{D}_u)$, whereas $V(\boldsymbol{D}_u)$, the variance of the indicator function of the two-dimensional projection design $\boldsymbol{D}_u$, is the base upon which the criteria of ave(χ^2), max(χ^2), $E(d^2)$, and max(d^2) are built.

A general version of Theorems 8.1 and 8.2 is the next result.

Theorem 10.21. *For a balanced s-level supersaturated design of n runs and m factors, we have that*

$$A_2 \geq \frac{m(s-1)(ms-m-n+1)}{2(n-1)}.$$

In addition, the lower bound is attained if and only if any two rows of the design have the same number of coincidences.

The lower bound in Theorem 10.21 was obtained by Yamada and Matsui (2002) in terms of ave(χ^2) and by Lu and Sun (2001) in terms of $E(d^2)$. We next consider the construction of s-level supersaturated designs that attain the lower bound.

It turns out that the methods of construction developed in Tang and Wu (1997) and in Lin (1993), as presented in Theorems 8.3 and 8.4 for two-level designs, can both be generalized to multi-level designs (Xu and Wu, 2005).

Let $\boldsymbol{D}_1, \ldots, \boldsymbol{D}_k$ be k saturated OA$(n, s^m, 2)$s, and $\boldsymbol{C}_j$ be an sth fraction of $\boldsymbol{D}_j$ obtained by taking all the rows of $\boldsymbol{D}_j$ that have a fixed level in the first column and then deleting this first column. Consider designs

$$\boldsymbol{D} = (\boldsymbol{D}_1, \ldots, \boldsymbol{D}_k) \text{ and } \boldsymbol{C} = (\boldsymbol{C}_1, \ldots, \boldsymbol{C}_k).$$

Theorem 10.22. *Designs $\boldsymbol{D}$ and $\boldsymbol{C}$ both achieve the lower bound in Theorem 10.21, and thus are A_2-optimal.*

Theorem 10.22 follows from Lemma 10.4, by which any two rows of a saturated OA$(n, s^m, 2)$ has a Hamming distance of $d_{pq} = n/s$, and therefore has $\delta_{pq} = m - n/s$ coincidences. This implies that the number of coincidences of any two rows is $k(m - n/s)$ for design $\boldsymbol{D}$ and is $k(m - n/s - 1)$ for design $\boldsymbol{C}$.

We know from Section 10.2 that a saturated OA$(n, s^m, 2)$ can be constructed for $n = s^l$ and $m = (s^l - 1)/(s - 1)$ for any positive integer $l \geq 2$ and any prime power s. Using k copies of such a saturated OA$(n, s^m, 2)$ as $\boldsymbol{D}_1, \ldots, \boldsymbol{D}_k$ in conjunction with Theorem 10.22, we obtain a design $\boldsymbol{D}$ of $n = s^l$ runs for $m = k(s^l - 1)/(s - 1)$ factors and a design $\boldsymbol{C}$ of $n = s^{l-1}$ runs for $m = k(s^l - s)/(s - 1)$ factors that attain the lower bound in Theorem 10.21, and are therefore A_2 optimal.

Similar to two-level supersaturated designs attaining the lower bound in Theorem 8.1, multi-level supersaturated designs attaining the lower bound in Theorem 10.21 can also be shown to be equivalent to resolvable balanced incomplete block designs (Lu, Hu, and Zheng, 2003). Such a connection allows one to consider a broader class of A_2-optimal supersaturated designs than those given in Theorem 10.22. This is helpful if one likes to use $\max_{|u|=2} A_u$ as a secondary criterion for further design selection.

The optimality results in Theorem 10.22 do not require that designs $\boldsymbol{D}$ and $\boldsymbol{C}$ have no fully aliased columns. Two columns are said to be fully aliased with each other if one can be obtained from the other by a permutation of the s levels. Obviously, we want to avoid fully aliased columns in a supersaturated design. The device of row permutation discussed in Chapter 8 can also be employed here to construct designs $\boldsymbol{D}$ and $\boldsymbol{C}$ without fully aliased columns. This can be done by starting with one saturated OA$(n, s^m, 2)$, as demonstrated by Lu and Sun (2001) through their successful construction of many supersaturated designs with 3-5 levels that have no fully aliased columns.

Example 10.4. *By permuting the rows of a given saturated $OA(9, 3^4, 2)$, Lu and Sun (2001) conducted a computer search of other $OA(9, 3^4, 2)$s with an eye on minimizing* $\max_{|u|=2} A_u$*, and found six such $OA(9, 3^4, 2)$s. Combining the columns of these seven $OA(9, 3^4, 2)$s gives a 3-level supersaturated design of 9 runs for 28 factors. This design attains the lower bound in Theorem 10.22 and is A_2-optimal. Furthermore, the design is also* $\max_{|u|=2} A_u$*-optimal.*

The theory of multi-level supersaturated designs culminated with the paper by Xu and Wu (2005). The key result of Xu and Wu (2005) is a general lower bound on the A_2 value.

Theorem 10.23. *For a balanced s-level supersaturated design of n runs and m factors, we have that*

$$A_2 \geq \frac{m(s-1)(ms-m-n+1)}{2(n-1)} + \frac{(n-1)s^2\eta(1-\eta)}{2n},$$

where

$$\eta = \frac{m(n-s)}{(n-1)s} - \left\lfloor \frac{m(n-s)}{(n-1)s} \right\rfloor,$$

and hence $0 \leq \eta < 1$. The lower bound is attained if and only if the numbers δ_{pq} of coincidences between row p and row q for all $p < q$ differ by at most one.

We give a combined proof of Theorems 10.21 and 10.23.

Proof. For balanced designs, we have that $A_1 = 0$. By Corollary 10.2, we obtain that $K_1 = m(n-s)/((n-1)s)$ and

$$K_2 = \big(2nA_2 + nm(m+s-1) - (ms)^2\big)/\big((n-1)s^2\big). \tag{10.21}$$

Let $\eta = K_1 - \lfloor K_1 \rfloor$, the fractional part of $K_1 = m(n-s)/((n-1)s)$. Now consider the problem of minimizing

$$K_2 = \binom{n}{2}^{-1} \sum_{1 \leq p < q \leq n} \delta_{pq}^2$$

over δ_{pq}s under the constraint that $K_1 = \binom{n}{2}^{-1} \sum_{1 \leq p<q \leq n} \delta_{pq}$. We see that K_2 is minimized if and only if δ_{pq}s take on one of the two values, $\lfloor K_1 \rfloor$ and $\lfloor K_1 \rfloor + 1$. Let f_1 and f_2 be the frequencies that δ_{pq}s are equal to $\lfloor K_1 \rfloor$ and $\lfloor K_1 \rfloor + 1$, respectively. We then have that

$$f_1 + f_2 = \binom{n}{2} \quad \text{and} \quad f_1 \lfloor K_1 \rfloor + f_2(\lfloor K_1 \rfloor + 1) = \binom{n}{2} K_1,$$

which implies that

$$f_1 = \binom{n}{2}(1-\eta) \quad \text{and} \quad f_2 = \binom{n}{2}\eta.$$

Therefore, the minimum value of K_2 is

$$\binom{n}{2}^{-1} \left(f_1 \lfloor K_1 \rfloor^2 + f_2\big(\lfloor K_1 \rfloor + 1\big)^2\right),$$

which simplifies to $K_1^2 + \eta(1-\eta)$. This shows that

$$K_2 \geq \left[\frac{m(n-s)}{(n-1)s}\right]^2 + \eta(1-\eta).$$

Now using (10.21), we obtain the lower bound for A_2 in Theorem 10.23.

Theorem 10.21 corresponds to the situation that K_1 is an integer, in which case $\eta = 0$ and $\delta_{pq} = K_1$ for all $p < q$. □

The next corollary is immediate from Theorems 10.22 and 10.23.

Corollary 10.4. *(i) Any design obtained by adding a balanced column to or deleting one column from $\boldsymbol{D}$ achieves the lower bound in Theorem 10.23, and is hence A_2-optimal. The same is also true for design $\boldsymbol{C}$.*
(ii) The design obtained by collecting the rows of two or more different sth fractions of a saturated $OA(n, s^m, 2)$ must also be A_2-optimal.

Xu and Wu (2005) provided many explicit constructions of A_2-optimal supersaturated designs that achieve the lower bound in Theorem 10.23. Their designs do not have fully aliased columns and can be thought of as special versions of designs $\boldsymbol{D}$ and $\boldsymbol{C}$ in Theorem 10.22 and those related designs in Corollary 10.4. To put it briefly, Xu and Wu (2005) constructed their designs by first dissecting Addelman-Kempthorne construction and then carefully manipulating and judiciously selecting different pieces.

As Xu and Wu (2005) pointed out, the designs attaining the lower bound in Theorem 10.23 all have the same distance distribution, and consequently are not just A_2-optimal; they all have minimum G_2-aberration.

Supersaturated designs with mixed levels have also drawn the attention of design researchers, and relevant work can be found in Yamada and Matsui (2002), Xu and Wu (2005), and Sun, Lin, and Liu (2011). In a supersaturated situation, it is hard to imagine that one would consider using more than two levels to study quantitative factors. With this in mind, it seems that of most practical importance are supersaturated designs with one or two factors having more than two levels to accommodate qualitative factors and all the other factors having two levels to accommodate quantitative factors. Future research effort may well be focused on this type of supersaturated designs.

10.9 Three-Level Orthogonal Arrays with Quantitative Factors

Three-level orthogonal arrays selected by the minimum G_2- and/or G-aberration criteria have good all-around properties, regardless of whether the factors under study are quantitative or qualitative. When the factors are all quantitative, one may wish to modify or enhance the aberration criteria. In the present section, we discuss two proposals from the design literature.

The first proposal is the minimum β-aberration criterion, advocated by Cheng and Ye (2004). For a three-level quantitative factor, each main effect has two orthogonal contrasts, the linear and quadratic effects. Consequently, the full model for an experiment with m three-level factors is a polynomial of $2m$ degrees. If we group the factorial effects by their polynomial degrees, then the full model under a design $\boldsymbol{D}$ as in (10.5) can be written as

$$\boldsymbol{\mu}(\boldsymbol{D}) = \beta_0 \mathbf{1}_n + \sum_{k=1}^{2m} \boldsymbol{P}_k(\boldsymbol{D})\boldsymbol{\beta}_k, \tag{10.22}$$

where $\boldsymbol{P}_k(\boldsymbol{D})$ is the model matrix corresponding to $\boldsymbol{\beta}_k$, β_0 is the intercept, $\boldsymbol{\beta}_1$ is the vector of all m linear main effects, $\boldsymbol{\beta}_2$ consists of all quadratic main effects and all linear-by-linear interactions, $\boldsymbol{\beta}_3$ consists of all linear-by-linear-by-linear interactions and all linear-by-quadratic interactions, and so on. Define

$$b_k = n^{-2}\left\|\boldsymbol{P}_k^T(\boldsymbol{D})\mathbf{1}_n\right\|^2,$$

where $\boldsymbol{P}_k(\boldsymbol{D})$ is given in (10.22) and $\mathbf{1}_n$ is the all-one vector. Then the minimum β-aberration is to sequentially minimize $b_1, b_2, \ldots, b_{2m}$ when selecting factorial designs. For orthogonal arrays of strength two, $b_1 = b_2 = 0$; so the criterion is to select designs by sequentially minimizing $b_3, b_4, \ldots, b_{2m}$. The rationale is simple: the terms of lower degrees are more important than those of higher degrees, and terms of the same degree are equally important.

Example 10.5. *Consider two orthogonal arrays of three levels:*

$$\boldsymbol{D}_1 = \begin{bmatrix} 0 & 0 & 0 \\ 0 & 1 & 2 \\ 0 & 2 & 1 \\ 1 & 0 & 2 \\ 1 & 1 & 1 \\ 1 & 2 & 0 \\ 2 & 0 & 1 \\ 2 & 1 & 0 \\ 2 & 2 & 2 \end{bmatrix}, \qquad \boldsymbol{D}_2 = \begin{bmatrix} 0 & 0 & 0 \\ 0 & 1 & 1 \\ 0 & 2 & 2 \\ 1 & 0 & 1 \\ 1 & 1 & 2 \\ 1 & 2 & 0 \\ 2 & 0 & 2 \\ 2 & 1 & 0 \\ 2 & 2 & 1 \end{bmatrix}.$$

For design $\boldsymbol{D}_1$, we have that

$$(b_3, b_4, b_5, b_6) = (0, 3/2, 0, 1/2),$$

and for design $\boldsymbol{D}_2$, we obtain that

$$(b_3, b_4, b_5, b_6) = (3/8, 3/8, 9/8, 1/8).$$

Therefore, design $\boldsymbol{D}_1$ has less β-aberration than design $\boldsymbol{D}_2$. Interestingly, $\boldsymbol{D}_1$ has $b_3 = 0$, showing that the linear-by-linear interactions are orthogonal to the linear main effects.

In the above example, the two arrays are actually isomorphic, as one can be obtained from the other by switching the two levels 1 and 2 in the third column. Yet, they are very different under the minimum β-aberration. Because of this, Cheng and Ye (2004) introduced a notion of geometric isomorphism for the study of minimum β-aberration designs. We refer to the original paper for the details.

To find the best three-level orthogonal arrays for quantitative factors, Xu, Cheng, and Wu (2004) proposed to enhance the minimum G_2- and G-aberration with the *projection-efficiency* criteria of Cheng and Wu (2001). The projection-efficiency criteria are based on the estimation capacity and information capacity sequences, which are three-level versions of those introduced in Section 8.5 for supersaturated designs. The motivation of Cheng and Wu (2001) is to use a single three-level design to identify important factors and then build a second-order model with only the important factors. In contrast, the traditional wisdom in response surface methodology is to use a first-order design to identify important factors and then use a second-order design to build a response surface model.

To achieve the two objectives of factor screening and response surface exploration, Cheng and Wu (2001) proposed using a three-level orthogonal array with large values of E_f and

D_f for $f = 3, \ldots, m$, where E_f is defined as the number of f-factor projection designs that allow estimation of the full second-order model, and D_f is the average D-efficiency of the f-factor projection designs. Then the projection-efficiency criteria are to formally select designs by first sequentially maximizing $E_3, E_4, \ldots, E_m$ and then sequentially maximizing $D_3, D_4, \ldots, D_m$.

Xu, Cheng, and Wu (2004) proposed a three-step procedure for the selection of three-level orthogonal arrays:

Step 1. First implement the weak G_2-aberration. To be specific, we find all orthogonal arrays that minimize A_3.

Step 2. Then implement the weak G-aberration. This means that from those arrays found in Step 1, we find all arrays using the G-aberration criterion by using only A_u values for $|u| = 3$.

Step 3. Finally, for each array obtained in Step 2, implement the projection-efficiency criteria to all the designs obtained by permuting the levels in each column of this given array.

Xu, Cheng, and Wu (2004) presented some optimal designs of 18 and 27 runs from using this three-step design selection strategy. They found that nonregular designs generally perform better than regular designs, and provided some constructions of nonregular three-level orthogonal arrays that perform well under their selection criteria.

Related in spirit to Xu, Cheng, and Wu (2004) is the work of Tsai, Gilmour, and Mead (2000), who proposed a criterion by averaging an approximation to the A_s-efficiencies over lower-dimensional projection designs.

10.10 Definitive Screening Designs

Introduced by Jones and Nachtsheim (2011), definitive screening designs are a class of three-level designs for studying quantitative factors. These designs are not orthogonal arrays. Under consideration is the full second-order model that consists of an intercept, linear main effects, quadratic main effects, and linear-by-linear two-factor interactions, as given by

$$Y = \beta_0 + \sum_i \beta_i x_i + \sum_i \beta_{ii} x_i^2 + \sum_{i<j} \beta_{ij} x_i x_j + \epsilon.$$

In a nutshell, definitive screening designs provide unbiased estimation of linear main effects in the presence of all second-order effects including quadratic main effects and linear-by-linear two-factor interactions, while being capable of separating quadratic main effects.

Consider the following two designs:

$$\boldsymbol{D}_1 = \begin{bmatrix} 0 & 1 & 1 & 1 \\ -1 & 0 & -1 & 1 \\ -1 & 1 & 0 & -1 \\ -1 & -1 & 1 & 0 \\ 0 & -1 & -1 & -1 \\ 1 & 0 & 1 & -1 \\ 1 & -1 & 0 & 1 \\ 1 & 1 & -1 & 0 \\ 0 & 0 & 0 & 0 \end{bmatrix}, \qquad \boldsymbol{D}_2 = \begin{bmatrix} -1 & -1 & -1 & -1 \\ -1 & -1 & 1 & 1 \\ -1 & 1 & -1 & 1 \\ -1 & 1 & 1 & -1 \\ 1 & -1 & -1 & 1 \\ 1 & -1 & 1 & -1 \\ 1 & 1 & -1 & -1 \\ 1 & 1 & 1 & 1 \\ 0 & 0 & 0 & 0 \end{bmatrix}.$$

Design $\boldsymbol{D}_1$ is a definitive screening design and design $\boldsymbol{D}_2$ is an OA$(8, 2^4, 3)$, a two-level orthogonal array of strength three, with an added center run. Both designs allow unbiased estimation of linear main effects in the presence of quadratic main effects and linear-by-linear two-factor interactions, as one can easily check that for both designs linear main effects are mutually orthogonal and are also orthogonal to all quadratic main effects and linear-by-linear two-factor interactions. However, the two designs are very different if we zoom in on quadratic main effects. For design $\boldsymbol{D}_1$, no quadratic main effect is fully aliased with another quadratic main effect, while for design $\boldsymbol{D}_2$ all quadratic main effects are fully aliased with one another. Indeed, separating quadratic main effects is the chief motivation behind the introduction of definitive screening designs. The price for this gain, when comparing $\boldsymbol{D}_1$ with $\boldsymbol{D}_2$, is the small loss of estimation efficiency for linear main effects due to two extra zeros in each design column.

Design $\boldsymbol{D}_1$ goes beyond just separating quadratic main effects; it actually allows estimation of all linear and quadratic main effects under the assumption that two-factor interactions are negligible. So, in terms of the ability of estimating linear and quadratic main effects, design $\boldsymbol{D}_1$ is on par with an OA$(9, 3^4, 2)$, a three-level orthogonal array, although the former is not as efficient as the latter in doing so.

Following the original proposal of Jones and Nachtsheim (2011), a definitive screening design has the following form

$$\boldsymbol{D} = \begin{pmatrix} \boldsymbol{C} \\ -\boldsymbol{C} \\ \boldsymbol{0} \end{pmatrix}, \tag{10.23}$$

where $\boldsymbol{C} = (c_{ij})_{m \times m}$ is a square matrix with entries $c_{ii} = 0$ and $c_{ij} = \pm 1$ for $i \neq j$, and $\boldsymbol{0}$ is a row vector of all zeros, with a further requirement that the columns of $\boldsymbol{C}$ are as orthogonal as possible. Jones and Nachtsheim (2011) conducted an algorithmic search of the best $\boldsymbol{C}$ for $m = 4, 5, \ldots, 30$, and they found an orthogonal $\boldsymbol{C}$ for $m = 4, 6, 8, 10$.

An orthogonal $\boldsymbol{C}$ is actually a *conference matrix*, which Xiao, Lin, and Bai (2012) used to construct their definitive screening designs. Formally, a conference matrix $\boldsymbol{C} = (c_{ij})_{m \times m}$ with entries $c_{ii} = 0$ and $c_{ij} = \pm 1$ for $i \neq j$ is an orthogonal square matrix, meaning that $\boldsymbol{C}^T\boldsymbol{C} = (m-1)\boldsymbol{I}_m$ where $\boldsymbol{I}_m$ is an identity matrix of order m.

Obviously, for $\boldsymbol{C}$ to be a conference matrix, it is necessary that m is even. It is currently known that a conference matrix of order m exists for all even $m \leq 64$ except for 22, 34, and 58 (Vazquez, Goos, and Schoen, 2020). A sufficient condition for a conference matrix to exist is that $m - 1$ is a prime power. Consider

$$\boldsymbol{C} = \begin{pmatrix} 0 & -\boldsymbol{1}_s^T \\ \boldsymbol{1}_s & \boldsymbol{Q} \end{pmatrix}, \tag{10.24}$$

where $\boldsymbol{Q}$ is defined in (2.7) of Chapter 2. Then one can easily check, using properties (i) and (ii) of $\boldsymbol{Q}$, that $\boldsymbol{C}$ given in (10.24) is a conference matrix. It is worth noting that property (iii) of $\boldsymbol{Q}$ is irrelevant for $\boldsymbol{C}$ to be a conference matrix, but it is important for Paley's two constructions of Hadamard matrices.

The next theorem presents the principal properties of design $\boldsymbol{D}$ in (10.23) constructed using a conference matrix $\boldsymbol{C}$.

Theorem 10.24. *Suppose that $\boldsymbol{C}$ is a conference matrix. Then the following properties hold for design $\boldsymbol{D}$.*
(i) linear main effects are mutually orthogonal and all are orthogonal to the grand mean,
(ii) linear main effects are orthogonal to all quadratic main effects and all linear-by-linear two-factor interactions, and
(iii) design $\boldsymbol{D}$ is a saturated design for estimating the grand mean, all linear, and all quadratic main effects.

Parts (i) and (ii) are obviously true. Only part (iii) requires a proof, which is a straightforward verification that the model matrix consisting of the grand mean, all linear effects, and all quadratic effects is nonsingular. We refer to Xiao, Lin, and Bai (2012) for details.

Two other properties of design $\boldsymbol{D}$ are also known. Although quadratic effects are not fully aliased with each other, severe aliasing is present between them, as can be seen by $J(\boldsymbol{d}_i^2, \boldsymbol{d}_j^2) = 2m-4$ for $i \neq j$, where $\boldsymbol{d}_i$ is the ith column of $\boldsymbol{D}$. On the other hand, the aliasing between quadratic effects and two-factor interactions is minimal, because $J(\boldsymbol{d}_i^2, \boldsymbol{d}_j\boldsymbol{d}_k) = 0$ if i is equal to j or k and $J(\boldsymbol{d}_i^2, \boldsymbol{d}_j\boldsymbol{d}_k) = \pm 2$ if i is equal to neither j nor k.

What cannot be deduced from $\boldsymbol{C}$ simply being a conference matrix are the aliasing properties between different two-factor interactions. Wang, Liu, and Lin (2022b) investigated the choice of a conference matrix for the construction of definitive screening designs and found that minimal aliasing between two-factor interactions in the resulting design is achieved by using Paley's conference matrices, which are those given in (10.24) and available whenever $m-1$ is a prime power.

Schoen, Eendebak, and Goos (2019) conducted a systematic investigation into the selection problem of definitive screening designs. They showed that projection designs onto two or three columns of a definitive screening design are essentially unique and therefore one only needs to consider the projection designs onto four columns in comparing different definitive screening designs. The projection designs onto four columns are determined by the J_4-characteristics, $J(\boldsymbol{d}_i, \boldsymbol{d}_j, \boldsymbol{d}_k, \boldsymbol{d}_l)$s for all $i < j < k < l$. Following the idea of minimum G-aberration, Schoen, Eendebak, and Goos (2019) proposed a criterion for the selection of definitive screening designs, which is to sequentially minimize the frequencies of the absolute values of J_4-characteristics sorted in descending order.

11

Mixed-Level Designs and Complicated Unit Structures

A Bayesian approach was used in Chapter 5 to derive a minimum G_2-aberration criterion for choosing good two-level designs with unstructured experimental units. It was shown in Chapter 7 how this approach can be applied to several cases where the experimental units have more complicated structures. In this chapter, we shall present a general theory for a broad class of structures of experimental units and mixed-level designs where different treatment factors may have different numbers of levels.

11.1 Introduction

A comparative experiment is conducted by assigning treatments to experimental units. The treatments may or may not have a structure. They have a *factorial structure* if each treatment is a combination of several factors, and the purpose of the experiment is to study how the response depends on these factors. Although this book is devoted to factorial designs, we note that the treatments may have other structures in different types of experiments.

Like treatments, the experimental units also may or may not have a structure. In completely randomized experiments, the experimental units are unstructured. In Chapter 7, we studied block, row-column, blocked row-column experiments, etc., where the experimental units have certain structures. We shall review some basic concepts and results on the factorial treatment structure in Section 11.2. Section 11.3 is devoted to a discussion of structures of experimental units, called *block structures*. We would prefer to call them unit structures, but "block structures" was first coined by Nelder (1965) and is a well established terminology in the literature. Nelder introduced *simple block structures* that cover most, but not all block structures encountered in practice. These structures and a more general notion of *orthogonal block structures* will be discussed in this chapter. In Section 11.4 we show how to determine the strata under an orthogonal block structure. Similar to what was done in Section 7.2.3, a Bayesian approach is used in Section 11.5 to derive a minimum G_2-aberration criterion for the selection of general mixed-level factorial designs for experiments with block structures more general than simple block structures. A necessary and sufficient condition for a design to have minimum G_2-aberration for all feasible prior stratum variances, extending Theorems 7.1, 7.2, and 7.3, is obtained. Some proofs are deferred to Section 11.6.

DOI: 10.1201/9781003371861-11

11.2 Factorial Treatment Structure

In a 2^m factorial experiment, $2^m - 1$ orthogonal treatment contrasts can be defined to represent factorial effects including main effects and interactions. In an $s_1 \times \cdots \times s_m$ experiment where the ith factor has s_i levels, orthogonal treatment contrasts representing factorial effects can be defined as follows:

Definition 11.1. *Two treatment contrasts are said to be orthogonal if their coefficient vectors are orthogonal. A contrast of treatment effects is said to represent the main effect of the ith factor if its coefficients only depend on the levels of that factor. It is said to represent the interaction of the ith and jth factors if it is orthogonal to all the main-effect contrasts and its coefficients only depend on the levels of the ith and jth factors. It is said to represent the interaction of factors $i_1, \ldots, i_k, k \geq 3$, if it is orthogonal to all the main-effect and k'-factor interaction contrasts with $k' < k$, and its coefficients only depend on the levels of factors $i_1, \ldots, i_k$.*

According to this definition, the dimension of the space of treatment contrasts representing the main effect of factor i is $s_i - 1$. Since the dimension of the space of treatment contrasts whose coefficients only depend on the levels of factors i and j is $s_i s_j - 1$, the space of treatment contrasts representing the interaction of these two factors has dimension $s_i s_j - 1 - (s_i - 1) - (s_j - 1) = (s_i - 1)(s_j - 1)$. Continuing this process or by induction, we conclude that the dimension of the space of treatment contrasts representing the interaction of factors $i_1, \ldots, i_k$ is $(s_{i_1} - 1) \cdots (s_{i_k} - 1)$.

For each $i = 1, \cdots, m$, let

$$\boldsymbol{P}_i = \begin{bmatrix} \boldsymbol{p}_0^i & \boldsymbol{p}_1^i & \cdots & \boldsymbol{p}_{s_i-1}^i \end{bmatrix} \tag{11.1}$$

be an $s_i \times s_i$ matrix such that its first column has constant entries, and the remaining columns are coefficient vectors of $s_i - 1$ mutually orthogonal contrasts of s_i levels. One common choice is to let the entries of $\boldsymbol{p}_j^i$ be the values of an orthogonal polynomial of degree j on s_i equally spaced points; see Section 5.7 of Cheng (2014a). For simplicity, we normalize the columns so that each has length 1. As an example, for $s_i = 2$ and 3, we can let $\boldsymbol{P}_i$ be

$$\begin{bmatrix} 1/\sqrt{2} & -1/\sqrt{2} \\ 1/\sqrt{2} & 1/\sqrt{2} \end{bmatrix}$$

and

$$\begin{bmatrix} 1/\sqrt{3} & -1/\sqrt{2} & 1/\sqrt{6} \\ 1/\sqrt{3} & 0 & -2/\sqrt{6} \\ 1/\sqrt{3} & 1/\sqrt{2} & 1/\sqrt{6} \end{bmatrix},$$

respectively.

Let

$$\boldsymbol{P} = \boldsymbol{P}_1 \otimes \cdots \otimes \boldsymbol{P}_m, \tag{11.2}$$

with $\boldsymbol{P}_i$ as in (11.1). Then $\boldsymbol{P}$ is an orthogonal matrix. The first column of $\boldsymbol{P}$ has constant entries, and the remaining $s_1 \cdots s_m - 1$ columns define a complete system of orthogonal contrasts representing the factorial effects of factors $1, \ldots, m$. In particular, the columns $\boldsymbol{p}_{j_1}^1 \otimes \cdots \otimes \boldsymbol{p}_{j_m}^m$, where $1 \leq j_l \leq s_l - 1$ and all the other j_k's are equal to 0, define $s_l - 1$ mutually orthogonal contrasts representing the main effect of factor l. For any $2 \leq k \leq m$ and any k factors $i_1, i_2, \ldots, i_k$, all the products $\boldsymbol{p}_{j_1}^1 \otimes \cdots \otimes \boldsymbol{p}_{j_m}^m$, where $1 \leq j_l \leq s_l - 1$ for $l = i_1, \ldots, i_k$, and all the other j_l's are 0, can be used as coefficient vectors of $(s_{i_1} - 1) \cdots (s_{i_k} - 1)$ mutually orthogonal treatment contrasts representing the interaction of factors $i_1, \ldots,$ and i_k.

Summarizing the above discussions, we conclude that, after the column vector with constant entries is deleted from the Kronecker product $\boldsymbol{P}_1 \otimes \cdots \otimes \boldsymbol{P}_m$, the remaining $s_1 \cdots s_m - 1$ columns define a complete system of orthogonal factorial effects of factors $1, \ldots, m$.

As in Section 5.2.4, let $\boldsymbol{\mu}$ be an $s_1 \ldots s_m \times 1$ vector whose components are effects $\mu(x_1, \ldots, x_m)$ of the $s_1 \ldots s_m$ treatment combinations $\boldsymbol{x} = (x_1, \ldots, x_m)$ and $\boldsymbol{\beta}$ be the $s_1 \ldots s_m \times 1$ vector consisting of the grand mean and all the factorial effects (main effects and interactions). When the components of $\boldsymbol{\mu}$ and $\boldsymbol{\beta}$ are arranged in the Yates order, we have

$$\boldsymbol{\mu} = \boldsymbol{P}\boldsymbol{\beta}. \tag{11.3}$$

The $s_1 \cdots s_m$ components of $\boldsymbol{\beta}$'s are divided into 2^m sets, corresponding to the 2^m subsets of $\{1, 2, \ldots, m\}$. The grand mean corresponds to the empty set $\emptyset$. To each nonempty subset $S = \{i_1, \ldots, i_k\}$ of $\{1, 2, \ldots, m\}$, there correspond $(s_{i_1} - 1) \cdots (s_{i_k} - 1)$ components of $\boldsymbol{\beta}$, representing the main effect of factor i_1 if $k = 1$ and the interaction of factors $i_1, \ldots, i_k$ if $k \geq 2$. We denote the vector of these components of $\boldsymbol{\beta}$ by $\boldsymbol{\beta}_S$.

Remark 11.1. We note that (11.3) has already appeared in (5.11), (7.16), and (10.4) except that there is a minor difference between the definition of matrix $\boldsymbol{P}$ in this chapter and that in Chapters 5, 7, and 10. The columns of $\boldsymbol{P}$ have squared length equal to $s_1 \cdots s_m$ in Chapters 5, 7, and 10, whereas they have length one in this chapter. The sole reason for this is the convenience of presentation.

11.3 Block Structures (Unit Structures)

We summarize below some concepts and results on unit factors and block structures. For more discussions of these topics and technical details including proofs, we refer the readers to Chapters 4 and 12 of Cheng (2014a).

The experimental units are partitioned into blocks in a block design, and they are partitioned into rows and columns in a row-column design. Mathematically, a factor F on a given set (of experimental units or treatments) can be defined as a partition of that set into nonoverlapping subsets, called F-classes, with all the units or treatments in the same set taking the same level of F. It is called a *uniform* factor if all the F-classes are of the same size. In this section, the discussions are presented in the context of experimental units. There are two "trivial" partitions that are relevant in our discussion: the partition of units into singletons, and the "partition" that puts all the units in one single class. We denote these two factors as E and U, respectively: E for the factor that partitions the units into singletons, and U for the factor with a single class, where E and U stand for *equality* and *universal*, respectively. Why these two trivial partitions (unit factors) are of interest will become clear as we continue. In fact, they have already appeared in Section 7.2.2 for labeling two "trivial" strata and stratum variances.

Given two factors F and G defined on the same set, we say that F is *finer* than G (G is coarser than F, or F is *nested* in G), denoted by $F \prec G$, if each F-class is a subset of some G-class, and $F \neq G$. We write $F \preceq G$ for "$F = G$ or $F \prec G$". We see that E is the finest and U is the coarsest of all factors on a given set.

We use the set $\{U, B, E\}$ to represent the *block structure* of a block design, where B is the factor that partitions the set of experimental units into blocks. For completely randomized designs, row-column designs, and blocked row-columns designs, the block structures are represented by $\{U, E\}$, $\{U, R, C, E\}$, and $\{U, B, R, C, E\}$, respectively. In general, a block structure is a set of factors on the same set of experimental units. The two trivial unit factors E and U are always included in any block structure.

Given two factors F_1 and F_2 on the same set, we define their *supremum* $F_1 \vee F_2$ as the finest factor G such that $F_1 \preceq G$ and $F_2 \preceq G$; that is, $F_1 \preceq F_1 \vee F_2$, $F_2 \preceq F_1 \vee F_2$, and if $F_1 \preceq G'$ and $F_2 \preceq G'$, then $F_1 \vee F_2 \preceq G'$. We also define the *infimum* of F_1 and F_2, denoted by $F_1 \wedge F_2$, as the coarsest factor G such that $G \preceq F_1$ and $G \preceq F_2$; that is, $F_1 \wedge F_2 \preceq F_1$, $F_1 \wedge F_2 \preceq F_2$, and if $G' \preceq F_1$ and $G' \preceq F_2$, then $G' \preceq F_1 \wedge F_2$. For example, under the block structure $\{U, B, R, C, E\}$ of a blocked row-column design, we have $R \vee C = B$ and $R \wedge C = E$.

The following result is easy to prove.

Proposition 11.1. *For any two factors F and G defined on the same set, each class of $F \wedge G$ is the intersection of an F-class and a G-class.*

There are several important matrices and vector spaces that are associated with each unit factor. Suppose there are n units and F is a factor with f levels. Then the relation between the units and the levels of F can be described by an $n \times f$ (0,1)-matrix $\boldsymbol{X}_F$, called the *incidence matrix* of F, such that the (i, j)th entry is 1 if and only if the jth level of F is assigned to unit i. For block, row, column, and treatment factors, this matrix tells us which block, row, and column a unit belongs to, and which treatment is assigned to that unit, respectively.

The matrix

$$\boldsymbol{R}_F = \boldsymbol{X}_F \boldsymbol{X}_F^T \tag{11.4}$$

is an $n \times n$ (0,1)-matrix such that the (i, j)th entry is equal to 1 if and only if the same level of F appears at units i and j (or, units i and j are in the same F-class), and can be called a *relation matrix*. It tells us whether two units are in the same block, row, and column, or whether they are assigned the same treatment. Note that such notations have appeared in expressions such as (7.10), (7.11), and elsewhere.

Let V_F be the column space of $\boldsymbol{X}_F$, i.e., the space generated by all the column vectors of $\boldsymbol{X}_F$. Clearly if $F_1 \preceq F_2$, then $V_{F_2} \subseteq V_{F_1}$. Also, it can be shown that for any two factors F_1 and F_2 on the same set,

$$V_{F_1 \vee F_2} = V_{F_1} \cap V_{F_2}. \tag{11.5}$$

Since each unit is in exactly one F-class, each row of $\boldsymbol{X}_F$ has exactly one entry equal to 1. It follows that the sum of all the columns of $\boldsymbol{X}_F$ is the column of all ones; thus each V_F is an f-dimensional space containing the one-dimensional space generated by the vector of 1's. This one-dimensional space is in fact V_U, where U is the universal factor with one single class. Therefore for any two different factors F_1 and F_2 on the same set of units, since both V_{F_1} and V_{F_2} contain V_U, they cannot be orthogonal. However, there is a very nice necessary and sufficient condition for F_1 and F_2 to have the property that $V_{F_1} \ominus V_U$ is orthogonal to $V_{F_2} \ominus V_U$. Given an f_1-level factor F_1 and an f_2-level factor F_2 on a set of n units, let n_{ij} be the number of units in the intersection of the ith F_1-class and the jth F_2-class. For example, if F_1 is the treatment factor and F_2 is the block factor, then n_{ij} is the number of times the ith treatment appears in the jth block. We say that F_1 and F_2 satisfy the *condition of proportional frequencies* if

$$n_{ij} = \frac{n_{i+} n_{+j}}{n} \text{ for all } i,\ j,$$

where $n_{i+} = \sum_{j=1}^{f_2} n_{ij}$ and $n_{+j} = \sum_{i=1}^{f_1} n_{ij}$. This means that n_{ij}/n_{i+} does not depend on i and n_{ij}/n_{+j} does not depend on j.

Then we have the following result:

Theorem 11.1. *For any two factors F_1 and F_2 on the same set, $V_{F_1} \ominus V_U$ and $V_{F_2} \ominus V_U$ are orthogonal if and only if F_1 and F_2 satisfy the condition of proportional frequencies.*

This result also appeared as Theorem 2.5 in Cheng (2014a). The readers are referred to a proof there.

It follows from Theorem 11.1 that F_1 and F_2 satisfy the condition of proportional frequencies if and only if $V_{F_1} \cap V_{F_2} = V_U$.

In general, we have the following definition of the orthogonality of two factors:

Definition 11.2. *Two factors F_1 and F_2 on the same set are said to be orthogonal if $V_{F_1} \ominus (V_{F_1} \cap V_{F_2})$ is orthogonal to $V_{F_2} \ominus (V_{F_1} \cap V_{F_2})$.*

By (11.5), two factors F_1 and F_2 on the same set are orthogonal to each other if and only if $V_{F_1} \ominus V_{F_1 \vee F_2}$ is orthogonal to $V_{F_2} \ominus V_{F_1 \vee F_2}$. It can be shown that this holds if and only if the condition of proportional frequencies holds in each $(F_1 \vee F_2)$-class. This means that for each $(F_1 \vee F_2)$-class H, if both the ith F_1-class and the jth F_2-class are contained in H, then

$$n_{ij} = \frac{n_{i+}n_{+j}}{|H|},$$

wherer $|H|$ is the size of H. For example, the two factors R and C in the block structure $b/(r \times c)$ do not have proportional frequencies since rows in one block do not share any unit with columns from another block, but in any block, each row shares one unit with every column.

The block structures $\{U, B, E\}$ and $\{U, R, C, E\}$ for block designs and row-column designs are the two simplest nontrivial block structures constructed by using the operations of *nesting* and *crossing*, respectively. In a block design, the position of a unit in a block is irrelevant to those in other blocks since a within-block randomization will be performed separately in each block. On the other hand, in a row-column design, two units in the same row (or column) remain in the same row (or column) after randomization. For b blocks of size k, the block structure is also written as b/k for short, and a short hand notation for the structure of r rows and c columns is $r \times c$.; here "/" and "×" stand for "nesting" and "crossing", respectively. The block structure of a blocked split-plot design with b blocks, w whole plots in each block, and s subplots in each whole plot involves two layers of nesting, and can be expressed as $(b/w)/s$ or $b/(w/s)$. That of a blocked row-column design with b blocks, each of which is split into r rows and c columns, is constructed via a combination of nesting and crossing and can be denoted as $b/(r \times c)$. All the block structures that can be obtained via iterations of nesting and crossing were introduced and called *simple block structures* by Nelder (1965).

Speed and Bailey (1982) introduced a more general class of block structures, which were called *orthogonal block structures* in Bailey (1985).

Definition 11.3. *A block structure $\mathfrak{B}$ satisfying the the following conditions is called an orthogonal block structure:*

1. *All the factors in $\mathfrak{B}$ are uniform;*
2. $E \in \mathfrak{B}$;
3. $U \in \mathfrak{B}$;
4. *if $F, G \in \mathfrak{B}$, then $F \wedge G \in \mathfrak{B}$;*
5. *if $F, G \in \mathfrak{B}$, then $F \vee G \in \mathfrak{B}$;*
6. *the factors in $\mathfrak{B}$ are pairwise orthogonal.*

It can be shown that all simple block structures are orthogonal block structures. A more detailed discussion of orthogonal block structures, including derivation of strata, will be discussed in the next section.

Example 11.1. *The two-stage experiment described in Example 7.1 is a row-column experiment, and hence the underlying block structure is a simple block structure: washer × dryer. If the experiment is conducted on two days, then the resulting block structure day/(washer× dryer) is still a simple block structure.*

Example 11.2. *Mee and Bates (1998) discussed designs of experiments with multiple processing stages in the fabrication of integrated circuits. Here, however, the experimental units do not necessarily have a simple block structure. Suppose there are 16 batches of material and the experiment is run in several stages. At each stage, the experimental units are divided into four groups of size four, with the same level of each treatment factor to be processed at that stage assigned to all the batches in the same group. This makes the factorial effects of these treatment factors confounded with group contrasts, a sacrifice in precision for convenience or cost considerations. Suppose there are two stages and each of the first-stage groups and each of the second-stage groups share exactly one batch. If the first-stage and second-stage partitions of the batches are considered as rows and columns, respectively, then the batches have a row-column layout and hence the block structure is the same as that in Example 7.1. Suppose there are three stages. One way of grouping the batches at the third stage is to use the numbers (or letters) in a* 4×4 *Latin square: each third-stage treatment factor takes the same level on the four batches corresponding to the same number (or letter) in the Latin square. Then we no longer have a simple block structure since it cannot be constructed by iterations of nesting and crossing operations. To have a simple block structure, one needs at least* $4 \times 4 \times 4 = 64$ *batches; however, only 16 batches are available. In this case, the block structure can be described by* $\{E, R, C, L, U\}$*, where* R*,* C*, and* L *are the partitions based on rows, columns, and numbers in the Latin square, respectively. It can easily be verified that this is an orthogonal block structure. Clearly conditions 1, 2, and 3 in Definition 11.3 are satisfied. These unit factors are also pairwise orthogonal; e.g., each pair of* R*,* C*, and* L *satisfies the condition of proportional frequencies. Furthermore, each pair of them has* U *as their supremum and* E *as their infimum. Therefore conditions 4, 5, and 6 are also satisfied.*

In the next example, we have a block structure that is neither a simple block structure nor an orthogonal block structure.

Example 11.3. *In both Examples 11.1 and 11.2, any two groups of units from two different stages share a common unit, a property that is crucial for the units to have an orthogonal block structure. This may not be possible if the number of available units is limited. In an experiment with two processing stages considered in Bingham, Sitter, Kelly, Moore, and Olivas (2008), the experimental units are required to be partitioned into eight groups at both stages. For every pair of groups from the two stages to share a unit, we need at least 64 units. However, only 32 units are available. Thus one needs to use a "half fraction" such as displayed overleaf.*

Let R *and* C *be the partitions of the 32 starred units into eight rows and eight columns of size four, respectively. Then the block structure of the 32 units can be represented by* $\mathfrak{B} = \{U, R, C, E\}$*. All these factors are uniform and orthogonal to one another. However,* $R \vee C$ *is the partition of the 32 units into the two* 4×4 *squares with stars, which can be considered as two "blocks" of size 16. Denote this partition by* B*. These two "pseudo blocks", however, are not part of the original blocking scheme. Since* $R \vee C$ *is not in* $\mathfrak{B}$*,* $\mathfrak{B}$ *is not an orthogonal block structure. On the other hand,* $\mathfrak{B}' = \{U, B, R, C, E\}$*, obtained by adding*

B to $\mathfrak{B}$, is an orthogonal bock structure. Under the kind of statistical models considered in Chapter 7, a random effect is present for each class of the partition induced by a unit factor. If we enlarge $\mathfrak{B}$ to $\mathfrak{B}'$ by adding the unit factor defined by the pseudo blocks, nothing is really changed including the statistical model, provided that we make $\sigma_B = 0$. Then the experiment can be treated as one with the orthogonal block structure $2/(4 \times 4)$, and thus all the results derived for orthogonal block structures, to be presented in the next section, can be applied. This is a very useful device for dealing with complications in the design and analysis of such experiments caused by a nonstandard block structure; see the discussions in Cheng and Tsai (2011).

*	*	*	*				
*	*	*	*				
*	*	*	*				
*	*	*	*				
				*	*	*	*
				*	*	*	*
				*	*	*	*
				*	*	*	*

11.4 Statistical Models and Strata for Orthogonal Block Structures

Statistical models for block designs, nested block designs (or blocked split-plot designs), and row-column designs have been given in (7.3)-(7.4), (7.27)-(7.28), and (7.36)-(7.37), respectively. Each of these includes a set of uncorrelated random effects for the levels of each factor in the block structure. Such models can be extended to any orthogonal block structure $\mathfrak{B}$ as follows.

Suppose there are n experimental units. For each $1 \le i \le n$, let y_i be the observation on the ith unit, $\phi(i) \in \{1, \ldots, t\}$ be the label of the treatment assigned to unit i, and for each $F \in \mathfrak{B}$, $F(i) \in \{1, \ldots, f\}$ be the label of the F-class (block, row, column, etc.) to which unit i belongs; that is, the level of F at the ith unit.

We assume that

$$y_i = \mu_{\phi(i)} + \sum_{F \in \mathfrak{B}} \gamma^F_{F(i)},$$

where the treatment effects $\mu_1, \ldots, \mu_t$ are unknown constants, and $\{\gamma^F_1, \ldots, \gamma^F_f\}_{F \in \mathfrak{B}}$ are uncorrelated random variables with $\mathrm{E}(\gamma^F_k) = 0$ and $\mathrm{var}(\gamma^F_k) = \sigma^2_F$ for all k. Then we can write this model in the matrix form

$$\boldsymbol{y} = \boldsymbol{X}_T \boldsymbol{\mu} + \sum_{F \in \mathfrak{B}} \boldsymbol{X}_F \boldsymbol{\gamma}^F, \tag{11.6}$$

where $\boldsymbol{\mu} = (\mu_1, \ldots, \mu_t)^T$, $\boldsymbol{\gamma}^F = (\gamma^F_1, \ldots, \gamma^F_f)^T$, and $\boldsymbol{X}_F$ is the incidence matrix between the units and factor F. Then

$$\mathrm{E}(\boldsymbol{y}) = \boldsymbol{X}_T \boldsymbol{\mu} \tag{11.7}$$

and

$$\boldsymbol{V} = \mathrm{cov}(\boldsymbol{y}) = \sum_{F \in \mathfrak{B}} \sigma^2_F \boldsymbol{X}_F \boldsymbol{X}^T_F = \sum_{F \in \mathfrak{B}} \sigma^2_F \boldsymbol{R}_F, \tag{11.8}$$

where the last equality follows from (11.4). In this form of the covariance matrix, we have one variance component for each of the factors that define the block structure. In the last equality in (11.8), we see that the covariance matrix can be expressed as a linear combination of the relation matrices of the factors.

One immediate consequence of (11.8) is that for any $1 \leq i, i' \leq n$,

$$\text{cov}(y_i, y_{i'}) = \sum_{G \in \mathfrak{B}:\text{units } i \text{ and } i' \text{ are in the same } G\text{-class}} \sigma_G^2. \tag{11.9}$$

Since the coarsest factor U has only one class, trivially for any $1 \leq i, i' \leq n$, units i and i' are in the same U-class; therefore there is at least one factor F in $\mathfrak{B}$ that has units i and i' in the same class. Let $F^{i,i'}$ be the infimum of all such factors. Then since $\mathfrak{B}$ is closed under the operation of $\wedge$, we have $F^{i,i'} \in \mathfrak{B}$. It is easy to see that $F^{i,i'}$ is the *finest* factor in $\mathfrak{B}$ such that units i and i' are in the same class: first of all, by Proposition 11.1, units i and i' are in the same $F^{i,i'}$-class; on the other hand, if G is a factor in $\mathfrak{B}$ such that units i and i' are in the same G-class, then since $F^{i,i'}$ is the infimum of all such factors, we have $F^{i,i'} \preceq G$.

In view of the observation in the previous paragraph, we can re-write (11.9) as

$$\text{cov}(y_i, y_{i'}) = \sum_{G \in \mathfrak{B}:\ F^{i,i'} \preceq G} \sigma_G^2. \tag{11.10}$$

In particular, for each $1 \leq i \leq n$,

$$\text{var}(y_i) = \sum_{G \in \mathfrak{B}} \sigma_G^2. \tag{11.11}$$

For each factor F in $\mathfrak{B}$, Let $\boldsymbol{A}_F$ be the $n \times n$ matrix such that

$$[\boldsymbol{A}_F]_{ii'} = \begin{cases} 1, & \text{if } F \text{ is the finest factor in } \mathfrak{B} \text{ such that units } i \text{ and } i' \\ & \text{are in the same } F\text{-class}, \\ 0, & \text{otherwise}, \end{cases} \tag{11.12}$$

where $[\boldsymbol{A}_F]_{ii'}$ is the (i, i')th entry of $\boldsymbol{A}_F$. That is, $[\boldsymbol{A}_F]_{ii'} = 1$ if and only if units i and i' are in the same F-class but are in different classes of any factor in $\mathfrak{B}$ that is finer than F. For example, under the block structure $\{U, B, R, C, E\}$ of a blocked row-column design, $[\boldsymbol{A}_B]_{i,i'}$ is a $(0, 1)$-matrix such that $[\boldsymbol{A}_B]_{i,i'}$ is equal to 1 if and only if units i and i' are in different rows and different columns of the same block, and $[\boldsymbol{A}_R]_{i,i'}$ is equal to 1 if and only if units i and i' are in the same row and different columns of the same block.

Recall that $\boldsymbol{R}_F$ is the (0,1)-matrix such that its (i, i')th entry is 1 if and only if units i and i' are in the same F-class. Since for any two units, there is exactly one factor that is the finest factor that have them in the same class, we must have

$$\boldsymbol{R}_F = \sum_{G \in \mathfrak{B}:G \preceq F} \boldsymbol{A}_G. \tag{11.13}$$

In particular, since $\boldsymbol{R}_U = \boldsymbol{J}$,

$$\sum_{F:F \in \mathfrak{B}} \boldsymbol{A}_F = \boldsymbol{J} \tag{11.14}$$

holds. For the trivial factor E at the bottom of the block structure, with one single unit in each class, clearly

$$\boldsymbol{A}_E = \boldsymbol{I}_n. \tag{11.15}$$

Suppose units i and i' are in the same F-class. Let G be the finest factor such that they are in the same G-class. Then we must have $G \preceq F$. Therefore for all $F \in \mathfrak{B}$, we have

$$\boldsymbol{A}_F = \boldsymbol{R}_F - \sum_{G \in \mathfrak{B}: G \prec F} \boldsymbol{A}_G. \tag{11.16}$$

At the bottom of the block structure, we have $\boldsymbol{A}_E = \boldsymbol{R}_E = \boldsymbol{I}_n$. Starting from E and applying (11.16) repeatedly, we can work up the hierarchy to write each $\boldsymbol{A}_F$ as a linear combination of the $\boldsymbol{R}_G$'s with $G \preceq F$. Eventually we can express $\boldsymbol{V}$ as a linear combination of the $\boldsymbol{A}_F$'s:

$$\boldsymbol{V} = \sum_{\mathcal{F}: \mathcal{F} \in \mathfrak{B}} \delta_{\mathcal{F}} \boldsymbol{A}_F, \tag{11.17}$$

where

$$\delta_F = \sum_{G \in \mathfrak{B}:\ F \preceq G} \sigma_G^2. \tag{11.18}$$

Write $\text{var}(y_i)$ in (11.11) as σ^2, and let

$$\rho_F = \delta_F / \sigma^2.$$

Then we have

$$\rho_F = \frac{\sum_{G \in \mathfrak{B}: F \preceq G} \sigma_G^2}{\sum_{G: G \in \mathfrak{B}} \sigma_G^2}. \tag{11.19}$$

It follows that

$$F_1 \prec F_2 \Rightarrow \rho_{F_1} \geq \rho_{F_2}. \tag{11.20}$$

If $\sigma_F^2 > 0$ for all F, then we have a strict inequality in (11.20).

Example 11.4. *Consider block designs with b blocks of size k. The block structure is $\mathfrak{B} = \{E, U, B\}$. Starting with*

$$\boldsymbol{A}_E = \boldsymbol{I}_n$$

in (11.15), where $n = bk$, and applying (11.16) with $F = B$, we have

$$\boldsymbol{A}_B = \boldsymbol{R}_B - \boldsymbol{A}_E = \boldsymbol{R}_B - \boldsymbol{I}_n.$$

Next, we move up to U. Applying (11.16) again but with $F - U$, we obtain

$$\begin{aligned} \boldsymbol{A}_U &= \boldsymbol{R}_U - \boldsymbol{A}_B - \boldsymbol{A}_E \\ &= \boldsymbol{R}_U - (\boldsymbol{R}_B - \boldsymbol{I}_n) - \boldsymbol{I}_n \\ &= \boldsymbol{R}_U - \boldsymbol{R}_B \\ &= \boldsymbol{J}_n - \boldsymbol{R}_B. \end{aligned}$$

Thus $\boldsymbol{V}$ is of the form

$$\delta_E \boldsymbol{I}_n + \delta_B(\boldsymbol{R}_B - \boldsymbol{I}_n) + \delta_U(\boldsymbol{J}_n - \boldsymbol{R}_B). \tag{11.21}$$

This is the form we saw in (7.10) with $\delta_E = \sigma^2, \delta_B = \rho_1\sigma^2$, and $\delta_U = \rho_2\sigma^2$. Here δ_E is the variance of each observation, δ_B is the covariance of the observations on two different units in the same block, and δ_U is the covariance of two observations in different blocks. We have also seen (11.20) in (7.5).

Example 11.5. *Consider row-column designs with r rows and c columns. As in Example 11.4, we start with $\boldsymbol{A}_E = \boldsymbol{I}_{rc}$, and repeatedly applying (11.16), first to R and C in any order (the order does not matter here since neither is nested in the other), and then to U. Then we have*

$$\boldsymbol{V} = \delta_E \boldsymbol{I}_{rc} + \delta_R(\boldsymbol{R}_R - \boldsymbol{I}_{rc}) + \delta_C(\boldsymbol{R}_C - \boldsymbol{I}_{rc}) + \delta_U(\boldsymbol{I}_{rc} - \boldsymbol{R}_R - \boldsymbol{R}_C + \boldsymbol{J}_{rc}). \tag{11.22}$$

This is the form in (7.39) with $\delta_E = \sigma^2, \delta_R = \rho_1\sigma^2, \delta_C = \rho_2\sigma^2$, and $\delta_U = \rho_3\sigma^2$. Here δ_E is the variance of each observation, δ_R is the covariance of the observations on two units in the same row but different columns, δ_C is the covariance of two observations in the same column but different rows, and δ_U is the covariance of two observations in different rows and different columns.

In the rest of this section, we derive the spectral decomposition of the covariance matrix in (11.17). We point out that so far we have used all the properties of orthogonal block structures as given in Definition 11.3 except closure with respect to the $\vee$ operation. This condition is needed here.

First, recall the space V_F associated with a factor F defined in Section 11.3. All the vectors in V_F have constant entries for all the components corresponding to the same level of F (say, units in the same block). Thus the orthogonal projection of an $n \times 1$ vector $\boldsymbol{z}$ onto V_F replaces all the entries of $\boldsymbol{z}$ that correspond to units having the same level of F with their average. Now since the (i, i')th entry of the $(0,1)$-relation matrix $\boldsymbol{R}_F$ is 1 if and only if the ith and i'th units have the same level of F, it follows that

$$\boldsymbol{P}_{V_F} = \frac{f}{n}\boldsymbol{R}_F, \tag{11.23}$$

where $\boldsymbol{P}_{V_F}$ is the orthogonal projection matrix onto V_F and f is the number of levels of F.

Theorem 11.2 below shows that under an orthogonal block structure $\mathfrak{B}$, R^n can be decomposed as an orthogonal sum of subspaces, with one subspace associated with each factor in $\mathfrak{B}$, and that each of these subspaces is an eigenspace of $\text{cov}(\boldsymbol{y})$ provided it is not a null space.

Theorem 11.2. *Let $\mathfrak{B}$ be an orthogonal block structure. For each factor $F \in \mathfrak{B}$, let*

$$W_F = V_F \ominus \left(\sum_{G \in \mathfrak{B}: F \prec G} V_G \right). \tag{11.24}$$

Then the spaces W_F, $F \in \mathfrak{B}$, are mutually orthogonal, and

$$V_F = \bigoplus_{G \in \mathfrak{B}: F \preceq G} W_G \tag{11.25}$$

for each $F \in \mathfrak{B}$. Applying this to E, we obtain

$$V_E = R^n = \bigoplus_{F: F \in \mathfrak{B}} W_F. \tag{11.26}$$

Furthermore, the covariance matrix $cov(\boldsymbol{y})$ as given in (11.17) can be expressed as

$$cov(\boldsymbol{y}) = \sum_{F: F \in \mathfrak{B}} \xi_F \boldsymbol{P}_{W_F}, \tag{11.27}$$

where

$$\xi_F = \sum_{G \in \mathfrak{B}: G \preceq F} \frac{n}{g}\sigma_G^2, \tag{11.28}$$

and $\boldsymbol{P}_{W_F}$ is the orthogonal projection matrix onto W_F.

Theorem 11.2 will be proved in Section 11.6. It is worth pointing out that the spaces W_F depend only on the block structure and do *not* depend on the entries of cov($\boldsymbol{y}$). It is possible that some of the W_F's may be the null space. Each nonnull W_F is an eigenspace of cov($\boldsymbol{y}$), and is called a *stratum*. The dimension of W_F is called its *degrees of freedom*.

By (11.24),

$$W_U = V_U, \text{ and } \dim(W_U) = 1.$$

We call V_U the mean stratum. The strata other than the mean stratum give an orthogonal decomposition of the space of all *unit contrasts*.

Each $n \times 1$ vector $\boldsymbol{c} \in W_F$, where $F \neq U$, must be orthogonal to the all-one vector. Thus $\boldsymbol{c}^T\boldsymbol{y}$ is a contrast. We have

$$\begin{aligned} \text{var}(\boldsymbol{c}^T\boldsymbol{y}) &= \boldsymbol{c}^T\text{cov}(\boldsymbol{y})\boldsymbol{c} \\ &= \xi_F\boldsymbol{c}^T\boldsymbol{P}_{W_F}\boldsymbol{c} \\ &= \xi_F\|\boldsymbol{c}\|^2. \end{aligned}$$

In particular, var($\boldsymbol{c}^T\boldsymbol{y}$) $= \xi_F$ if $\|\boldsymbol{c}\| = 1$. The eigenvalues ξ_F are called *stratum variances.* An immediate consequence of (11.28) is that

$$F_1 \prec F_2 \Rightarrow \xi_{F_1} \leq \xi_{F_2}. \tag{11.29}$$

Thus ξ_E is the smallest among all the stratum variances. If $\sigma_F^2 > 0$ for all $F \neq U$, then we have strict inequalities in (11.29).

To determine the strata, we point out that the stratum W_U corresponding to the universal factor U at the top of the block structure is the one-dimensional space consisting of all the vectors with constant entries. From W_U, We can obtain the other strata by repeatedly applying (11.24) while going down the block structure hierarchy. The degrees of freedom can be determined similarly.

For simple block structures, Nelder (1965) provided a very simple and explicit recipe for determining the strata, degrees of freedom, and projections onto the strata.

Examples 11.4 and 11.5 revisited. Again consider block designs with b blocks of size k. We start from U with $W_U = V_U$, the one-dimensional space spanned by the $bk \times 1$ vector of 1's. The associated projection matrix is $\boldsymbol{P}_{W_U} = \boldsymbol{J}/bk$, which produces the overall mean. The unit factor immediately below U in the block structure hierarchy is B; so $W_B = V_B \ominus W_U$, with dimension $b-1$ and projection matrix $\boldsymbol{P}_{W_B} = \boldsymbol{P}_{V_B} - \boldsymbol{P}_{W_U}$, which produces deviations of block means from the overall mean. Then from the unit factor E at the bottom, we obtain $W_E = V_E \ominus (W_U \oplus W_B)$ with dimension $bk - (b-1) - 1 = b(k-1)$ and projection matrix $\boldsymbol{P}_{W_E} = \boldsymbol{P}_{V_E} - \boldsymbol{P}_{W_B} - \boldsymbol{P}_{W_U} = \boldsymbol{I}_{bk} - \boldsymbol{P}_{V_B}$, which produces deviations of individual observations from the block means. The two stratum variances ξ_E and ξ_B are, respectively, intra- and interblock variances. By (11.29), $\xi_E \leq \xi_B$. We have seen this in (7.15).

The strata for row-column designs can be determined in the same way. We start from U, then work with R and C in either order (since neither R nor C is nested in the other), and then end with E. In this case, (7.40) follows from (11.29).

11.5 Minimum G_2-Aberration

We are now ready to present a general theory of minimum G_2-aberration for mixed-level designs with orthogonal block structures using the same Bayesian framework adopted in

Chapters 5 and 7. In fact, the previous results for two-level designs and relatively simple block structures are covered as special cases.

In (11.6), for an $s_1 \times \cdots \times s_m$ factorial experiment, $\boldsymbol{\mu}$ is the $s_1 \cdots s_m \times 1$ vector of treatment effects. By (11.3), we can write it as $\boldsymbol{P\beta}$, where $\boldsymbol{P}$ is as in (11.2) and $\boldsymbol{\beta}$ is the $s_1 \cdots s_m \times 1$ vector of the grand mean and the values of a system of $s_1 \cdots s_m - 1$ mutually orthogonal treatment contrasts representing factorial effects of the m factors. Let $\boldsymbol{y}$ be the $n \times 1$ vector of observations obtained by conducting a fractional factorial experiment of run size n. Then the mean of $\boldsymbol{y}$, $\boldsymbol{X}_T\boldsymbol{\mu}$, as in (11.7), can be written as

$$\boldsymbol{X}_T\boldsymbol{\mu} = \boldsymbol{X}_T\boldsymbol{P\beta} = \boldsymbol{U\beta}, \tag{11.30}$$

where $\boldsymbol{U}$ is the submatrix of $\boldsymbol{P}$ consisting of the n rows of $\boldsymbol{P}$ corresponding to the treatment combinations chosen for experimentation.

As in Section 5.2.4, suppose that the factorial effects (components of $\boldsymbol{\beta}$) are independent and normally distributed with zero means, and that, given $\boldsymbol{\beta}$, the observations $\boldsymbol{y}$ on the selected treatment combinations are normally distributed. Then, by (11.7), (11.8), and (11.30), we have

$$\boldsymbol{y}|\boldsymbol{\beta} \sim \mathrm{N}(\boldsymbol{U\beta}, \sum_{F:F\in\mathfrak{B}} \sigma_F^2 \boldsymbol{R}_F). \tag{11.31}$$

By Theorem 11.2,

$$\sum_{F:F\in\mathfrak{B}} \sigma_F^2 \boldsymbol{R}_F = \sum_{F:F\in\mathfrak{B}} \xi_F \boldsymbol{P}_{W_F}. \tag{11.32}$$

In the following, we write the matrix in (11.32) as $\boldsymbol{V}$.

As in Chapter 7, in what follows we use $\boldsymbol{\Sigma}_{\boldsymbol{\beta}}$ to denote the prior covariance matrix $\mathrm{cov}(\boldsymbol{\beta})$ of $\boldsymbol{\beta}$. Then $\boldsymbol{\Sigma}_{\boldsymbol{\beta}}$ is a diagonal matrix with variances of the components of $\boldsymbol{\beta}$ as diagonal entries. The prior variance of each treatment factorial effect measures its importance. The more important a factorial effect is, the larger is its prior variance.

We also assume that for any given $S = \{i_1, \ldots, i_k\}$, all the components of $\boldsymbol{\beta}_S$ have the same variance, which we denote as v_S. This amounts to saying that, for $k = 1$, all the main-effect contrasts of factor i_1 are equally important and, for $k > 2$, all the k-factor interaction contrasts of factors $i_1, \ldots, i_k$ are equally important. As in (5.16), under the usual hierarchical assumption, we would have

$$S \subset S' \Rightarrow v_S \geq v_{S'}. \tag{11.33}$$

Under a more stringent version of the hierarchical assumption that factorial effects of the same order are equally important and lower-order effects are more important than higher-order ones, we would assume that the variance of each component of $\boldsymbol{\beta}$ only depends on the number of factors it involves, and that the fewer factors it involves, the greater is the prior variance. In this case, denote the prior variance of $\boldsymbol{\beta}_S$ as v_k if $|S| = k$; then

$$k > k' \Rightarrow v_k \leq v_{k'}. \tag{11.34}$$

More generally, the treatment factorial effects may be divided into, say L groups $G_1, \ldots G_L$, such that the effects in the same group are equally important and, for $i < j$, the effects in group i are more important than those in group j. Then there is a common prior variance v_i for the effects in the ith group, and we have

$$i < j \Rightarrow v_i > v_j. \tag{11.35}$$

For the scenario in the previous paragraph, we have $L = m$, and each group consists of the factorial effects $\boldsymbol{\beta}_S$ with the same $|S|$. For experiments with requirement sets discussed in

Section 6.3, if the requirement set consists of the main effects and some interactions, then the first group consists of the required effects, and the other effects are split into groups according to the usual effect hierarchy. In certain experiments, there may be departures from the usual hierarchical assumption. For example, in robust parameter designs for quality improvement, interactions of control factors and noise factors may be more important than the main effects of noise factors.

Using the $(M.S)$-criterion as a surrogate for the D-criterion, repeating the same argument as in Section 7.2.3, we can show that a good surrogate for the Bayesian D-criterion is to first maximize $\text{tr}\left(\boldsymbol{\Sigma}_{\boldsymbol{\beta}}\boldsymbol{U}^T\boldsymbol{V}^{-1}\boldsymbol{U}\right)$ and then minimize $\text{tr}\left(\boldsymbol{\Sigma}_{\boldsymbol{\beta}}\boldsymbol{U}^T\boldsymbol{V}^{-1}\boldsymbol{U}\right)^2$ among those that maximize $\text{tr}\left(\boldsymbol{\Sigma}_{\boldsymbol{\beta}}\boldsymbol{U}^T\boldsymbol{V}^{-1}\boldsymbol{U}\right)$.

In Section 7.2.3 where minimum G_2-aberration for two-level designs and some simple block structures were considered, one important step was the derivation of (7.21). For (7.21) to hold, a crucial fact is that $\|\boldsymbol{u}_S\|^2$ is a constant for all S, where $\boldsymbol{u}_S$ is the column of $\boldsymbol{U}$ associated with S. This is trivial for two-level designs, since in this case all the entries of $\boldsymbol{u}_S$ are 1 or -1; thus $\|\boldsymbol{u}_S\|^2 = n$. It is no longer true for mixed-level designs, but the following can be established.

For each $S \subseteq \{1, \ldots, m\}$, let $\boldsymbol{U}_S$ be the submatrix of $\boldsymbol{U}$ consisting of the columns that are associated with $\boldsymbol{\beta}_S$. For example, if S contains three factors, one of which has two levels and both of the other two factors have three levels, then $\boldsymbol{U}_S$ consists of $(2-1)(3-1)(3-1) = 4$ columns of $\boldsymbol{U}$. Then we have the following result:

Lemma 11.1. *For any $S \subseteq \{1, \ldots, m\}$, tr$(\boldsymbol{U}_S^T\boldsymbol{U}_S)$ is a constant that does not depend on the choices of the n-run design and orthogonal bases to form the columns of $\boldsymbol{P}$.*

Note that for two-level designs, each $\boldsymbol{U}_S$ has only one column, all of whose entries are $1/\sqrt{n}$ or $-1/\sqrt{n}$. In this case $\boldsymbol{U}_S = \boldsymbol{u}_S/\sqrt{n}$, and we have $\text{tr}(\boldsymbol{U}_S^T\boldsymbol{U}_S) = 1$.

We shall defer the proof of Lemma 11.1 to Section 11.6, and proceed, with Lemma 11.1 as a tool, to derive the minimum G_2-aberration criterion.

Following exactly the same argument employed after (7.19) in Section 7.2.3, except that now we may have more than three strata, we obtain

$$\begin{aligned}\text{tr}\left(\boldsymbol{\Sigma}_{\boldsymbol{\beta}}\boldsymbol{U}^T\boldsymbol{V}^{-1}\boldsymbol{U}\right) &= \sum_{F\in\mathfrak{B}} \frac{1}{\xi_F}\text{tr}\left(\boldsymbol{\Sigma}_{\boldsymbol{\beta}}\boldsymbol{U}^T\boldsymbol{P}_{W_F}\boldsymbol{U}\right)\\ &= \sum_{F\in\mathfrak{B}} \frac{1}{\xi_F}\text{tr}\left[\boldsymbol{\Sigma}_{\boldsymbol{\beta}}^{1/2}(\boldsymbol{P}_{W_F}\boldsymbol{U}\boldsymbol{\Sigma}_{\boldsymbol{\beta}}^{1/2})^T(\boldsymbol{P}_{W_F}\boldsymbol{U})\right]\\ &= \sum_{F\in\mathfrak{B}} \frac{1}{\xi_F}\text{tr}\left[(\boldsymbol{P}_{W_F}\boldsymbol{U}\boldsymbol{\Sigma}_{\boldsymbol{\beta}}^{1/2})^T(\boldsymbol{P}_{W_F}\boldsymbol{U}\boldsymbol{\Sigma}_{\boldsymbol{\beta}}^{1/2})\right].\end{aligned}$$

This leads to

$$\text{tr}\left(\boldsymbol{\Sigma}_{\boldsymbol{\beta}}\boldsymbol{U}^T\boldsymbol{V}^{-1}\boldsymbol{U}\right) = \sum_{F\in\mathfrak{B}} \frac{1}{\xi_F}\left[\sum_{S\subseteq\{1,\ldots,m\}} v_S\text{tr}\left(\boldsymbol{U}_S^T\boldsymbol{P}_{W_F}\boldsymbol{U}_S\right)\right]. \tag{11.36}$$

By Lemma 11.1, $\frac{1}{\xi_E}\left[\sum_{S\subseteq\{1,\ldots,m\}} v_S\text{tr}\left(\boldsymbol{U}_S^T\boldsymbol{U}_S\right)\right]$ is a constant. Since $\sum_{F\in\mathfrak{B}}\boldsymbol{P}_{W_F} = \boldsymbol{I}$, it follows that $\sum_{F\in\mathfrak{B}}\frac{1}{\xi_E}\left[\sum_{S\subseteq\{1,\ldots,m\}} v_S\text{tr}\left(\boldsymbol{U}_S^T\boldsymbol{P}_{W_F}\boldsymbol{U}_S\right)\right]$ is a constant. Subtracting (11.36) from this constant, we see that the maximization of $\mathrm{tr}\left(\boldsymbol{\Sigma}_\beta \boldsymbol{U}^T \boldsymbol{V}^{-1}\boldsymbol{U}\right)$, the first step of Bayesian (M.S.)-optimality, is equivalent to the minimization of

$$\sum_{F\in\mathfrak{B},F\neq E}\left(\frac{1}{\xi_E}-\frac{1}{\xi_F}\right)\left[\sum_{S\subseteq\{1,\ldots,m\}} v_S \mathrm{tr}\left(\boldsymbol{U}_S^T\boldsymbol{P}_{W_F}\boldsymbol{U}_S\right)\right]. \tag{11.37}$$

When v_S only depends on $|S|$, this reduces to the minimization of

$$\sum_{F\in\mathfrak{B},F\neq E}\left(\frac{1}{\xi_E}-\frac{1}{\xi_F}\right)\left[\sum_{S\subseteq\{1,\ldots,m\}} v_{|S|} \mathrm{tr}\left(\boldsymbol{U}_S^T\boldsymbol{P}_{W_F}\boldsymbol{U}_S\right)\right];$$

that is, the minimization of

$$\sum_{F\in\mathfrak{B},F\neq E}\left(\frac{1}{\xi_E}-\frac{1}{\xi_F}\right)\left[\sum_{k=1}^{m} v_k \sum_{S\subseteq\{1,\ldots,m\},|S|=k} \mathrm{tr}\left(\boldsymbol{U}_S^T\boldsymbol{P}_{W_F}\boldsymbol{U}_S\right)\right]. \tag{11.38}$$

If $v_k >> v_{k'}$ for $k < k'$, then a good surrogate for minimizing (11.37) is the sequential minimization of the terms in

$$W_\xi = \left(\sum_{F\in\mathfrak{B},F\neq E}\left(\frac{1}{\xi_E}-\frac{1}{\xi_F}\right)B_{1,F}(d),\ldots,\sum_{F\in\mathfrak{B},F\neq E}\left(\frac{1}{\xi_E}-\frac{1}{\xi_F}\right)B_{m,F}(d)\right), \tag{11.39}$$

where

$$B_{k,F}(d) = \sum_{S:S\subseteq\{1,\ldots,m\},|S|=k} \mathrm{tr}(\boldsymbol{U}_S^T\boldsymbol{P}_{W_F}\boldsymbol{U}_S). \tag{11.40}$$

For unstructured units, $\mathfrak{B} = \{U, E\}$. In this case, there is only one unit factor (U) that is different from E. Thus (11.39) reduces to

$$W_\xi = \left(\left(\frac{1}{\xi_E}-\frac{1}{\xi_U}\right)B_{1,U}(d),\ldots,\left(\frac{1}{\xi_E}-\frac{1}{\xi_U}\right)B_{m,U}(d)\right).$$

Except for the constant multiple $\frac{1}{\xi_E}-\frac{1}{\xi_U}$, this is Xu and Wu's (2001) extension of the wordlength pattern introduced by Tang and Deng (1999) for two-level nonregular designs (as discussed in Chapter 5) to mixed-level designs.

For two-level block designs, nested block designs, and row-column designs, (11.39) reduces to (7.22), (7.33), and (7.41), respectively. Again, one problem in practical applications is that the stratum variances ξ_F's are unknown except the constraint that $F \prec F' \Rightarrow \xi_F \leq \xi_{F'}$. Furthermore, the derivation that leads to (11.39) relies on the strong assumption that $v_k >> v_{k'}$ for $k < k'$. Under the weaker hierarchical assumption that $S \subset S' \Rightarrow v_S \geq v_{S'}$, one also faces the problem of not knowing prior variances of the treatment factorial effects. We shall extend Theorems 7.1, 7.2, and 7.3 (for two-level block designs, nested block designs, and row-column designs, respectively) to a general necessary and sufficient condition for mixed-level designs with orthogonal block structures to maximize $\mathrm{tr}\left(\boldsymbol{\Sigma}_\beta \boldsymbol{U}^T \boldsymbol{V}^{-1}\boldsymbol{U}\right)$ for all *feasible* ξ_F's and v_S's. Here the ξ_F's are said to be feasible if they satisfy (11.29), and the v_S's are feasible if more important $\boldsymbol{\beta}_S$'s have larger v_S's, e.g., (11.33) and (11.34) under the two versions of hierarchical assumption mentioned earlier, or (11.35). Such a result reduces verification of Bayesian M-optimality (or a minimum G_2-aberration surrogate) to certain extreme cases without having to know the v_S (or v_k) and ξ_F values.

We now present several results for the minimization of (11.37) and sequential minimization of the terms of (11.39). The first result is concerned with the simpler case of (11.39), which is a direct extension of Theorems 7.1, 7.2, and 7.3.

Theorem 11.3. *A necessary and sufficient condition for a design d to sequentially minimize the terms of W_ξ in (11.39) for all feasible ξ's is that it sequentially minimizes the terms of*

$$\left(\sum_{F\in\mathfrak{G}} B_{1,F}(d), \cdots, \sum_{F\in\mathfrak{G}} B_{m,F}(d)\right) \tag{11.41}$$

for all nonempty $\mathfrak{G} \subseteq \mathfrak{B} \setminus \{E\}$ such that

$$F \in \mathfrak{G}, F' \in \mathfrak{B}, \text{ and } F \prec F' \Rightarrow F' \in \mathfrak{G}. \tag{11.42}$$

Imposing the stronger hierarchical assumption that $v_k >> v_{k'}$ for $k < k'$ eliminates the need to know actual values of the prior variances of treatment factorial effects. In this case, Theorem 11.3 says that in order to check the Bayesian M-optimality, one only needs to verify minimum aberration with respect to the generalized wordlength pattern (11.41) for certain block structures. The block structures that need to be considered are those satisfying the hierarchical condition that if a certain unit factor is included in the block structure, then all the coarser unit factors (those with larger stratum variances) should also be present. So, e.g., for block designs, we need to check minimum aberration with respect to (11.41) for the block structures $\{U\}$ and $\{B, U\}$, leading to the wordlength patterns (7.23) and (7.24), respectively. For blocked split-plot designs, we have to consider the three block structures $\{U\}, \{B, U\}$, and $\{W, B, U\}$, and end up with the three wordlength patterns in Theorem 7.2. For row-column designs, we need to consider the block structures $\{U\}, \{R, U\}, \{C, U\}$, and $\{R, C, U\}$, which yield the four wordlength patterns in Theorem 7.3.

The same argument as in the proof of Theorem 11.3 can be used to show the following result for the case where v_S only depends on $|S|$, and $|S| \le |S'| \Rightarrow v_S \ge v_{S'}$.

Theorem 11.4. *For given v_k values satisfying $k < k' \Rightarrow v_k \ge v_{k'}$, a necessary and sufficient condition for a design to be Bayesian M-optimal for all feasible ξ_F's is that it minimizes*

$$\sum_{F\in\mathfrak{G}} \sum_{k=1}^{m} v_k B_{k,F}(d) \tag{11.43}$$

for all nonempty $\mathfrak{G} \subseteq \mathfrak{B} \setminus \{E\}$ such that

$$F \in \mathfrak{G}, F' \in \mathfrak{B}, \text{ and } F \prec F' \Rightarrow F' \in \mathfrak{G}. \tag{11.44}$$

We note that Theorem 11.3 also follows from Theorem 11.4. Theorem 11.4 addresses directly the question of Bayesian M-optimality, instead of employing a minimum aberration type surrogate. To apply this result, however, we need to know the v_k values, or as in the case of Theorem 11.3, resort to the strong hierarchical condition that $k < k' \Rightarrow v_k >> v_{k'}$.

The following result provides a necessary and sufficient condition for the Bayesian M-optimality under all feasible stratum variances and prior variances of factorial effects without making the stronger hierarchical assumption that $k < k' \Rightarrow v_k >> v_{k'}$.

Theorem 11.5. *A necessary and sufficient condition for a design to be Bayesian M-optimal (that is, it maximizes $tr\left(\mathbf{\Sigma}_\beta \mathbf{U}^T \mathbf{V}^{-1} \mathbf{U}\right)$) for all feasible v_S's and ξ_F's is that it minimizes*

$$\sum_{S\in\mathfrak{S}} \sum_{F\in\mathfrak{G}} tr\left[\mathbf{U}_S^T \mathbf{P}_{W_F} \mathbf{U}_S\right] \tag{11.45}$$

for all nonempty subsets $\mathfrak{S}$ of $2^{\{1,\ldots,m\}}$ and $\mathfrak{G} \subseteq \mathfrak{B} \setminus \{E\}$ such that

$$S \in \mathfrak{S}, S' \text{ is a nonempty subset of } 2^{\{1,\ldots,m\}}, \text{ and } v_{S'} \ge v_S \Rightarrow S' \in \mathfrak{S} \tag{11.46}$$

and

$$F \in \mathfrak{G}, F' \in \mathfrak{B}, \text{ and } F \prec F' \Rightarrow F' \in \mathfrak{G}, \tag{11.47}$$

where $2^{\{1,\ldots,m\}}$ *is the set of all the subsets of* $\{1,\ldots,m\}$.

If we have the hierarchical assumption that $S' \subset S \Rightarrow v_{S'} \geq v_S$, then in (11.46), "$v_{S'} \geq v_S \Rightarrow S' \in \mathfrak{S}$" can be replaced by "$S' \subset S \Rightarrow S' \in \mathfrak{S}$".

A price to pay for using Theorem 11.5 is that one needs to check some combinations of treatment and block structures – not only the block structures required in Theorem 11.3 and Theorem 11.4, but also the treatment structures prescribed in (11.46). This makes the result more difficult to apply in practice than Theorem 11.3 and Theorem 11.4 because much more configurations need to be checked.

In applications of Theorems 11.3, 11.4, and 11.5, if there is only one nonisomorphic design that is the best for all the block/treatment structures that are required to check, then it is the only Bayesian M-optimal design for all feasible stratum variances and prior variances of treatment factorial effects, and thus is Bayesian $(M.S)$-optimal. If there is more than one such M-optimal design, then we need to discriminate them according to the S-criterion. It may happen that no design is M-optimal for all feasible stratum variances and prior variances of treatment factorial effects. In this case Theorems 11.3, 11.4, and 11.5 are useful for ruling out the so-called *inadmissible* designs, those that are dominated by some other designs and need not be considered further. Here a design d_1 is inadmissible if there is another design d_2 such that d_2 is at least as good as d_1 for all all possible combinations of treatment and block structures prescribed in (11.46) and (11.47), and d_1 is worse than d_2 with respect to at least one such combination.

Remark 11.2. It is easy to see that the condition in Theorem 11.3 is necessary. For any $\mathfrak{G} \subseteq \mathfrak{B} \setminus \{E\}$ that satisfies (11.42), let $\xi_F = a$ for all F such that $F \in \mathfrak{G}$, and $\xi_G = b$ for all G such that $G \notin \mathfrak{G}$, where $b < a$, Then these ξ values are feasible. Furthermore, in this case, the terms in W_ξ are those in (11.41) multiplied by $1/b - 1/a$. If a design sequentially minimizes the terms in W_ξ for all feasible ξ's, then it must also sequentially minimize the terms in (11.41). Thus we only need to prove the sufficiency part. This is also true for both Theorems 11.4 and 11.5.

We will defer the proof of the sufficiency part of Theorem 11.3 to Section 11.6. Theorem 11.4 can be proved in a similar way. The proof of Theorem 11.5 is more complicated, but it applies a method similar to what is used in the proof of Theorem 11.3 to both the treatment and block structures and thus is omitted. The readers are referred to Chang (2022) for details.

Remark 11.3. A condition with a similar flavor (in fact, equivalent) to (11.42) first appeared in Theorem 4.1 of Cheng and Tsai (2011), where the selection of regular fractional factorial designs with orthogonal block structures was considered. They adopted the model robustness criterion of information capacity discussed in Section 5.2.3 of this book, and derived a surrogate criterion based on aliasing patterns of the competing designs. This surrogate criterion also depends on the stratum variances; so a device like what is used in Theorem 11.3 was developed to rule out inadmissible designs without having to know the stratum variances. The same condition also appeared in Theorem 5.1 of Chang and Cheng (2018), where since the stronger hierarchical condition ($k < k' \Rightarrow v_k >> v_{k'}$) was imposed, there is no need to check the minimization of (11.43) for the various treatment structures specified in (11.46). While Chang and Cheng (2018) only considered two-level designs, Theorem 11.5 (due to Chang (2022)), in addition to a less stringent hierarchical assumption, covers mixed-level designs as well. Lemma 11.1 is also due to Chang (2022). Chang and Cheng (2018) introduced an aberration-type criterion to assess multi-stratum factorial designs. The computation of wordlength patterns relies on model matrices of different orders of factorial effects, resulting in a significant computational load. Building on the concept of moment aberration proposed by Xu (2003) for factorial designs with unstructured experimental units, Chang (2024) developed an orthogonal moment aberration criterion for multi-stratum factorial designs using kernel functions. This new approach requires row-wise computation, thereby significantly reducing the computational load. Additionally, it offers flexibility, allowing users to select appropriate kernel functions for their specific needs.

Remark 11.4. As in Remarks 7.1, 7.2, and 7.3, checking the minimization of (11.45) in Theorem 11.5 for the block structures $\mathfrak{G}$ satisfying (11.47) amounts to considering, for each such $\mathfrak{G}$, the models with $\xi_F = \xi_U = \infty$ for all $F \in \mathfrak{G}$ and $\xi_F = \xi_E$ for all $F \notin \mathfrak{G}$; that is, comparing block designs with the $\mathfrak{G}$-classes as blocks under models with fixed block effects. For example, for experiments with the block structure $b/(r \times c)$, to apply Theorem 11.5 (or 11.3) to determine admissible designs, one has to compare designs under the setting of brc unstructured units, b blocks of size rc, br blocks of size c, and bc blocks of size r, with fixed block effects in all the block design settings.

11.6 Proofs

Proof of Theorem 11.2:

If $F \prec G$, then by (11.24), it is clear that W_F and W_G are orthogonal. Otherwise, we must have $G \prec F \vee G$; then again by (11.24),

$$W_G \subseteq V_G \ominus V_{F \vee G} \tag{11.48}$$

and

$$W_F \subseteq V_F. \tag{11.49}$$

Since F and G are orthogonal factors,

$$V_G \ominus V_{F \vee G} \text{ and } V_F \text{ are orthogonal.} \tag{11.50}$$

The orthogonality of W_F and W_G then follows immediately from (11.48), (11.49), and (11.50).

For (11.25), we see that it holds if there is no factor G that is coarser than F, since $W_F = V_F$ in this case. We prove the general case by induction. Suppose (11.25) holds for all the factors G such that $F \prec G$. Then

$$\sum_{G \in \mathfrak{B}: F \prec G} V_G = \sum_{G \in \mathfrak{B}: F \prec G} \left[\bigoplus_{H \in \mathfrak{B}: G \preceq H} W_H \right] = \bigoplus_{G \in \mathfrak{B}: F \prec G} W_G.$$

This together with (11.24) imply that

$$V_F = W_F \oplus \sum_{G \in \mathfrak{B}: F \prec G} V_G = \bigoplus_{G \in \mathfrak{B}: F \preceq G} W_G.$$

For (11.27), let g be the number of levels of G; then we have

$$\begin{aligned}
\operatorname{cov}(\boldsymbol{y}) &= \sum_{G:G\in\mathfrak{B}} \sigma_G^2 \boldsymbol{R}_G \\
&= \sum_{G:G\in\mathfrak{B}} \sigma_G^2 \frac{n}{g} \boldsymbol{P}_{V_G} \\
&= \sum_{G:G\in\mathfrak{B}} \sigma_G^2 \frac{n}{g} \left[\sum_{F\in\mathfrak{B}:G\preceq F} \boldsymbol{P}_{W_F} \right] \\
&= \sum_{F:F\in\mathfrak{B}} \left[\sum_{G\in\mathfrak{B}:G\preceq F} \sigma_G^2 \frac{n}{g} \right] \boldsymbol{P}_{W_F},
\end{aligned}$$

where the second equality follows from (11.23) and the third equality follows from (11.25). Therefore each nonnull $W_F, F \in \mathfrak{B}$, is an eigenspace of $\operatorname{cov}(\boldsymbol{y})$ with eigenvalue

$$\xi_F = \sum_{G\in\mathfrak{B}:G\preceq F} \frac{n}{g}\sigma_G^2. \tag{11.51}$$

This completes the proof.

Proof of Lemma 11.1:

We first show that for any given n-run design, $\operatorname{tr}(\boldsymbol{U}_S^T\boldsymbol{U}_S)$ does not depend on the choices of orthogonal bases to form the columns of $\boldsymbol{P}$. Given two different full model matrices $\boldsymbol{P}$ and $\boldsymbol{P}^*$ with orthonormal columns, let $\boldsymbol{U}$ and $\boldsymbol{U}^*$ be the corresponding submatrices consisting of the rows associated with the n runs in the design. Furthermore, let $\boldsymbol{P}_S, \boldsymbol{P}_S^*, \boldsymbol{U}_S$, and $\boldsymbol{U}_S^*$ be the corresponding submatrices of $\boldsymbol{P}, \boldsymbol{P}^*, \boldsymbol{U}$, and $\boldsymbol{U}^*$ consisting of the columns associated with S. Since there is an orthogonal matrix $\boldsymbol{O}$ such that $\boldsymbol{P}_S = \boldsymbol{P}_S^*\boldsymbol{O}$, we have $\boldsymbol{U}_S = \boldsymbol{U}_S^*\boldsymbol{O}$, and thus $\boldsymbol{U}_S\boldsymbol{U}_S^T = (\boldsymbol{U}_S^*\boldsymbol{O})(\boldsymbol{U}_S^*\boldsymbol{O})^T = \boldsymbol{U}_S^*\boldsymbol{U}_S^{*T}$.

To show that $\operatorname{tr}(\boldsymbol{U}_S^T\boldsymbol{U}_S)$ does not depend on choices of designs, we first note that $\operatorname{tr}(\boldsymbol{U}_S^T\boldsymbol{U}_S) = \operatorname{tr}(\boldsymbol{U}_S\boldsymbol{U}_S^T)$, the sum of the squared lengths of the rows of $\boldsymbol{U}_S$, which are n rows of $\boldsymbol{P}_S$. Thus it is sufficient to show that all the rows of $\boldsymbol{P}_S$ have the same length. This can be seen by induction on $|S|$, the size of S. For the case where $|S| = 1$, without loss of generality, we may assume that $S = \{1\}$. Then

$$\left[\frac{1}{\sqrt{s_1\cdots s_m}}\mathbf{1}_{s_1\cdots s_m}, \boldsymbol{P}_S\right] = \boldsymbol{P}_1 \otimes \frac{1}{\sqrt{s_2}}\mathbf{1}_{s_2} \otimes \cdots \otimes \frac{1}{\sqrt{s_m}}\mathbf{1}_{s_m}, \tag{11.52}$$

where $\mathbf{1}_k$ is the $k \times 1$ vector of 1's. We have

$$\begin{aligned}
&\left[\frac{1}{\sqrt{s_1\cdots s_m}}\mathbf{1}_{s_1\cdots s_m}, \boldsymbol{P}_S\right]\left[\frac{1}{\sqrt{s_1\cdots s_m}}\mathbf{1}_{s_1\cdots s_m}, \boldsymbol{P}_S\right]^T \\
&= \left[\boldsymbol{P}_1 \otimes \frac{1}{\sqrt{s_2}}\mathbf{1}_{s_2} \otimes \cdots \otimes \frac{1}{\sqrt{s_m}}\mathbf{1}_{s_m}\right]\left[\boldsymbol{P}_1 \otimes \frac{1}{\sqrt{s_2}}\mathbf{1}_{s_2} \otimes \cdots \otimes \frac{1}{\sqrt{s_m}}\mathbf{1}_{s_m}\right]^T \\
&= \left(\boldsymbol{P}_1\boldsymbol{P}_1^T\right) \otimes \frac{\boldsymbol{J}_{s_2\cdots s_m}}{s_2\cdots s_m} \\
&= \frac{\boldsymbol{I}_{s_1} \otimes \boldsymbol{J}_{s_2\cdots s_m}}{s_2\cdots s_m}.
\end{aligned}$$

Since the ith diagonal of $\boldsymbol{P}_S\boldsymbol{P}_S^T$ is the squared length of the ith row of $\boldsymbol{P}_S$, it follows that all the rows of $\boldsymbol{P}_S$ have the same length.

Now suppose that for all the S's with $|S| < k$, the rows of $\boldsymbol{P}_S$ have the same length. We show that for all the S's with $|S| = k$, all the rows of $\boldsymbol{P}_S$ also have the same length. Again, without loss of generality, assume that $S = \{1, \ldots, k\}$. Then, similar to (11.52), we have

$$\left[\frac{1}{\sqrt{s_1 \cdots s_m}}\mathbf{1}_{s_1 \cdots s_m}, \boldsymbol{P}_{\{1\}} \ldots, \boldsymbol{P}_{\{1,\ldots,k\}}\right] = \boldsymbol{P}_1 \otimes \cdots \otimes \boldsymbol{P}_k \otimes \frac{1}{\sqrt{s_{k+1}}}\mathbf{1}_{s_{k+1}} \otimes \cdots \otimes \frac{1}{\sqrt{s_m}}\mathbf{1}_{s_m}.$$

Then since

$$[\boldsymbol{P}_1 \otimes \cdots \otimes \boldsymbol{P}_k]\,[\boldsymbol{P}_1 \otimes \cdots \otimes \boldsymbol{P}_k]^T = \boldsymbol{I}_{s_1 \cdots s_k}$$

and

$$\left[\frac{1}{\sqrt{s_{k+1} \cdots s_m}}\mathbf{1}_{s_{k+1} \cdots s_m}\right]\left[\frac{1}{\sqrt{s_{k+1} \cdots s_m}}\mathbf{1}_{s_{k+1} \cdots s_m}\right]^T = \frac{1}{s_{k+1} \cdots s_m}\boldsymbol{J}_{s_{k+1} \cdots s_m},$$

we have

$$\left[\frac{1}{\sqrt{s_1 \cdots s_m}}\mathbf{1}_{s_1 \cdots s_m}, \boldsymbol{P}_{\{1\}}, \ldots, \boldsymbol{P}_{\{1,\ldots,k\}}\right]\left[\frac{1}{\sqrt{s_1 \cdots s_m}}\mathbf{1}_{s_1 \cdots s_m}, \boldsymbol{P}_{\{1\}}, \ldots, \boldsymbol{P}_{\{1,\ldots,k\}}\right]^T$$
$$= \frac{1}{s_{k+1} \cdots s_m}\left[\boldsymbol{I}_{s_1 \cdots s_k} \otimes \boldsymbol{J}_{s_{k+1} \cdots s_m}\right].$$

From this identity we see that $\boldsymbol{P}_{\{1,\ldots,k\}}\boldsymbol{P}^T_{\{1,\ldots,k\}}$ can be expressed in terms of $\boldsymbol{P}_{S'}\boldsymbol{P}^T_{S'}$'s with $|S'| < k$. It then follows from the induction hypothesis that all rows of $\boldsymbol{P}_S = \boldsymbol{P}_{\{1,\ldots,k\}}$ have the same length.

Proof of Theorem 11.3:

It is enough to prove the sufficiency; see Remark 11.2.

We have to show that if a design sequentially minimizes the terms in (11.41) for all the subsets $\mathfrak{G}$ of $\mathfrak{B} \setminus \{E\}$ satisfying (11.42), then it sequentially minimizes the terms in (11.39) for all feasible ξ's.

For each $F \neq E$, let $\eta_F = \frac{1}{\xi_E} - \frac{1}{\xi_F}$ and η be the vector of the η_F's with $F \neq E$. We say that η is feasible if the η_F's are nonnegative and $F \preceq F' \Rightarrow \eta_F \leq \eta_{F'}$. Clearly η is feasible if and only if ξ is feasible. We will prove the following result, which is more general than what is stated in the theroem: for any subset $\mathfrak{B}'$ of $\mathfrak{B} \setminus \{E\}$, if d^* sequentially minimizes the terms in

$$\left(\sum_{F \in \mathfrak{G}} B_{1,F}(d), \cdots, \sum_{F \in \mathfrak{G}} B_{m,F}(d)\right) \tag{11.53}$$

for all subsets $\mathfrak{G}$ of $\mathfrak{B}'$ such that

$$F \in \mathfrak{G}, F' \in \mathfrak{B}' \text{ and } F \prec F' \Rightarrow F' \in \mathfrak{G}, \tag{11.54}$$

then d^* sequentially minimizes the terms of

$$\left(\sum_{F \in \mathfrak{B}'} \eta_F B_{1,F}(d), \ldots, \sum_{F \in \mathfrak{B}'} \eta_F B_{m,F}(d)\right). \tag{11.55}$$

for all feasible η's. If this is true for all subsets $\mathfrak{B}'$ of $\mathfrak{B} \setminus \{E\}$ satisfying (11.42), then in particular it holds for $\mathfrak{B}' = \mathfrak{B} \setminus \{E\}$; the conclusion of Theorem 11.3 then follows immediately.

We will prove the result mentioned in the previous paragraph by mathematical induction on the number of unit factors in $\mathfrak{B}'$. The case where $\mathfrak{B}'$ consists of one single unit factor is trivial. Suppose that it holds for all subsets $\mathfrak{B}'$ of $\mathfrak{B} \setminus \{E\}$ containing fewer than s unit factors, where $s \geq 2$. Then we need to show that if $|\mathfrak{B}'| = s$ and d^* sequentially minimizes the terms of (11.53) for all $\mathfrak{G}$ satisfying (11.54), then it sequentially minimizes the terms of (11.55) for all feasible η's.

Under the given assumption, since (11.54) is trivially satisfied when $\mathfrak{G}$ is chosen to be $\mathfrak{B}'$, we have that d^* sequentially minimizes the terms of (11.53) with $\mathfrak{G} = \mathfrak{B}'$; that is, it sequentially minimizes the terms of

$$\left(\sum_{F\in\mathfrak{B}'} B_{1,F}(d), \cdots, \sum_{F\in\mathfrak{B}'} B_{m,F}(d)\right). \tag{11.56}$$

If all the η_F's with $F \in \mathfrak{B}'$ are equal, then the terms of (11.55) are constant multiples of those of (11.56). It follows that d^* sequentially minimizes the terms of (11.55).

On the other hand, suppose not all the η_F's with $F \in \mathfrak{B}'$ are equal. Let η' be the smallest such η_F and let $\mathfrak{B}^* = \{F \in \mathfrak{B}' : \eta_F > \eta'\}$. Then $\mathfrak{B}^*$ is nonempty and $|\mathfrak{B}^*| < s$. For each $F \in \mathfrak{B}^*$, let $\eta_F^* = \eta_F - \eta'$; then $\eta_F^* > 0$. Also, $\eta_F^* \leq \eta_{F'}^*$ if $F \prec F'$ and $F, F' \in \mathfrak{B}^*$. Furthermore,

$$\sum_{F\in\mathfrak{B}'} \eta_F B_{k,F}(d) = \sum_{F\in\mathfrak{B}^*} \eta_F^* B_{k,F}(d) + \eta' \sum_{F\in\mathfrak{B}'} B_{k,F}(d). \tag{11.57}$$

We have that d^* sequentially minimizes the terms of (11.56); thus, by (11.57), it suffices to show that d^* sequentially minimizes the terms of

$$\left(\sum_{F\in\mathfrak{B}^*} \eta_F^* B_{1,F}(d), \cdots, \sum_{F\in\mathfrak{B}^*} \eta_F^* B_{m,F}(d)\right).$$

Since $|\mathfrak{B}^*| < s$, by the induction hypothesis, it remains to show that d^* sequentially minimizes the terms of (11.53) for all subsets $\mathfrak{G}$ of $\mathfrak{B}^*$ satisfying the following condition:

$$F \in \mathfrak{G}, F' \in \mathfrak{B}^* \text{ and } F \prec F' \Rightarrow F' \in \mathfrak{G}. \tag{11.58}$$

Suppose a subset $\mathfrak{G}$ of $\mathfrak{B}^*$ satisfies (11.58). By the assumption on $\mathfrak{B}'$, d^* sequentially minimizes the terms of (11.53) as long as $\mathfrak{G}$ also satisfies (11.54). That is, given $F \in \mathfrak{G}$, $F' \in \mathfrak{B}'$, and $F \prec F'$, it is enough to show that $F' \in \mathfrak{G}$. Because $F \in \mathfrak{G} \subseteq \mathfrak{B}^*$, by the definition of $\mathfrak{B}^*$, we have $\eta_F > \eta'$. Also, $\eta_{F'} \geq \eta_F$ since $F \prec F'$. Thus $\eta_{F'} \geq \eta_F > \eta'$, which implies that $F' \in \mathfrak{B}^*$. Then by (11.58), $F' \in \mathfrak{G}$. This completes the proof.

12

Space-Filling Designs

In this chapter we provide a concise treatment of space-filling designs, a class of designs suitable for running computer experiments. The design literature on this subject is vast and diverse. Space-filling designs deserve a separate book by themselves if a full treatment is attempted. Our goal is to present those topics that are most related to factorial designs, so as to be consistent with the theme of this book. After an introductory section, we introduce various approaches to the construction of space-filling designs and discuss related design selection criteria. We then present some major results on the existence and construction of orthogonal designs, of which orthogonal Latin hypercube designs are special cases. OA-based designs and their advantages over general balanced designs are discussed next, and adequate attention is given to strong orthogonal arrays, a special class of OA-based designs. The remaining items on the agenda are maximin distance designs, nested space-filling designs, and designs for both qualitative and quantitative factors.

12.1 Introduction

Computer models are powerful tools that enable researchers to investigate complex systems from almost every imaginable field of studies in natural sciences, engineering, social sciences, and humanities. While some computer models are stochastic, others are deterministic. We concern ourselves with deterministic computer models. A deterministic computer model is essentially a mathematical function, so complex that it is represented by a computer code written by subject specialists. To users, a computer model is a black box – one can obtain an output for a given input by running the computer code without the need of knowing the details of how the code was written. In many applications, the computer code representing a computer model is expensive to run. It is therefore desirable to build a cheaper surrogate model in such situations. Computer experiments are concerned with the building of a statistical surrogate model based on the data consisting of a set of carefully selected inputs and the corresponding outputs from running the computer code. Though not always, a Gaussian process model is employed to connect the output variable with the input variables, and the method of kriging is used to predict the output at an unobserved site of input variables.

Designing a computer experiment, namely a judicious selection of input settings to run the computer code, is a crucial step in the process of building a statistical surrogate model. No matter how elaborate and sophisticated the model building process is, a statistical model contains no more information than what the data are able to offer. In the past three decades since the dawn of the design and analysis of computer experiments as a subject of study (Sacks, Welch, Mitchell, and Wynn, 1989), space-filling designs have been accepted far and wide as appropriate designs for computer experiments. A space-filling design refers very broadly to any design for distributing points (observation sites) in the

DOI: 10.1201/9781003371861-12

experimental region in some uniform fashion. It is sensible to use space-filling designs in computer experiments because any interesting feature in the relationship between the output variable and the input variables is equally likely to occur in all portions of the experimental region, owing to the unknown and complex nature of the true model. Additionally, a design having some uniformity in distributing the points allows the fitting of diverse models and tends to minimize the bias due to inadequacy of the fitted model; see Box and Draper (1959) and Sacks and Ylvisaker (1984).

We now introduce some notation and basic concepts. We consider designs of n runs for m factors with s levels. The s levels are denoted by $0, 1, \dots, s-1$, unless stated otherwise. All designs discussed in this chapter are balanced, meaning that the s levels occur equally often in every design column. A balanced design is also called a U-type design, and is in fact an orthogonal array of strength one. We use $\mathrm{U}(n, s^m)$ to denote a balanced design of n runs for m factors, each having s levels. In the theory of factorial design, experimenters typically use designs with a few levels, two or three levels the most common, to facilitate the estimation of models with various factorial effects. The situation is different for space-filling designs, where experimenters are concerned with building statistical surrogate models and their preferred choices are designs with many levels. A design with many levels not only allows one to more fully explore each factor and to avoid replication of design points in low dimensional projections but also permits a variety of models to be fitted. A Latin hypercube design has as many levels as experimental runs. In our notation, a balanced design $\mathrm{U}(n, s^m)$ is a Latin hypercube if $s = n$, and we use $\mathrm{LH}(n, m)$ to denote a Latin hypercube of n runs for m factors.

For simplicity, we only consider the situation where the design region is rectangular, which we can assume to be the m-dimensional unit cube $[0, 1]^m$. Let $\boldsymbol{D} = (d_{ij})_{n\times m}$ be a given $\mathrm{U}(n, s^m)$. Then one can use the rows of

$$\boldsymbol{X} = (x_{ij})_{n\times m} \quad \text{with } x_{ij} = (d_{ij} + 0.5)/s$$

to obtain the design points within the unit cube. Alternatively, one can also take

$$x_{ij} = (d_{ij} + U_{ij})/s,$$

where the U_{ij}'s are generated independently from the uniform distribution on the unit interval $[0, 1]$. The former uses the center of a cell identified by a row of $\boldsymbol{D}$, and the latter randomly takes a point from that cell.

For general references on design and analysis of computer experiments, we refer to the books by Santner, Williams, and Notz (2003), Fang, Li, and Sudjianto (2005), and Gramacy (2020). A broad review of space-filling designs is available in Lin and Tang (2015).

12.2 Approaches to Constructing Space-Filling Designs

Many ideas are useful for distributing design points uniformly in the experimental region. We discuss some of them through four approaches to the construction of space-filling designs.

A simple approach is to use the correlation. We recall that the correlation between two column vectors $\boldsymbol{x} = (x_1, \dots, x_n)^T$ and $\boldsymbol{y} = (y_1, \dots, y_n)^T$ is

$$\rho(\boldsymbol{x}, \boldsymbol{y}) = \frac{\sum_{i=1}^n (x_i - \bar{x})(y_i - \bar{y})}{\left[\sum_{i=1}^n (x_i - \bar{x})^2\right]^{1/2} \left[\sum_{i=1}^n (y_i - \bar{y})^2\right]^{1/2}},$$

where, for example, $\bar{x}$ is the mean of $x_1, \ldots, x_n$. Consider a balanced design $\boldsymbol{D}$, a $\mathrm{U}(n, s^m)$. Let ρ_{jk} be the correlation between the jth and kth columns. Then design $\boldsymbol{D}$ is said to be orthogonal if $\rho_{jk} = 0$ for all $j < k$, and it is said to be nearly orthogonal if its correlations are made small in some way. Formally, one can use the minimization of

$$\frac{1}{\binom{m}{2}} \sum_{j<k} \rho_{jk}^2 \quad \text{or} \quad \max_{j<k} \rho_{jk}^2$$

as the design selection criterion to produce a design that has small correlations. An orthogonal design is useful in its own right, as it allows uncorrelated estimation of main effects under the main-effect model. We can also regard orthogonality as a stepping stone to the construction of space-filling designs. If a design is such that its points are uniformly scattered in the design region, then the projections of the design points onto any two dimensions should also be uniform in some way, and, to say the least, should not exhibit any apparent pattern. This must result in a low correlation. In other words, one can always find a good space-filling design from the collection of designs that are orthogonal or nearly orthogonal. Consideration of Latin hypercube designs with low correlations goes back to Iman and Conover (1982) and Owen (1994). The commonly used correlation as given above is actually the linear correlation. Tang (1998) proposed an algorithm for reducing quadratic correlations in Latin hypercube designs. Butler (2001) constructed Latin hypercube designs that are orthogonal with respect to trigonometric functions. We devote Section 12.3 to the construction of orthogonal designs and orthogonal Latin hypercubes.

Perhaps, the most fruitful approach to the construction of space-filling designs is to make use of orthogonal arrays. Latin hypercube designs, introduced by McKay, Beckman, and Conover (1979), are essentially orthogonal arrays of strength one. The other two early papers along this line are Owen (1992) and Tang (1993). Owen (1992) used randomized orthogonal arrays directly as space-filling designs, whereas Tang (1993) constructed Latin hypercube designs using orthogonal arrays. Orthogonal array-based designs, or simply OA-based designs, achieve uniformity of design points in low-dimensional projections by matching up empirical probabilities with theoretical ones over various regions of interest. Strong orthogonal arrays offer even better projection properties than ordinary orthogonal arrays (He and Tang, 2013), and will therefore receive special attention in our discussion of OA-based designs. Minimum G_2- and G-aberration criteria are helpful for finding the best orthogonal arrays due to their projection justification given in Section 10.4. However, because space-filling designs generally need to have many levels, the aberration criteria are not directly useful. For example, the two aberration criteria cannot distinguish Latin hypercube designs; more precisely all Latin hypercube designs of n runs and m factors have the same performance under the minimum G_2- or G-aberration criteria. Instead of directly using the aberration criteria for selecting designs, we can use minimum G_2- or G-aberration orthogonal arrays with a smaller number of levels to construct OA-based designs with more levels. Section 12.4 provides a general discussion on OA-based designs, and Sections 12.5 and 12.6 give an adequate treatment of space-filling designs based on strong orthogonal arrays.

Using distance-based criteria is another popular approach to constructing space-filling designs. Let $\boldsymbol{D} = (d_{ij})_{n \times m}$ be a $\mathrm{U}(n, s^m)$. Then

$$\delta_{uv} = \left(\sum_{j=1}^{m} |d_{uj} - d_{vj}|^p \right)^{1/p}$$

is the L_p distance between the uth and vth runs. Commonly used are L_1 and L_2 distances, corresponding to $p = 1$ and $p = 2$, respectively. A space-filling design needs to spread out its points in the design region and therefore should have large δ_{uv}'s for all $u < v$. There are many ways one can make all δ_{uv}'s large. The simplest is to use the maximin distance criterion (Johnson, Moore, and Ylvisaker, 1990), which is to maximize $\min_{u<v} \delta_{uv}$. In our setting, a $\mathrm{U}(n, s^m)$ is a maximin distance design if it maximizes $\min_{u<v} \delta_{uv}$ among all $\mathrm{U}(n, s^m)$s. A generalization of the maximin distance criterion is to sequentially minimize the frequencies of the distinct δ_{uv} values arranged in ascending order (Morris and Mitchell, 1995). Other possibilities for making all δ_{uv}s large will be seen later in the chapter. Besides the maximin distance criterion, Johnson, Moore, and Ylvisaker (1990) also introduced the minimax distance criterion. This criterion seeks designs by minimizing the distance of the most remote point in the design region from the design points. Although the idea is attractive, the minimax distance criterion seldom gets used in practice because it involves points outside the design points and hence is much harder to implement than the maximin distance criterion. We provide some results in Section 12.7 on the construction of maximin distance designs.

Space-filling designs can be constructed as well by making the empirical distribution of design points as close as possible to the uniform distribution on the design region $[0, 1]^m$. If cumulative distribution functions are utilized to represent the empirical distribution and theoretical uniform distribution, then this idea gives rise to the criteria of star discrepancy and L_p star discrepancy. These criteria measure the difference between the empirical and theoretical cumulative distribution functions, which are defined with reference to one particular vertex $(0, \ldots, 0)$ of the cube $[0, 1]^m$. To overcome this shortcoming, the centered L_2 and wrap-around L_2 discrepancies (Hickernell, 1998) and a mixture discrepancy (Zhou, Fang, and Ning, 2013) were proposed. Here we give a detailed expression for the centered L_2 discrepancy:

$$\mathrm{CD} = \left(\frac{13}{12}\right)^m - \frac{2}{n}\sum_{i=1}^{n}\prod_{k=1}^{m}\left(1 + \frac{1}{2}\left|\frac{2d_{ik}+1-s}{2s}\right| - \frac{1}{2}\left|\frac{2d_{ik}+1-s}{2s}\right|^2\right)$$
$$+ \frac{1}{n^2}\sum_{i=1}^{n}\sum_{j=1}^{n}\prod_{k=1}^{m}\left(1 + \frac{1}{2}\left|\frac{2d_{ik}+1-s}{2s}\right| + \frac{1}{2}\left|\frac{2d_{jk}+1-s}{2s}\right| - \left|\frac{d_{ik}-d_{jk}}{2s}\right|\right).$$

We will not provide further discussion on discrepancy-based criteria, except for some occasional references to them. Instead, we refer the reader to the treatise of Fang, Liu, Qin, and Zhou (2018) for an extensive discussion.

Different types of space-filling criteria are not unrelated to each other. The connections between them were explored by Wang, Yang, and Xu (2018b), Sun, Wang, and Xu (2019), and Wang, Sun, and Xu (2022c).

12.3 Orthogonal Designs and Orthogonal Latin Hypercubes

In this section, we examine the construction of orthogonal designs and orthogonal Latin hypercubes. The literature on this topic is rich; so we only present some major results. Recall that a balanced design $\mathrm{U}(n, s^m)$ with levels denoted by $0, 1, \ldots, s-1$ is said to be orthogonal if any two of its columns have zero correlation. For the discussion of orthogonal designs, it is convenient to use centered levels

$$u - (s-1)/2 \quad \text{for } u = 0, 1, \ldots, s-1,$$

which we do for the present section. Then a design $\mathrm{U}(n, s^m)$, $\boldsymbol{D} = (\boldsymbol{d}_1, \ldots, \boldsymbol{d}_m)$, is orthogonal if any two of its columns are orthogonal; that is, $J(\boldsymbol{d}_i, \boldsymbol{d}_j) = \boldsymbol{d}_i^T \boldsymbol{d}_j = 0$ for any $i < j$. We use $\mathrm{OD}(n, s^m)$ to denote an orthogonal design. When $s = n$, an $\mathrm{OD}(n, s^m)$ becomes an orthogonal Latin hypercube, in which case we use the notation $\mathrm{OLH}(n, m)$.

For a smooth reading of this section, we stress that all designs in the entire section have centered levels $u - (s-1)/2$ for $u = 0, 1, \ldots, s-1$, unless stated otherwise.

12.3.1 Existence and basic methods of construction

A balanced design $\mathrm{U}(n, s^m)$ is an orthogonal array of strength one. It is therefore necessary that $n = \lambda s$ for some positive integer λ. To be an orthogonal design, further conditions are required of λ and s. The next theorem provides a complete characterization for the existence and non-existence of orthogonal designs.

Theorem 12.1. *An $OD(n, s^m)$, an n-run orthogonal design for m s-level factors with $n \geq 4$ and $m = 2$ exists if and only if λ is even or $s \neq 4k+2$ for any integer $k \geq 1$.*

Equivalently, Theorem 12.1 states that an $\mathrm{OD}(n, s^m)$ with $n \geq 4$ and $m = 2$ does not exist if and only if λ is odd and $s = 4k+2$ for some integer $k \geq 1$. When $\lambda = 1$, an $\mathrm{OD}(n, s^m)$ becomes an orthogonal Latin hypercube $\mathrm{OLH}(n, m)$. Applying Theorem 12.1 to this situation leads to the next corollary.

Corollary 12.1. *An $OLH(n, m)$ with $n \geq 4$ and $m = 2$ exists if and only if $n \neq 4k+2$ for any integer $k \geq 1$.*

Corollary 12.1 was presented and proved in Lin, Bingham, Sitter, and Tang (2010). The general version as given in Theorem 12.1 appeared in Karunanayaka and Tang (2018), but they only presented and proved the necessity part. We now give a proof of the sufficiency part of Theorem 12.1.

If $s \neq 4k+2$, then an $\mathrm{OLH}(s, 2)$ exists according to Corollary 12.1. Then for any $\lambda \geq 1$, we obtain an $\mathrm{OD}(\lambda s, 2)$ by using λ copies of the $\mathrm{OLH}(s, 2)$. It remains to show that when $s = 4k+2$, an $\mathrm{OD}(\lambda s, s^2)$ exists for even λ. It is sufficient to establish this for $\lambda = 2$. Let $(\boldsymbol{a}, \boldsymbol{b})$ be a Latin hypercube with s runs and two columns. Then $(\boldsymbol{c}, \boldsymbol{d})$ is an $\mathrm{OD}(2s, s^2)$ with $\lambda = 2$, where $\boldsymbol{c} = (\boldsymbol{a}^T, \boldsymbol{a}^T)^T$ and $\boldsymbol{d} = (\boldsymbol{b}^T, -\boldsymbol{b}^T)^T$.

We next present two basic methods of constructing orthogonal designs. Both of these methods can be used to construct large orthogonal designs from small ones.

Theorem 12.2. *Let $\boldsymbol{A} = (a_{ij})_{n_1 \times m_1}$ be a matrix with ± 1 entries and orthogonal columns, and $\boldsymbol{D}_0$ be an $OD(n_2, s^{m_2})$. Then $\boldsymbol{D} = \boldsymbol{A} \otimes \boldsymbol{D}_0$ is an $OD(n_1 n_2, s^{m_1 m_2})$.*

Theorem 12.2 follows immediately from an application of Lemma 4.2. Suppose that $\boldsymbol{D}_0$ is an $\mathrm{OD}(n_2, s^{m_2})$. If we take

$$\boldsymbol{A} = \begin{pmatrix} 1 & 1 \\ 1 & -1 \end{pmatrix}$$

in Theorem 12.2, then

$$\boldsymbol{D} = \boldsymbol{A} \otimes \boldsymbol{D}_0 = \begin{pmatrix} \boldsymbol{D}_0 & \boldsymbol{D}_0 \\ \boldsymbol{D}_0 & -\boldsymbol{D}_0 \end{pmatrix}$$

is an $\mathrm{OD}(2n_2, s^{2m_2})$. The method in Theorem 12.2 was proposed by Bingham, Sitter, and Tang (2009), who also gave two generalizations of this basic tensor product method. The generalizations are useful if one is interested in finding orthogonal designs with other desirable properties. For details, we refer to Bingham, Sitter, and Tang (2009). Here we are content with simple illustrations of the generalizations. Let $\boldsymbol{D}_1$ and $\boldsymbol{D}_2$ both be $\mathrm{OD}(n_2, s^{m_2})$s. Then the two generalizations give the following two orthogonal designs:

$$\begin{pmatrix} \boldsymbol{D}_1 & \boldsymbol{D}_2 \\ \boldsymbol{D}_1 & -\boldsymbol{D}_2 \end{pmatrix} \text{ and } \begin{pmatrix} \boldsymbol{D}_1 & \boldsymbol{D}_1 \\ \boldsymbol{D}_2 & -\boldsymbol{D}_2 \end{pmatrix},$$

both being $\mathrm{OD}(2n_2, s^{2m_2})$s.

Example 12.1. *Consider the following* $OD(10, 5^6)$*:*

$$D_0 = \begin{bmatrix} 0 & -2 & -2 & -2 & -1 & 0 \\ 1 & -2 & -1 & 2 & 0 & -1 \\ -1 & -1 & 0 & 1 & 1 & 1 \\ -1 & -1 & 1 & -1 & 0 & 1 \\ 0 & 0 & 2 & -2 & 2 & -2 \\ -2 & 1 & 0 & 1 & 1 & 2 \\ 2 & 1 & 1 & -1 & -2 & 2 \\ -2 & 2 & -1 & 0 & -2 & -2 \\ 1 & 0 & 2 & 2 & -1 & -1 \\ 2 & 2 & -2 & 0 & 2 & 0 \end{bmatrix},$$

found by Karunanayaka and Tang (2018) using a computer search. Let $\boldsymbol{A}$ *be a Hadamard matrix of order* k*. Taking the* $OD(10, 5^6)$ *as* $\boldsymbol{D}_0$ *in Theorem 12.2, we obtain design* $\boldsymbol{D} = \boldsymbol{A} \otimes \boldsymbol{D}_0$*, which is an* $OD(10k, 5^{6k})$*. For* $k = 2, 4,$ *and* 8*, we obtain an* $OD(20, 5^{12})$*, an* $OD(40, 5^{24})$*, and an* $OD(80, 5^{48})$*, respectively.*

Remark 12.1. *Bingham, Sitter, and Tang (2009) also considered designs with higher-order orthogonality. An* $OD(n, s^m)$*,* $\boldsymbol{D} = (\boldsymbol{d}_1, \ldots, \boldsymbol{d}_m)$*, is said to be 3-orthogonal if* $J(\boldsymbol{d}_i, \boldsymbol{d}_j, \boldsymbol{d}_k) = 0$ *for all* i, j, k*. Then it can be shown that design* $\boldsymbol{D} = \boldsymbol{A} \otimes \boldsymbol{D}_0$ *is 3-orthogonal if*

(i) $\boldsymbol{A}$ *has orthogonal columns and* $\boldsymbol{D}_0$ *is 3-orthogonal, or*

(ii) $\boldsymbol{A}$ *is an orthogonal array of strength 3 and* $\boldsymbol{D}_0$ *is orthogonal.*

The second basic method is similar to the first method in that it can also be used to construct large orthogonal designs from small ones, but the two methods are quite different otherwise.

Theorem 12.3. *Let* $\boldsymbol{A}$ *be an* $OA(n, s^{m_1}, 2)$*,* $\boldsymbol{D}_0$ *be an* $OD(s, p^{m_2})$*, and* $\boldsymbol{D}$ *be the design obtained by replacing the* u*th level in each column of* $\boldsymbol{A}$ *with the* u*th row of* $\boldsymbol{D}_0$*. Then* $\boldsymbol{D}$ *is an* $OD(n, p^{m_1 m_2})$*.*

We may take the columns of $\boldsymbol{D}_0$ as a set of orthogonal contrasts for s treatments. If these orthogonal contrasts are used for each factor of s levels, then $\boldsymbol{D}$ is simply the model matrix under design $\boldsymbol{A}$ for the main-effect model with the all-one column removed. The orthogonality of $\boldsymbol{D}$ is hence obvious.

Example 12.2. *Let* $\boldsymbol{A}$ *be an* $OA(25, 5^6, 2)$ *and*

$$D_0 = \begin{bmatrix} -2 & -1 \\ -1 & 2 \\ 0 & 0 \\ 1 & -2 \\ 2 & 1 \end{bmatrix},$$

an $OLH(5, 2)$*. After replacing the five levels of* $\boldsymbol{A}$ *with the five rows of* $\boldsymbol{D}_0$*, we obtain an* $OD(25, 5^{12})$*.*

For the study of orthogonal Latin hypercubes, more general than the existence problem is the determination of the largest m for an OLH(n, m) to exist for given n. Let m^* be the largest such m. Then Corollary 12.1 says that $m^* = 1$ if $n \geq 4$ has the form $n = 4k + 2$, and $m^* \geq 2$ otherwise. Lin, Bingham, Sitter, and Tang (2010) provided stronger results. One of their results is that $m^* \geq 6$ for all $n = 16k + j$ with $k \geq 1$ and $j \neq 2, 6, 10, 14$.

12.3.2 Rotation methods

Orthogonal designs constructed in Section 12.3.1 may have a small number of levels. Rotation methods allow orthogonal designs with more levels, orthogonal Latin hypercubes in particular, to be constructed from orthogonal designs with fewer levels. Many papers have been written on this subject, including Steinberg and Lin (2006), Pang, Liu, and Lin (2009), Lin, Mukerjee, and Tang (2009), Sun and Tang (2017a), and Sun and Tang (2017b). Instead of discussing the methods in their general form, our presentation concentrates on three most important situations.

The first situation is rotation in groups of two columns. Its spirit is reflected in the next lemma.

Lemma 12.1. *Suppose that*

$$\boldsymbol{C} = (\boldsymbol{C}_1, \ldots, \boldsymbol{C}_k) \quad \textit{is an } OD(n, p^{2k}),$$

an orthogonal design with p levels, where each $\boldsymbol{C}_j$ is an $OA(n, p^2, 2)$, a two-column orthogonal array of strength two. Let

$$\boldsymbol{R} = \begin{pmatrix} 1 & -p \\ p & 1 \end{pmatrix}. \tag{12.1}$$

Then the design

$$\boldsymbol{D} = (\boldsymbol{C}_1\boldsymbol{R}, \ldots, \boldsymbol{C}_k\boldsymbol{R})$$

is an $OD\big(n, (p^2)^{2k}\big)$, an orthogonal design with p^2 levels.

The orthogonality of design $\boldsymbol{D}$ is obvious because of the orthogonality of $\boldsymbol{C}$ and $\boldsymbol{R}$. One only needs to show that $\boldsymbol{D}$ has p^2 equally spaced centered levels, which only requires a simple verification. To apply Lemma 12.1, we need to create an orthogonal design $\boldsymbol{C}$ with the required structure that $\boldsymbol{C}$ itself is an orthogonal design and its columns are grouped in pairs, with each pair forming an orthogonal array of strength two.

Let $\boldsymbol{A}$ be an OA$(n, s^{m_1}, 2)$, $\boldsymbol{B}$ be an OD(s, p^{m_2}), and $\boldsymbol{A}_i = (\boldsymbol{a}_{i1}, \ldots, \boldsymbol{a}_{im_2})$ be the matrix obtained from the ith column of $\boldsymbol{A}$ by replacing the uth level with the uth row of $\boldsymbol{B}$, where $u = 1, \ldots, s$. Then according to Theorem 12.3, $(\boldsymbol{A}_1, \ldots, \boldsymbol{A}_{m_1})$ is an OD$(n, p^{m_1 m_2})$.

Arrange the columns of

$$(\boldsymbol{A}_1, \ldots, \boldsymbol{A}_{m_1}) = (\boldsymbol{a}_{11}, \boldsymbol{a}_{12}, \ldots, \boldsymbol{a}_{1m_2}; \boldsymbol{a}_{21}, \boldsymbol{a}_{22}, \ldots, \boldsymbol{a}_{2m_2}; \ldots; \boldsymbol{a}_{m_1 1}, \boldsymbol{a}_{m_1 2}, \ldots, \boldsymbol{a}_{m_1 m_2})$$

in the order of

$$\boldsymbol{a}_{11}, \boldsymbol{a}_{21}, \ldots, \boldsymbol{a}_{m_1 1}; \boldsymbol{a}_{12}, \boldsymbol{a}_{22}, \ldots, \boldsymbol{a}_{m_1 2}; \ldots; \boldsymbol{a}_{1m_2}, \boldsymbol{a}_{2m_2}, \ldots, \boldsymbol{a}_{m_1 m_2}.$$

We then successively collect two columns at a time as $\boldsymbol{C}_1$, $\boldsymbol{C}_2$, and so on, each of which must be an OA$(n, p^2, 2)$. If $m_1 m_2$ is even, then all the $m_1 m_2$ columns are collected; otherwise, there is one leftover column. In any case, we obtain a $\boldsymbol{C} = (\boldsymbol{C}_1, \ldots, \boldsymbol{C}_k)$, an OD$(n, p^{2k})$, where each $\boldsymbol{C}_j$ is an OA$(n, p^2, 2)$ and $k = \lfloor m_1 m_2/2 \rfloor$. Then by Lemma 12.1, design $\boldsymbol{D} = (\boldsymbol{C}_1\boldsymbol{R}, \ldots, \boldsymbol{C}_k\boldsymbol{R})$ is an OD$\big(n, (p^2)^{2k}\big)$, where $\boldsymbol{R}$ is as in (12.1). We summarize the above construction in a theorem.

Theorem 12.4. *Let $\boldsymbol{A}$ be an $OA(n, s^{m_1}, 2)$ and $\boldsymbol{B}$ be an $OD(s, p^{m_2})$. Then design $\boldsymbol{D}$ constructed above using $\boldsymbol{A}$ and $\boldsymbol{B}$ is an $OD\big(n, (p^2)^{2k}\big)$, where $k = \lfloor m_1 m_2/2 \rfloor$. In particular, $\boldsymbol{D}$ is an orthogonal Latin hypercube if $\boldsymbol{A}$ is an orthogonal array of index unity and $\boldsymbol{B}$ is an orthogonal Latin hypercube.*

The method of rotation in groups of two columns as detailed in Lemma 12.1 and Theorem 12.4 was first developed in Lin, Mukerjee, and Tang (2009), and later refined in Sun and Tang (2017a) and Karunanayaka and Tang (2018).

Example 12.3. *Let* $\boldsymbol{A}$ *be an* $OA(25, 5^6, 2)$ *and* $\boldsymbol{B}$ *be the* $OLH(5, 2)$ *in Example 12.2. Then design* $\boldsymbol{D}$ *obtained by applying Theorem 12.4 is an orthogonal design* $OD(25, 25^{12})$, *which is an* $OLH(25, 12)$.

We next consider rotations in groups of four columns, as in the following lemma.

Lemma 12.2. *Suppose that*

$$\boldsymbol{C} = (\boldsymbol{C}_1, \ldots, \boldsymbol{C}_k) \quad \text{is an } OD(n, p^{4k}),$$

where each $\boldsymbol{C}_j$ *is an* $OA(n, p^4, 4)$, *a four-column orthogonal array of strength four. Then the design*

$$\boldsymbol{D} = (\boldsymbol{C}_1\boldsymbol{R}, \ldots, \boldsymbol{C}_k\boldsymbol{R}) \quad \text{is an } OD\big(n, (p^4)^{4k}\big),$$

where

$$\boldsymbol{R} = \begin{pmatrix} p^3 & -p^2 & -p & 1 \\ p^2 & p^3 & -1 & -p \\ p & -1 & p^3 & -p^2 \\ 1 & p & p^2 & p^3 \end{pmatrix}.$$

Through rotation, Lemma 12.2 converts an orthogonal design with p levels into one with p^4 levels. To apply Lemma 12.2, we need to create an orthogonal design $\boldsymbol{C}$ with the structure that its columns are in groups of four, each being a strength-four orthogonal array.

Let $\boldsymbol{A}$ be an $\mathrm{OA}(n, s^{m_1}, 2)$ and $\boldsymbol{B}$ an $\mathrm{OA}(s, p^{m_2}, 2)$. In each column of $\boldsymbol{A}$, if we replace the uth level, for $u = 1, \ldots, s$, by the uth row of $\boldsymbol{B}$, then the resulting matrix $\boldsymbol{E}$ must be an $\mathrm{OA}(n, p^{m_1 m_2}, 2)$. In fact, design $\boldsymbol{E}$ has another property useful for our purpose. Write $\boldsymbol{E}$ as

$$\boldsymbol{E} = (\boldsymbol{E}_1, \boldsymbol{E}_2, \ldots, \boldsymbol{E}_{m_1}), \tag{12.2}$$

where $\boldsymbol{E}_i$ represents the ith group of m_2 columns arising from replacing the levels in the ith column of $\boldsymbol{A}$ by the rows of $\boldsymbol{B}$. Then we have the following result.

Lemma 12.3. *Any four columns of* $\boldsymbol{E}$, *obtained by taking two columns from one group* $\boldsymbol{E}_{i_1}$ *and two columns from another group* $\boldsymbol{E}_{i_2}$, *where* $i_1 \neq i_2$, *must form an* $OA(n, p^4, 4)$, *a strength-four array.*

This result is simple and quite intuitive; no proof is required. Lemma 12.3 says that although $\boldsymbol{E}$ is an orthogonal array of strength two, many of its four-column subarrays in fact have strength four. Using Lemma 12.3, we can re-organize the columns of $\boldsymbol{E}$ into sets of four columns, each being an $\mathrm{OA}(n, p^4, 4)$. To ensure there are no leftover columns, let both m_2 and $m_1 m_2/2$ be even, and write $m_2 = 2k$. This means that m_1 can be any integer if m_2 is a multiple of 4, and m_1 has to be even if m_2 is even but not a multiple of 4. For the $m_2 = 2k$ columns in each group $\boldsymbol{E}_i$, we form k pairs of columns by writing $\boldsymbol{E}_i$ as

$$\boldsymbol{E}_i = (\boldsymbol{P}_{i1}, \boldsymbol{P}_{i2}, \ldots, \boldsymbol{P}_{ik}), \text{ for } i = 1, \ldots, m_1,$$

where each pair $\boldsymbol{P}_{ij}$ is composed of two columns from group $\boldsymbol{E}_i$, and altogether $\boldsymbol{P}_{i1}, \boldsymbol{P}_{i2}, \ldots, \boldsymbol{P}_{ik}$ exhaust all the $m_2 = 2k$ columns of $\boldsymbol{E}_i$. Now order the $m_1 k$ pairs as

$$\boldsymbol{P}_{11}, \boldsymbol{P}_{21}, \ldots, \boldsymbol{P}_{m_1 1}, \boldsymbol{P}_{12}, \boldsymbol{P}_{22}, \ldots, \boldsymbol{P}_{m_1 2}, \ldots, \boldsymbol{P}_{1k}, \boldsymbol{P}_{2k}, \ldots, \boldsymbol{P}_{m_1 k}. \tag{12.3}$$

Since any two successive pairs in the above list are from different groups, if we take two pairs at a time in the order given in (12.3), then we obtain q sets of four columns, each of which is an OA$(n, p^4, 4)$, where $q = m_1k/2 = m_1m_2/4$. Let these sets be $\boldsymbol{C}_1, \boldsymbol{C}_2, \ldots, \boldsymbol{C}_q$, and further, let

$$\boldsymbol{C} = (\boldsymbol{C}_1, \boldsymbol{C}_2, \ldots, \boldsymbol{C}_q). \tag{12.4}$$

It is seen that $\boldsymbol{C}$ and $\boldsymbol{E}$ consist of the same set of m_1m_2 columns. While the columns in $\boldsymbol{E}$ are in m_1 groups of size m_2 with the property as described in Lemma 12.3, the columns of $\boldsymbol{C}$ are in q sets of size 4, each being an OA$(n, p^4, 4)$, where $q = m_1m_2/4$. According to Lemma 12.2, the design

$$\boldsymbol{D} = (\boldsymbol{C}_1\boldsymbol{R}, \boldsymbol{C}_2\boldsymbol{R}, \ldots, \boldsymbol{C}_q\boldsymbol{R}) \tag{12.5}$$

is an OD$(n, (p^4)^{4q})$. We summarize the above construction in a theorem.

Theorem 12.5. *Let $\boldsymbol{A}$ be an $OA(n, s^{m_1}, 2)$ and $\boldsymbol{B}$ be an $OA(s, p^{m_2}, 2)$. Then design $\boldsymbol{D}$ constructed using $\boldsymbol{A}$ and $\boldsymbol{B}$ as given in (12.5) is an $OD(n, (p^4)^{m_1m_2})$. It is an $OLH(n, m_1m_2)$ if both $\boldsymbol{A}$ and $\boldsymbol{B}$ are arrays of index unity.*

Unlike the method of Steinberg and Lin (2006) and Pang, Liu, and Lin (2009), the method presented here, which was developed in Sun and Tang (2017a), requires no use of finite fields and projective geometries. In particular, it does not require s and p to be primes or prime powers.

Example 12.4. *If we take $\boldsymbol{A}$ to be an $OA(81, 9^{10}, 2)$ and $\boldsymbol{B}$ to be an $OA(9, 3^4, 2)$, then design $\boldsymbol{D}$ in Theorem 12.5 is an orthogonal Latin hypercube $OLH(81, 40)$. If we take $\boldsymbol{A}$ to be an $OA(32, 4^8, 2)$ and $\boldsymbol{B}$ to be an $OA(4, 2^2, 2)$, then design $\boldsymbol{D}$ in Theorem 12.5 is an orthogonal design $OD(32, 16^{16})$.*

Corollary 12.2. *Let $\boldsymbol{A}$ and $\boldsymbol{B}$ be as in Theorem 12.5. Suppose that an $OLH(p, m')$ is available. Then an $OD(n, (p^4)^m)$ with $m = m_1m_2m'$ can be constructed.*

Replacing each column of the OLH(p, m') with a copy of $\boldsymbol{D}$, altogether, we can construct m' copies of $\boldsymbol{D}$ with all the columns being mutually orthogonal. This establishes Corollary 12.2.

Example 12.5. *Let $\boldsymbol{A}$ be an $OA(625, 25^{26}, 2)$ and $\boldsymbol{B}$ be an $OA(25, 5^6, 2)$. Theorem 12.5 allows the construction of an $OLH(625, 156)$. Using an $OLH(5, 2)$, we obtain an $OLH(625, 312)$.*

The third situation we consider for the rotation method is also rotation in groups of four columns, but it converts an orthogonal design with p levels into one with p^3 levels.

Lemma 12.4. *Suppose that*

$$\boldsymbol{C} = (\boldsymbol{C}_1, \ldots, \boldsymbol{C}_k) \quad \textit{is an } OD(n, p^{4k}),$$

where each $\boldsymbol{C}_j$ is an $OA(n, p^4, 3)$, a strength-three orthogonal array for four factors. Then the design

$$\boldsymbol{D} = (\boldsymbol{C}_1\boldsymbol{R}, \ldots, \boldsymbol{C}_k\boldsymbol{R}) \quad \textit{is an } OD(n, (p^3)^{4k}),$$

where

$$\boldsymbol{R} = \begin{pmatrix} p^2 & -p & -1 & 0 \\ p & p^2 & 0 & 1 \\ 1 & 0 & p^2 & -p \\ 0 & -1 & p & p^2 \end{pmatrix}.$$

The rotation method in Lemma 12.4 is similar to that in Lemma 12.2 in that for both methods, the orthogonal design $\boldsymbol{C}$ to be rotated has its columns divided into groups of four. But there are two differences. The first is the difference in rotation matrices. The other is that while each group of four columns is required to have strength four in Lemma 12.2, it is only required to have strength three in Lemma 12.4.

We need to create an orthogonal design $\boldsymbol{C}$, required by Lemma 12.4. Let $\boldsymbol{A}$ be an $\mathrm{OA}(n', s^{m_1}, 2)$ and $\boldsymbol{B}$ be an $\mathrm{OD}(s, p^{m_2})$. As in the construction for the rotation in groups of two columns preceding Theorem 12.4, we can obtain an $\mathrm{OD}(n', p^{2k})$ with $k = \lfloor m_1 m_2/2 \rfloor$, where the $2k$ columns are in k pairs, with each pair being an $\mathrm{OA}(n', p^2, 2)$. We now denote this design by $\boldsymbol{E} = (\boldsymbol{E}_1, \ldots, \boldsymbol{E}_k)$ with each $\boldsymbol{E}_j$ being an $\mathrm{OA}(n', p^2, 2)$. Obviously,

$$\boldsymbol{F}_j = \boldsymbol{E}_j + (p-1)/2$$

is also an $\mathrm{OA}(n', p^2, 2)$, but has levels $0, 1, \ldots, p-1$. Let p be a prime and define

$$\boldsymbol{G}_j = \begin{pmatrix} \boldsymbol{F}_j & \boldsymbol{F}_j \\ \boldsymbol{F}_j & 1+\boldsymbol{F}_j \\ \vdots & \vdots \\ \boldsymbol{F}_j & p-1+\boldsymbol{F}_j \end{pmatrix},$$

where all calculations are modulo p. Then $\boldsymbol{G}_j$ is an $\mathrm{OA}(pn', p^4, 3)$, a strength-three orthogonal array of four columns with levels $0, 1, \ldots, p-1$. Furthermore,

$$\boldsymbol{C}_j = \boldsymbol{G}_j - (p-1)/2$$

is also an $\mathrm{OA}(pn', p^4, 3)$, but with centered levels. By Lemma 12.4, design $\boldsymbol{D} = (\boldsymbol{C}_1\boldsymbol{R}, \ldots, \boldsymbol{C}_k\boldsymbol{R})$ is an $\mathrm{OD}(n, (p^3)^{4k})$, where $\boldsymbol{R}$ is given in Lemma 12.4. We summarize the above development in a theorem.

Theorem 12.6. *Let $\boldsymbol{A}$ be an $OA(n', s^{m_1}, 2)$ and $\boldsymbol{B}$ be an $OD(s, p^{m_2})$. Then design $\boldsymbol{D}$ in the above construction is an $OD(n, (p^3)^{4k})$, where $n = pn'$ and $k = \lfloor m_1 m_2/2 \rfloor$. Design $\boldsymbol{D}$ is an orthogonal Latin hypercube if $n' = s^2$ and $s = p$.*

The method embodied in Lemma 12.4 and Theorem 12.6 is a special case of a general rotation method developed in Sun and Tang (2017b).

Example 12.6. *Let $\boldsymbol{A}$ be an $OA(9, 3^4, 2)$ and $\boldsymbol{B}$ be an $OD(3, 3^1)$. Here $\boldsymbol{B} = (-1, 0, 1)^T$ is an orthogonal design with one column. Then Theorem 12.6 allows the construction of an $OD(27, 27^8)$, or an $OLH(27, 8)$.*

12.3.3 Other methods

Sun, Liu, and Lin (2009) presented a method of constructing two orthogonal Latin hypercubes with 2^c factors for any $c \geq 1$. Let $\boldsymbol{X}^*$ denote the matrix obtained by switching the signs in the top half of a matrix $\boldsymbol{X}$ with an even number of rows. Their method proceeds as follows.

First let

$$\boldsymbol{S}_1 = \begin{pmatrix} 1 & 1 \\ 1 & -1 \end{pmatrix} \quad \text{and} \quad \boldsymbol{T}_1 = \begin{pmatrix} 1 & 2 \\ 2 & -1 \end{pmatrix}.$$

For $c \geq 2$, let

$$\boldsymbol{S}_c = \begin{pmatrix} \boldsymbol{S}_{c-1} & -\boldsymbol{S}^*_{c-1} \\ \boldsymbol{S}_{c-1} & \boldsymbol{S}^*_{c-1} \end{pmatrix} \quad \text{and} \quad \boldsymbol{T}_c = \begin{pmatrix} \boldsymbol{T}_{c-1} & -(\boldsymbol{T}^*_{c-1} + 2^{c-1}\boldsymbol{S}^*_{c-1}) \\ \boldsymbol{T}_{c-1} + 2^{c-1}\boldsymbol{S}_{c-1} & \boldsymbol{T}^*_{c-1} \end{pmatrix}.$$

Finally, obtain

$$\boldsymbol{L}_c = \begin{pmatrix} \boldsymbol{T}_c \\ \boldsymbol{0} \\ -\boldsymbol{T}_c \end{pmatrix} \quad \text{and} \quad \boldsymbol{K}_c = \begin{pmatrix} \boldsymbol{T}_c - \boldsymbol{S}_c/2 \\ -(\boldsymbol{T}_c - \boldsymbol{S}_c/2) \end{pmatrix}.$$

Theorem 12.7. *Design $\boldsymbol{L}_c$ is an OLH$(2^{c+1}+1, 2^c)$ and $\boldsymbol{K}_c$ is an OLH$(2^{c+1}, 2^c)$. They are in fact 3-orthogonal.*

Lin, Bingham, Sitter, and Tang (2010) proposed a method for constructing orthogonal Latin hypercubes using

$$\boldsymbol{L} = \boldsymbol{A} \otimes \boldsymbol{B} + n_2 \boldsymbol{E} \otimes \boldsymbol{F},$$

where $\boldsymbol{A}_{n_1 \times m_1}$ and $\boldsymbol{F}_{n_2 \times m_2}$ are matrices of ± 1 with orthogonal columns, and $\boldsymbol{B}_{n_2 \times m_2}$ and $\boldsymbol{E}_{n_1 \times m_1}$ are orthogonal Latin hypercubes. They showed that the resulting $\boldsymbol{L}$ is an orthogonal Latin hypercube of $n_1 n_2$ runs for $m_1 m_2$ factors, if some further conditions on $\boldsymbol{A}, \boldsymbol{B}, \boldsymbol{E}$, and $\boldsymbol{F}$ are met. One application of this result gives rise to the following theorem.

Theorem 12.8. *Suppose that an OLH(n, m) is available, where n is a multiple of 4, such that a Hadamard matrix of order n exists. Then the following orthogonal Latin hypercubes, an OLH$(2n, m)$, an OLH$(4n, 2m)$, an OLH$(8n, 4m)$, and an OLH$(16n, 8m)$, can all be constructed.*

12.4 OA-Based Designs

Let $\boldsymbol{x}_1, \ldots, \boldsymbol{x}_n$ be a set of n points in the design region, assumed to be the m-dimensional unit cube $[0,1)^m$. One way to distribute the design points uniformly is to make the empirical distribution based on the design points as close as possible to the theoretical uniform distribution, with both distributions defined on $[0,1)^m$ and represented by the cumulative distribution functions. This idea gives rise to various discrepancy-based criteria for the selection of space-filling designs.

Another idea of making the design points uniformly distributed in $[0,1)^m$ is to match up the empirical probabilities with the theoretical ones. Let $R_1, \ldots, R_k$ form an equal-volume partition of $[0,1)^m$, meaning that $R_1, \ldots, R_k$ are mutually exclusive and exhaustive with each R_j having a volume of $1/k$. Then a set of n design points, $\boldsymbol{x}_1, \ldots, \boldsymbol{x}_n$, is said to be uniformly distributed in $[0,1)^m$ with respect to the equal-volume partition

$$[0,1)^m = R_1 \cup \cdots \cup R_k,$$

if each R_j contains exactly n/k design points, where k is a divisor of n. As a result, the empirical probability of each R_j, namely the proportion of points in each R_j, is $(n/k)/n = 1/k$, a perfect match with the theoretical probability given by the uniform distribution. It goes without saying that we prefer a design with its points uniformly distributed in $[0,1)^m$ with respect to various meaningful equal-volume partitions. This idea of making design points uniformly distributed in the design region spawns a class of space-filling designs, namely orthogonal array-based designs or OA-based designs for short, if we focus on low-dimensional projections.

Being orthogonal arrays of strength one, Latin hypercubes are the simplest OA-based designs. For notational brevity, first consider a Latin hypercube $\boldsymbol{D} = (d_{ij})$ with n runs for $m = 2$ factors, where each column of $\boldsymbol{D}$ is a permutation of $0, 1, \ldots, n-1$. Based on this Latin hypercube, n design points $\boldsymbol{x}_1, \ldots, \boldsymbol{x}_n$ in the unit square $[0,1)^2$ are given by the rows of $\boldsymbol{X} = (x_{ij})$, where $x_{ij} = (d_{ij} + 0.5)/n$. We see that the n design points $\boldsymbol{x}_1, \ldots, \boldsymbol{x}_n$ are uniform with respect to both of the following two equal-volume partitions of $[0,1)^2$:

$$[0,1)^2 = \bigcup_{i=1}^{n} \left(\left[\frac{i-1}{n}, \frac{i}{n}\right) \times [0,1) \right) \tag{12.6}$$

and

$$[0,1)^2 = \bigcup_{i=1}^{n} \left([0,1) \times \left[\frac{i-1}{n}, \frac{i}{n}\right) \right). \tag{12.7}$$

To put it in another way, when projected onto any dimension, the design points $\boldsymbol{x}_1, \ldots, \boldsymbol{x}_n$ are uniform with respect to the equal-volume partition $[0,1) = \bigcup_{i=1}^{n} \big[(i-1)/n, i/n\big)$ in that each interval $\big[(i-1)/n, i/n\big)$ contains exactly one point. The design literature often uses the sentence that the design achieves the maximum stratification in each dimension to describe this property. In general, an n-run Latin hypercube for m factors therefore achieves the maximum stratification in each of the m dimensions. The discussion above equally applies if the design points are obtained via $x_{ij} = (d_{ij} + U_{ij})/n$, where the U_{ij}'s are independently and identically distributed as a uniform distribution on $[0,1)$. We will make no further mention of this possibility from now on.

Now consider an $\mathrm{OA}(n, s^m, 2)$, $\boldsymbol{D} = (d_{ij})$. Again we first take $m = 2$ for easy exposition. Then the n design points $\boldsymbol{x}_1, \ldots, \boldsymbol{x}_n$ in the unit square $[0,1)^2$ are given by the rows of $\boldsymbol{X} = (x_{ij})$ where $x_{ij} = (d_{ij} + 0.5)/s$. We see that $\boldsymbol{x}_1, \ldots, \boldsymbol{x}_n$ are uniform with respect to the following equal-volume partition of $[0,1)^2$:

$$[0,1)^2 = \bigcup_{i=1}^{s} \bigcup_{j=1}^{s} \left(\left[\frac{i-1}{s}, \frac{i}{s}\right) \times \left[\frac{j-1}{s}, \frac{j}{s}\right) \right). \tag{12.8}$$

We say that the design achieves stratifications on $s \times s$ grids. In general, an $\mathrm{OA}(n, s^m, 2)$ achieves stratifications on $s \times s$ grids in every two-dimensional projection.

From an $\mathrm{OA}(n, s^m, 2)$ $\boldsymbol{D}'$, one can construct an OA-based Latin hypercube $\boldsymbol{D}$. This is done as follows: in each column of $\boldsymbol{D}'$, we replace the $\alpha = n/s$ occurrences of level j by any permutation of

$$\alpha j, \alpha j + 1, \ldots, \alpha j + \alpha - 1, \quad \text{for } j = 0, 1, \ldots, s-1.$$

The resulting design $\boldsymbol{D} = (d_{ij})$ is a Latin hypercube. As before, $\boldsymbol{D}$ gives rise to the design points $\boldsymbol{x}_1, \ldots, \boldsymbol{x}_n$ in $[0,1)^m$, which are the rows of $\boldsymbol{X} = (x_{ij})$ with $x_{ij} = (d_{ij} + 0.5)/n$. Then we see that these design points achieve the maximum stratification when projected onto any dimension and also achieve stratifications on $s \times s$ grids in every two-dimensional projection. If $\boldsymbol{D}'$ is an $\mathrm{OA}(n, s^m, 3)$, an array of strength 3, then the resulting Latin hypercube $\boldsymbol{D}$ achieves stratifications on $s \times s \times s$ grids in every three-dimensional projection in addition to achieving the maximum stratification in every dimension.

Example 12.7. *We give a small example to illustrate OA-based Latin hypercubes. In the following, $\boldsymbol{D}$ is a Latin hypercube based on an orthogonal array $\boldsymbol{D}'$, with $\boldsymbol{X}$ being the corresponding design scaled into $[0,1)^2$:*

$$\boldsymbol{D}' = \begin{pmatrix} 0 & 0 \\ 0 & 1 \\ 1 & 0 \\ 1 & 1 \end{pmatrix}, \quad \boldsymbol{D} = \begin{pmatrix} 0 & 1 \\ 1 & 3 \\ 2 & 0 \\ 3 & 2 \end{pmatrix}, \quad \textit{and} \quad \boldsymbol{X} = \begin{pmatrix} 1/8 & 3/8 \\ 3/8 & 7/8 \\ 5/8 & 1/8 \\ 7/8 & 5/8 \end{pmatrix}.$$

The four design points, given by the rows of $\boldsymbol{X}$, are displayed in Figure 12.1, viewed through three different equal-volume partitions given in (12.6), (12.7), and (12.8).

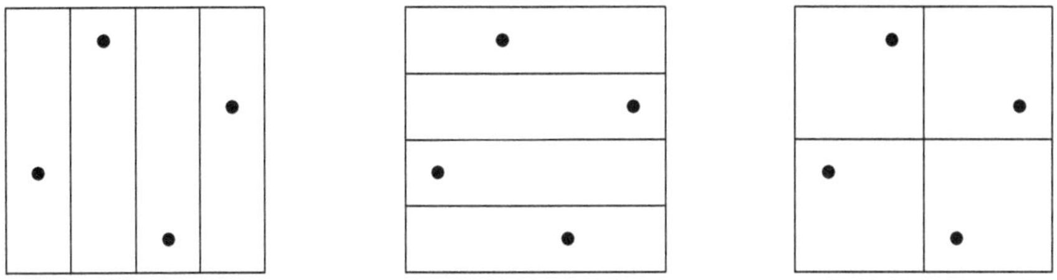

FIGURE 12.1
An OA-based Latin hypercube viewed through three different equal-volume partitions

OA-based Latin hypercubes are superior members in the whole class of Latin hypercubes. Chen and Tang (2022a) considered a general problem of evaluating OA-based designs in the whole class of balanced designs under a large collection of design criteria. In broad strokes, we provide a description of their result. Starting with an OA$(n, s^m, 2)$, by level permutation and level expansion one can obtain a class of balanced q-level designs, where $q = \alpha s$ with $\alpha \geq 2$ being an integer. Specifically, this is done by first permuting the s levels in each column of the OA$(n, s^m, 2)$, and then replacing the n/s occurrences of level $j = 0, 1, \ldots, s-1$ by any permutation of

$$(\alpha j, \ldots, \alpha j, \alpha j+1, \ldots, \alpha j+1, \ldots, \alpha j+\alpha-1, \ldots, \alpha j+\alpha-1),$$

where each distinct entry occurs n/q times. These OA-based designs are OA-based Latin hypercubes if $q = n$. Consider a general criterion

$$\chi(\boldsymbol{D}) = \frac{1}{m(m-1)} \sum_{u \neq v} \gamma(\boldsymbol{D}_{uv}),$$

where $\boldsymbol{D} = (d_{ij})$ denotes any balanced n-run design, $\boldsymbol{D}_{uv}$ is its projection design onto columns u and v, and

$$\gamma(\boldsymbol{D}_{uv}) = \gamma_0 + \frac{\gamma_1}{n} \sum_{i=1}^{n} g(d_{iu}) g(d_{iv}) + \frac{\gamma_2}{n^2} \sum_{i=1}^{n} \sum_{j=1}^{n} f(d_{iu}, d_{ju}) f(d_{iv}, d_{jv}),$$

where γ_0 and γ_1 are arbitrary real constants, f and g are arbitrary real-valued functions, and we only require that $\gamma_2 > 0$. The general criterion $\chi(\boldsymbol{D})$ defined above represents a very broad collection of criteria. In particular, with appropriate choices of constants γ_0, γ_1, and γ_2 as well as functions f and g, $\chi(\boldsymbol{D})$ gives rise to the following criteria:

(i) *the orthogonality criterion*

$$\rho(\boldsymbol{D}) = \frac{1}{\binom{m}{2}} \sum_{u<v} \rho_{uv}^2,$$

where ρ_{uv} is the correlation between columns u and v;

(ii) *a distance criterion*

$$\kappa(\boldsymbol{D}) = \frac{1}{\binom{n}{2}} \sum_{i<j} (\Delta_{ij} - \bar{\Delta})^2,$$

where Δ_{ij} is the squared L_2-distance between rows i and j, $\bar{\Delta}$ is the average, and $\kappa(\boldsymbol{D})$ is the variance of these pairwise distances;

(iii) *the uniform projection criterion* (Sun, Wang, and Xu, 2019)

$$\psi(\boldsymbol{D}) = \frac{1}{\binom{m}{2}} \sum_{u<v} \mathrm{CD}(\boldsymbol{D}_{uv}),$$

where $\mathrm{CD}(\boldsymbol{D}_{uv})$ is the centered L_2 discrepancy of the projection design $\boldsymbol{D}_{uv}$, introduced in Section 12.2.

We refer to Chen and Tang (2022a) for details regarding how to produce $\rho(\boldsymbol{D})$, $\kappa(\boldsymbol{D})$, and $\psi(\boldsymbol{D})$ from $\chi(\boldsymbol{D})$.

Let $\bar{\chi}_U$ be the average of the χ values of all n-run balanced designs (or U-type designs) for m q-level factors, and $\bar{\chi}_{OA}$ be the average of the χ values of all OA-based designs obtained from an $\mathrm{OA}(n, s^m, 2)$ by permuting the s levels in each column and then expanding the s levels to q levels. The following result was obtained by Chen and Tang (2022a).

Theorem 12.9. *The values of $\bar{\chi}_U$ and $\bar{\chi}_{OA}$ are constants, only depending on n, s, q, γ_2, and the f function. In addition, we have that*

$$\bar{\chi}_U - \bar{\chi}_{OA} > 0.$$

Theorem 12.9 shows that OA-based designs are superior members in the whole class of balanced designs in the sense that on average a random OA-based design is better than a random balanced design. Worth noting in the theorem is that $\bar{\chi}_{OA}$ does not depend on the particular $\mathrm{OA}(n, s^m, 2)$ used for constructing OA-based designs and that $\bar{\chi}_U$, $\bar{\chi}_{OA}$, and hence $\bar{\chi}_U - \bar{\chi}_{OA}$ do not depend on m, γ_0, γ_1, and the g function.

The χ criterion introduced earlier arises from evaluating all two-dimensional projection designs. Under the χ criterion, all $\mathrm{OA}(n, s^m, 2)$s perform the same in terms of the qualities of the classes of OA-based designs they generate. If a more general design criterion is used, different $\mathrm{OA}(n, s^m, 2)$s may behave differently. The line of research regarding the choice of orthogonal arrays for constructing OA-based designs via level permutation and level expansion was carried out by Zhou and Xu (2014), Xiao and Xu (2018), Jiang and Ai (2017), and Wang, Wang, Yuan, and Xiao (2021). Again in broad strokes, we present their findings.

Let $\boldsymbol{D}$ denote a q-level balanced design. Consider a general design criterion

$$\phi(\boldsymbol{D}) = \frac{1}{n^2} \sum_{i=1}^{n} \sum_{j=1}^{n} F(\boldsymbol{x}_i, \boldsymbol{x}_j),$$

where $\boldsymbol{x}_i$ is the ith row of $\boldsymbol{X} = (\boldsymbol{D} + 0.5)/q$, $F(\boldsymbol{x}_i, \boldsymbol{x}_j) = \prod_{k=1}^{m} f(x_{ik}, x_{jk})$, and $f(x, y)$ is nonnegative and further satisfies that

$$f(x, x) + f(y, y) > f(x, y) + f(y, x) \text{ for any } x, y \in [0, 1] \text{ and } x \neq y.$$

This gives a class of criteria that includes a general distance criterion that can be thought of as an extension of the maximin distance criterion. With proper choices of f, the ϕ criterion can also give rise to several discrepancy criteria including the centered L_2 and wrap-around L_2 discrepancies. We refer to Zhou and Xu (2014) for details.

Now consider all q-level balanced designs obtained by first applying level permutation and then level expansion to a given s-level balanced design $\boldsymbol{D}$. Let $\bar{\phi}(\boldsymbol{D})$ be the average of the ϕ values of all these q-level designs.

Theorem 12.10. *We have that*

$$\bar{\phi}(\boldsymbol{D}) = a + \sum_{j=2}^{m} b^j A_j(\boldsymbol{D}),$$

where a and b $(0 < b < 1)$ are constants, depending on n, m, s, q, and the f function, and $(A_2(\boldsymbol{D}), \ldots, A_m(\boldsymbol{D}))$ is the wordlength pattern of design $\boldsymbol{D}$.

Theorem 12.10 shows that minimum G_2-aberration designs with fewer levels should perform very well when used to construct designs with more levels via level permutation and level expansion. In particular, minimum G_2-aberration orthogonal arrays are excellent choices for the construction of OA-based designs.

In view of the projection justification of minimum G_2-aberration discussed in Section 10.4, Theorems 12.9 and 12.10 are not very surprising. Computer experiments call for designs with many levels and orthogonal arrays with many levels are often too expensive for researchers to afford. Using orthogonal arrays with a small number of levels to construct OA-based designs with many levels provides a compromised solution.

Latin hypercube sampling and OA-based Latin hypercube sampling are sampling methods that select a random Latin hypercube and a random OA-based Latin hypercube, respectively, for Monte Carlo integration. Stein (1987), Owen (1992), and Tang (1993) showed that these sampling methods achieve a reduction in the Monte Carlo variance compared with simple random sampling.

12.5 Strong Orthogonal Arrays

We have been using the notation $\mathrm{OA}(n, s^m, t)$ to denote a strength-t orthogonal array of n runs for m s-level factors. An equally popular notation for this array is $\mathrm{OA}(n, m, s, t)$, which we now adopt in the rest of the chapter. Correspondingly for asymmetric orthogonal arrays, we will use $\mathrm{OA}(n, m, s_1 \times \cdots \times s_m, t)$ instead of the simpler $\mathrm{OA}(n, s_1 \times \cdots \times s_m, t)$. This change in notation is to facilitate our discussion of strong orthogonal arrays.

We have seen in Section 12.4 that orthogonal arrays are useful for constructing space-filling designs. Strong orthogonal arrays, introduced by He and Tang (2013), offer an even better class of designs in terms of their stratification properties in lower dimensions.

Definition 12.1. *An $n \times m$ matrix with levels from $\{0, 1, \ldots, s^t - 1\}$ is called a strong orthogonal array of size n, m factors, s^t levels, and strength t if any subarray of g columns for any g with $1 \leq g \leq t$ can be collapsed into an $OA(n, g, s^{u_1} \times s^{u_2} \times \cdots \times s^{u_g}, g)$ for any positive integers $u_1, \ldots, u_g$ with $u_1 + \cdots + u_g = t$, where collapsing into s^{u_j} levels is done by $\lfloor a/s^{t-u_j} \rfloor$ for $a = 0, 1, \ldots, s^t - 1$. We use $SOA(n, m, s^t, t)$ to denote such an array.*

Taking $g = t$ and $u_1 = \cdots = u_t = 1$ in the definition, we see that an $\mathrm{SOA}(n, m, s^t, t)$ can be collapsed into an $\mathrm{OA}(n, m, s, t)$. This shows that $n = \lambda s^t$ for an integer λ. We call λ the index of the strong orthogonal array.

An $\mathrm{SOA}(8, 3, 8, 3)$ is displayed below:

$$\begin{bmatrix} 0 & 0 & 0 \\ 2 & 3 & 6 \\ 3 & 6 & 2 \\ 1 & 5 & 4 \\ 6 & 2 & 3 \\ 4 & 1 & 5 \\ 5 & 4 & 1 \\ 7 & 7 & 7 \end{bmatrix}.$$

It can be checked that

(i) the whole array becomes an OA(8, 3, 2, 3) after the eight levels are collapsed into two levels according to $\lfloor a/4 \rfloor = 0$ for $a = 0, 1, 2, 3$ and $\lfloor a/4 \rfloor = 1$ for $a = 4, 5, 6, 7$;

(ii) any subarray of two columns can be collapsed into an $\text{OA}(8, 2, 2 \times 4, 2)$ as well as an $\text{OA}(8, 2, 4 \times 2, 2)$, where collapsing into two levels is done by $\lfloor a/4 \rfloor$ and collapsing into four levels is done using $\lfloor a/2 \rfloor$; and

(iii) any subarray of one column is an OA(8, 1, 8, 1).

Strong orthogonal arrays (SOAs) are motivated by the desire of having better stratification properties in lower dimensions than ordinary orthogonal arrays. The idea of these arrays is not entirely new and in fact has an origin in quasi-Monte Carlo methods. An elementary interval in base s is an interval in $[0, 1)^m$ of the form

$$\prod_{j=1}^{m} \left[\frac{c_j}{s^{d_j}}, \frac{c_j + 1}{s^{d_j}} \right),$$

where nonnegative integers c_j and d_j satisfy $0 \le c_j < s^{d_j}$. For $0 \le w \le k$, a (w, k, m)-net in base s is a set of s^k points in $[0, 1)^m$ such that every elementary interval in base s of volume s^{w-k} contains exactly s^w points. Nets and related sequences were first defined by Sobol' (1967) for $s = 2$ and later by Niederreiter (1987) for general s in the literature of quasi-Monte Carlo methods.

Previously in Section 12.4, we have used equal-volume partitions of the design region $[0, 1)^m$ to measure the uniformity of design points. The definition of a (w, k, m)-net using elementary intervals is just a mathematically concise way of doing the same. We illustrate this through a simple example. Take $w = 0$, $k = 2$, and $m = 2$ in the definition of nets. Then a $(0, 2, 2)$-net in base s is a set of s^2 points in $[0, 1)^2$ such that every elementary interval of volume $s^{w-k} = 1/s^2$ contains exactly one point. Elementary intervals of volume $1/s^2$ must be one of the three types:

$$\left[\frac{i}{s^2}, \frac{i+1}{s^2} \right) \times [0, 1) \text{ for } i = 0, 1, \ldots, s^2 - 1,$$

$$[0, 1) \times \left[\frac{j}{s^2}, \frac{j+1}{s^2} \right) \text{ for } j = 0, 1, \ldots, s^2 - 1, \text{ and}$$

$$\left[\frac{i}{s}, \frac{i+1}{s} \right) \times \left[\frac{j}{s}, \frac{j+1}{s} \right) \text{ for } i = 0, 1, \ldots, s - 1 \text{ and } j = 0, 1, \ldots, s - 1.$$

We see that the s^2 elementary intervals of each type form an equal-volume partition of $[0, 1)^2$, with respect to which the points are uniformly distributed. The relationship between strong orthogonal arrays and (w, k, m)-nets in base s is in the next theorem.

Theorem 12.11. *If $\lambda = s^w$ for an integer w, then the existence of an $\text{SOA}(\lambda s^t, m, s^t, t)$ is equivalent to that of a (w, k, m)-net in base s, where $k = w + t$.*

Proof. Let $\boldsymbol{A} = (a_{ij})$ be an $\text{SOA}(n, m, s^t, t)$. Then we can easily check that the set of $n = \lambda s^t = s^{w+t}$ points,

$$(x_{i1}, \ldots, x_{im}) \quad \text{for } i = 1, \ldots, n, \text{ where } x_{ij} = (a_{ij} + 0.5)/s^t,$$

is a $(w, w+t, m)$-net in base s. Conversely, suppose that a set of $n = s^k$ points $(x_{i1}, \ldots, x_{im})$ for $i = 1, \ldots, n = s^k$ in $[0, 1)^m$ is a (w, k, m)-net in base s. Then it can be verified that the array

$$\boldsymbol{A} = (a_{ij}) \quad \text{where } a_{ij} = \lfloor x_{ij} s^{k-w} \rfloor,$$

is an $\text{SOA}(s^k, m, s^{k-w}, k - w)$. □

SOAs are defined without restricting the index to be a power of s; as a result they provide a more general concept than (w, k, m)-nets. Another major advantage of SOAs is that they are formulated in design language. This not only makes them easily accessible to design researchers and practitioners but also creates an opportunity for the development of new types of designs. In the next section, we will discuss several variants of SOAs that would not seem possible without the notion of SOAs.

Lawrence (1996) showed that nets can be completely characterized by generalized orthogonal arrays. As generalizations of nets, SOAs can also be characterized by generalized orthogonal arrays. Let $\boldsymbol{B}$ be an $n \times (tm)$ matrix with entries from $\{0, 1, \ldots, s-1\}$, where the tm columns $\boldsymbol{b}$'s are put into m groups of t columns each, i.e.

$$\boldsymbol{B} = (\boldsymbol{B}_1, \ldots, \boldsymbol{B}_m) \quad \text{where } \boldsymbol{B}_i = (\boldsymbol{b}_{i1}, \ldots, \boldsymbol{b}_{it}).$$

Then $\boldsymbol{B}$ is called a generalized orthogonal array of size n, m constraints, s levels, and strength t if the matrix $\boldsymbol{B}^*$ consisting of t columns $\boldsymbol{b}_{ij}$'s, where $i = i_1, \ldots, i_g$ and $j = 1, \ldots, u_i$, is an orthogonal array of strength t for any $1 \le g \le t$, any $1 \le i_1 < \cdots < i_g \le m$, and any positive integers u's with $u_1 + \cdots + u_g = t$. Note that $\boldsymbol{B}^*$ consists of the first u_1 columns from $\boldsymbol{B}_{i_1}$, the first u_2 columns from $\boldsymbol{B}_{i_2}$, and so on. We use GOA(n, m, s, t) to denote such an array.

We give a simple illustrative example. Let $\boldsymbol{a}_1, \boldsymbol{a}_2, \boldsymbol{a}_3, \boldsymbol{a}_4$ be the four columns of an OA$(8, 4, 2, 3)$. Then $\boldsymbol{B} = (\boldsymbol{B}_1, \boldsymbol{B}_2, \boldsymbol{B}_3)$ is a GOA$(8, 3, 2, 3)$, where $\boldsymbol{B}_1 = (\boldsymbol{a}_1, \boldsymbol{a}_4, \boldsymbol{a}_2)$, $\boldsymbol{B}_2 = (\boldsymbol{a}_2, \boldsymbol{a}_4, \boldsymbol{a}_3)$, and $\boldsymbol{B}_3 = (\boldsymbol{a}_3, \boldsymbol{a}_4, \boldsymbol{a}_1)$.

For an ordinary orthogonal array OA(n, m, s, t), any submatrix of t columns is an orthogonal array of strength t. For a generalized orthogonal array GOA(n, m, s, t), which has a total of mt columns, only a special collection of submatrices of t columns is required to have this property. Within each of the m groups of columns, the selection of columns has to be done in an orderly fashion and no column is allowed to be skipped. This is why generalized orthogonal arrays are also called ordered orthogonal arrays (Mullen and Schmid, 1996).

SOAs can be completely characterized by generalized orthogonal arrays, as shown in the next result.

Theorem 12.12. *Let* $\boldsymbol{B} = \{(\boldsymbol{b}_{11}, \ldots, \boldsymbol{b}_{1t}); \ldots; (\boldsymbol{b}_{m1}, \ldots, \boldsymbol{b}_{mt})\}$ *be a GOA*(n, m, s, t). *Then* $\boldsymbol{A} = (\boldsymbol{a}_1, \ldots, \boldsymbol{a}_m)$ *is an SOA*(n, m, s^t, t), *where*

$$\boldsymbol{a}_i = \sum_{j=1}^{t} \boldsymbol{b}_{ij} s^{t-j}. \tag{12.9}$$

Conversely, if $\boldsymbol{A} = (\boldsymbol{a}_1, \ldots, \boldsymbol{a}_m)$ *is an SOA*(n, m, s^t, t), *then*

$$\boldsymbol{B} = \{(\boldsymbol{b}_{11}, \ldots, \boldsymbol{b}_{1t}); \ldots; (\boldsymbol{b}_{m1}, \ldots, \boldsymbol{b}_{mt})\}$$

is a GOA(n, m, s, t), *where* $\boldsymbol{b}_{i1}, \ldots, \boldsymbol{b}_{it}$ *for each given* $i = 1, \ldots, m$ *are uniquely determined by* $\boldsymbol{a}_i$ *through (12.9).*

Theorem 12.12 is a more general version of a result in Lawrence (1996). The next result, also from Lawrence (1996), gives a construction of generalized orthogonal arrays from ordinary orthogonal arrays.

Lemma 12.5. *If an OA*(n, m, s, t) *exists, then a GOA*(n, m', s, t) *can be constructed, where*

$$m' = \begin{cases} \lfloor m/e \rfloor & \text{if } t = 2e \text{ is even,} \\ \lfloor (m-1)/e \rfloor & \text{if } t = 2e+1 \text{ is odd.} \end{cases}$$

Proof. Let $\boldsymbol{C} = (\boldsymbol{c}_1, \ldots, \boldsymbol{c}_m)$ be an OA(n, m, s, t). For the most useful cases $t = 2, 3, 4$, we present in detail the construction of a GOA(n, m', s, t),

$$\boldsymbol{B} = \{(\boldsymbol{b}_{11}, \ldots, \boldsymbol{b}_{1t}); \ldots; (\boldsymbol{b}_{m'1}, \ldots, \boldsymbol{b}_{m't})\},$$

where m' is as in Lemma 12.5. This should be sufficient for the reader to see the generality of the method.

For $t = 2$, we have $m' = m$. The $\boldsymbol{b}_{ij}$'s in

$$\boldsymbol{B} = \{(\boldsymbol{b}_{11}, \boldsymbol{b}_{12}); \ldots; (\boldsymbol{b}_{m1}, \boldsymbol{b}_{m2})\}$$

are obtained by taking

$$(\boldsymbol{b}_{11}, \ldots, \boldsymbol{b}_{m1}) = (\boldsymbol{c}_1, \ldots, \boldsymbol{c}_m) \quad \text{and} \quad (\boldsymbol{b}_{12}, \ldots, \boldsymbol{b}_{m2}) = (\boldsymbol{c}_2, \ldots, \boldsymbol{c}_m, \boldsymbol{c}_1).$$

For $t = 3$, $m' = m - 1$. We obtain

$$\boldsymbol{B} = \{(\boldsymbol{b}_{11}, \boldsymbol{b}_{12}, \boldsymbol{b}_{13}); \ldots; (\boldsymbol{b}_{m'1}, \boldsymbol{b}_{m'2}, \boldsymbol{b}_{m'3})\}$$

by taking

$$(\boldsymbol{b}_{11}, \ldots, \boldsymbol{b}_{m'1}) = (\boldsymbol{c}_1, \ldots, \boldsymbol{c}_{m'}), \quad (\boldsymbol{b}_{12}, \ldots, \boldsymbol{b}_{m'2}) = (\boldsymbol{c}_m, \ldots, \boldsymbol{c}_m),$$

and

$$(\boldsymbol{b}_{13}, \ldots, \boldsymbol{b}_{m'3}) = (\boldsymbol{c}_2, \ldots, \boldsymbol{c}_{m'}, \boldsymbol{c}_1).$$

For $t = 4$, we have $m' = \lfloor m/2 \rfloor$. We obtain

$$\boldsymbol{B} = \{(\boldsymbol{b}_{11}, \boldsymbol{b}_{12}, \boldsymbol{b}_{13}, \boldsymbol{b}_{14}); \ldots; (\boldsymbol{b}_{m'1}, \boldsymbol{b}_{m'2}, \boldsymbol{b}_{m'3}, \boldsymbol{b}_{m'4})\}$$

by taking

$$(\boldsymbol{b}_{11}, \ldots, \boldsymbol{b}_{m'1}) = (\boldsymbol{c}_1, \ldots, \boldsymbol{c}_{m'}), \quad (\boldsymbol{b}_{12}, \ldots, \boldsymbol{b}_{m'2}) = (\boldsymbol{c}_{m'+1}, \ldots, \boldsymbol{c}_{2m'}),$$

$$(\boldsymbol{b}_{13}, \ldots, \boldsymbol{b}_{m'3}) = (\boldsymbol{c}_{m'+2}, \ldots, \boldsymbol{c}_{2m'}, \boldsymbol{c}_{m'+1}), \quad \text{and } (\boldsymbol{b}_{14}, \ldots, \boldsymbol{b}_{m'4}) = (\boldsymbol{c}_2, \ldots, \boldsymbol{c}_{m'}, \boldsymbol{c}_1).$$

□

Theorem 12.12 and Lemma 12.5 together imply that strong orthogonal arrays can be constructed from ordinary orthogonal arrays.

Theorem 12.13. *If an OA(n, m, s, t) exists, then an SOA(n, m', s^t, t) can be constructed for*

$$m' = \begin{cases} \lfloor m/e \rfloor & \text{if } t = 2e \text{ is even,} \\ \lfloor (m-1)/e \rfloor & \text{if } t = 2e+1 \text{ is odd.} \end{cases}$$

In the same way as constructing Latin hypercubes from an ordinary orthogonal array, one can construct Latin hypercubes from an SOA(n, m, s^t, t) by expanding its s^t levels into n levels. The Latin hypercubes generated this way are called SOA-based Latin hypercubes. An SOA(n, m, s^t, t) promises stratification of design points on $s^{u_1} \times \cdots \times s^{u_g}$ grids for any positive u_j's with $u_1 + \cdots + u_g = t$ when projected onto any $g \leq t$ dimensions. A Latin hypercube based on an SOA(n, m, s^t, t) further maximizes the stratification of design points for any one-dimensional projection, while retaining all the stratification properties the SOA can offer.

A Latin hypercube based on an SOA$(n, m, s^2, 2)$ achieves stratifications on $s \times s$ grids in any two-dimensional projection, and so does a Latin hypercube based on an OA$(n, m, s, 2)$. Consequently, Latin hypercubes based on an OA$(n, m, s, 2)$ do as well as those based on an SOA$(n, m, s^2, 2)$ in terms of both one- and two-dimensional space-filling properties. For higher strengths, SOA-based Latin hypercubes are better than OA-based Latin hypercubes. Consider two Latin hypercubes, one based on an OA$(n, m, s, 3)$ and the other based on an SOA$(n, m, s^3, 3)$. Both achieve stratifications on $s \times s \times s$ grids in any three-dimensional projection. But the SOA-based achieves stratifications on $s^2 \times s$ and $s \times s^2$ grids in any two-dimensional projection, while the OA-based only promises stratifications on $s \times s$ grids for two-dimensional projections.

From Definition 12.1, the existence of an SOA(n, m, s^t, t) implies that of an OA(n, m, s, t). By Theorem 12.13, the existence of an OA$(n, m, s, 2)$ implies that of an SOA$(n, m, s^2, 2)$. This equivalence between strong orthogonal arrays and ordinary orthogonal arrays for strength two is responsible for why SOA-based Latin hypercubes are not better than OA-based Latin hypercubes in the case of strength two. For strength three, Theorem 12.13 states that the existence of an OA$(n, m, s, 3)$ implies that of an SOA$(n, m-1, s^3, 3)$, showing that one column is lost when constructing an SOA of strength three from an ordinary orthogonal array of strength three. However, there are situations where an OA$(n, m, s, 3)$, even when it is not a subarray of an OA$(n, m+1, s, 3)$, allows the construction of an SOA$(n, m, s^3, 3)$. We discuss this next.

The rest of the section is devoted to some results on SOAs of strength three. An orthogonal array OA(n, m, s, t) is said to be *embeddable* if it can be obtained by deleting one column from an OA$(n, m+1, s, t)$. For a given OA$(n, m, s, 3)$, the s levels in the first column divide the whole array into s subarrays, all of which are OA$(n/s, m-1, s, 2)$'s if their first columns are deleted. These s arrays of strength two are obtained by branching the first column. Branching any other column also produces s orthogonal arrays of strength two. In total, ms such arrays of strength two can be obtained, and they are called the children of the given OA$(n, m, s, 3)$. An OA$(n, m, s, 3)$ is said to be *semi-embeddable* if all of its ms children are embeddable. It is immediate that if an OA$(n, m, s, 3)$ is embeddable, then it must be semi-embeddable. The converse is not always true, however. Some examples will be given later. The usefulness of semi-embeddability is in the next result.

Theorem 12.14. *An SOA$(n, m, s^3, 3)$ exists if and only if a semi-embeddable OA$(n, m, s, 3)$ exists.*

We refer to He and Tang (2014) for a proof and details on how to construct an SOA$(n, m, s^3, 3)$ from a semi-embeddable OA$(n, m, s, 3)$. Theorem 12.12 is a general result, providing a fundamental characterization of a strong orthogonal array through a generalized orthogonal array. For strong orthogonal arrays of strength three, Theorem 12.14 offers a more direct and penetrating characterization. A result in Bierbrauer, Edel, and Schmid (2002) can be regarded as a linear version of Theorem 12.14. The following lemma has also been established by these authors albeit in different terminology.

Lemma 12.6. *A linear orthogonal array OA$(s^k, m, s, 3)$ is semi-embeddable, provided that $m \leq (s^{k-1} - 1)/(s - 1)$.*

The Bush construction, as in Theorem 10.7, gives a linear OA$(s^3, s+1, s, 3)$, which can be embedded into an OA$(s^3, s+2, s, 3)$ when s is an even prime power. For an odd prime power s, this OA$(s^3, s+1, s, 3)$ is not embeddable but, according to Lemma 12.6, it is semi-embeddable. Therefore, an SOA$(s^3, s+1, s^3, 3)$ can always be constructed whenever s is a prime power. Consider a linear OA$(s^4, s^2+1, s, 3)$ based on an ovoid, as given in Theorem 10.8. This array satisfies the condition in Lemma 12.6 and is therefore semi-embeddable. As a result, an $\text{SOA}(s^4, s^2+1, s^3, 3)$ can be constructed. We summarize the above discussion in the next result.

Theorem 12.15. *For any prime power s, an $SOA(s^3, s+1, s^3, 3)$ and an $SOA(s^4, s^2+1, s^3, 3)$ can be constructed.*

There are five nonisomorphic $\text{OA}(54, 5, 3, 3)$s (Hedayat, Seiden, and Stufken, 1997), which are all non-embeddable. He and Tang (2014) found that two of them are semi-embeddable, using which they constructed two $\text{SOA}(54, 5, 27, 3)$s.

Tian and Xu (2022) and Tian and Xu (2024) introduced and studied a space-filling pattern for strong orthogonal arrays, based on which a minimum aberration type criterion was proposed for assessing and selecting strong orthogonal arrays and, more generally, space-filling designs. The R package SOAs of Grömping (2022) provides a practical implementation of various construction methods of SOAs in the present section and of their variants to be discussed in the next section. A unifying overview of various construction methods of SOAs was given in Grömping (2023), who argued for using the term *stratum* orthogonal arrays in place of strong orthogonal arrays.

12.6 Variants of Strong Orthogonal Arrays

However attractive SOAs may be when compared with ordinary orthogonal arrays, their potential as space-filling designs cannot be fully realized if they are confined to the original definition. As discussed in Section 12.5, SOAs of strength 2 do not really improve OAs of strength 2 when it comes down to constructing Latin hypercubes. On the other hand, SOAs of strength 4 or higher are too costly to use, because, by Theorem 12.13, they are even more expensive than OAs of strength four, which are already fairly expensive. These arguments leave SOAs of strength 3 as the only practically viable choices in most applications. The goal of the present section is to discuss several variants of SOAs that significantly broaden the application scope of the SOA concept.

12.6.1 SOAs of strength 2+

SOAs of strength 2+ were introduced and studied by He, Cheng, and Tang (2018). As the name suggests, SOAs of strength 2+ are SOAs of strength 2 with an extra space-filling property.

Definition 12.2. *An $n \times m$ matrix with entries from $\{0, 1, \ldots, s^2-1\}$ is called a strong orthogonal array of strength 2+, with n runs and m factors of s^2 levels, if any subarray of two columns can be collapsed into an $OA(n, 2, s^2 \times s, 2)$ and an $OA(n, 2, s \times s^2, 2)$. We denote this array by $SOA(n, m, s^2, 2+)$.*

While an $\text{SOA}(n, m, s^2, 2)$ only promises stratifications on $s \times s$ grids in two-dimensional projections, an $\text{SOA}(n, m, s^2, 2+)$ achieves stratifications on $s^2 \times s$ and $s \times s^2$ grids in two-dimensional projections. This property of stratification on finer grids in two-dimensional projections is what makes an $\text{SOA}(n, m, s^3, 3)$ better than an $\text{OA}(n, m, s, 3)$.

Example 12.8. *An $SOA(16, 10, 4, 2+)$ is given below:*

2	2	2	2	2	2	0	0	0	0
2	2	0	2	0	0	1	2	2	2
2	0	2	0	2	1	2	1	2	2
2	0	0	0	0	3	3	3	0	0
0	2	2	1	1	2	2	2	1	2
0	2	0	1	3	0	3	0	3	0
0	0	2	3	1	1	0	3	3	0
0	0	0	3	3	3	1	1	1	2
1	1	1	2	2	2	2	2	2	1
1	1	3	2	0	0	3	0	0	3
1	3	1	0	2	1	0	3	0	3
1	3	3	0	0	3	1	1	2	1
3	1	1	1	1	2	0	0	3	3
3	1	3	1	3	0	1	2	1	1
3	3	1	3	1	1	2	1	1	1
3	3	3	3	3	3	3	3	3	3

One can check that any subarray of two columns is collapsible into an $OA(16, 2, 4\times 2, 2)$ *and an* $OA(16, 2, 2\times 4, 2)$. *An* $SOA(16, m, 8, 3)$ *enjoys the same two-dimensional space-filling property but, by an application of Theorem 12.14, exists only for* $m \leq 7$.

Theorem 12.12 shows that SOAs of strength t can be completely characterized by generalized orthogonal arrays of strength t. A similar result can also be established for SOAs of strength 2+.

Lemma 12.7. *An* $SOA(n, m, s^2, 2+)$*, say* $\boldsymbol{D}$*, exists if and only if there exist two arrays*

$$\boldsymbol{A} = (\boldsymbol{a}_1, \ldots, \boldsymbol{a}_m) \quad \text{and} \quad \boldsymbol{B} = (\boldsymbol{b}_1, \ldots, \boldsymbol{b}_m),$$

where $\boldsymbol{A}$ *is an* $OA(n, m, s, 2)$ *and* $\boldsymbol{B}$ *is an* $OA(n, m, s, 1)$*, such that* $(\boldsymbol{a}_j, \boldsymbol{a}_k, \boldsymbol{b}_k)$ *is an* $OA(n, 3, s, 3)$ *for any* $j \neq k$. *The three arrays are related by*

$$\boldsymbol{D} = s\boldsymbol{A} + \boldsymbol{B}.$$

In Lemma 12.7, it is important to note that while array $\boldsymbol{D}$ has levels $0, 1, \ldots, s^2 - 1$, both arrays $\boldsymbol{A}$ and $\boldsymbol{B}$ have levels $0, 1, \ldots, s-1$. To apply Lemma 12.7, we need to find two arrays $\boldsymbol{A}$ and $\boldsymbol{B}$ with the required properties. We first consider the use of regular $\boldsymbol{A}$ and $\boldsymbol{B}$, meaning that the columns of $\boldsymbol{A}$ and $\boldsymbol{B}$ are to be selected from a saturated regular design.

For $s = 2$, we use ± 1 to denote the two levels for convenience. To obtain an SOA$(n, m, 4, 2+)$ from $\boldsymbol{A}$ and $\boldsymbol{B}$ with levels ± 1, we need to convert $\boldsymbol{A}$ and $\boldsymbol{B}$ beforehand into arrays with levels 0 and 1 via $(\boldsymbol{A}+1)/2$ and $(\boldsymbol{B}+1)/2$. This shows that we should use

$$\boldsymbol{D} = 2(\boldsymbol{A}+1)/2 + (\boldsymbol{B}+1)/2 = \boldsymbol{A} + \boldsymbol{B}/2 + 3/2 \tag{12.10}$$

to construct an SOA$(n, m, 4, 2+)$ from $\boldsymbol{A}$ and $\boldsymbol{B}$.

Theorem 12.16. *Suppose that an SOA of strength 2+ is constructed through (12.10), where the columns of* $\boldsymbol{A}$ *and* $\boldsymbol{B}$ *are to be selected from* $\boldsymbol{S}$*, a saturated two-level regular design. Then it is necessary and sufficient that* $\bar{\boldsymbol{A}}$ *is a second order saturated design, where* $\bar{\boldsymbol{A}} = \boldsymbol{S} \setminus \boldsymbol{A}$ *consists of all the columns of* $\boldsymbol{S}$ *that are not in* $\boldsymbol{A}$.

Second-order saturated (SOS) designs were previously discussed in Chapter 6. The result in Theorem 12.16 establishes a somewhat surprising connection between SOAs of strength

TABLE 12.1
The smallest number m'_k of factors for a regular minimal SOS design, the largest number m_k of factors for an SOA of strength 2+ based on regular designs, and the largest number m''_k of factors for an SOA of strength 3 based on regular designs.

k	$n = 2^k$	m'_k	m_k	m''_k
4	16	5	10	7
5	32	9	22	15
6	64	13	50	31
7	128	19	108	63
8	256	[25, 26]	[229, 230]	127
9	512	[34, 39]	[472, 477]	255
10	1024	[47, 51]	[972, 976]	511

2+ and SOS designs. We omit the proof of Theorem 12.16. Instead, we show how to construct an SOA of strength 2+ for a given $\boldsymbol{A}$ when $\bar{\boldsymbol{A}}$ is an SOS design. Because $\bar{\boldsymbol{A}}$ is SOS, for any column $\boldsymbol{a}_i$ of $\boldsymbol{A}$ we must have that $\boldsymbol{a}_i = \boldsymbol{b}_i\boldsymbol{b}'_i$ for some $\boldsymbol{b}_i$ and $\boldsymbol{b}'_i$ in $\bar{\boldsymbol{A}}$. We simply take $\boldsymbol{B} = (\boldsymbol{b}_1, \ldots, \boldsymbol{b}_m)$, and then obtain $\boldsymbol{D}$ via (12.10).

Any regular SOS design can be used to construct an SOA of strength 2+. Theorem 6.12 is a general result, providing four minimal SOS designs constructed from two Hadamard matrices $\boldsymbol{H}_{n_1}$ and $\boldsymbol{H}_{n_2}$. We now consider regular versions of these SOS designs. Suppose that $\boldsymbol{H}_{n_1}$ with $n_1 = 2^{k_1}$ and $\boldsymbol{H}_{n_2}$ with $n_2 = 2^{k_2}$ are of Sylvester type, i.e., they are obtained by adding an all-ones column to saturated regular designs based on k_1 and k_2 independent columns, respectively. Then $\boldsymbol{D}_1$ in Theorem 6.12 is a minimal regular SOS design with $n = 2^k$ runs and $2^{k_1} + 2^{k_2} - 2$ factors, and $\boldsymbol{D}_2$, $\boldsymbol{D}_3$, and $\boldsymbol{D}_4$ in Theorem 6.12 are minimal regular SOS designs with $n = 2^k$ runs and $2^{k_1} + 2^{k_2} - 3$ factors, where $k = k_1 + k_2$. All these SOS designs can be used to construct SOAs of strength 2+. From $\boldsymbol{D}_1$, we can obtain an SOA$(2^k, m, 4, 2+)$ with $m = 2^k - 2^{k_1} - 2^{k_2} + 1$ factors, and from $\boldsymbol{D}_2$, $\boldsymbol{D}_3$, and $\boldsymbol{D}_4$ we can obtain SOA$(2^k, m, 4, 2+)$s with $m = 2^k - 2^{k_1} - 2^{k_2} + 2$ factors. If we want to construct an SOA$(2^k, m, 4, 2+)$ with the largest m from these SOS designs, for given k, we can take $k_1 = \lfloor k/2 \rfloor$ and $k_2 = k - \lfloor k/2 \rfloor$, and use design $\boldsymbol{D}_2$ to obtain an SOA$(2^k, m, 4, 2+)$ with

$$m = 2^k - 2^{\lfloor k/2 \rfloor} - 2^{k - \lfloor k/2 \rfloor} + 2,$$

which equals 10, 22, 50, and 106 for $k = 4, 5, 6$, and 7, respectively.

To construct an SOA$(2^k, m, 4, 2+)$ with the largest number m_k of factors for given k, we need to use the smallest minimal SOS design. Smaller minimal SOS designs do exist for 128 or more runs; see Cheng, He, and Tang (2021) for a discussion of two general constructions of these SOS designs, originally presented in Gabidulin, Davydov, and Tombak (1991) and Davydov, Marcugini, and Pambianco (2006) in the literature of coding theory and projective geometry.

Computer search results of Davydov, Marcugini, and Pambianco (2006) give lower and upper bounds on m'_k, the size of the smallest regular minimal SOS design for $4 \leq k \leq 10$. Since $m'_k + m_k = 2^k - 1$, the bounds on m'_k can then be used to obtain bounds on m_k, the greatest m such that an SOA$(2^k, m, 4, 2+)$ based on regular designs exists. The results are presented in Table 12.1, where for $4 \leq k \leq 7$, the m'_k and m_k values are exact and for $8 \leq k \leq 10$, the best known lower and upper bounds are given (e.g., $25 \leq m'_8 \leq 26$ and $229 \leq m_8 \leq 230$). For comparison, Table 12.1 also gives m''_k, the greatest m for an SOA$(2^k, m, 8, 3)$ to exist. By an application of Theorem 12.14, we have that $m''_k = 2^{k-1} - 1$.

Shi and Tang (2019b) considered the design selection problem for SOAs of strength 2 and 2+. Their idea is to find the SOAs of strength 2 that are closest to SOAs of strength 2+ and the SOAs of strength 2+ that are closest to SOAs of strength 3. This study provides useful space-filling designs when SOAs of strength 2+ do not exist and when SOAs of strength 2+ exist but SOAs of strength 3 do not. One of their results generalizes Theorem 12.16 as it shows that the best $\boldsymbol{A}$ to use in (12.10) is such that $\bar{\boldsymbol{A}}$ maximizes the number of degrees of freedom. See Remark 6.3 in Section 6.5 for a definition of the number of degrees of freedom. Another interesting result in Shi and Tang (2019b) is that the regular SOS $\boldsymbol{D}_1$ in Theorem 6.12 produces the best SOA of strength 2+ among all SOAs of strength 2+ that can be constructed from SOS designs of the same size.

A characterization similar to Theorem 12.16 can also be obtained for constructing SOA$(n, m, s^2, 2+)$ for $s \geq 3$ using regular $\boldsymbol{A}$ and $\boldsymbol{B}$. We refer to He, Cheng, and Tang (2018) for details. From this characterization, the next result can be established.

Theorem 12.17. *For any $k \geq 3$ and any prime power $s \geq 3$, an SOA$(s^k, m, s^2, 2+)$ with*

$$m = \frac{s^k - 1}{s - 1} - \frac{(s-1)^k - 1}{s - 2}$$

can be constructed.

To see the advantage of SOAs of strength 2+ over SOAs of strength 3, consider $k = 3$ and $k = 4$. For $k = 3$, by Theorem 12.15, an SOA$(s^3, m, s^3, 3)$ for $m = s + 1$ factors is available, whereas by Theorem 12.17, an SOA$(s^3, m, s^2, 2+)$ can be constructed for

$$m = \frac{s^3 - 1}{s - 1} - \frac{(s-1)^3 - 1}{s - 2} = 2s$$

factors. Similarly, for $k = 4$, an SOA$(s^4, m, s^3, 3)$ for $m = s^2 + 1$ factors is available, whereas an SOA$(s^4, m, s^2, 2+)$ can be constructed for

$$m = \frac{s^4 - 1}{s - 1} - \frac{(s-1)^4 - 1}{s - 2} = 3s^2 - s + 1$$

factors.

The construction of SOA$(n, m, s^2, 2+)$s using nonregular $\boldsymbol{A}$ and $\boldsymbol{B}$ in Lemma 12.7 was also considered by Cheng, He, and Tang (2021). For $s = 2$, a result similar to, but more complicated than, Theorem 12.16 was established. We omit the details here. At any rate, Cheng, He, and Tang (2021) showed that the four minimal SOS designs in Theorem 6.12 can all be used to construct SOAs of strength 2+. These SOAs of strength 2+ have run sizes that are multiples of 16. For example, if we take $n_1 = 4$ and $n_2 = 12$ in Theorem 6.12, then design $\boldsymbol{D}_2$ is a minimal SOS design with 48 runs for 13 factors, which allows the construction of an SOA$(48, 34, 4, 2+)$. For run sizes that are not multiples of 16, SOAs of strength 2+ may not be available. Chen and Tang (2022b) investigated the problem of how to construct the best space-filling designs using two-level nonregular designs through the notion of non-empty-cell designs, and presented their results for designs of $n = 16, 20, 24, 28, 32, 36$, and 40 runs. Results on the construction of SOA$(n, m, s^2, 2+)$s for $s \geq 3$ using nonregular $\boldsymbol{A}$ and $\boldsymbol{B}$ are quite limited, although some progress has been made in Jiang, Wang, and Wang (2021) and Chen and Tang (2022b).

SOAs of strength 2+ are justifiable under two-dimensional space-filling criteria based on the correlation, distance, and discrepancy measures. For a detailed discussion on these justification results, we refer to Chen and Tang (2022a) and Sun and Tang (2023).

12.6.2 SOAs of strength 3+

As has been discussed previously, SOAs of strength 2+ are those SOAs of strength 2 that share a stratification property of SOAs of strength 3. In a similar vein, we would like SOAs of strength 3 to have some stratification properties that only SOAs of strength 4 can offer. Naturally, these better SOAs of strength 3 can all be called SOAs of strength 3+. The present situation, however, is more complicated.

An SOA$(n, m, s^3, 3)$ achieves stratifications on $s \times s \times s$ grids in all three-dimensional projections and on $s \times s^2$ and $s^2 \times s$ grids in all two-dimensional projections. An SOA$(n, m, s^4, 4)$ is more space-filling, achieving

(α) stratifications on $s^2 \times s^2$ grids in all two-dimensional projections,

(β) stratifications on $s^2 \times s \times s$, $s \times s^2 \times s$, and $s \times s \times s^2$ grids in all three-dimensional projections, and

(γ) stratifications on $s^3 \times s$ and $s \times s^3$ grids in all two-dimensional projections.

An SOA$(n, m, s^4, 4)$ has two more space-filling properties: (δ) stratification on a set of s^4 intervals in all one-dimensional projections and (ϵ) stratifications on $s \times s \times s \times s$ grids in all four-dimensional projections. These two properties are not very interesting for SOAs of strength 3. Property δ is not interesting at all because Latin hypercubes based on an SOA$(n, m, s^3, 3)$ can achieve the maximum stratification in all one-dimensional projections. Property ϵ is not very interesting because an SOA$(n, m, s^3, 3)$ with this property implies the existence of an OA$(n, m, s, 4)$, requiring a large run size n for a given number m of factors.

We present some construction results for SOAs of strength 3 that possess some or all of properties α, β, and γ from using regular two-level designs. This line of studies was started by Shi and Tang (2020) and further enriched by Shi and Xu (2024). We present three major results from the two papers.

Theorem 12.18. *We have that for* $n = 2^k$ *with* $k \geq 4$,

(i) an SOA$(n, m, 8, 3)$ *with properties* α *and* β *can be constructed for* $m = n/4$ *factors, and*

(ii) this SOA$(n, n/4, 8, 3)$ *has the maximum number of factors among all SOA*$(n, m, 8, 3)$ *with properties* α *and* β.

Theorem 12.19. *With the exception of* $k = 5$, *we have that for* $n = 2^k$,

(i) an SOA$(n, m, 8, 3)$ *with properties* α *and* γ *can be constructed for every* $k \geq 4$ *and for* $m = 5n/16$ *factors, and*

(ii) this SOA$(n, 5n/16, 8, 3)$ *has the maximum number of factors among all SOA*$(n, m, 8, 3)$ *with properties* α *and* γ.

Theorem 12.20. *We have that for* $n = 2^k$ *with* $k \geq 4$,

(i) an SOA$(n, m, 8, 3)$ *with properties* α, β, *and* γ *can be constructed for* $m = n/4 - 1$ *factors, and*

(ii) this SOA$(n, n/4-1, 8, 3)$ *has the maximum number of factors among all SOA*$(n, m, 8, 3)$ *with properties* α, β, *and* γ.

Theorem 12.18 was obtained by Shi and Tang (2020) and Theorem 12.19 was obtained by Shi and Xu (2024). Part (i) of Theorem 12.20 is due to Shi and Tang (2020) and part (ii) of Theorem 12.20 is due to Shi and Xu (2024). We omit the proofs and detailed constructions for the above results. Interested readers may wish to read these two original papers on SOAs of strength 3+.

One may wonder if it is possible to construct SOAs of strength 3 with more factors if only one of the three properties α, β, and γ is required. Shi and Tang (2020) showed that an SOA$(n = 2^k, m, 8, 3)$ with property α must have $m \leq 5n/16$ and that an SOA$(n = 2^k, m, 8, 3)$ with property β must have $m \leq n/4$. Therefore, it is impossible to obtain an SOA$(n = 2^k, m, 8, 3)$ with property α that has more factors than that in Theorem 12.19 and an SOA$(n = 2^k, m, 8, 3)$ with property β that has more factors than that in Theorem 12.18. On the other hand, it is possible to construct an SOA$(n = 2^k, m, 8, 3)$ with property γ that has more factors than that in Theorem 12.19.

Apparently missing from the above theorems and discussion are SOA$(n = 2^k, m, 8, 3)$s with properties β and γ. From the aforementioned result that an SOA$(n = 2^k, m, 8, 3)$ with property β must have $m \leq n/4$, and by Theorem 12.20, we can deduce that an SOA$(n = 2^k, m, 8, 3)$ with properties β and γ exists for $m = n/4 - 1$ factors but not for $m \geq n/4 + 1$ factors. It remains to be settled if an SOA$(n = 2^k, m, 8, 3)$ with properties β and γ can be found for $m = n/4$ factors.

Example 12.9. *The following array (transposed)*

$$
\begin{array}{cccccccccccccccccccccccccccccccc}
7&6&6&7&6&7&7&6&2&3&3&2&3&2&2&3&0&1&1&0&1&0&0&1&5&4&4&5&4&5&5&4\\
7&0&4&3&4&3&7&0&2&5&1&6&1&6&2&5&0&7&3&4&3&4&0&7&5&2&6&1&6&1&5&2\\
7&4&2&1&4&7&1&2&2&1&7&4&1&2&4&7&0&3&5&6&3&0&6&5&5&6&0&3&6&5&3&0\\
7&2&0&5&6&3&1&4&2&7&5&0&3&6&4&1&0&5&7&2&1&4&6&3&5&0&2&7&4&1&3&6\\
7&4&6&5&2&1&3&0&2&1&3&0&7&4&6&5&0&3&1&2&5&6&4&7&5&6&4&7&0&3&1&2\\
7&2&4&1&0&5&3&6&2&7&1&4&5&0&6&3&0&5&3&6&7&2&4&1&5&0&6&3&2&7&1&4\\
7&6&2&3&0&1&5&4&2&3&7&6&5&4&0&1&0&1&5&4&7&6&2&3&5&4&0&1&2&3&7&6
\end{array}
$$

is an $SOA(32, 7, 8, 3)$ with all of the properties α, β, and γ. This array has property α because any subarray of two columns becomes an $OA(32, 2, 4, 2)$ when the eight levels are collapsed into four levels. The array has property β since any subarray of three columns becomes an $OA(32, 3, 4 \times 2 \times 2, 3)$, an $OA(32, 3, 2 \times 4 \times 2, 3)$, or an $OA(32, 3, 2 \times 2 \times 4, 3)$ when the levels of one factor are collapsed into four levels, and the levels of the other two factors are collapsed into two levels. It has property γ because any subarray of two columns becomes an $OA(32, 2, 2 \times 8, 2)$ or an $OA(32, 2, 8 \times 2, 2)$ when the levels of one factor are collapsed into two levels.

Shi and Xu (2024) introduced a stratification pattern for strong orthogonal arrays, which is a refinement of the space-filling pattern introduced in Tian and Xu (2022). As shown in Shi and Xu (2024), the stratification pattern provides a useful tool for studying and optimizing SOAs of strength 3.

12.6.3 Other variants of SOAs

For starters, there is a line of research that aims at adding column-orthogonality to various strong orthogonal arrays that have been discussed so far. Many papers have been written on the topic, some of which are Liu and Liu (2015), Zhou and Tang (2019) and Li, Liu, and Yang (2022b).

The main focus of this subsection is to give a brief account of *mappable nearly orthogonal arrays* (MNOAs), which are similar to SOAs but have a slightly different flavour, and can therefore be regarded as cousins of SOAs. For a fuller treatment, we refer to Mukerjee, Sun, and Tang (2014), who introduced MNOAs and conducted a systematic investigation into their properties and constructions.

Let $\boldsymbol{A}$ be an OA$(n, s^{m_1}, 2)$ and $\boldsymbol{B}$ an OA$(s, p^{m_2}, 2)$. Furthermore, let $\boldsymbol{C}$ be a Latin hypercube based on $\boldsymbol{B}$, and $\boldsymbol{D}_j$ be obtained from the jth column of $\boldsymbol{A}$ by replacing the uth level, for $u = 1, \ldots, s$, by the uth row of $\boldsymbol{C}$. Then

$$\boldsymbol{D} = (\boldsymbol{D}_1, \ldots, \boldsymbol{D}_{m_1})$$

is an s-level design of n runs with $m = m_1 m_2$ factors that are divided into m_1 groups of size m_2. Design $\boldsymbol{D}$ is called an MNOA and has the following properties:

(a) the whole array can be mapped or collapsed into an $\mathrm{OA}(n, m, p, 2)$, a fully orthogonal array of p levels, and

(b) any two columns from different groups $\boldsymbol{D}_i$ and $\boldsymbol{D}_j$ with $i \neq j$ form an $\mathrm{OA}(n, 2, s, 2)$, an s-level orthogonal array.

These two properties imply that design $\boldsymbol{D}$ achieves stratifications on $p \times p$ grids in all two-dimensional projections, and achieves stratifications on finer $s \times s$ grids in all those two-dimensional projections with the two factors coming from different groups. The proportion of the two-dimensional projections with the better stratification property is

$$\pi = 1 - m_1 \binom{m_2}{2} \Big/ \binom{m_1 m_2}{2} = 1 - m_2/(m_1 m_2 - 1),$$

which can be used to evaluate an MNOA.

To demonstrate the usefulness of MNOAs, we consider the following example. If we take an $\mathrm{OA}(81, 10, 9, 2)$ as $\boldsymbol{A}$ and an $\mathrm{OA}(9, 4, 3, 2)$ as $\boldsymbol{B}$, then we obtain an MNOA $\boldsymbol{D}$, a 9-level design of 81 runs with 40 factors, divided into 10 groups of size 4. Design $\boldsymbol{D}$ achieves stratifications on 3×3 grids in all two-dimensional projections just as an $\mathrm{OA}(81, 40, 3, 2)$ does. This design also achieves stratifications on 9×9 grids in 720 out of all 780 two-dimensional projections, having a π value equal to 12/13. If one insists on the use of a design achieving stratifications on 9×9 grids in all two-dimensional projections, then one has to consider a 9-level orthogonal array of 81 runs, but such an array only exists for 10 or fewer factors.

The research on MNOAs has been followed up by a number of papers. A notable one is Wang, Yang, and Liu (2022a), who proposed a class of arrays by mixing together the ideas of MNOAs and SOAs.

12.7 Maximin Distance Designs

Let $\boldsymbol{D} = (x_{ij})_{n \times m}$ be a $\mathrm{U}(n, s^m)$, an n-run balanced design for m s-level factors. The L_p-distance between the uth and vth design points $\boldsymbol{x}_u = (x_{u1}, \ldots, x_{um})$ and $\boldsymbol{x}_v = (x_{v1}, \ldots, x_{vm})$ is

$$d_p(\boldsymbol{x}_u, \boldsymbol{x}_v) = \left(\sum_{j=1}^{m} |x_{uj} - x_{vj}|^p \right)^{1/p}.$$

We would like to have a design that has large $d_p(\boldsymbol{x}_u, \boldsymbol{x}_v)$ for all $u < v$. One simple way to achieve this is to use the maximin distance criterion that requires the maximization of the *separation distance* of design $\boldsymbol{D}$ defined as

$$d_p(\boldsymbol{D}) = \min_{1 \leq u < v \leq n} d_p(\boldsymbol{x}_u, \boldsymbol{x}_v).$$

A design that maximizes the separation distance is called a maximin distance design.

Finding a maximin distance design from among all $U(n, s^m)$s is not a simple matter. One popular approach is to employ computational algorithms such as those of simulated annealing (Morris and Mitchell, 1995; Joseph and Hung, 2008), iterated local search (Grosso, Jamali, and Locatelli, 2009), particle swarm (Moon, Dean, and Santner, 2011; Chen, Hsieh, Hung, and Wang, 2013) and threshold-accepting (Xiao and Xu, 2018). Algorithmic search of maximin distance designs is applicable to all scenarios, at least in principle, without restrictions on the run size and number of factors. Perhaps a more important advantage of an algorithmic search is its flexibility in the choice of a design criterion. For example, instead of the maximin distance criterion defined above, Morris and Mitchell (1995) considered a more sophisticated criterion that seeks designs by minimizing

$$\left(\sum_{1 \le u < v \le n} \frac{1}{[d_p(\boldsymbol{x}_u, \boldsymbol{x}_v)]^q} \right)^{1/q},$$

for some positive integer q. When a sufficiently large q is used, this provides an easy implementation of the criterion of sequentially minimizing the frequencies of the distinct $d_p(\boldsymbol{x}_u, \boldsymbol{x}_v)$ values arranged in an ascending order, which is a refined version of the maximin distance criterion. Alternatively, Joseph, Gul, and Ba (2015) proposed a criterion of minimizing

$$\left(\frac{1}{\binom{n}{2}} \sum_{1 \le u < v \le n} \frac{1}{\prod_{j=1}^{m} (x_{uj} - x_{vj})^2} \right)^{1/m}.$$

This criterion has a virtue in that it penalizes any pair of close points, whether they are close in the whole design region or are close in any low dimensional projection.

Available to generate maximin distance designs are the R packages SLHD (Ba, 2013) and lhs (Carnell, 2020). SLHD can also be used to generate *sliced* Latin hypercube designs, which will be discussed in the next section.

A salient fact with algorithmic searches is that computational algorithms always have limits on the sizes of the problem they can handle no matter how smartly computer codes are written and how powerful computers become. Theoretical results are indispensable if one wants to gain insights into the nature of maximin distance designs and to have available a construction method for large design problems. However, theoretical investigations of maximin distance designs are challenging. So far, satisfactory solutions have been obtained only for two-dimensional problems. van Dam, Husslage, Den Hertog, and Melissen (2007) constructed maximin distance two-dimensional Latin hypercubes for any run size n when the L_∞ and L_1 distances are considered, and for run size $n \le 70$ when the L_2 distance is used.

In the remainder of this section, we present some general theoretical results on maximin distance designs. For the convenience of theoretical investigation, we use the L_p distance measure without taking the pth root; that is, we re-define the L_p distance between two design points $\boldsymbol{x}_u = (x_{u1}, \ldots, x_{um})$ and $\boldsymbol{x}_v = (x_{v1}, \ldots, x_{vm})$ as

$$d_p(\boldsymbol{x}_u, \boldsymbol{x}_v) = \sum_{j=1}^{m} |x_{uj} - x_{vj}|^p,$$

and correspondingly we re-define the separation distance of design $\boldsymbol{D}$ as $d_p(\boldsymbol{D}) = \min_{1 \le u < v \le n} d_p(\boldsymbol{x}_u, \boldsymbol{x}_v)$ using the newly defined $d_p(\boldsymbol{x}_u, \boldsymbol{x}_v)$.

For a $U(n, s^m)$, one can easily show that the average L_p distance between all pairs of points is

$$d_{p,\text{ave}} = \frac{mC_p}{n^2 - n}, \text{ where } C_p = \textstyle\sum_{u \ne v} |x_{uj} - x_{vj}|^p \text{ for any fixed } j.$$

Obviously, C_p is a constant for balanced designs. Thus $d_{p,\text{ave}}$ is also a constant, depending only on n, m, s, and p. It is obvious that, for any design $\boldsymbol{D}$, we must have

$$d_p(\boldsymbol{D}) \leq \lfloor d_{p,\text{ave}} \rfloor,$$

which gives an upper bound for the separation distance. With just some calculations, we obtain that

$$d_1(\boldsymbol{D}) \leq \lfloor n(s^2-1)m/(3ns-3s) \rfloor$$

for the L_1 distance, and

$$d_2(\boldsymbol{D}) \leq \lfloor n(s^2-1)m/(6n-6) \rfloor$$

for the L_2 distance. These upper bounds on the separation distances were obtained by van Dam, Rennen, and Husslage (2009) and Zhou and Xu (2015). To assess the performance of a design under the maximin distance criterion, one can use a distance efficiency defined by $d_p(\boldsymbol{D})/\lfloor d_{p,\text{ave}} \rfloor$, as suggested by Wang, Xiao, and Xu (2018a). We use the simple version defined as

$$d_{p,\text{eff}}(\boldsymbol{D}) = \frac{d_p(\boldsymbol{D})}{d_{p,\text{ave}}}.$$

If a design has that $d_{p,\text{eff}}(\boldsymbol{D}) = 1$, then $d_p(\boldsymbol{x}_u, \boldsymbol{x}_v)$ is a constant across all $u \neq v$, in which case the design is said to be equidistant. An equidistant design must be a maximin distance design. On the other hand, a maximin distance design need not have $d_{p,\text{eff}}(\boldsymbol{D}) = 1$, but if a design has a $d_{p,\text{eff}}(\boldsymbol{D})$ value close to 1, then it is at least nearly optimal under the maximin distance criterion.

The following theorem collects some of the important construction results on optimal or nearly optimal Latin hypercubes under the maximin distance criterion. Recall that LH(n, m) denotes a Latin hypercube of n runs for m factors.

Theorem 12.21. *The following assertions hold.*

(i) If $2s+1$ is a prime, then an LH(s, s) can be constructed to have $d_{1,\text{eff}} = 1$.

(ii) If $2s-1$ is a prime, then an LH$(s, s-1)$ can be constructed to have $d_{1,\text{eff}} = s/(s+1)$.

(iii) If s is a prime, then an LH$(s, s-1)$ can be constructed to have

$$d_{1,\text{eff}} \geq 1 - 2/(3s^2-3)^{1/2}.$$

(iv) An LH$(2^k, 2^{k-1})$ can be constructed to have $d_{2,\text{eff}} \geq 1 - 1/2^k$ for $k \geq 2$.

Parts (i), (ii), and (iii) were established by Wang, Xiao, and Xu (2018a) and part (iv) was obtained by Zhou, Yang, and Liu (2020). For detailed constructions, we refer to these papers. We also note that parts (i), (ii), and (iii) are about the L_1 distance whereas part (iv) is about the L_2 distance.

Li, Liu, and Tang (2021) presented a method of constructing large maximin distance designs from small ones. In the process of construction, the Hamming distance plays a supporting role. For a U(n, s^m) design $\boldsymbol{D} = (x_{ij})_{n\times m}$, the Hamming distance $d_H(\boldsymbol{x}_u, \boldsymbol{x}_v)$ between $\boldsymbol{x}_u$ and $\boldsymbol{x}_v$, the uth and vth rows of $\boldsymbol{D}$, is the number of components where they differ. The Hamming separation distance of $\boldsymbol{D}$, denoted by $d_H(\boldsymbol{D})$, is the minimum Hamming distance between all pairs of design points. We can easily find the average of the Hamming distances among all pairs of rows, which is given by

$$d_{H,\text{ave}} = n(s-1)m/((n-1)s),$$

a constant. Similarly, we define

$$d_{H,\text{eff}}(\boldsymbol{D}) = d_H(\boldsymbol{D})/d_{H,\text{ave}}$$

as the Hamming distance efficiency of $\boldsymbol{D}$.

Let $\boldsymbol{A}$ be a $\mathrm{U}(n, s^{m_1})$ and $\boldsymbol{B}$ a $\mathrm{U}(s, q^{m_2})$. From $\boldsymbol{A}$ and $\boldsymbol{B}$, we construct a design $\boldsymbol{D}$ by replacing the uth level of $\boldsymbol{A}$ by the uth row of $\boldsymbol{B}$ for $u = 1, \ldots, s$. Then $\boldsymbol{D}$ is a $\mathrm{U}(n, q^{m_1 m_2})$. The following theorem establishes the distance properties of design $\boldsymbol{D}$ in relation to those of $\boldsymbol{A}$ and $\boldsymbol{B}$.

Theorem 12.22. *We have that*

(i) $d_p(\boldsymbol{D}) \geq d_H(\boldsymbol{A})\, d_p(\boldsymbol{B})$, *and*

(ii) $d_{p,\mathrm{eff}}(\boldsymbol{D}) \geq d_{H,\mathrm{eff}}(\boldsymbol{A})\, d_{p,\mathrm{eff}}(\boldsymbol{B})$.

(iii) If $\boldsymbol{B}$ *is equidistant, then the equalities in (i) and (ii) are both attained.*

We refer to Li, Liu, and Tang (2021) for a proof of Theorem 12.22. While the Hamming distance of design $\boldsymbol{A}$ enters the picture, the distance measure for designs $\boldsymbol{B}$ and $\boldsymbol{D}$ is the L_p distance for any $p \geq 1$. From the proof, we see that Theorem 12.22 is valid for any additive distance measure, which refers to a distance of the form $d(\boldsymbol{x}, \boldsymbol{y}) = \sum_{j=1}^{m} d(x_j, y_j)$ for $\boldsymbol{x} = (x_1, \ldots, x_m)$ and $\boldsymbol{y} = (y_1, \ldots, y_m)$. One can therefore construct a good large design $\boldsymbol{D}$ from a good small design $\boldsymbol{B}$ using the L_1, L_2 or any additive distance, provided that $\boldsymbol{A}$ is good under the Hamming distance.

If we choose $\boldsymbol{A}$ to be a saturated orthogonal array $\mathrm{OA}(n, s^{m_1})$, then it is equidistant with $d_H(\boldsymbol{A}) = n/s$ according to Cheng (2014, Theorem 8.6). The Rao-Hamming construction gives a saturated $\mathrm{OA}(n, s^{m_1})$ with $n = s^k$ and $m_1 = (n-1)/(s-1)$, and is applicable whenever s is a prime power. By taking the designs in Theorem 12.21 as $\boldsymbol{B}$, we obtain the next theorem.

Theorem 12.23. *Suppose that s is a prime power and $k \geq 2$ is an integer. Then the following assertions hold.*

(i) If $2s+1$ is a prime, then a $U(n, s^m)$ with $n = s^k$ and $m = s(s^k-1)/(s-1)$ can be constructed to have $d_{1,\mathrm{eff}} = 1$.

(ii) If $2s-1$ is a prime, then a $U(n, s^m)$ with $n = s^k$ and $m = s^k - 1$ can be constructed to have $d_{1,\mathrm{eff}} \geq s/(s+1)$.

(iii) If s is a prime, then a $U(n, s^m)$ with $n = s^k$ and $m = s^k - 1$ can be constructed to have $d_{1,\mathrm{eff}} \geq 1 - 2/(3s^2-3)^{1/2}$.

(iv) A $U(n, s^m)$ with $s = 2^{k_2}$, $n = 2^{k_1 k_2}$, and $m = 2^{k_2-1}(2^{k_1 k_2}-1)/(2^{k_2}-1)$ can be constructed to have $d_{2,\mathrm{eff}} \geq 1 - 1/2^{k_2}$ for all $k_1 > 2$ and $k_2 > 2$.

Taking $s = 11$ and $k = 2$ in parts (i) and (iii) of Theorem 12.23, we obtain a $\mathrm{U}(121, 11^{132})$ with $d_{1,\mathrm{eff}} = 1$, and a $\mathrm{U}(121, 11^{120})$ with $d_{1,\mathrm{eff}} \geq 0.975$. Taking $s = 9$ and $k = 2$ in part (ii) of Theorem 12.23, we obtain a $\mathrm{U}(81, 9^{80})$ that has $d_{1,\mathrm{eff}} \geq 0.90$. Taking $k_1 = 2$ and $k_2 = 3$ in part (iv) of Theorem 12.23, we obtain a $\mathrm{U}(64, 8^{36})$ with $d_{2,\mathrm{eff}} \geq 7/8$.

12.8 Other Topics in Space-Filling Designs

In this section, we give a brief discussion on two types of space-filling designs that have not been covered in the previous sections. Nested space-filling designs are of one type, and they were introduced for computer experiments that can be conducted with two levels of accuracy. The other type of designs is custom-made for computer experiments that involve both qualitative and quantitative factors.

12.8.1 Nested space-filling designs

Nested space-filling designs were introduced for computer experiments that can be conducted at two different levels of accuracy (Qian, Seepsad, Joseph, Allen, and Wu, 2006; Qian, Tang, and Wu, 2009b). In short, a nested space-filling design refers to a pair of space-filling designs with one design nested within the other. Nested space-filling designs are useful for other purposes as well. Rennen, Husslage, Van Dam, and Hertog (2010) has a discussion on their usefulness for dealing with training and testing sets, linking parameters, and sequential evaluations.

Finding nested space-filling designs appears to be simple, at least in principle. One can start with a large space-filling design and then identify the best subdesign using some space-filling criterion. Or alternatively, one can also start with a small space-filling design and then add some points so that the resulting whole design is also space-filling. The problem becomes non-trivial if low dimensional space-filling properties are to be stressed. Here we present two construction results from Qian, Tang, and Wu (2009b). For further developments on nested space-filling designs, we refer to Qian, Ai, and Wu (2009a), Rennen, Husslage, Van Dam, and Hertog (2010), He and Qian (2011), Wang and Li (2013), Sun, Liu, and Qian (2014), and Tsai (2017).

We use $\{\boldsymbol{D}_1, \boldsymbol{D}_2\}$ to denote a nested space-filling design, where $\boldsymbol{D}_1$ is a $\mathrm{U}(n_1, s^m)$ and $\boldsymbol{D}_2$ consists of n_2 rows of $\boldsymbol{D}_1$ with $n_2 < n_1$. We require that both $\boldsymbol{D}_1$ and $\boldsymbol{D}_2$ be space-filling. Since $\boldsymbol{D}_1$ has more points than $\boldsymbol{D}_2$, naturally we can make $\boldsymbol{D}_1$ more space-filling than $\boldsymbol{D}_2$.

Our first construction is simple, producing a design pair $\{\boldsymbol{D}_1, \boldsymbol{D}_2\}$ such that $\boldsymbol{D}_1$ is space-filling in all three-dimensional projections while $\boldsymbol{D}_2$ is space-filling in all two-dimensional projections. The basic idea is to obtain $\{\boldsymbol{D}_1, \boldsymbol{D}_2\}$ by branching a column of an orthogonal array of strength 3. Specifically, suppose that $\boldsymbol{A}_1$ is an $\mathrm{OA}(n, m+1, s, 3)$. Let $\boldsymbol{A}_2$ be the subarray of $\boldsymbol{A}_1$ consisting of all the rows that have the first entry being a fixed level, say level 0. Let $\boldsymbol{D}_1$ and $\boldsymbol{D}_2$ be obtained, respectively, from $\boldsymbol{A}_1$ and $\boldsymbol{A}_2$ by deleting their first columns.

Theorem 12.24. *Given an $OA(n, m+1, s, 3)$, we can construct a nested space-filling design $\{\boldsymbol{D}_1, \boldsymbol{D}_2\}$ with the properties that*

(i) $\boldsymbol{D}_1$ is an $OA(n, m, s, 3)$ and thus achieves stratifications on $s \times s \times s$ grids in all three-dimensional projections, and

(ii) $\boldsymbol{D}_2$ is an $OA(n/s, m, s, 2)$ and thus achieves stratifications on $s \times s$ grids in all two-dimensional projections.

We give an example. Using an $\mathrm{OA}(64, 6, 4, 3)$ as $\boldsymbol{A}_1$, we can then obtain $\{\boldsymbol{D}_1, \boldsymbol{D}_2\}$ where $\boldsymbol{D}_1$ is an $\mathrm{OA}(64, 5, 4, 3)$ and $\boldsymbol{D}_2$ is an $\mathrm{OA}(16, 5, 4, 2)$. The simple method of constructing nested space-filling designs described above can be generalized in two ways. If we use an $\mathrm{OA}(n, m+1, s, t)$ as the original array, then $\boldsymbol{D}_1$ is an $\mathrm{OA}(n, m, s, t)$ and $\boldsymbol{D}_2$ is an $\mathrm{OA}(n/s, m, s, t-1)$. By branching the first two columns of an $\mathrm{OA}(n, m+2, s, t)$ together, we can obtain $\{\boldsymbol{D}_1, \boldsymbol{D}_2\}$ such that $\boldsymbol{D}_1$ is an $\mathrm{OA}(n, m, s, t)$ and $\boldsymbol{D}_2$ is an $\mathrm{OA}(n/s^2, m, s, t-2)$.

The next result gives a construction of $\{\boldsymbol{D}_1, \boldsymbol{D}_2\}$ such that both $\boldsymbol{D}_1$ and $\boldsymbol{D}_2$ are space-filling in all two-dimensional projections, but $\boldsymbol{D}_1$ does so on finer grids than $\boldsymbol{D}_2$.

Theorem 12.25. *Let $s_1 = p^{u_1}$ and $s_2 = p^{u_2}$ be two powers of the same prime p with $2u_2 \le u_1 + 1$. Then a nested space-filling design pair $\{\boldsymbol{D}_1, \boldsymbol{D}_2\}$ can be constructed, where*

(i) $\boldsymbol{D}_1$ is an $OA(s_1^k, m, s_1, 2)$ with $m = (s_2^k - 1)/(s_2 - 1)$, thereby achieving stratifications on $s_1 \times s_1$ grids in all two-dimensional projections, and

(ii) $\boldsymbol{D}_2$ *consists of* $n_2 = s_2^k$ *rows of* $\boldsymbol{D}_1$ *and becomes an* $OA(s_2^k, m, s_2, 2)$ *if the* s_1 *levels are collapsed into* s_2 *levels, thereby achieving stratifications on* $s_2 \times s_2$ *grids in all two-dimensional projections.*

We refer to the original paper Qian, Tang, and Wu (2009b) for detailed constructions of $\boldsymbol{D}_1$ and $\boldsymbol{D}_2$. Essentially, $\boldsymbol{D}_1$ is obtained by a judicious selection of $m = (s_2^k - 1)/(s_2 - 1)$ columns from a saturated $\text{OA}(s_1^k, (s_1^k - 1)/(s_1 - 1), s_1, 2)$ given by the Rao-Hamming construction, and $\boldsymbol{D}_2$ is obtained by a further judicious selection of s_2^k rows from $\boldsymbol{D}_1$.

We give an example. Let $p = 2$, $u_1 = 3$, and $u_2 = 2$. So $s_1 = 8$ and $s_2 = 4$. The condition $2u_2 \leq u_1 + 1$ is satisfied. Taking $k = 2$ in Theorem 12.25, we obtain a nested space-filling design pair $\{\boldsymbol{D}_1, \boldsymbol{D}_2\}$, where $\boldsymbol{D}_1$ is an $\text{OA}(64, 5, 8, 2)$ and $\boldsymbol{D}_2$, consisting of 16 rows of $\boldsymbol{D}_1$, becomes an $\text{OA}(16, 5, 4, 2)$ if the eight levels are collapsed into four levels. One may find it interesting to compare this nested design with what is given right after Theorem 12.24.

If desired, one can construct a Latin hypercube from $\{\boldsymbol{D}_1, \boldsymbol{D}_2\}$. Then the resulting design pair $\{\boldsymbol{E}_1, \boldsymbol{E}_2\}$ retains the space-filling properties of $\boldsymbol{D}_1$ and $\boldsymbol{D}_2$ while achieving the maximum stratification in one-dimensional projections.

12.8.2 Designs with both qualitative and quantitative factors

Our discussion on space-filling designs thus far has been assuming that all studied factors are quantitative. Computer experiments, however, often involve some qualitative factors besides quantitative factors (Schmidt, Cruz, and Iyengar, 2005). The idea of *a sliced space-filling design* (Qian and Wu, 2009) stems from the need for a design that is not just space-filling across all quantitative factors, but also space-filling for each level combination of qualitative factors. Formally, a sliced space-filling design is a space-filling design $\boldsymbol{D}$ that can be divided into several slices $\boldsymbol{D}_1, \ldots, \boldsymbol{D}_l$, i.e.

$$\boldsymbol{D} = (\boldsymbol{D}_1^T, \ldots, \boldsymbol{D}_l^T)^T,$$

such that each slice $\boldsymbol{D}_j$ is also space-filling. When running the experiment, we associate each slice of the design with a level combination of the qualitative factors.

Sliced space-filling designs are related to nested space-filling designs; the former requires that all slices be space-filling while the latter only requires that one slice be space-filling. Because of this connection, Theorems 12.24 and 12.25 have their counterparts for the construction of sliced space-filling designs.

Theorem 12.26. *From an* $OA(n, m+1, s, 3)$, *a sliced space-filling design*

$$\boldsymbol{D} = (\boldsymbol{D}_1^T, \ldots, \boldsymbol{D}_s^T)^T$$

can be constructed such that

(i) $\boldsymbol{D}$ *is an* $OA(n, m, s, 3)$*; and*

(ii) each $\boldsymbol{D}_j$ *is an* $OA(n/s, m, s, 2)$ *for* $j = 1, \ldots, s$.

Theorem 12.27. *Let* $s_1 = p^{u_1}$ *and* $s_2 = p^{u_2}$ *be two powers of the same prime* p *with* $2u_2 \leq u_1 + 1$. *Then a sliced space-filling design*

$$\boldsymbol{D} = (\boldsymbol{D}_1^T, \ldots, \boldsymbol{D}_l^T)^T,$$

where $l = (s_1/s_2)^k$, *can be constructed such that*

(i) $\boldsymbol{D}$ *is an* $OA(s_1^k, m, s_1, 2)$ *with* $m = (s_2^k - 1)/(s_2 - 1)$*, and*

(ii) each $\boldsymbol{D}_j$ *consists of* $n_2 = s_2^k$ *rows of* $\boldsymbol{D}$ *and becomes an* $OA(s_2^k, m, s_2, 2)$ *if the* s_1 *levels are collapsed into* s_2 *levels.*

We look at an example. For $p = 2$, $u_1 = 3$, and $u_2 = 2$, and hence $s_1 = 8$ and $s_2 = 4$, the condition $2u_2 \leq u_1 + 1$ is satisfied. Taking $k = 2$ in Theorem 12.27, we obtain a sliced space-filling design

$$\boldsymbol{D} = (\boldsymbol{D}_1^T, \ldots, \boldsymbol{D}_4^T)^T,$$

where $\boldsymbol{D}$ is an OA(64, 5, 8, 2) and each $\boldsymbol{D}_j$ becomes an OA(16, 5, 4, 2) if the eight levels are collapsed into four levels. The four slices can be the four levels of one qualitative factor or the four level combinations of two two-level qualitative factors.

Again, if a sliced space-filling design with more levels is to be used, then one can construct a Latin hypercube by expanding the levels of the design $\boldsymbol{D}$ in Theorems 12.26 and 12.27. For more results and further developments on sliced space-filling designs, we refer to Ai, Jiang, and Li (2014) and Ba, Myers, and Brenneman (2015).

When multiple qualitative factors are studied in a computer experiment, the total number of level combinations becomes large. In this case, a sliced space-filling design needs to have a large number of slices, which results in a very large design. To handle this, Deng, Hung, and Lin (2015) proposed *marginally coupled designs.* A design denoted by $\boldsymbol{D} = (\boldsymbol{D}_1, \boldsymbol{D}_2)$, where $\boldsymbol{D}_1$ is for qualitative factors and $\boldsymbol{D}_2$ is a Latin hypercube for quantitative factors, is called a marginally coupled design if, for each level of any qualitative factor, the corresponding design points of quantitative factors also form a Latin hypercube design. We present one result from He, Lin, and Sun (2017) who investigated the construction of marginally coupled designs with an extra property that for each level of any qualitative factor, the corresponding design points of quantitative factors achieve stratifications in all two-dimensional projections in addition to forming a Latin hypercube.

Theorem 12.28. *Let* $\boldsymbol{A}$ *be an* $OA(n, s^m q^1, 2)$ *where* $q = n/s$ *and* $\boldsymbol{B}$ *be an* $OA(q, k, s_1, 2)$*. From* $\boldsymbol{A}$ *and* $\boldsymbol{B}$*, a marginally coupled design* $(\boldsymbol{D}_1, \boldsymbol{D}_2)$ *of* n *runs, where* $\boldsymbol{D}_1$ *has* m *qualitative factors of* s *levels and* $\boldsymbol{D}_2$ *is a Latin hypercube for* k *quantitative factors, can be constructed to have the property that the subdesign of* $\boldsymbol{D}_2$ *corresponding to each level of any factor in* $\boldsymbol{D}_1$ *achieves stratifications on* $s_1 \times s_1$ *grids in all two-dimensional projections.*

If we take $\boldsymbol{A}$ to be an $\text{OA}(8, 2^4 4^1, 2)$ and $\boldsymbol{B}$ to be an OA(4, 3, 2, 2). Then we can construct an eight-run marginally coupled design $\boldsymbol{D} = (\boldsymbol{D}_1, \boldsymbol{D}_2)$ for four qualitative and three quantitative factors. This design has the property that for each level of any qualitative factor, the four design points from the quantitative factors achieve stratifications on 2×2 grids in all two-dimensional projections. The design is displayed below.

$\boldsymbol{D}_1$				$\boldsymbol{D}_2$		
0	0	0	0	5	7	2
1	1	1	1	4	6	3
0	0	1	1	6	2	4
1	1	0	0	7	3	5
0	1	0	1	0	4	6
1	0	1	0	1	5	7
0	1	1	0	2	1	1
1	0	0	1	3	0	0

A

Appendix: Background Knowledge

This appendix chapter provides some background knowledge for reading the book. Since our book is about the theory of *nonregular* factorial designs, basic knowledge of *regular* factorial designs, though not absolutely necessary, will be helpful for the readers. We first give a rundown of the most important concepts and results on regular two-level designs. Finite fields and finite geometries are next topics on the agenda. These mathematical tools are essential for a proper understanding of the theory of regular multi-level designs, which we review next. In the final section, we provide an introduction to error-correcting codes, the theory of which has intersected multiple times with that of factorial designs.

A.1 Regular Two-Level Factorial Designs

This section is a brief account of regular two-level designs. The purpose is to provide the reader with an easy access to the fundamentals of regular two-level designs that one may need while reading this book. We strive for a concise presentation that covers a variety of important topics, with the aim of providing a sufficient background for studying nonregular designs. A detailed treatment of most of the topics can be found in Box, Hunter, and Hunter (1978) and Wu and Hamada (2000). A full and advanced coverage is given by Cheng (2014a). Regular multi-level designs will be reviewed in Section A.4.

Two-level designs deserve a separate treatment for at least three reasons. The foremost of these is that one can fully understand and appreciate regular two-level designs without a knowledge of finite fields and projective geometries. Another is the convenience and intuition offered by the coding of two levels with ± 1, as they can be used directly in writing down a model matrix, of which the columns correspond to factorial effects. The last but definitely not least is that research results are not just simpler, but also richer for two-level designs.

A.1.1 Generators, defining relation, and aliasing pattern

To study m factors, each having two levels, a full factorial requires 2^m runs. This run size is too large for most applications, which necessitates the use of fractional factorials. Finney (1945) was the first who systematically studied such designs, and it is remarkable that some of his terminology and notation is still in use today. Elementary concepts in fractional factorials such as those of generators, defining relation, and aliasing patterns are best illustrated using an example.

DOI: 10.1201/9781003371861-A

Suppose that five factors are to be studied in a scientific investigation. A full factorial design requires 32 runs. Consider a fractional factorial design of 8 runs given below:

$$\boldsymbol{D} = \begin{pmatrix} -1 & -1 & -1 & 1 & 1 \\ 1 & -1 & -1 & -1 & 1 \\ -1 & 1 & -1 & -1 & -1 \\ 1 & 1 & -1 & 1 & -1 \\ -1 & -1 & 1 & 1 & -1 \\ 1 & -1 & 1 & -1 & -1 \\ -1 & 1 & 1 & -1 & 1 \\ 1 & 1 & 1 & 1 & 1 \end{pmatrix}.$$

Let $\boldsymbol{d}_j$ be the jth column of $\boldsymbol{D}$, so $\boldsymbol{D} = (\boldsymbol{d}_1, \boldsymbol{d}_2, \boldsymbol{d}_3, \boldsymbol{d}_4, \boldsymbol{d}_5)$. The first three columns $(\boldsymbol{d}_1, \boldsymbol{d}_2, \boldsymbol{d}_3)$ form a full factorial for three factors. Column $\boldsymbol{d}_4$ is the componentwise product of columns $\boldsymbol{d}_1$ and $\boldsymbol{d}_2$; that is, $\boldsymbol{d}_4 = \boldsymbol{d}_1\boldsymbol{d}_2$. Column $\boldsymbol{d}_5$ is obtained by $\boldsymbol{d}_5 = \boldsymbol{d}_2\boldsymbol{d}_3$. The two equations $\boldsymbol{d}_4 = \boldsymbol{d}_1\boldsymbol{d}_2$ and $\boldsymbol{d}_5 = \boldsymbol{d}_2\boldsymbol{d}_3$ are called *generators* of the design. Equivalently, they can be written as $\boldsymbol{1}_8 = \boldsymbol{d}_1\boldsymbol{d}_2\boldsymbol{d}_4$ and $\boldsymbol{1}_8 = \boldsymbol{d}_2\boldsymbol{d}_3\boldsymbol{d}_5$, where $\boldsymbol{1}_8$ represents a column of 8 plus ones. Together, these two relations imply another one: $\boldsymbol{1}_8 = \boldsymbol{d}_1\boldsymbol{d}_3\boldsymbol{d}_4\boldsymbol{d}_5$. We then have

$$\boldsymbol{1}_8 = \boldsymbol{d}_1\boldsymbol{d}_2\boldsymbol{d}_4 = \boldsymbol{d}_2\boldsymbol{d}_3\boldsymbol{d}_5 = \boldsymbol{d}_1\boldsymbol{d}_3\boldsymbol{d}_4\boldsymbol{d}_5,$$

which is called the *defining relation*. Following the tradition, factor labels $\boldsymbol{d}_1, \boldsymbol{d}_2, \boldsymbol{d}_3, \boldsymbol{d}_4$, and $\boldsymbol{d}_5$ are called *letters*. Then combinations of letters such as $\boldsymbol{d}_1\boldsymbol{d}_2\boldsymbol{d}_3$, $\boldsymbol{d}_1\boldsymbol{d}_2\boldsymbol{d}_4$, $\boldsymbol{d}_1\boldsymbol{d}_3\boldsymbol{d}_4\boldsymbol{d}_5$, and $\boldsymbol{d}_2\boldsymbol{d}_3\boldsymbol{d}_4\boldsymbol{d}_5$ are called *words*. The number of letters in a word is its length. The defining relation of the design contains three words, $\boldsymbol{d}_1\boldsymbol{d}_2\boldsymbol{d}_4$, $\boldsymbol{d}_2\boldsymbol{d}_3\boldsymbol{d}_5$, and $\boldsymbol{d}_1\boldsymbol{d}_3\boldsymbol{d}_4\boldsymbol{d}_5$. They are called *defining words*.

Multiplying by $\boldsymbol{d}_1$ on all terms of the defining relation, we obtain that

$$\boldsymbol{d}_1 = \boldsymbol{d}_2\boldsymbol{d}_4 = \boldsymbol{d}_1\boldsymbol{d}_2\boldsymbol{d}_3\boldsymbol{d}_5 = \boldsymbol{d}_3\boldsymbol{d}_4\boldsymbol{d}_5.$$

This means that the same column consists of coefficients of a contrast that estimates all these effects; actually each estimates the sum of four factorial effects. Similarly, we have

$$\boldsymbol{d}_2 = \boldsymbol{d}_1\boldsymbol{d}_4 = \boldsymbol{d}_3\boldsymbol{d}_5 = \boldsymbol{d}_1\boldsymbol{d}_2\boldsymbol{d}_3\boldsymbol{d}_4\boldsymbol{d}_5.$$

Continuing this process for the other effects, altogether we obtain the full *aliasing pattern*:

$$\begin{aligned} &\boldsymbol{1}_8 = \boldsymbol{d}_1\boldsymbol{d}_2\boldsymbol{d}_4 = \boldsymbol{d}_2\boldsymbol{d}_3\boldsymbol{d}_5 = \boldsymbol{d}_1\boldsymbol{d}_3\boldsymbol{d}_4\boldsymbol{d}_5, && \boldsymbol{d}_1 = \boldsymbol{d}_2\boldsymbol{d}_4 = \boldsymbol{d}_1\boldsymbol{d}_2\boldsymbol{d}_3\boldsymbol{d}_5 = \boldsymbol{d}_3\boldsymbol{d}_4\boldsymbol{d}_5, \\ &\boldsymbol{d}_2 = \boldsymbol{d}_1\boldsymbol{d}_4 = \boldsymbol{d}_3\boldsymbol{d}_5 = \boldsymbol{d}_1\boldsymbol{d}_2\boldsymbol{d}_3\boldsymbol{d}_4\boldsymbol{d}_5, && \boldsymbol{d}_3 = \boldsymbol{d}_1\boldsymbol{d}_2\boldsymbol{d}_3\boldsymbol{d}_4 = \boldsymbol{d}_2\boldsymbol{d}_5 = \boldsymbol{d}_1\boldsymbol{d}_4\boldsymbol{d}_5, \\ &\boldsymbol{d}_4 = \boldsymbol{d}_1\boldsymbol{d}_2 = \boldsymbol{d}_2\boldsymbol{d}_3\boldsymbol{d}_4\boldsymbol{d}_5 = \boldsymbol{d}_1\boldsymbol{d}_3\boldsymbol{d}_5, && \boldsymbol{d}_5 = \boldsymbol{d}_1\boldsymbol{d}_2\boldsymbol{d}_4\boldsymbol{d}_5 = \boldsymbol{d}_2\boldsymbol{d}_3 = \boldsymbol{d}_1\boldsymbol{d}_3\boldsymbol{d}_4, \\ &\boldsymbol{d}_1\boldsymbol{d}_3 = \boldsymbol{d}_2\boldsymbol{d}_3\boldsymbol{d}_4 = \boldsymbol{d}_1\boldsymbol{d}_2\boldsymbol{d}_5 = \boldsymbol{d}_4\boldsymbol{d}_5, && \boldsymbol{d}_1\boldsymbol{d}_2\boldsymbol{d}_3 = \boldsymbol{d}_3\boldsymbol{d}_4 = \boldsymbol{d}_1\boldsymbol{d}_5 = \boldsymbol{d}_2\boldsymbol{d}_4\boldsymbol{d}_5. \end{aligned}$$

The grand mean, corresponding to $\boldsymbol{1}_8$, is not a treatment contrast, but for convenience of discussion, let us call it a factorial effect (of order 0). With this convention, we see that the aliasing pattern divides the 32 factorial effects into 8 groups of size 4. The effects in the same group are aliased with one another and those from different groups are mutually orthogonal. Each group containing the effects that are aliased with one another is called an *alias set*. Thus, the aliasing pattern of a fractional factorial provides a clear picture of effect aliasing in partitioning all the factorial effects into alias sets.

In general, starting with a full factorial for k factors and using q independent (mutually orthogonal) interaction columns to define additional factors, we obtain a design of 2^k runs for $m = k + q$ factors. Such a design is often called a 2^{m-q} design. A 2^{m-q} design has q independent generators, giving rise to q independent defining words. Multiplying these defining words in all possible ways, we obtain the defining relation, which contains $2^q - 1$ defining words. From the defining relation, we can then obtain the aliasing pattern, which partitions the 2^m effects into 2^k alias sets, each containing 2^q effects such that the effects in the same alias set are aliased with one another and those from different groups are mutually orthogonal. This complete picture of the effect aliasing afforded by the aliasing pattern is worth documenting as a theorem.

Theorem A.1. *The aliasing pattern of a 2^{m-q} design provides a partition of all 2^m factorial effects (the grand mean included) into 2^k alias sets ($k = m - q$) of size 2^q such that the effects in the same alias set are aliased with one another, and those from different groups are mutually orthogonal.*

Aliasing is an unavoidable consequence of fractionation. The usefulness of fractional factorials is derived from effect sparsity, which says that only a few out of a large number of effects are important. With this in mind, if all but one effect in each alias set are negligible, then all the nonnegligible effects can be estimated. The requirement set problem in fractional factorial designs, as pioneered by Greenfield (1976) and later systematically studied by Franklin and Bailey (1977), is to find a design such that all the required effects come from different alias sets.

A.1.2 Resolution

While all the properties of a 2^{m-q} design are completely determined by its defining relation through the aliasing pattern, one would like to have something simpler and powerful enough to capture the design's main characteristics. This is offered by the concept of resolution.

The *resolution* of a 2^{m-q} design is defined as the length of the shortest words in its defining relation. If a 2^{m-q} design has resolution R, we say it is a 2_R^{m-q} design. The resolution captures the aliasing properties of factorial effects of lower orders. In a 2_R^{m-q} design, no effect involving r factors ($r < R$) is aliased with any effect involving fewer than $R - r$ factors, although there exist two effects, one of order r and the other of order $R-r$, that are aliased with each other.

The resolution of a 2^{m-q} design has an important implication in its projection designs.

Theorem A.2. *A 2_R^{m-q} design is an orthogonal array of strength $R - 1$. In other words, any subdesign of $R - 1$ or fewer factors is a full factorial or several replicates thereof.*

The resolution of a 2^{m-q} design also has an implication in estimability of models containing lower order effects. We use $\lfloor x \rfloor$ and $\lceil x \rceil$ to denote the floor and ceiling functions, respectively.

Theorem A.3. *A 2_R^{m-q} design allows the estimation of any model consisting of effects of order $\lfloor (R-1)/2 \rfloor$ or lower, provided that effects of order $\lceil (R-1)/2 \rceil + 1$ or higher are negligible.*

It is useful to specialize Theorem A.3 for designs of resolution III, IV, and V, as they are the most useful regular factorial designs. According to Theorem A.3, we have that

- a 2^{m-q} design of resolution III allows the estimation of all main effects, provided that all interaction effects are negligible;
- a 2^{m-q} design of resolution IV allows the estimation of all main effects, provided that all interaction effects involving three or more factors are negligible; and

- a 2^{m-q} design of resolution V allows the estimation of all main effects and all two-factor interactions, provided that all interaction effects involving three or more factors are negligible.

The construction of resolution III designs is easy. Starting with a full factorial for k factors, any interaction column can be used to define an additional factor. One obtains a resolution III design of 2^k runs for $m = 2^k - 1$ factors if all the interaction columns are used to define additional factors. This design is a saturated resolution III design, as the main-effect columns use up all the degrees of freedom.

Resolution IV designs are also easy to construct. This will be dealt with in the next subsection.

The construction of resolution V designs is considerably harder compared to that of resolution III and IV designs. For designs of 16, 32, 64, and 128 runs, the maximum number of factors for a regular design to have resolution V is 5, 6, 8, and 11, respectively. Textbooks such as Box, Hunter, and Hunter (1978) and Wu and Hamada (2000) have details on the construction of these resolution V designs. The best general construction method of resolution V designs is via BCH codes (Hedayat, Sloane, and Stufken, 1999, Section 5.4).

A.1.3 Resolution IV designs and clear two-factor interactions

Standard textbooks give two equivalent constructions of resolution IV designs. Starting with a full factorial for k factors, if we use all the interactions involving odd numbers of factors to define additional factors, then a resolution IV design is obtained for 2^{k-1} factors. This resolution IV design is called a saturated resolution IV design, because there does not exist a resolution IV design of 2^k runs for more than 2^{k-1} factors. An equivalent construction is by folding over a saturated resolution III design. Adding a column of all ones to a saturated resolution III design of 2^{k-1} runs gives a square matrix $\boldsymbol{H}$ of order 2^{k-1}. Then the foldover design $\boldsymbol{D} = (\boldsymbol{H}^T, -\boldsymbol{H}^T)^T$ is a resolution IV design of 2^k runs with 2^{k-1} factors. Resolution IV designs with fewer than 2^{k-1} factors can be obtained by deleting some factors from the resolution IV design above.

Theorem A.4. *For given $n = 2^k$ runs, a resolution IV design can be constructed for m factors provided that $m \leq n/2$.*

The resolution IV designs obtained above are called even designs because the lengths of all defining words must be even. This raises the question of whether or not even designs are the only resolution IV designs. The answer is positive but only for designs of $n = 2^k$ runs with more than $5n/16$ factors. When the number of factors is $5n/16$ or less, non-even designs of resolution IV, called even/odd designs, of $n = 2^k$ runs can be constructed, one of which is obtained by repeated doubling of a 2_V^{5-1} design generated by $\boldsymbol{d}_5 = \boldsymbol{d}_1\boldsymbol{d}_2\boldsymbol{d}_3\boldsymbol{d}_4$. Let $\boldsymbol{D}_0$ denote this 2_V^{5-1} design. Then its double

$$\boldsymbol{D} = \begin{pmatrix} \boldsymbol{D}_0 & \boldsymbol{D}_0 \\ \boldsymbol{D}_0 & -\boldsymbol{D}_0 \end{pmatrix}$$

is an even/odd resolution IV design of 32 runs for 10 factors. Doubling $\boldsymbol{D}$ again, we obtain an even/odd resolution IV design of 64 runs for 20 factors. We summarize the above discussion in the next theorem.

Theorem A.5. *For $n = 2^k$ runs, we have that*

(i) resolution IV designs with $m > 5n/16$ factors must be even designs; and

(ii) even/odd resolution IV designs can be constructed for $m \leq 5n/16$ factors.

Theorem A.5 was presented in Chen and Cheng (2006); the original results are from studies on caps in finite projective geometries (Bruen, Haddad, and Wehlau, 1998).

A more general problem than the existence of even/odd designs is the study of maximal resolution IV designs. A resolution IV design is said to be *maximal* if it becomes a resolution III design whenever any factor is added. We have just seen two families of such designs. One is the saturated resolution IV designs of $n = 2^k$ runs for $m = n/2$ factors, and the other is the designs with $n = 2^k$ runs for $m = 5n/16$ factors obtained by repeated doubling of $\boldsymbol{D}_0$.

Obviously, any resolution IV design is either maximal in itself or a projection design of a maximal resolution IV design. If a complete collection of maximal resolution IV designs is obtained, then all resolution IV designs can be obtained by taking projections of the maximal resolution IV designs. Chen and Cheng (2006) provided a complete characterization of maximal resolution IV designs of n runs for $n/4+1$ or more factors, again based on results from finite projective geometries. Here we give a simple version of the results.

Theorem A.6. *For $n = 2^k$ with $k \geq 4$, let $m \geq n/4 + 1$. We have that*

(i) there exists a maximal resolution IV design for $m = n/4 + 1$ factors; and

(ii) for $m > n/4 + 1$, there exists a maximal resolution IV design of n runs for m factors if and only if $m \in \{n/2, 5n/16, 9n/32, 17n/64, 33n/128, \ldots\}$.

Resolution IV designs allow the estimation of all main effects under the assumption that all interactions involving three or more factors are negligible. Some two-factor interactions are aliased with one another in a resolution IV design, but the presence of nonnegligible two-factor interactions does not affect the estimation of main effects. This robust property makes resolution IV designs special.

The robust property of resolution IV designs just discussed is greatly enhanced when they are coupled with *clear two-factor interactions*. Suppose that interactions involving three or more factors are all negligible. Then a two-factor interaction is said to be clear if it is not aliased with any main effect and any other two-factor interaction. Now suppose that a resolution IV design has some clear two-factor interactions. Then these clear two-factor interactions along with all main effects can be estimated without assuming the negligibility of other two-factor interactions. Clear two-factor interactions do not always exist. A necessary and sufficient condition was discovered by Chen and Hedayat (1998).

Theorem A.7. *For a resolution IV design of $n = 2^k$ runs with m factors to have clear two-factor interactions, it is necessary and sufficient that $m \geq n/4 + 1$.*

Clear two-factor interactions can also be defined for resolution III designs but are less useful. For more details on clear two-factor interactions, we refer to Chen and Hedayat (1998) and Tang, Ma, Ingram, and Wang (2002).

A.1.4 Minimum aberration and other topics

The foundation on which most design selection criteria are built is the *hierarchical principle* of factorial effects, which states that (i) effects of lower orders are more important than those of higher orders and (ii) effects of the same order are equally important. This is also called the hierarchical assumption. The nomenclature "principle" is preferred to "assumption" as the idea has a similar flavor to that of the Pareto principle. The hierarchical principle of factorial effects is a reflection of the philosophy in scientific modeling that simple models should take precedence over complex ones.

The resolution provides a simple characterization for a regular factorial design in terms of its projection designs and the estimability of lower order factorial effects, and can therefore be used as a design selection criterion. According to the resolution criterion, designs of higher resolutions are better than those of lower resolutions, which makes a lot of sense as a design of higher resolution accomplishes more than a design of lower resolution in terms of both projection designs and the estimability of lower order factorial effects. As a selection criterion, the resolution is intuitive and easy to use. The only complaint one may possibly have about it is its crudeness when used as a selection criterion. So many designs have the same resolution. Which one to choose then? This question leads to the introduction of the minimum aberration criterion by Fries and Hunter (1980), perhaps the most celebrated concept in regular fractional factorials that has occupied the center stage in this area of research for more than four decades.

To define the minimum aberration criterion, we recall the defining words of a regular fractional factorial, which are the words in its defining relation. Let A_j be the number of defining words of length j for $j = 3, \ldots, m$. The vector $(A_3, \ldots, A_m)$ is the word length pattern. The resolution is given by the smallest integer R such that $A_R > 0$. The minimum aberration criterion is then defined as the one that selects designs by sequentially minimizing A_3, A_4, and so on. Designs found using this criterion are called minimum aberration designs. To be more specific, according to this criterion, competing designs are first compared using their A_3 values and those with the smallest A_3 value are selected. The selected designs are further discriminated by their A_4 values and those with the smallest A_4 values are identified. The process is continued until a unique design is found or the last entry A_m is used for comparison. Clearly, the minimum aberration criterion is a refinement of the resolution criterion. Consequently, minimum aberration designs must have the highest resolutions.

The minimum aberration criterion can be justified statistically from a number of angles. Here we briefly discuss two of them. Let ν_j be the number of j-factor interactions that are aliased with the main effects, for $j = 2, 3, \ldots, m$. Tang and Deng (1999) obtained the next result. Note that $A_1 = A_2 = 0$.

Theorem A.8. *We have that*

$$\nu_j = (j+1)A_{j+1} + (m-j+1)A_{j-1}$$

for $j = 2, \ldots, m-1$, *and* $\nu_m = A_{m-1}$. *Therefore, sequentially minimizing* $A_3, A_4, \ldots, A_m$ *is equivalent to sequentially minimizing* $\nu_2, \nu_3, \ldots, \nu_m$.

Minimum aberration designs have another statistical justification in having better estimation capacity. Let E_j be the number of estimable models consisting of all main effects and j two-factor interactions, and $(E_1, E_2, \ldots)$ is called the estimation capacity sequence. Cheng, Steinberg, and Sun (1999) showed that the minimum aberration criterion is a good surrogate for the estimation capacity criterion that selects designs with large E_j values for all j.

Finding minimum aberration designs is not a simple matter. For small run sizes, this can be done by looking up the complete catalogs of nonisomorphic designs provided by Chen, Sun, and Wu (1993) and Xu (2009). Chen and Wu (1991) and Chen (1992) obtained the first set of theoretical results, allowing the construction of minimum aberration 2^{m-q} designs for any m and $q = 1, \ldots, 5$. To date, the most powerful method for the construction of minimum aberration designs remains to be that using a complementary design theory. The theory was first developed by Chen and Hedayat (1996) and Tang and Wu (1996), and further refined and extended by Butler (2003a) and Xu and Cheng (2008). To present the basic complementary design theory, consider a regular saturated factorial design of 2^k runs for $2^k - 1$ factor. Let $\boldsymbol{D}$ be a design obtained by selecting m columns from such a saturated design and $\bar{\boldsymbol{D}}$ be its complementary design, which consists of the remaining columns.

Theorem A.9. *We have that*

$$A_j(\boldsymbol{D}) = a_0 + \sum_{i=3}^{j} a_i A_i(\bar{\boldsymbol{D}}),$$

where $a_j = (-1)^j$ *and* $a_0, a_3, \ldots, a_j$ *are all constants, depending only on* j, m, *and* k.

Theorem A.9 implies that one can find a minimum aberration design by finding its complementary design through sequentially minimizing $-A_3(\bar{\boldsymbol{D}}), A_4(\bar{\boldsymbol{D}}), -A_5(\bar{\boldsymbol{D}})$, and so on.

Blocking is a powerful device that allows the elimination of a known systematic variation in the experimental material. Although an unequivocal aberration-type criterion can be formulated for the selection of the best blocking schemes for full factorials–see Sun, Wu, and Chen (1997) and Tang (2007) for details, it is not clear how to best define an appropriate aberration criterion for the simultaneous selection of fractions and blocking schemes. A general Bayesian approach is presented in Chapter 11.

A.2 Introduction to Finite Fields

This section provides a brief introduction to finite fields, also called Galois fields in honor of Evariste Galois (1811–1832). Although finite fields are not a must for reading this book, some basic knowledge will be very helpful for better understanding certain topics such as multi-level regular designs and Paley's constructions of Hadamard matrices. For design researchers who want to know more about finite fields, we recommend Appendix A of Hedayat, Sloane, and Stufken (1999) or Chapter 2 of Vanstone and Van Oorschot (2013). Readers who wish for a more complete coverage may want to read Lidl and Niederreiter (1994).

A.2.1 Fields

The notion of a field is abstracted from our number systems in which we can perform the usual addition, subtraction, multiplication, and division for any two numbers (except that 0 cannot be in the denominator). Three familiar fields are all rational numbers, all real numbers, and all complex numbers. More concretely, all rational numbers form a field with respect to the usual addition, subtraction, multiplication, and division. So are all real numbers and all complex numbers. A formal definition is given below.

Definition A.1. *A field* F *is a set of elements which is closed under two binary operations, denoted by* $+$ *and* $\times$, *such that the following properties are satisfied for all* $a, b, c \in F$:

(i) $(a + b) + c = a + (b + c)$

(ii) $a + b = b + a$

(iii) there exists an element $0 \in F$ *such that* $a + 0 = a$

(iv) there exists an element $-a \in F$ *such that* $a + (-a) = 0$

(v) $(a \times b) \times c = a \times (b \times c)$

(vi) $a \times b = b \times a$

(vii) there exists an element $1 \in F$ *such that* $1 \times a = a$

(viii) for each $a \neq 0$*, there exists an element* $a^{-1} \in F$ *such that* $a^{-1} \times a = 1$

(ix) $a \times (b + c) = a \times b + a \times c$.

We refer to $+$ as addition and $\times$ as multiplication. Subtraction is defined as $a - b = a + (-b)$ and division is defined as $a/b = a \times b^{-1}$. Then we see that a field is a set of elements in which we can perform addition, subtraction, multiplication, and division for any two elements except that $a/0$ is not allowed. The addition obeys associative and commutative laws; so does the multiplication. The multiplication is distributive with respect to the addition. Furthermore, both additive and multiplicative identities, denoted by 0 and 1, respectively, need to exist.

For convenience, we simply write $a \times b$ as ab. As mentioned earlier, the set of rational numbers is a field with respect to the usual addition and multiplication. So are the set of real numbers and that of complex numbers.

A less familiar field is the set of numbers given by

$$F = \{a + b\sqrt{2} \mid a, b \text{ are rational numbers}\}$$

with respect to the usual addition and multiplication. All requirements but (viii) for F to be a field are obviously satisfied. To verify (viii), for every nonzero number $a + b\sqrt{2}$ in F where a, b cannot be both zero, we see that its inverse

$$(a + b\sqrt{2})^{-1} = \frac{1}{a + b\sqrt{2}} = \frac{a - b\sqrt{2}}{a^2 - 2b^2}$$

still belongs to F.

Another example of a field is the set of all numbers of the form

$$a + b\sqrt{2} + c\sqrt{3} + d\sqrt{6},$$

where a, b, c, and d are rational numbers.

Examples of sets of numbers that do not form fields are abundant. The set of positive integers is not a field, since it does not even have 0. The set of all integers is not a field, because $1/n$ is not an integer for any integer n with $|n| \geq 2$. The set of all irrational numbers is not a field, because it is not closed under the addition ($\sqrt{2} - \sqrt{2} = 0$ is rational).

Remark A.1. *A set of elements is called a* group *if it is closed under a binary operation, which follows the associative law and admits an inverse for any element. It is called an Abelian group if the binary operation is commutative. To put it simply, a field is an Abelian group under the addition, and the set of nonzero elements is an Abelian group under the multiplication, where the addition and multiplication are connected through the distributive law.*

A.2.2 Finite fields

A *finite field* is a field that has a finite number of elements. This subsection provides results on the existence and construction of finite fields.

Theorem A.10. *For every prime* p*, the set* $\{0, 1, \ldots, p - 1\}$ *of residues modulo* p *is a finite field under the standard addition and multiplication modulo* p*. We denote this field by* $GF(p)$.

TABLE A.1
Addition table for $GF(5)$.

+	0	1	2	3	4
0	0	1	2	3	4
1	1	2	3	4	0
2	2	3	4	0	1
3	3	4	0	1	2
4	4	0	1	2	3

TABLE A.2
Multiplication table for $GF(5)$.

×	0	1	2	3	4
0	0	0	0	0	0
1	0	1	2	3	4
2	0	2	4	1	3
3	0	3	1	4	2
4	0	4	3	2	1

Example A.1. *Consider* $GF(5) = \{0, 1, 2, 3, 4\}$. *The addition and multiplication tables are given in Tables A.1 and A.2.*

Using the tables, we obtain that $-1 = 4$ *as* $1 + 4 = 0$, *and that* $2^{-1} = 3$ *as* $2 \times 3 = 1$.

We now consider polynomials with coefficients all from $GF(p)$. Let

$$g(x) = a_0 + a_1x + \cdots + a_{k-1}x^{k-1} + x^k$$

be a monic polynomial of degree k. For any polynomial $f(x)$, there are unique polynomials $q(x)$ and $r(x)$ such that

$$f(x) = q(x)g(x) + r(x),$$

where the degree of $r(x)$ is lower than k. Then $r(x)$ is called the residue of $f(x)$ modulo $g(x)$. We say that $g(x)$ is *irreducible* if it cannot be factorized into a product of two polynomials with their degrees both lower than k.

Theorem A.11. *If* $g(x)$ *is irreducible, then the set of residues modulo* $g(x)$ *as given by*

$$\{b_0 + b_1x + \cdots + b_{k-1}x^{k-1} \mid b_0, b_1, \ldots, b_{k-1} \in GF(p)\}$$

is a finite field under the operations of polynomial addition and multiplication modulo $g(x)$. *We denote this field by* $GF(p^k)$.

Example A.2. *One can check that* $g(x) = x^2 + x + 1$ *is irreducible over* $GF(2) = \{0, 1\}$. *We therefore obtain a* $GF(4) = \{0, 1, x, x + 1\}$ *with addition and multiplication modulo* $g(x) = x^2 + x + 1$. *The addition and multiplication tables of this field are given in Tables A.3 and A.4.*

Using these tables, we obtain that $-x = x$ *as* $x + x = 0$ *and* $x^{-1} = x + 1$ *as* $x(x + 1) = 1$.

The next result completes the construction of finite fields.

Theorem A.12. *For every positive integer* $k \geq 1$, *there exists an irreducible monic polynomial of degree* k. *This implies that a* $GF(p^k)$ *can be constructed for every positive integer* k *and every prime* p.

TABLE A.3
Addition table for $GF(4)$.

$+$	0	1	x	$x+1$
0	0	1	x	$x+1$
1	1	0	$x+1$	x
x	x	$x+1$	0	1
$x+1$	$x+1$	x	1	0

TABLE A.4
Multiplication table for $GF(4)$.

times	0	1	x	$x+1$
0	0	0	0	0
1	0	1	x	$x+1$
x	0	x	$x+1$	1
$x+1$	0	$x+1$	1	x

Are there other finite fields than those given in Theorem A.12? The question is settled in the next theorem.

Theorem A.13. *Every finite field is isomorphic to a $GF(p^k)$ given in Theorem A.12 for some prime p and some positive integer k.*

Two isomorphic fields are essentially the same. To put it briefly, two finite fields are isomorphic if there is a one-to-one mapping between them that preserves the operations of addition and multiplication. Because of Theorem A.13, essentially there is only one $GF(p^k)$. When referring to it, we can simply use $GF(p^k)$ without the need of an article "a" or "the".

A.2.3 Primitive elements and primitive polynomials

A *primitive element* in $GF(p^k)$ is an element a that can generate all the non-zero elements through its powers. More precisely, an element a in $GF(p^k)$ is called a primitive element if the non-zero elements of $GF(p^k)$ can be written as $a, a^2, \ldots, a^{s-1} = 1$ where $s = p^k$. In group-theoretic language, the multiplicative group formed by the non-zero elements in $GF(p^k)$ is cyclic.

This representation of the non-zero elements of a finite field facilitates multiplicative calculations, because one can calculate the product easily using $a^m a^n = a^{m+n}$. For example, in $GF(5)$, 2 is primitive. In fact, we have that $2^2 = 4$, $2^3 = 3$, and $2^4 = 1$ in $GF(5)$. Therefore $2^2 \times 2^3 = 2^5 = 2^1$ because $2^4 = 1$. The representation using powers of a primitive element for a finite field is akin to the polar representation of complex numbers.

Theorem A.14. *There exists a primitive element in every $GF(p^k)$.*

For $k \geq 2$, the elements of $GF(p^k)$ are polynomials over $GF(p)$, with the addition and multiplication defined using an irreducible polynomial. It would be very convenient if x, the simplest polynomial, is primitive in $GF(p^k)$. An irreducible polynomial is called a *primitive polynomial* if x is a primitive element in $GF(p^k)$.

Theorem A.15. *There exists a primitive polynomial for every $k \geq 2$.*

In Example A.2, the irreducible $x^2 + x + 1$ is a primitive polynomial, because $x, x^2 = x + 1, x^3 = 1$ are the three non-zero elements of $GF(4)$.

Table A.5 provides primitive elements in small $GF(p)$s. Table A.6 gives a list of primitive polynomials for small $GF(p^k)$s.

TABLE A.5
Primitive elements a in small $GF(p)$s.

p	3	5	7	11	13	17	19	23	29	31	37	41	43	47
a	2	2	3	2	2	3	2	5	2	3	2	6	3	5

TABLE A.6
Primitive polynomials for small $GF(p^k)$s.

p	k	primitive polynomial
2	2	x^2+x+1
2	3	x^3+x+1
2	4	x^4+x+1
2	5	x^5+x^2+1
3	2	x^2+x+2
3	3	x^3+2x+1
3	4	x^4+x+2
5	2	x^2+x+2
5	3	x^3+3x+2
7	2	x^2+x+3
7	3	x^3+3x+2
11	2	x^2+x+7
13	2	x^2+x+2

In $GF(s)$ where s is a prime power, a nonzero element b is said to be a quadratic residue if $b=c^2$ for some $c \in GF(s)$. This concept is relevant to Paley's constructions of Hadamard matrices, discussed in Section 2.5.

Theorem A.16. *The following assertions hold.*

(i) For even s, all nonzero elements of $GF(s)$ are quadratic residues.

(ii) For odd s, half of all nonzero elements of $GF(s)$ are quadratic residues and half are not. Let a be a primitive element. Then $a^2, a^4, \ldots, a^{s-1}$ are quadratic residues and $a, a^3, \ldots, a^{s-2}$ are not.

A.3 Finite Geometries

Let F be a field. Consider

$$F^m = \{\boldsymbol{x} = (x_1, \ldots, x_m) \mid x_j \in F \text{ for } j = 1, \ldots, m\},$$

the set of all vectors of length n with components from F. For two vectors $\boldsymbol{x} = (x_1, \ldots, x_m)$ and $\boldsymbol{y} = (y_1, \ldots, y_m)$ in F^m and a scalar $a \in F$, we define the vector addition as

$$\boldsymbol{x} + \boldsymbol{y} = (x_1 + y_1, \ldots, x_m + y_m),$$

and the scalar multiplication as

$$a\boldsymbol{x} = (ax_1, \ldots, ax_m).$$

Then F^m is an m-dimensional vector space over F.

If F is the real field R, then $F^m = R^m$ is the usual m-dimensional Euclidean space over R.

A.3.1 Finite Euclidean geometries

From this point on, we take F to be a finite field $GF(s)$, where s is a prime power. Then F^m is an m-dimensional Euclidean space over the finite field F, and it is also called an m-dimensional Euclidean geometry over F. This is a *finite Euclidean geometry* since the total number of points (vectors) in the geometry is s^m.

Almost all concepts and results one has learned from a linear algebra course about R^m are valid for a finite Euclidean geometry. We discuss a few of these that are most relevant to the design theory.

A set of vectors, $\boldsymbol{x}_1, \ldots, \boldsymbol{x}_k$, in F^m is said to be linearly independent if we must have $a_1 = \cdots = a_k = 0$ in order for $a_1\boldsymbol{x}_1 + \cdots + a_k\boldsymbol{x}_k = \boldsymbol{0}$ to hold with $a_j \in F$. If $\boldsymbol{x}_1, \ldots, \boldsymbol{x}_m$ in F^m are linearly independent, then they form a basis, meaning that for any $\boldsymbol{y}$ in F^m, there are unique $a_1, \ldots, a_m$ in F such that $\boldsymbol{y} = a_1\boldsymbol{x}_1 + \cdots + a_m\boldsymbol{x}_m$.

A subset U of F^m is a linear subspace if it is closed under the vector addition and scalar multiplication, and it has dimension k if the vectors in U can be generated from a set of k linearly independent vectors $\boldsymbol{x}_1, \ldots, \boldsymbol{x}_k$ as given below:

$$U = \{a_1\boldsymbol{x}_1 + \cdots + a_k\boldsymbol{x}_k \mid a_j \in F\}. \tag{A.1}$$

For two vectors $\boldsymbol{x} = (x_1, \ldots, x_m)$ and $\boldsymbol{y} = (y_1, \ldots, y_m)$, their inner product is defined as

$$\langle \boldsymbol{x}, \boldsymbol{y} \rangle = x_1y_1 + \cdots + x_my_m.$$

If $\langle \boldsymbol{x}, \boldsymbol{y} \rangle = 0$, we say that $\boldsymbol{x}$ and $\boldsymbol{y}$ are orthogonal. Let $\boldsymbol{A} = (a_{ij})_{l\times m}$ be an $l \times m$ matrix with entries all from F such that the row vectors are linearly independent. Then the set of vectors given by

$$U = \{\boldsymbol{x} \mid \boldsymbol{A}\boldsymbol{x}^T = \boldsymbol{0}\} \tag{A.2}$$

is a linear subspace of dimension $m - l$, which consists of all vectors that are orthogonal to the rows of $\boldsymbol{A}$. If we take all linear combinations of the rows of $\boldsymbol{A}$, we obtain another linear subspace, which is the orthogonal complement of the linear subspace U in (A.2).

We have seen that a linear subspace can either be generated using its own basis as in (A.1) or be defined through a basis of its orthogonal complement as in (A.2).

A.3.2 Finite projective geometries

Finite projective geometries can be derived from finite Euclidean geometries. This is done as follows. Consider all nonzero vectors in F^m. Two nonzero vectors $\boldsymbol{x} = (x_1, \ldots, x_m)$ and $\boldsymbol{y} = (y_1, \ldots, y_m)$ are said to be equivalent if $(x_1, \ldots, x_m) = a(y_1, \ldots, y_m)$ for some nonzero $a \in F$. If we regard each set of equivalent vectors as one point, we obtain an $(m-1)$-dimensional finite projective geometry, which is denoted by $PG(m-1, s)$. A $PG(m-1, s)$ has $(s^m - 1)/(s - 1)$ points in total, since each set of equivalent nonzero vectors has $s - 1$ members. We may use any vector in an equivalence set as a representative for a point in the $PG(m-1, s)$.

Suppose that $\boldsymbol{x}_1, \ldots, \boldsymbol{x}_m$ are a basis in F^m. Then the points in the corresponding $PG(m-1, s)$ can be obtained by using all linear combinations

$$a_1\boldsymbol{x}_1 + \cdots + a_m\boldsymbol{x}_m$$

with the restrictions that $a_1, \ldots, a_m$ are not all zero and that the first nonzero a_j is set to be 1.

Corresponding to a k-dimensional linear subspace of F^m is a projective geometry $PG(k-1, s)$, which is called an $(k-1)$-*flat* in $PG(m-1, s)$. An $(k-1)$-flat has $(s^k-1)/(s-1)$ points. A 1-flat, called a line, has $s+1$ points.

Again suppose that $\boldsymbol{x}_1, \ldots, \boldsymbol{x}_k$ are linearly independent in F^m. Then a $(k-1)$-flat is given by all linear combinations

$$a_1\boldsymbol{x}_1 + \cdots + a_k\boldsymbol{x}_k$$

such that $a_1, \ldots, a_k$ are not all zero and the first nonzero a_j is set to be 1. A 1-flat or a line is given by

$$\boldsymbol{x}_2,\ \boldsymbol{x}_1 + \alpha_0\boldsymbol{x}_2,\ \boldsymbol{x}_1 + \alpha_1\boldsymbol{x}_2,\ \ldots,\ \boldsymbol{x}_1 + \alpha_{s-1}\boldsymbol{x}_2,$$

where we have listed the elements of F as $F = \{\alpha_0, \alpha_1, \ldots, \alpha_{s-1}\}$.

Three points are said to be collinear if they are on the same line. A set of points is called a *cap* if no three of them are collinear.

A.4 Regular s-Level Factorial Designs

This section gives a brief review of regular s-level factorial designs. We only cover those topics that are crucial to the readers of this book. For a detailed treatment of regular three-level designs, we refer to standard textbooks such as Wu and Hamada (2000) and Montgomery (2017). For a general treatment of regular s-level designs, we refer to Cheng (2014a).

A.4.1 s^{m-q} designs as finite geometries

Let F be a finite field of order s, where s is a prime power. Consider a full factorial $\boldsymbol{C}$ for k s-level factors, which is an $s^k \times k$ matrix consisting of all s^k level combinations. One way to write down this full factorial is as follows. Let

$$\boldsymbol{x}_1 = (1, 0, 0, \ldots, 0),\ \boldsymbol{x}_2 = (0, 1, 0, \ldots, 0),\ \ldots,\ \boldsymbol{x}_k = (0, 0, 0, \ldots, 1). \tag{A.3}$$

Then the s^k runs of the full factorial $\boldsymbol{C}$ are all the linear combinations

$$a_1\boldsymbol{x}_1 + \cdots + a_k\boldsymbol{x}_k$$

with coefficients $a_1, \ldots, a_k$ from F. Let

$$\boldsymbol{C} = (\boldsymbol{u}_1, \ldots, \boldsymbol{u}_k),$$

where $\boldsymbol{u}_j$ is the jth column of the full factorial $\boldsymbol{C}$. We refer to $\boldsymbol{u}_j$ as the main-effect column of the jth factor. A linear combination

$$b_1\boldsymbol{u}_1 + \cdots + b_k\boldsymbol{u}_k$$

of the main-effect columns is an interaction column. If there are l nonzero b_js in the linear combination, then it is an l-factor interaction column, which is a main-effect column if $l = 1$. Two interaction columns $b_1\boldsymbol{u}_1 + \cdots + b_k\boldsymbol{u}_k$ and $b_1'\boldsymbol{u}_1 + \cdots + b_k'\boldsymbol{u}_k$ are equivalent if $(b_1, \ldots, b_k) = a(b_1', \ldots, b_k')$ for some a in F. Two equivalent interaction columns define the same $(s-1)$-dimensional contrast space, so we only need to choose one from a set of equivalent interaction columns. One choice for a full set of nonequivalent interaction columns is to use all linear combinations

$$b_1\boldsymbol{u}_1 + \cdots + b_k\boldsymbol{u}_k$$

with the restrictions that the b_js are not all zeros and the first nonzero b_j is set to be 1. There are in total $(s^k - 1)/(s - 1)$ such linear combinations. Collecting all these columns gives a matrix $\boldsymbol{S}$ of s^k rows and $(s^k - 1)/(s - 1)$ columns. We see that the columns of $\boldsymbol{S}$ are the points in the projective geometry $PG(k - 1)$. In fact, $\boldsymbol{S}$ is a saturated orthogonal array of strength two, and the construction method is called the Rao-Hamming construction.

If we take m columns ($m > k$) from $\boldsymbol{S}$, we obtain a regular s^{m-q} design where $q = m - k$. Let $\boldsymbol{D}$ denote an s^{m-q} design. Further let

$$\boldsymbol{x}_1^*,\ \boldsymbol{x}_2^*,\ \ldots,\ \boldsymbol{x}_k^*$$

denote the k rows of $\boldsymbol{D}$ that correspond to $\boldsymbol{x}_1, \ldots, \boldsymbol{x}_k$ in (A.3). Then we see that the rows of $\boldsymbol{D}$ are given by taking all the linear combinations of $\boldsymbol{x}_1^*, \boldsymbol{x}_2^*, \ldots, \boldsymbol{x}_k^*$. We summarize the above discussion in a theorem.

Theorem A.17. *Let $\boldsymbol{D}$ be an s^{m-q} design. Then we have that*

(i) the rows of design $\boldsymbol{D}$ form a k-dimensional linear subspace of the m-dimensional Euclidean geometry F^m; and

(ii) the columns of design $\boldsymbol{D}$ are a set of distinct points in the $(k-1)$-dimensional projective geometry $PG(k - 1, s)$.

A.4.2 Defining relation, resolution, and minimum aberration

Let $\boldsymbol{D}$ be an s^{m-q} design. We know from Theorem A.17 that the rows of $\boldsymbol{D}$ form a k-dimensional linear subspace of the m-dimensional linear space F^m. The orthogonal complement of this k-dimensional linear subspace has dimension q and contains s^q vectors. Collecting the s^q vectors as rows, we obtain a matrix, denote by $\boldsymbol{D}^{\perp}$. We see that $\boldsymbol{D}^{\perp}$ is the orthogonal complement of $\boldsymbol{D}$, or we may simply say that $\boldsymbol{D}^{\perp}$ is the dual of $\boldsymbol{D}$.

The dual $\boldsymbol{D}^{\perp}$ gives the *defining relation* of design $\boldsymbol{D}$. More precisely, Any row $\boldsymbol{x} \neq (0, \ldots, 0)$ of $\boldsymbol{D}^{\perp}$ is called a *defining word*. Two defining words $\boldsymbol{x} = (x_1, \ldots, x_m)$ and $\boldsymbol{y} = (y_1, \ldots, y_m)$ are equivalent if $(x_1, \ldots, x_m) = \lambda(y_1, \ldots, y_m)$ for some λ in F. The defining relation collects a complete set of $(s^q - 1)/(s - 1)$ nonequivalent defining words. We see that the defining relation consists of all the points in the projective geometry derived from $\boldsymbol{D}^{\perp}$.

A defining word is said to have length j if it contains j nonzero entries. The resolution of design $\boldsymbol{D}$ is the length of the shortest defining word. Let A_j be the number of defining words of length j. The vector $(A_1, \ldots, A_m)$ is the word length pattern. The minimum aberration criterion is defined as one that selects designs by sequentially minimizing $A_1, \ldots, A_m$.

The columns of an s^{m-q} design are a set of distinct points in $PG(k - 1, s)$; see Theorem A.17. Such an s^{m-q} design must have resolution III or higher. However, if we allow a point in $PG(k - 1, s)$ to be selected more than once, the resulting design has resolution II or higher. For an s^{m-q} design to have resolution IV or higher, no three points can be collinear. In other words, an s^{m-q} design of resolution IV or higher is a cap in $PG(k - 1, s)$.

When viewed as an orthogonal array, an s^{m-q} design is a linear orthogonal array because its rows form a linear subspace over F. The strength of the array is determined by the resolution of the design.

Theorem A.18. *An s^{m-q} design of resolution R is a linear orthogonal array of strength R–1.*

Let $\boldsymbol{D}$ be an s^{m-q} design and $\bar{\boldsymbol{D}}$ be its complement, which consists of the columns in the saturated design $\boldsymbol{S}$ that are not selected by $\boldsymbol{D}$. In the language of projective geometries, $\boldsymbol{D}$ is a set of m points and $\bar{\boldsymbol{D}}$ contains all the points of $PG(k-1,s)$ outside $\boldsymbol{D}$. The following result was established by Suen, Chen, and Wu (1997) using a result from coding theory.

Theorem A.19. *We have that*

$$A_j(\boldsymbol{D}) = \sum_{i=0}^{j} c_{ji} A_i(\bar{\boldsymbol{D}})$$

where $c_{jj} = (-1)^j$ and the other c_{ji}'s are all real constants depending only on n, m, and s.

Theorem A.19 shows that finding a minimum aberration design $\boldsymbol{D}$ from a saturated regular s-level design $\boldsymbol{S}$ is equivalent to finding a complementary design $\bar{\boldsymbol{D}}$ that sequentially minimizes $-A_3(\bar{\boldsymbol{D}}), A_4(\bar{\boldsymbol{D}}), -A_5(\bar{\boldsymbol{D}})$, and so on.

We conclude this section with an illustrative example.

Example A.3. *Consider the following three-level design:*

$$\boldsymbol{D} = \begin{pmatrix} 0 & 0 & 0 & 0 \\ 0 & 1 & 1 & 2 \\ 0 & 2 & 2 & 1 \\ 1 & 0 & 1 & 1 \\ 1 & 1 & 2 & 0 \\ 1 & 2 & 0 & 2 \\ 2 & 0 & 2 & 2 \\ 2 & 1 & 0 & 1 \\ 2 & 2 & 1 & 0 \end{pmatrix}.$$

This is a 3^{4-2} design, which has 9 runs for four 3-level factors. The rows of design $\boldsymbol{D}$ is a linear subspace of F^4, where $F = \{0,1,2\}$ is a finite field of order 3. The linear subspace has a basis consisting of two row vectors

$$(0,1,1,2) \ \ \textit{and} \ \ (1,0,1,1);$$

so the rows of $\boldsymbol{D}$ can be obtained by taking all linear combinations of these two vectors.

Let $\boldsymbol{D} = (\boldsymbol{u}_1, \boldsymbol{u}_2, \boldsymbol{u}_3, \boldsymbol{u}_4)$, where $\boldsymbol{u}_j$ is the jth column of $\boldsymbol{D}$. The two columns $\boldsymbol{u}_1$ and $\boldsymbol{u}_2$ give a full factorial for two factors. Columns $\boldsymbol{u}_3$ and $\boldsymbol{u}_4$ are generated by

$$\boldsymbol{u}_3 = \boldsymbol{u}_1 + \boldsymbol{u}_2 \ \ \textit{and} \ \ \boldsymbol{u}_4 = \boldsymbol{u}_1 + 2\boldsymbol{u}_2, \tag{A.4}$$

with all calculations performed modulo 3. The four columns $\boldsymbol{u}_1, \boldsymbol{u}_2, \boldsymbol{u}_3, \boldsymbol{u}_4$ can be regarded as the four points in a one-dimensional projective geometry $PG(1,3)$. The equations in (A.4) can be equivalently written as

$$\boldsymbol{u}_1 + \boldsymbol{u}_2 + 2\boldsymbol{u}_3 = \boldsymbol{0} \ \ \textit{and} \ \ \boldsymbol{u}_1 + 2\boldsymbol{u}_2 + 2\boldsymbol{u}_4 = \boldsymbol{0}, \tag{A.5}$$

where $\boldsymbol{0}$ denotes a column vector of all zeros. The coefficient vectors

$$\boldsymbol{e}_1 = (1,1,2,0) \ \ \textit{and} \ \ \boldsymbol{e}_2 = (1,2,0,2)$$

of the two equations in (A.5) form a basis in $\boldsymbol{D}^{\perp}$. We can obtain $\boldsymbol{D}^{\perp}$ by taking all linear combinations $a_1\boldsymbol{e}_1 + a_2\boldsymbol{e}_2$ with $a_1, a_2 \in F$. Each nonzero row vector in $\boldsymbol{D}^{\perp}$ is a defining word. There are four nonequivalent defining words, given by

$$\boldsymbol{e}_1 = (1,1,2,0), \ \ \boldsymbol{e}_2 = (1,2,0,2), \ \ \boldsymbol{e}_1 + \boldsymbol{e}_2 = (2,0,2,2), \ \ \boldsymbol{e}_1 + 2\boldsymbol{e}_2 = (0,2,2,1).$$

A simpler set of nonequivalent defining words are

$$(1,1,2,0),\quad (1,2,0,2),\quad (1,0,1,1),\quad (0,1,1,2).$$

In terms of the design columns, the four defining words mean that

$$\boldsymbol{u}_1 + \boldsymbol{u}_2 + 2\boldsymbol{u}_3 = \boldsymbol{u}_1 + 2\boldsymbol{u}_2 + 2\boldsymbol{u}_4 = \boldsymbol{u}_1 + \boldsymbol{u}_3 + \boldsymbol{u}_4 = \boldsymbol{u}_2 + \boldsymbol{u}_3 + 2\boldsymbol{u}_4 = \boldsymbol{0}.$$

In standard textbooks, the above four equations are written symbolically as

$$I = ABC^2 = AB^2D^2 = ACD = BCD^2.$$

Since all the defining words have length 3, the design is of resolution III.

Remark A.2. *As a linear subspace of F^m, an s^{m-q} design can be defined as one that collects as its runs all row vectors $\boldsymbol{x}$ satisfying*

$$\boldsymbol{A}\boldsymbol{x}^T = \boldsymbol{0},$$

where the rows of $\boldsymbol{A}$ are q independent defining words and $\boldsymbol{0}$ is a column vector of all zeros. A broader definition of an s^{m-q} design is given by all row vectors $\boldsymbol{x}$ satisfying

$$\boldsymbol{A}\boldsymbol{x}^T = \boldsymbol{b},$$

where $\boldsymbol{b}$ is a column vector of length q with entries from F. For any $\boldsymbol{b}$, the solution set is called a flat. Flats for the same $\boldsymbol{A}$ but different $\boldsymbol{b}$s are said to be parallel. A parallel-flat design is one that combines several parallel flats as one design.

A.5 Introduction to Error-Correcting Codes

Some concepts and results from coding theory are alluded to several times in this book. The brief introduction to the theory of error-correcting codes given here is meant to be useful for our readers. For textbooks on the subject, we refer to Vanstone and Van Oorschot (2013) and Pless (2011). For comprehensive coverage, we refer to MacWilliams and Sloane (1977).

A.5.1 Fundamental concepts

Let F be a finite field of order s. Consider row vectors of length m with entries from F. The *Hamming weight* of a row vector $\boldsymbol{x} = (x_1, \ldots, x_m)$, denoted by $w(\boldsymbol{x})$, is the number of nonzero entries in $\boldsymbol{x}$. The *Hamming distance* between two vectors $\boldsymbol{x} = (x_1, \ldots, x_m)$ and $\boldsymbol{y} = (y_1, \ldots, y_m)$, denoted by $d(\boldsymbol{x}, \boldsymbol{y})$, is the number of entries where the two vectors differ. Obviously, we have that

$$d(\boldsymbol{x}, \boldsymbol{y}) = w(\boldsymbol{x} - \boldsymbol{y}).$$

A *code* is a collection of row vectors with entries from F, and can be represented by an $n \times m$ matrix $\boldsymbol{C} = (c_{ij})_{n\times m}$ with $c_{ij} \in F$. If $s = 2$, $\boldsymbol{C}$ is a binary code. A row vector of $\boldsymbol{C}$ is called a *codeword*; so $\boldsymbol{C}$ has n codewords. The distance of the code $\boldsymbol{C}$ is the minimum distance between two distinct codewords; that is,

$$d(\boldsymbol{C}) = \min_{\boldsymbol{x}\neq\boldsymbol{y}} d(\boldsymbol{x}, \boldsymbol{y}),$$

where the minimization is over all pairs of distinct codewords $\boldsymbol{x}$ and $\boldsymbol{y}$.

We use a code to transmit messages over a communication channel. When a codeword representing a message is sent through a noisy channel, errors may occur in some components of the codeword, causing the received vector to be possibly different from the sent codeword. A decoding strategy is then used to recover the true codeword from the received vector. The *nearest neighbor decoding* strategy decodes a received vector to the codeword that is closest with respect to the Hamming distance. This strategy is intuitively appealing, and can also be formally justified by *maximum likelihood decoding*.

Suppose a code has distance d, meaning that any two distinct codewords are apart by at least distance d. If no more than $\lfloor (d-1)/2 \rfloor$ errors occur when a codeword is sent through a noisy channel, the nearest neighbor decoding strategy can always find the correct codeword. We say that the code can *correct* $\lfloor (d-1)/2 \rfloor$ errors, and write this as a theorem.

Theorem A.20. *A code with distance d can correct $\lfloor (d-1)/2 \rfloor$ errors.*

According to Theorem A.20, a code with distance 3 can correct one error and a code with distance 5 can correct two errors. Consider a code with distance 4. It can correct one error by Theorem A.20. If two errors occur, there may be two or more codewords that have distance 2 from the received vector. In this case, the nearest neighbor decoding strategy cannot identify the correct codeword, but we do know that two errors have occurred. Therefore, a code with distance 4 can *correct* one error and *detect* two errors.

A code is called *linear* if its codewords form a linear subspace over F, and called *nonlinear* otherwise. The codewords of a linear code can be generated using its *generator matrix* $\boldsymbol{G}_{k\times m}$ via

$$\boldsymbol{x} = \boldsymbol{u}\boldsymbol{G} \quad \text{for all } \boldsymbol{u} = (u_1, \ldots, u_k) \text{ with } u_j \in F.$$

The resulting code has $n = s^k$ codewords. The rows of $\boldsymbol{G}$ are independent codewords.

The orthogonal complement $\boldsymbol{C}^{\perp}$ of a linear code $\boldsymbol{C}$ is called the *dual code* of $\boldsymbol{C}$. This means that the codewords of $\boldsymbol{C}$ are orthogonal to those of $\boldsymbol{C}^{\perp}$. Let $\boldsymbol{H}$ be the generator matrix of $\boldsymbol{C}^{\perp}$. Then $\boldsymbol{H}$ is the *parity check matrix* of $\boldsymbol{C}$. The codewords of $\boldsymbol{C}$ are all the row vectors $\boldsymbol{x}$ satisfying

$$\boldsymbol{H}\boldsymbol{x}^T = \boldsymbol{0}.$$

We see that a code $\boldsymbol{C}$ can be specified either by its generator matrix $\boldsymbol{G}$ or by its parity check matrix $\boldsymbol{H}$.

For any two codewords $\boldsymbol{x}$ and $\boldsymbol{y}$ of a linear code $\boldsymbol{C}$, $\boldsymbol{x} - \boldsymbol{y}$ must also a codeword in $\boldsymbol{C}$. Therefore

$$d(\boldsymbol{C}) = \min\{w(\boldsymbol{x}) : \boldsymbol{x} \in \boldsymbol{C}, \boldsymbol{x} \neq \boldsymbol{0}\}.$$

That is, the distance of $\boldsymbol{C}$ is the Hamming weight of the lightest nonzero codeword in $\boldsymbol{C}$. The distance of $\boldsymbol{C}^{\perp}$ is called the *dual distance* of $\boldsymbol{C}$.

In light of the above discussion, when a linear $\boldsymbol{C}$ is viewed as a regular s-level factorial design, the codewords of $\boldsymbol{C}^{\perp}$ are defining words of design $\boldsymbol{C}$. This gives the following result.

Theorem A.21. *Let $d^{\perp}$ be the dual distance of a linear code $\boldsymbol{C}$. Then as a regular design, $\boldsymbol{C}$ has resolution $d^{\perp}$.*

A.5.2 MacWilliams theorem and MacWilliams transform

We first continue our discussion on linear codes. Let $\boldsymbol{C}$ be a linear code and B_j be the number of codewords with Hamming weight j. Then $(B_0, B_1, \ldots, B_m)$ is called the *weight distribution* of $\boldsymbol{C}$. We use $(B_0^{\perp}, B_1^{\perp}, \ldots, B_m^{\perp})$ to denote the weight distribution of the dual code $\boldsymbol{C}^{\perp}$. The MacWilliams theorem establishes a relationship between $(B_0, B_1, \ldots, B_m)$ and $(B_0^{\perp}, B_1^{\perp}, \ldots, B_m^{\perp})$. A concise form of the relationship is provided through the *weight enumerators* of the two codes.

We define the weight enumerator of $\boldsymbol{C}$ as

$$W_{\boldsymbol{C}}(x,y)=\sum_{j=0}^{m}B_j x^{m-j}y^j.$$

The MacWilliam theorem shows that there is a simple relationship between the weight enumerator of $\boldsymbol{C}$ and that of $\boldsymbol{C}^{\perp}$.

Theorem A.22. *We have that*

$$W_{\boldsymbol{C}^{\perp}}(x,y)=\frac{1}{n}W_{\boldsymbol{C}}(x+(s-1)y,x-y). \tag{A.6}$$

Both sides of (A.6) are homogeneous polynomials. By matching the coefficients of both sides, we obtain the following explicit relationship between the two weight distributions.

Corollary A.1. *We have that*

$$B_j^{\perp}=n^{-1}\sum_{i=0}^{m}B_iP_j(i;m,s)$$

for $j=0,1,\ldots,m$, *where*

$$P_j(x;m,s)=\sum_{i=0}^{j}(-1)^i(s-1)^{j-i}\binom{x}{i}\binom{m-x}{j-i} \tag{A.7}$$

are the Krawtchouk polynomials.

As mentioned at the end of Section A.5.1, when $\boldsymbol{C}$ is viewed as a regular factorial design, the rows of $\boldsymbol{C}^{\perp}$ are defining words of design $\boldsymbol{C}$. We see that the word length pattern $(A_1,\ldots,A_m)$ of design $\boldsymbol{C}$ is related to the weight distribution of the dual code $\boldsymbol{C}^{\perp}$ via

$$A_j=(s-1)^{-1}B_j^{\perp}.$$

Making use of this connection between the word length pattern and the weight distribution, Suen, Chen, and Wu (1997) derived a relation between the word length pattern of a regular s-level design and that of its complementary design; see Theorem A.19.

We now consider a general code $\boldsymbol{C}$, linear or nonlinear. Let

nB_j be the number of ordered pairs of codewords that are at distance j.

The vector $(B_0,B_1,\ldots,B_m)$ is called the *distance distribution* of $\boldsymbol{C}$. If $\boldsymbol{C}$ is linear, then the distance distribution is the same as the weight distribution.

If $\boldsymbol{C}$ is nonlinear, there is no such thing as its dual code. But we can still define

$$B_j^{\perp}=n^{-1}\sum_{i=0}^{m}B_iP_j(i;m,s)$$

for $j=0,1,\ldots,m$, where $P_j(x;m,s)$ is as in (A.7). We call $(B_0^{\perp},B_1^{\perp},\ldots,B_m^{\perp})$ the *MacWilliams transform* of the distance distribution $(B_0,B_1,\ldots,B_m)$.

As shown in Delsarte (1973), the MacWilliams transform of the distance distribution for a nonlinear code has the same properties as the weight distribution of the dual code for a linear code, based on which Delsarte (1973) derived the powerful *linear programming bound* for the size of a code.

The MacWilliams transform of the distance distribution is relevant to nonregular designs. As discussed in Section 10.3, it is the same as the generalized word length pattern of a nonregular design.

References

S. Addelman and O. Kempthorne. Some main-effect plans and orthogonal arrays of strength two. *Annals of Mathematical Statistics*, 32:1167–1176, 1961.

M. Ai and R. Zhang. Projection justification of generalized minimum aberration for asymmetrical fractional factorial designs. *Metrika*, 60:279–285, 2004.

M. Ai, L. Kang, and V. R. Joseph. Bayesian optimal blocking of factorial designs. *Journal of Statistical Planning and Inference*, 139:3319–3328, 2009.

M. Ai, B. Jiang, and K. Li. Construction of sliced space-filling designs based on balanced sliced orthogonal arrays. *Statistica Sinica*, 24:1685–1702, 2014.

S. Ba. SLHD: maximin-distance (sliced) latin hypercube designs. R package version 1.2-1, available at `http://cran.r-project.org/package=SLHD`, 2013.

S. Ba, W. R. Myers, and W. A. Brenneman. Optimal sliced Latin hypercube designs. *Technometrics*, 57:479–487, 2015.

R. Bailey. The decomposition of treatment degrees of freedom in quantitative factorial experiments. *Journal of the Royal Statistical Society: Series B (Methodological)*, 44:63–70, 1982.

R. Bailey, F. Gilchrist, and H. Patterson. Identification of effects and confounding patterns in factorial designs. *Biometrika*, 64:347–354, 1977.

R. A. Bailey. Factorial design and Abelian groups. *Linear Algebra and its Applications*, 70: 349–368, 1985.

N. Balakrishnan and P. Yang. Connections between the resolutions of general two-level factorial designs. *Annals of the Institute of Statistical Mathematics*, 58:609, 2006.

T. Banerjee and R. Mukerjee. Optimal factorial designs for cDNA microarray experiments. *Annals of Applied Statistics*, 2:366–385, 2008.

J. Bierbrauer, Y. Edel, and W. C. Schmid. Coding-theoretic constructions for (t, m, s)-nets and ordered orthogonal arrays. *Journal of Combinatorial Designs*, 10:403–418, 2002.

D. Bingham and R. Sitter. Minimum-aberration two-level fractional factorial split-plot designs. *Technometrics*, 41:62–70, 1999a.

D. Bingham and R. Sitter. Some theoretical results for fractional factorial split-plot designs. *Annals of Statistics*, 27:1240–1255, 1999b.

D. Bingham and R. Sitter. Design issues in fractional factorial split-plot experiments. *Journal of Quality Technology*, 33:2–15, 2001.

D. Bingham, R. Sitter, E. Kelly, L. Moore, and J. D. Olivas. Factorial designs with multiple levels of randomization. *Statistica Sinica*, 18:493–513, 2008.

D. Bingham, R. R. Sitter, and B. Tang. Orthogonal and nearly orthogonal designs for computer experiments. *Biometrika*, 96:51–65, 2009.

R. M. Block and R. W. Mee. Second order saturated resolution IV designs. *J. Stat. Theory Appl*, 2:96–112, 2003.

K. H. Booth and D. R. Cox. Some systematic supersaturated designs. *Technometrics*, 4: 489–495, 1962.

R. Bose and K. Bush. Orthogonal arrays of strength two and three. *Annals of Mathematical Statistics*, 23:508–524, 1952.

R. C. Bose. Mathematical theory of the symmetrical factorial design. *Sankhyā: The Indian Journal of Statistics*, 8:107–166, 1947.

G. E. Box and N. R. Draper. A basis for the selection of a response surface design. *Journal of the American Statistical Association*, 54:622–654, 1959.

G. E. Box and J. S. Hunter. The 2^{k-p} fractional factorial designs. *Technometrics*, 3:311–351, 1961.

G. E. Box and J. Tyssedal. Projective properties of certain orthogonal arrays. *Biometrika*, 83:950–955, 1996.

G. E. Box, W. H. Hunter, and S. Hunter. *Statistics for Experimenters*, volume 664. John Wiley & Sons, 1978.

A. A. Bruen, L. Haddad, and D. L. Wehlau. Binary codes and caps. *Journal of Combinatorial Designs*, 6:275–284, 1998.

D. A. Bulutoglu and C.-S. Cheng. Hidden projection properties of some nonregular fractional factorial designs and their applications. *Annals of Statistics*, 31:1012–1026, 2003.

D. A. Bulutoglu and C.-S. Cheng. Construction of E(s^2)-optimal supersaturated designs. *Annals of Statistics*, 32:1662–1678, 2004.

D. A. Bulutoglu and D. M. Kaziska. A counterexample to Beder's conjectures about Hadamard matrices. *Journal of Statistical Planning and Inference*, 139:3381–3383, 2009.

D. A. Bulutoglu and K. J. Ryan. Algorithms for finding generalized minimum aberration designs. *Journal of Complexity*, 31:577–589, 2015.

K. Bush. Orthogonal arrays of index unity. *Annals of Mathematical Statistics*, 23:426–434, 1952.

N. A. Butler. Optimal and orthogonal latin hypercube designs for computer experiments. *Biometrika*, 88:847–857, 2001.

N. A. Butler. Some theory for constructing minimum aberration fractional factorial designs. *Biometrika*, 90:233–238, 2003a.

N. A. Butler. Minimum aberration construction results for nonregular two-level fractional factorial designs. *Biometrika*, 90:891–898, 2003b.

N. A. Butler. Minimum G_2-aberration properties of two-level foldover designs. *Statistics & probability letters*, 67:121–132, 2004.

N. A. Butler. Generalised minimum aberration construction results for symmetrical orthogonal arrays. *Biometrika*, 92:485–491, 2005.

N. A. Butler. Optimal blocking of two-level factorial designs. *Biometrika*, 93:289–302, 2006.

N. A. Butler. Results for two-level fractional factorial designs of resolution IV or more. *Journal of Statistical Planning and Inference*, 137:317–323, 2007.

N. A. Butler, R. Mead, K. M. Eskridge, and S. G. Gilmour. A general method of constructing E(s^2)-optimal supersaturated designs. *Journal of the Royal Statistical Society: Series B (Statistical Methodology)*, 63:621–632, 2001.

M. Caboara, G. Pistone, E. Riccomagno, and H. P. Wynn. The fan of an experimental design. Unpublished manuscript, 1999.

R. Carnell. lhs: Latin hypercube samples. R package version 1.0.2, available at `https://CRAN.R-project.org/package=lhs`, 2020.

I. M. Chakravarti. Fractional replication in asymmetrical factorial designs and partially balanced arrays. *Sankhyā: The Indian Journal of Statistics (1933-1960)*, 17:143–164, 1956.

M.-C. Chang. *Effect Aliasing in Gaussian Random Field Models.* PhD thesis, National Tsing Hua University, 2015.

M.-C. Chang. A unified framework for minimum aberration. *Statistica Sinica*, 32:251–269, 2022.

M.-C. Chang. Orthogonalized moment aberration for multi-stratum factorial designs. Unpublished manuscript, 2024.

M.-C. Chang and C.-S. Cheng. A Bayesian approach to the selection of two-level multi-stratum factorial designs. *Annals of Statistics*, 46:1779–1806, 2018.

G. Chen and B. Tang. A study of orthogonal array-based designs under a broad class of space-filling criteria. *Annals of Statistics*, 50:2925–2949, 2022a.

G. Chen and B. Tang. Using nonregular designs to generate space-filling designs. *Journal of Statistical Planning and Inference*, 221:201–211, 2022b.

G. Chen and B. Tang. Minimum aberration factorial designs under a mixed parametrization. *Statistica Sinica, accepted*, 2024.

G. Chen, C. Shi, and B. Tang. Nonregular designs from Paley's Hadamard matrices: Generalized resolution, projectivity and hidden projection property. *Electronic Journal of Statistics*, 17:2120–2138, 2023.

H. Chen. Some projective properties of fractional factorial designs. *Statistics & Probability Letters*, 40:185–188, 1998.

H. Chen and C.-S. Cheng. Theory of optimal blocking of 2^{n-m} designs. *Annals of Statistics*, 27:1948–1973, 1999.

H. Chen and C.-S. Cheng. Aberration, estimation capacity and estimation index. *Statistica Sinica*, 14:203–215, 2004.

H. Chen and C.-S. Cheng. Doubling and projection: a method of constructing two-level designs of resolution IV. *Annals of Statistics*, 34:546–558, 2006.

H. Chen and A. Hedayat. 2^{n-l} designs with weak minimum aberration. *Annals of Statistics*, 24:2536–2548, 1996.

H. Chen and A. Hedayat. 2^{n-m} designs with resolution III or IV containing clear two-factor interactions. *Journal of Statistical Planning and Inference*, 75:147–158, 1998.

J. Chen. Some results on 2^{n-k} fractional factorial designs and search for minimum aberration designs. *Annals of Statistics*, 20:2124–2141, 1992.

J. Chen and C. Wu. Some results on s^{n-k} fractional factorial designs with minimum aberration or optimal moments. *Annals of Statistics*, 19:1028–1041, 1991.

J. Chen, D. Sun, and C. Wu. A catalogue of two-level and three-level fractional factorial designs with small runs. *International Statistical Review/Revue Internationale de Statistique*, 61:131–145, 1993.

R.-B. Chen, D.-N. Hsieh, Y. Hung, and W. Wang. Optimizing Latin hypercube designs by particle swarm. *Statistics and Computing*, 23:663–676, 2013.

C.-S. Cheng. Optimality of certain asymmetrical experimental designs. *Annals of Statistics*, 6:1239–1261, 1978.

C.-S. Cheng. Optimality of some weighing and 2^n fractional factorial designs. *Annals of Statistics*, 8:436–446, 1980.

C.-S. Cheng. An application of the kiefer-wolfowitz equivalence theorem to a problem in hadamard transform optics. *Annals of Statistics*, 15:1593–1603, 1987.

C.-S. Cheng. On the optimality of (M,S)-optimal designs in large systems. *Sankhyā: The Indian Journal of Statistics, Series A*, 54:117–125, 1992.

C.-S. Cheng. Some projection properties of orthogonal arrays. *Annals of Statistics*, 23: 1223–1233, 1995a.

C.-S. Cheng. Complete class results for the moment matrices of designs over permutation-invariant sets. *Annals of Statistics*, 23:41–54, 1995b.

C.-S. Cheng. E(s^2)-optimal supersaturated designs. *Statistica Sinica*, 7:929–939, 1997.

C.-S. Cheng. Some hidden projection properties of orthogonal arrays with strength three. *Biometrika*, 85:491–495, 1998.

C.-S. Cheng. *Theory of Factorial Design: Single- and Multi-Stratum Experiments.* CRC Press, 2014a.

C.-S. Cheng. Optimal biased weighing designs and two-level main-effect plans. *Journal of Statistical Theory and Practice*, 8:83–99, 2014b.

C.-S. Cheng and M.-H. Kao. Optimal experimental designs for fMRI via circulant biased weighing designs. *Annals of Statistics*, 43:2565–2587, 2015.

C.-S. Cheng and B. Tang. Upper bounds on the number of columns in supersaturated designs. *Biometrika*, 88:1169–1174, 2001.

C.-S. Cheng and B. Tang. A general theory of minimum aberration and its applications. *Annals of Statistics*, 33:944–958, 2005.

C.-S. Cheng and P.-W. Tsai. Multistratum fractional factorial designs. *Statistica Sinica*, 21:1001–1021, 2011.

C.-S. Cheng, D. M. Steinberg, and D. X. Sun. Minimum aberration and model robustness for two-level fractional factorial designs. *Journal of the Royal Statistical Society: Series B (Statistical Methodology)*, 61:85–93, 1999.

C.-S. Cheng, L.-Y. Deng, and B. Tang. Generalized minimum aberration and design efficiency for nonregular fractional factorial designs. *Statistica Sinica*, 12:991–1000, 2002.

C.-S. Cheng, R. W. Mee, and O. Yee. Second order saturated orthogonal arrays of strength three. *Statistica Sinica*, 18:105–119, 2008.

C.-S. Cheng, M.-H. Kao, and F. K. H. Phoa. Optimal and efficient designs for functional brain imaging experiments. *Journal of Statistical Planning and Inference*, 181:71–80, 2017.

C.-S. Cheng, A. Das, R. Singh, and P.-W. Tsai. E(s^2)-and UE(s^2)-optimal supersaturated designs. *Journal of Statistical Planning and Inference*, 196:105–114, 2018.

C.-S. Cheng, Y. He, and B. Tang. Minimal second order saturated designs and their applications to space-filling designs. *Statistica Sinica*, 31:867–890, 2021.

S.-W. Cheng and C. Wu. Factor screening and response surface exploration. *Statistica Sinica*, 11:553–580, 2001.

S. W. Cheng and C. J. Wu. Choice of optimal blocking schemes in two-level and three-level designs. *Technometrics*, 44:269–277, 2002.

S.-W. Cheng and K. Q. Ye. Geometric isomorphism and minimum aberration for factorial designs with quantitative factors. *Annals of Statistics*, 32:2168–2185, 2004.

S.-W. Cheng, W. Li, and K. Q. Ye. Blocked nonregular two-level factorial designs. *Technometrics*, 46:269–279, 2004.

C. J. Colbourn. Combinatorial aspects of covering arrays. *Le Matematiche*, 59:125–172, 2004.

C. Daniel. Use of half-normal plots in interpreting factorial two-level experiments. *Technometrics*, 1:311–341, 1959.

A. Das, A. Dey, L.-Y. Chan, and K. Chatterjee. On E(s^2)-optimal supersaturated designs. *Journal of Statistical Planning and Inference*, 138:3749–3757, 2008.

A. A. Davydov, S. Marcugini, and F. Pambianco. Minimal 1-saturating sets and complete caps in binary projective spaces. *Journal of Combinatorial Theory, Series A*, 113:647–663, 2006.

P. Delsarte. An algebraic approach to the association schemes of coding theory. *Philips Res. Rep. Suppl.*, 10:vi+–97, 1973.

L.-Y. Deng and B. Tang. Generalized resolution and minimum aberration criteria for Plackett-Burman and other nonregular factorial designs. *Statistica Sinica*, 9:1071–1082, 1999.

L.-Y. Deng and B. Tang. Design selection and classification for Hadamard matrices using generalized minimum aberration criteria. *Technometrics*, 44:173–184, 2002.

L.-Y. Deng, D. K. Lin, and J. Wang. A resolution rank criterion for supersaturated designs. *Statistica Sinica*, 9:605–610, 1999.

X. Deng, Y. Hung, and C. D. Lin. Design for computer experiments with qualitative and quantitative factors. *Statistica Sinica*, 25:1567–1581, 2015.

A. Dey and R. Mukerjee. *Fractional factorial plans.* John Wiley & Sons, 1999.

A. Dey and C.-Y. Suen. Optimal fractional factorial plans for main effects and specified two-factor interactions: a projective geometric approach. *Annals of Statistics*, 30:1512–1523, 2002.

R. Dorfman. The detection of defective members of large populations. *Annals of mathematical statistics*, 14:436–440, 1943.

W. DuMouchel and B. Jones. A simple Bayesian modification of D-optimal designs to reduce dependence on an assumed model. *Technometrics*, 36:37–47, 1994.

J. Eccleston and A. Hedayat. On the theory of connected designs: characterization and optimality. *Annals of Statistics*, 2:1238–1255, 1974.

H. Evangelaras. Construction of minimum generalized aberration two-level orthogonal arrays. *Electronic Journal of Statistics*, 9:2689–2705, 2015.

H. Evangelaras, C. Koukouvinos, A. Dean, and C. Dingus. Projection properties of certain three level orthogonal arrays. *Metrika*, 62:241–257, 2005.

H. Evangelaras, C. Koukouvinos, and E. Lappas. Further contributions to nonisomorphic two level orthogonal arrays. *Journal of Statistical Planning and Inference*, 137:2080–2086, 2007.

H. Evangelaras, C. Koukouvinos, and E. Lappas. 27-run nonisomorphic three level orthogonal arrays: Identification, evaluation and projection properties. *Utilitas Mathematica*, 84, 2011.

K. Fang, M.-Q. Liu, H. Qin, and Y.-D. Zhou. *Theory and Application of Uniform Experimental Designs*, volume 221. Springer, 2018.

K.-T. Fang, D. K. Lin, and C.-X. Ma. On the construction of multi-level supersaturated designs. *Journal of Statistical Planning and Inference*, 86:239–252, 2000.

K.-T. Fang, R. Li, and A. Sudjianto. *Design and Modeling for Computer Experiments.* Chapman and Hall/CRC, 2005.

D. Finney. The fractional replication of factorial arrangement. *Annals of Eugenics*, 12: 291–301, 1945.

R. Fontana, G. Pistone, and M. P. Rogantin. Classification of two-level factorial fractions. *Journal of Statistical Planning and Inference*, 87:149–172, 2000.

M. Franklin. Selecting defining contrasts and confounded effects in p^{n-m} factorial experiment. *Technometrics*, 27:165–172, 1985.

M. Franklin and R. Bailey. Selection of defining contrasts and confounded effects in two-level experiments. *Journal of the Royal Statistical Society: Series C (Applied Statistics)*, 26:321–326, 1977.

A. Fries and W. G. Hunter. Minimum aberration 2^{k-p} designs. *Technometrics*, 22:601–608, 1980.

E. M. Gabidulin, A. A. Davydov, and L. M. Tombak. Linear codes with covering radius 2 and other new covering codes. *IEEE Transactions on Information Theory*, 37:219–224, 1991.

Z. Galil and J. Kiefer. D-optimum weighing designs. *Annals of Statistics*, 8:1293–1306, 1980.

S. Ghosh. Sequential assembly of fractions in factorial experiments. *Handbook of Statistics*, 13:407–435, 1996.

G. Glonek and P. Solomon. Factorial and time course designs for cDNA microarray experiments. *Biostatistics*, 5:89–111, 2004.

R. B. Gramacy. *Surrogates: Gaussian Process Modeling, Design, and Optimization for the Applied Sciences.* Chapman and Hall/CRC, 2020.

A. Greenfield. Selection of defining contrasts in two-level experiments. *Journal of the Royal Statistical Society: Series C (Applied Statistics)*, 25:64–67, 1976.

U. Grömping. A note on dominating fractional factorial two-level designs with clear two-factor interactions. *Technometrics*, 56:42–45, 2014.

U. Grömping. An algorithm for blocking regular fractional factorial 2-level designs with clear two-factor interactions. *Computational Statistics & Data Analysis*, 153:107059, 2021.

U. Grömping. R package SOAs: creation of stratum orthogonal arrays, version 1.2. In *R Core Team (2022). R: A Language and Environment for Statistical Computing.* R Foundation for Statistical Computing Vienna, Austria, 2022.

U. Grömping. A unifying implementation of stratum (aka strong) orthogonal arrays. *Computational Statistics & Data Analysis*, 183:107739, 2023.

U. Grömping and H. Xu. Generalized resolution for orthogonal arrays. *Annals of Statistics*, 42:918–939, 2014.

A. Grosso, A. Jamali, and M. Locatelli. Finding maximin Latin hypercube designs by iterated local search heuristics. *European Journal of Operational Research*, 197:541–547, 2009.

A. J. Gutman, D. A. Bulutoglu, and E. D. White. Supersaturated designs with the maximum number of factors for a given resolution-rank. *Journal of Statistical Theory and Practice*, 9:419–435, 2015.

M. Harwit and N. J. A. Sloane. *Hadamard transform optics.* New York: Academic Press, 1979.

X. He and P. Z. Qian. Nested orthogonal array-based latin hypercube designs. *Biometrika*, 98:721–731, 2011.

Y. He and B. Tang. Strong orthogonal arrays and associated Latin hypercubes for computer experiments. *Biometrika*, 100:254–260, 2013.

Y. He and B. Tang. A characterization of strong orthogonal arrays of strength three. *Annals of Statistics*, 42:1347–1360, 2014.

Y. He, C. D. Lin, and F. Sun. On construction of marginally coupled designs. *Statistica Sinica*, 27:665–683, 2017.

Y. He, C.-S. Cheng, and B. Tang. Strong orthogonal arrays of strength two plus. *Annals of Statistics*, 46:457–468, 2018.

A. Hedayat, K. Pu, and J. Stufken. On the construction of asymmetrical orthogonal arrays. *Annals of Statistics*, 20:2142–2152, 1992.

A. Hedayat, E. Seiden, and J. Stufken. On the maximal number of factors and the enumeration of 3-symbol orthogonal arrays of strength 3 and index 2. *Journal of Statistical Planning and Inference*, 58:43–63, 1997.

A. S. Hedayat, N. J. A. Sloane, and J. Stufken. *Orthogonal arrays: theory and applications.* Springer Science & Business Media, 1999.

F. Hickernell. A generalized discrepancy and quadrature error bound. *Mathematics of Computation*, 67:299–322, 1998.

K. J. Horadam. *Hadamard Matrices and Their Applications.* Princeton University Press, 2007.

H. Hotelling. Some improvements in weighing and other experimental techniques. *Annals of Mathematical Statistics*, 15:297–306, 1944.

P. Huang, D. Chen, and J. O. Voelkel. Minimum-aberration two-level split-plot designs. *Technometrics*, 40:314–326, 1998.

S. Huda and R. Mukerjee. Two-level minimum aberration designs in $n = 2 \pmod 4$ runs. *Journal of Statistical Planning and Inference*, 143:1872–1879, 2013.

R. L. Iman and W.-J. Conover. A distribution-free approach to inducing rank correlation among input variables. *Communications in Statistics-Simulation and Computation*, 11: 311–334, 1982.

D. Ingram and B. Tang. Minimum G-aberration design construction and design tables for 24 runs. *Journal of Quality Technology*, 37:101–114, 2005.

M. Jacroux. Blocking in two-level non-regular fractional factorial designs. *Journal of Statistical Planning and Inference*, 139:1215–1220, 2009.

M. Jacroux, C. Wong, and J. Masaro. On the optimality of chemical balance weighing designs. *Journal of Statistical Planning and Inference*, 8:231–240, 1983.

B. Jiang and M. Ai. Construction of uniform U-designs. *Journal of Statistical Planning and Inference*, 181:1–10, 2017.

B. Jiang, Z. Wang, and Y. Wang. Strong orthogonal arrays of strength two-plus based on the Addelman-Kempthorne method. *Statistics & Probability Letters*, 175:109114, 2021.

P. W. John. Three-quarter replicates of 2^4 and 2^5 designs. *Biometrics*, 17:319–321, 1961.

P. W. John. Three quarter replicates of 2^n designs. *Biometrics*, 18:172–184, 1962.

M. E. Johnson, L. M. Moore, and D. Ylvisaker. Minimax and maximin distance designs. *Journal of Statistical Planning and Inference*, 26:131–148, 1990.

B. Jones and D. Majumdar. Optimal supersaturated designs. *Journal of the American Statistical Association*, 109:1592–1600, 2014.

B. Jones and C. J. Nachtsheim. A class of three-level designs for definitive screening in the presence of second-order effects. *Journal of Quality Technology*, 43:1–15, 2011.

B. Jones, R. Lekivetz, D. Majumdar, C. J. Nachtsheim, and J. W. Stallrich. Construction, properties, and analysis of group-orthogonal supersaturated designs. *Technometrics*, 62: 403–414, 2020.

B. A. Jones, W. Li, C. J. Nachtsheim, and K. Q. Ye. Model-robust supersaturated and partially supersaturated designs. *Journal of Statistical Planning and Inference*, 139:45–53, 2009.

V. R. Joseph. A Bayesian approach to the design and analysis of fractionated experiments. *Technometrics*, 48:219–229, 2006.

V. R. Joseph and J. D. Delaney. Functionally induced priors for the analysis of experiments. *Technometrics*, 49:1–11, 2007.

V. R. Joseph and Y. Hung. Orthogonal-maximin Latin hypercube designs. *Statistica Sinica*, 18:171–186, 2008.

V. R. Joseph, M. Ai, and C. J. Wu. Bayesian-inspired minimum aberration two-and four-level designs. *Biometrika*, 96:95–106, 2009.

V. R. Joseph, E. Gul, and S. Ba. Maximum projection designs for computer experiments. *Biometrika*, 102:371–380, 2015.

L. Kang and V. R. Joseph. Bayesian optimal single arrays for robust parameter design. *Technometrics*, 51:250–261, 2009.

M.-H. Kao. Universally optimal fMRI designs for comparing hemodynamic response functions. *Statistica Sinica*, 25:499–506, 2015.

R. C. Karunanayaka and B. Tang. Compromise designs under baseline parameterization. *Journal of Statistical Planning and Inference*, 190:32–38, 2017.

R. C. Karunanayaka and B. Tang. On the existence and constructions of orthogonal designs. *Australian & New Zealand Journal of Statistics*, 60:471–480, 2018.

T. I. Katsaounis and A. M. Dean. A survey and evaluation of methods for determination of combinatorial equivalence of factorial designs. *Journal of Statistical Planning and Inference*, 138:245–258, 2008.

W. Ke and B. Tang. Selecting 2^{m-p} designs using a minimum aberration criterion when some two-factor interactions are important. *Technometrics*, 45:352–360, 2003.

W. Ke, B. Tang, and H. Wu. Compromise plans with clear two-factor interactions. *Statistica Sinica*, 15:709–715, 2005.

K. F. Kerr. Efficient 2^k factorial designs for blocks of size 2 with microarray applications. *Journal of Quality Technology*, 38:309–318, 2006.

M. K. Kerr. Bayesian optimal fractional factorials. *Statistica Sinica*, 11:605–630, 2001.

H. Kharaghani and B. Tayfeh-Rezaie. Hadamard matrices of order 32. *Journal of Combinatorial Designs*, 21:212–221, 2013.

J. Kiefer. General equivalence theory for optimum designs (approximate theory). *Annals of Statistics*, 2:849–879, 1974.

J. Kiefer. Construction and optimality of generalized Youden designs II. In *A Survey of Statistical Designs and Linear Models*, pages 333–353. North Holland, 1975.

J. Kiefer and J. Wolfowitz. The equivalence of two extremum problems. *Canadian Journal of Mathematics*, 12:363–366, 1960.

J. Lawrence, R. N. Kacker, Y. Lei, D. R. Kuhn, and M. Forbes. A survey of binary covering arrays. *Electronic Journal of Combinatorics*, 18:84–84, 2011.

K. M. Lawrence. A combinatorial characterization of (t, m, s)-nets in base b. *Journal of Combinatorial Designs*, 4:275–293, 1996.

R. Lekivetz and J. Morgan. Covering arrays: Using prior information for construction, evaluation and to facilitate fault localization. *Journal of Statistical Theory and Practice*, 14:1–20, 2020.

R. Lekivetz and B. Tang. Robust designs through partially clear two-factor interactions. *Biometrika*, 98:733–739, 2011.

R. A. Lekivetz. *Optimal Factorial Designs with Robust Properties*. PhD thesis, Simon Fraser University, 2011.

R. V. Lenth. Quick and easy analysis of unreplicated factorials. *Technometrics*, 31:469–473, 1989.

S. Lewis and A. Dean. Detection of interactions in experiments on large numbers of factors. *Journal of the Royal Statistical Society Series B: Statistical Methodology*, 63:633–672, 2001.

P. Li, A. Miller, and B. Tang. Algorithmic search for baseline minimum aberration designs. *Journal of Statistical Planning and Inference*, 149:172–182, 2014.

W. Li, M.-Q. Liu, and B. Tang. A method of constructing maximin distance designs. *Biometrika*, 108:845–855, 2021.

W. Li, M.-Q. Liu, and B. Tang. A systematic construction of compromise designs under baseline parameterization. *Journal of Statistical Planning and Inference*, 219:33–42, 2022a.

W. Li, M.-Q. Liu, and J.-F. Yang. Construction of column-orthogonal strong orthogonal arrays. *Statistical Papers*, 63:515–530, 2022b.

C. Liao, H. Iyer, and D. Vecchia. Construction of orthogonal two-level designs of user-specified resolution where $n \neq 2^k$. *Technometrics*, 38:342–353, 1996.

R. Lidl and H. Niederreiter. *Introduction to Finite Fields and Their Applications*. Cambridge University Press, 1994.

C. D. Lin and B. Tang. Latin hypercubes and space-filling designs. In *Handbook of Design and Analysis of Experiments*, pages 613–646. Chapman and Hall/CRC, 2015.

C. D. Lin, A. Miller, and R. Sitter. Folded over non-orthogonal designs. *Journal of Statistical Planning and Inference*, 138:3107–3124, 2008.

C. D. Lin, R. Mukerjee, and B. Tang. Construction of orthogonal and nearly orthogonal Latin hypercubes. *Biometrika*, 96:243–247, 2009.

C. D. Lin, D. Bingham, R. R. Sitter, and B. Tang. A new and flexible method for constructing designs for computer experiments. *Annals of Statistics*, 38:1460–1477, 2010.

D. K. Lin. A new class of supersaturated designs. *Technometrics*, 35:28–31, 1993.

D. K. Lin. Generating systematic supersaturated designs. *Technometrics*, 37:213–225, 1995.

H. Liu and M.-Q. Liu. Column-orthogonal strong orthogonal arrays and sliced strong orthogonal arrays. *Statistica Sinica*, 25:1713–1734, 2015.

M. Liu and R. Zhang. Construction of E(s^2)-optimal supersaturated designs using cyclic BIBDs. *Journal of Statistical Planning and Inference*, 91:139–150, 2000.

X. Lu and Y. Sun. Supersaturated design with more than two levels. *Chinese Annals of Mathematics*, 22:183–194, 2001.

X. Lu, W. Hu, and Y. Zheng. A systematical procedure in the construction of multi-level supersaturated design. *Journal of Statistical Planning and Inference*, 115:287–310, 2003.

C.-X. Ma and K.-T. Fang. A note on generalized aberration in factorial designs. *Metrika*, 53:85–93, 2001.

F. J. MacWilliams and N. J. A. Sloane. *The Theory of Error-Correcting Codes*, volume 16. Elsevier, 1977.

B. H. Margolin. Results on factorial designs of resolution IV for the 2^n and 2^n3^m series. *Technometrics*, 11:431–444, 1969a.

B. H. Margolin. Resolution IV fractional factorial designs. *Journal of the Royal Statistical Society: Series B (Methodological)*, 31:514–523, 1969b.

M. McKay, R. Beckman, and W. Conover. A comparison of three methods for selecting values of input variables in the analysis of output from a computer code. *Technometrics*, 21:239–245, 1979.

R. W. Mee and R. L. Bates. Split-lot designs: Experiments for multistage batch processes. *Technometrics*, 40:127–140, 1998.

A. Miller. Strip-plot configurations of fractional factorials. *Technometrics*, 39:153–161, 1997.

A. Miller and R. Sitter. Choosing columns from the 12-run Plackett–Burman design. *Statistics & Probability Letters*, 67:193–201, 2004.

A. Miller and R. R. Sitter. Using the folded-over 12-run Plackett—Burman design to consider interactions. *Technometrics*, 43:44–55, 2001.

A. Miller and B. Tang. Minimal dependent sets for evaluating supersaturated designs. *Statistica Sinica*, 22:1273–1285, 2012.

A. Miller and B. Tang. Finding MDS-optimal supersaturated designs using computer searches. *Journal of Statistical Theory and Practice*, 7:703–712, 2013.

A. Miller and B. Tang. Using regular fractions of two-level designs to find baseline designs. *Statistica Sinica*, 26:745–759, 2016.

T. J. Mitchell, M. D. Morris, and D. Ylvisaker. Two-level fractional factorials and Bayesian prediction. *Statistica Sinica*, 4:559–573, 1995.

D. C. Montgomery. *Design and Analysis of Experiments.* John Wiley & Sons, 2017.

H. Moon, A. Dean, and T. Santner. Algorithms for generating maximin Latin hypercube and orthogonal designs. *Journal of Statistical Theory and Practice*, 5:81–98, 2011.

L. B. Morales and G. Vega. On the enumeration of E(s^2)-optimal and minimax-optimal k-circulant supersaturated designs. *Journal of Combinatorial Designs*, 22:149–160, 2014.

L. B. Morales, D. A. Bulutoglu, and K. T. Arasu. The maximum number of columns in supersaturated designs with $s_{\max} = 2$. *Journal of Combinatorial Designs*, 27:448–472, 2019.

J. Morgan, S. Ghosh, and A. M. Dean. J.N. Srivastava and experimental design. *Journal of Statistical Planning and Inference*, 144:3–18, 2014.

S. Moriguti. Optimality of orthogonal designs. *Rep. Statist. Appl. Res. Un. Japan Sci. Engrs*, 3:75–98, 1954.

M. D. Morris and T. J. Mitchell. Exploratory designs for computational experiments. *Journal of Statistical Planning and Inference*, 43:381–402, 1995.

R. Mukerjee and K.-T. Fang. Fractional factorial split-plot designs with minimum aberration and maximum estimation capacity. *Statistica Sinica*, 12:885–903, 2002.

R. Mukerjee and B. Tang. Optimal fractions of two-level factorials under a baseline parameterization. *Biometrika*, 99:71–84, 2012.

R. Mukerjee and B. Tang. A complementary set theory for quaternary code designs. *Annals of Statistics*, 41:2768–2785, 2013.

R. Mukerjee and B. Tang. Optimal two-level regular designs under baseline parametrization via cosets and minimum moment aberration. *Statistica Sinica*, 26:1001–1019, 2016.

R. Mukerjee and C. J. Wu. On the existence of saturated and nearly saturated asymmetrical orthogonal arrays. *Annals of Statistics*, 23:2102–2115, 1995.

R. Mukerjee, F. Sun, and B. Tang. Nearly orthogonal arrays mappable into fully orthogonal arrays. *Biometrika*, 101:957–963, 2014.

G. L. Mullen and W. C. Schmid. An equivalence between (t, m, s)-nets and strongly orthogonal hypercubes. *Journal of Combinatorial Theory, Series A*, 76:164–174, 1996.

J. A. Nelder. The analysis of randomized experiments with orthogonal block structure. I. Block structure and the null analysis of variance. *Proceedings of the Royal Society of London. Series A. Mathematical and Physical Sciences*, 283:147–162, 1965.

N.-K. Nguyen. An algorithmic approach to constructing supersaturated designs. *Technometrics*, 38:69–73, 1996.

N.-K. Nguyen and C.-S. Cheng. New E(s^2)-optimal supersaturated designs constructed from incomplete block designs. *Technometrics*, 50:26–31, 2008.

H. Niederreiter. Point sets and sequences with small discrepancy. *Monatshefte für Mathematik*, 104:273–337, 1987.

A. W. Nordstrom and J. P. Robinson. An optimum nonlinear code. *Information and Control*, 11:613–616, 1967.

A. B. Owen. Orthogonal arrays for computer experiments, integration and visualization. *Statistica Sinica*, 2:439–452, 1992.

A. B. Owen. Controlling correlations in Latin hypercube samples. *Journal of the American Statistical Association*, 89:1517–1522, 1994.

R. E. Paley. On orthogonal matrices. *Journal of Mathematics and Physics*, 12:311–320, 1933.

F. Pang and M.-Q. Liu. Indicator function based on complex contrasts and its application in general factorial designs. *Journal of Statistical Planning and Inference*, 140:189–197, 2010.

F. Pang, M.-Q. Liu, and D. K. Lin. A construction method for orthogonal Latin hypercube designs with prime power levels. *Statistica Sinica*, 19:1721–1728, 2009.

S. Pang, J. Wang, D. K. Lin, and M.-Q. Liu. Construction of mixed orthogonal arrays with high strength. *Annals of Statistics*, 49:2870–2884, 2021.

F. K. Phoa and H. Xu. Quarter-fraction factorial designs constructed via quaternary codes. *Annals of Statistics*, 37:2561–2581, 2009.

G. Pistone and H. P. Wynn. Generalised confounding with Gröbner bases. *Biometrika*, 83: 653–666, 1996.

R. L. Plackett and J. P. Burman. The design of optimum multifactorial experiments. *Biometrika*, 33:305–325, 1946.

V. Pless. *Introduction to The Theory of Error-Correcting Codes*. John Wiley & Sons, 2011.

F. Pukelsheim. On linear regression designs which maximize information. *Journal of Statistical Planning and Inference*, 4:339–364, 1980.

F. Pukelsheim. *Optimal Design of Experiments*. Wiley, New York, 1993.

P. Z. Qian and C. J. Wu. Sliced space-filling designs. *Biometrika*, 96:945–956, 2009.

P. Z. Qian, M. Ai, and C. J. Wu. Construction of nested space-filling designs. *Annals of Statistics*, 37:3616–3643, 2009a.

P. Z. Qian, B. Tang, and C. J. Wu. Nested space-filling designs for computer experiments with two levels of accuracy. *Statistica Sinica*, 19:287–300, 2009b.

Z. Qian, C. C. Seepsad, V. Joseph, J. K. Allen, and C. Wu. Building surrogate models based on detailed and approximate simulations. *Journal of Mechanical Design*, 128:668–677, 2006.

X. Qu. Statistical properties of Rechtschaffner designs. *Journal of Statistical Planning and Inference*, 137:2156–2164, 2007.

J. Rafter and E. Seiden. Contributions to the theory and construction of balanced arrays. *Annals of Statistics*, 2:1256–1273, 1974.

C. R. Rao. Factorial experiments derivable from combinatorial arrangements of arrays. *Supplement to the Journal of the Royal Statistical Society*, 9:128–139, 1947.

R. Rechtschaffner. Saturated fractions of 2^n and 3^n factorial designs. *Technometrics*, 9: 569–575, 1967.

G. Rennen, B. Husslage, E. R. Van Dam, and D. D. Hertog. Nested maximin Latin hypercube designs. *Structural and Multidisciplinary Optimization*, 41:371–395, 2010.

K. J. Ryan and D. A. Bulutoglu. E(s^2)-optimal supersaturated designs with good minimax properties. *Journal of Statistical Planning and Inference*, 137:2250–2262, 2007.

J. Sacks and D. Ylvisaker. Some model robust designs in regression. *Annals of Statistics*, 12:1324–1348, 1984.

J. Sacks, W. Welch, T. Mitchell, and H. Wynn. Design and analysis of computer experiments. *Statistical Science*, 4:409–435, 1989.

T. J. Santner, B. J. Williams, and W. I. Notz. *The Design and Analysis of Computer Experiments*, volume 1. Springer, 2003.

Y. Sathe and R. Shenoy. *A*-optimal weighing designs when $n \equiv 3 \pmod 4$. *Annals of Statistics*, 17:1906–1915, 1989.

F. Satterthwaite. Random balance experimentation. *Technometrics*, 1:111–137, 1959.

R. R. Schmidt, E. E. Cruz, and M. Iyengar. Challenges of data center thermal management. *IBM Journal of Research and Development*, 49:709–723, 2005.

E. D. Schoen. All orthogonal arrays with 18 runs. *Quality and Reliability Engineering International*, 25:467–480, 2009.

E. D. Schoen, P. T. Eendebak, and M. V. Nguyen. Complete enumeration of pure-level and mixed-level orthogonal arrays. *Journal of Combinatorial Designs*, 18:123–140, 2010.

E. D. Schoen, B. Sartono, and P. Goos. Optimal blocking for general resolution-3 designs. *Journal of Quality Technology*, 45:166–187, 2013.

E. D. Schoen, N. Vo-Thanh, and P. Goos. Two-level orthogonal screening designs with 24, 28, 32, and 36 runs. *Journal of the American Statistical Association*, 112:1354–1369, 2017.

E. D. Schoen, P. T. Eendebak, and P. Goos. A classification criterion for definitive screening designs. *Annals of Statistics*, 47:1179–1202, 2019.

E. Seiden and R. Zemach. On orthogonal arrays. *Annals of Mathematical Statistics*, 37: 1355–1370, 1966.

K. R. Shah and B. K. Sinha. Weighing designs. In *Theory of Optimal Designs*, pages 141–163. Springer, 1989.

C. Shi and B. Tang. Designs from good Hadamard matrices. *Bernoulli*, 24:661–671, 2018.

C. Shi and B. Tang. Supersaturated designs robust to two-factor interactions. *Journal of Statistical Planning and Inference*, 200:119–128, 2019a.

C. Shi and B. Tang. Design selection for strong orthogonal arrays. *Canadian Journal of Statistics*, 47:302–314, 2019b.

C. Shi and B. Tang. Construction results for strong orthogonal arrays of strength three. *Bernoulli*, 26:418–431, 2020.

C. Shi and B. Tang. On construction of nonregular two-level factorial designs with maximum generalized resolution. *Statistica Sinica*, 33:593–607, 2023.

C. Shi and H. Xu. A projection space-filling criterion and related optimality results. *Journal of the American Statistical Association, accepted*, 2024.

A. C. Shoemaker, K.-L. Tsui, and C. J. Wu. Economical experimentation methods for robust design. *Technometrics*, 33:415–427, 1991.

R. R. Sitter, J. Chen, and M. Feder. Fractional resolution and minimum aberration in blocked 2^{n-k} designs. *Technometrics*, 39:382–390, 1997.

I. M. Sobol'. On the distribution of points in a cube and the approximate evaluation of integrals. *Zhurnal Vychislitel'noi Matematiki i Matematicheskoi Fiziki*, 7:784–802, 1967.

T. Speed and R. A. Bailey. On a class of association schemes derived from lattices of equivalence relations. In *Algebraic Structures and Applications*, pages 55–74. Marcel Dekker, 1982.

J. Srivastava. Designs for searching non-negligible effects. In *A Survey of Statistical Design and Linear Models*, pages 507–520. North-Holland, 1975.

J. Srivastava and D. Chopra. Balanced optimal 2^m fractional factorial designs of resolution V, $m \leq 6$. *Technometrics*, 13:257–269, 1971.

J. Srivastava and S. Ghosh. Balanced 2^m factorial designs of resolution V which allow search and estimation of one extra unknown effect, $4 \leq m \leq 8$. *Communications in Statistics-Theory and Methods*, 6:141–166, 1977.

J. Srivastava and B. Gupta. Main effect plan for 2^m factorials which allow search and estimation of one unknown effect. *Journal of Statistical Planning and Inference*, 3:259–265, 1979.

J. Srivastava and D. Throop. Orthogonal arrays obtainable as solutions to linear equations over finite fields. *Linear Algebra and Its Applications*, 127:283–300, 1990.

J. N. Srivastava and J. Li. Orthogonal designs of parallel flats type. *Journal of Statistical Planning and Inference*, 53:261–283, 1996.

M. Stein. Large sample properties of simulations using Latin hypercube sampling. *Technometrics*, 29:143–151, 1987.

D. M. Steinberg and D. K. Lin. A construction method for orthogonal Latin hypercube designs. *Biometrika*, 93:279–288, 2006.

J. Stufken and B. Tang. Complete enumeration of two-level orthogonal arrays of strength d with d+ 2 constraints. *Annals of Statistics*, 35:793–814, 2007.

C.-Y. Suen, H. Chen, and C. Wu. Some identities on q^{n-m} designs with application to minimum aberration designs. *Annals of Statistics*, 25:1176–1188, 1997.

C.-Y. Sun and B. Tang. Relationship between orthogonal and baseline parameterizations and its application to design constructions. *Statistica Sinica*, 32:239–250, 2022.

C.-Y. Sun and B. Tang. Uniform projection designs and strong orthogonal arrays. *Journal of the American Statistical Association*, 118:417–423, 2023.

D. X. Sun. *Estimation Capacity and Related Topics in Experimental Designs*. PhD thesis, University of Waterloo, 1993.

D. X. Sun, C. Wu, and Y. Chen. Optimal blocking schemes for 2^n and 2^{n-p} designs. *Technometrics*, 39:298–307, 1997.

D. X. Sun, W. Li, and K. Q. Ye. Algorithmic construction of catalogs of non-isomorphic two-level orthogonal designs for economic run sizes. *Statistics and Applications*, 6:141–155, 2008.

F. Sun and B. Tang. A method of constructing space-filling orthogonal designs. *Journal of the American Statistical Association*, 112:683–689, 2017a.

F. Sun and B. Tang. A general rotation method for orthogonal Latin hypercubes. *Biometrika*, 104:465–472, 2017b.

F. Sun, M.-Q. Liu, and D. K. Lin. Construction of orthogonal Latin hypercube designs. *Biometrika*, 96:971–974, 2009.

F. Sun, D. K. Lin, and M.-Q. Liu. On construction of optimal mixed-level supersaturated designs. *Annals of Statistics*, 39:1310–1333, 2011.

F. Sun, M.-Q. Liu, and P. Z. Qian. On the construction of nested space-filling designs. *Annals of Statistics*, 42:1394–1425, 2014.

F. Sun, Y. Wang, and H. Xu. Uniform projection designs. *Annals of Statistics*, 47:641–661, 2019.

B. Tang. Orthogonal array-based Latin hypercubes. *Journal of the American Statistical Association*, 88:1392–1397, 1993.

B. Tang. Selecting Latin hypercubes using correlation criteria. *Statistica Sinica*, 8:965–977, 1998.

B. Tang. Theory of J-characteristics for fractional factorial designs and projection justification of minimum G_2-aberration. *Biometrika*, 88:401–407, 2001.

B. Tang. Orthogonal arrays robust to nonnegligible two-factor interactions. *Biometrika*, 93:137–146, 2006.

B. Tang. Construction results on minimum aberration blocking schemes for 2^m designs. *Journal of Statistical Planning and Inference*, 137:2355–2361, 2007.

B. Tang and L.-Y. Deng. Minimum G_2-aberration for nonregular fractional factorial designs. *Annals of Statistics*, 27:1914–1926, 1999.

B. Tang and L.-Y. Deng. Construction of generalized minimum aberration designs of 3, 4 and 5 factors. *Journal of Statistical Planning and Inference*, 113:335–340, 2003.

B. Tang and C. Wu. Characterization of minimum aberration 2^{n-k} designs in terms of their complementary designs. *Annals of Statistics*, 24:2549–2559, 1996.

B. Tang and C. Wu. A method for constructing supersaturated designs and its $E(s^2)$-optimality. *Canadian Journal of Statistics*, 25:191–201, 1997.

B. Tang and J. Zhou. Existence and construction of two-level orthogonal arrays for estimating main effects and some specified two-factor interactions. *Statistica Sinica*, 19: 1193–1201, 2009.

B. Tang and J. Zhou. D-optimal two-level orthogonal arrays for estimating main effects and some specified two-factor interactions. *Metrika*, 76:325–337, 2013.

B. Tang, F. Ma, D. Ingram, and H. Wang. Bounds on the maximum number of clear two-factor interactions for 2^{m-p} designs of resolution III and IV. *Canadian Journal of Statistics*, 30:127–136, 2002.

Y. Tian and H. Xu. A minimum aberration-type criterion for selecting space-filling designs. *Biometrika*, 109:489–501, 2022.

Y. Tian and H. Xu. Stratification pattern enumerator and its applications. *Journal of the Royal Statistical Society Series B: Statistical Methodology*, 86:364–385, 2024.

R. D. Tobias. Saturated second-order two-level designs: an empirical approach. *SAS Technicall Report*, 1996.

P. W. Tsai, S. G. Gilmour, and R. Mead. Projective three-level main effects designs robust to model uncertainty. *Biometrika*, 87:467–475, 2000.

S.-F. Tsai. Construction of nested orthogonal arrays of parallel-flats type. *Journal of Statistical Planning and Inference*, 181:41–51, 2017.

E. R. van Dam, B. Husslage, D. Den Hertog, and H. Melissen. Maximin Latin hypercube designs in two dimensions. *Operations Research*, 55:158–169, 2007.

E. R. van Dam, G. Rennen, and B. Husslage. Bounds for maximin Latin hypercube designs. *Operations Research*, 57:595–608, 2009.

S. A. Vanstone and P. C. Van Oorschot. *An Introduction to Error Correcting Codes With Applications*, volume 71. Springer Science & Business Media, 2013.

A. R. Vazquez, P. Goos, and E. D. Schoen. Projections of definitive screening designs by dropping columns: Selection and evaluation. *Technometrics*, 62:37–47, 2020.

E. Verheiden. Integral and rational completions of combinatorial matrices. *Journal of Combinatorial Theory, Series A*, 25:267–276, 1978.

A. Vine, S. Lewis, and A. Dean. Two-stage group screening in the presence of noise factors and unequal probabilities of active effects. *Statistica Sinica*, 15:871–888, 2005.

C. Wang, J. Yang, and M.-Q. Liu. Construction of strong group-orthogonal arrays. *Statistica Sinica*, 32:1225–1243, 2022a.

J. Wang and C. Wu. An approach to the construction of asymmetrical orthogonal arrays. *Journal of the American Statistical Association*, 86:450–456, 1991.

J. Wang and C. J. Wu. A hidden projection property of Plackett-Burman and related designs. *Statistica Sinica*, 5:235–250, 1995.

K. Wang and Y. Li. Constructions of nested orthogonal arrays. *Journal of Combinatorial Designs*, 21:464–477, 2013.

L. Wang, Q. Xiao, and H. Xu. Optimal maximin L_1-distance Latin hypercube designs based on good lattice point designs. *Annals of Statistics*, 46:3741–3766, 2018a.

Y. Wang, J. Yang, and H. Xu. On the connection between maximin distance designs and orthogonal designs. *Biometrika*, 105:471–477, 2018b.

Y. Wang, F. Wang, Y. Yuan, and Q. Xiao. Connecting U-type designs before and after level permutations and expansions. *Journal of Statistical Theory and Practice*, 15:1–19, 2021.

Y. Wang, S. Liu, and D. K. Lin. On definitive screening designs using Paley's conference matrices. *Statistics & Probability Letters*, 181:109267, 2022b.

Y. Wang, F. Sun, and H. Xu. On design orthogonality, maximin distance, and projection uniformity for computer experiments. *Journal of the American Statistical Association*, 117:375–385, 2022c.

G. Watson. A study of the group screening method. *Technometrics*, 3:371–388, 1961.

S. Webb. Non-orthogonal designs of even resolution. *Technometrics*, 10:291–299, 1968.

J. Williamson. Hadamard's determinant theorem and the sum of four squares. *Duke Mathematical Journal*, 11:65, 1944.

C. Wu. Construction of 2^m4^n designs via a grouping scheme. *Annals of Statistics*, 17: 1880–1885, 1989.

C. Wu. Construction of supersaturated designs through partially aliased interactions. *Biometrika*, 80:661–669, 1993.

C. Wu, R. Zhang, and R. Wang. Construction of asymmetrical orthogonal arrays of the type $OA(s^k, s^m(s^{r_1})^{n_1}...(s^{r_t})^{n_t})$. *Statistica Sinica*, 2:203–219, 1992.

C. J. Wu and Y. Chen. A graph-aided method for planning two-level experiments when certain interactions are important. *Technometrics*, 34:162–175, 1992.

C. J. Wu and M. S. Hamada. *Experiments: Planning, Analysis, and Optimization*. John Wiley & Sons, 2000.

H. Wu and C. Wu. Clear two-factor interactions and minimum aberration. *Annals of Statistics*, 30:1496–1511, 2002.

H. Wu, R. Mee, and B. Tang. Fractional factorial designs with admissible sets of clear two-factor interactions. *Technometrics*, 54:191–197, 2012.

L. Xiao, D. K. Lin, and F. Bai. Constructing definitive screening designs using conference matrices. *Journal of Quality Technology*, 44:2–8, 2012.

Q. Xiao and H. Xu. Construction of maximin distance designs via level permutation and expansion. *Statistica Sinica*, 28:1395–1414, 2018.

H. Xu. An algorithm for constructing orthogonal and nearly-orthogonal arrays with mixed levels and small runs. *Technometrics*, 44:356–368, 2002.

H. Xu. Minimum moment aberration for nonregular designs and supersaturated designs. *Statistica Sinica*, 13:691–708, 2003.

H. Xu. A catalogue of three-level regular fractional factorial designs. *Metrika*, 62:259–281, 2005.

H. Xu. Algorithmic construction of efficient fractional factorial designs with large run sizes. *Technometrics*, 51:262–277, 2009.

H. Xu and C.-S. Cheng. A complementary design theory for doubling. *Annals of statistics*, 36:445–457, 2008.

H. Xu and A. Wong. Two-level nonregular designs from quaternary linear codes. *Statistica Sinica*, 17:1191–1213, 2007.

H. Xu and C. Wu. Generalized minimum aberration for asymmetrical fractional factorial designs. *Annals of Statistics*, 29:1066–1077, 2001.

H. Xu and C. Wu. Construction of optimal multi-level supersaturated designs. *Annals of Statistics*, 33:2811–2836, 2005.

H. Xu, S.-W. Cheng, and C. J. Wu. Optimal projective three-level designs for factor screening and interaction detection. *Technometrics*, 46:280–292, 2004.

S. Yamada and D. K. Lin. Three-level supersaturated designs. *Statistics & Probability Letters*, 45:31–39, 1999.

S. Yamada and T. Matsui. Optimality of mixed-level supersaturated designs. *Journal of Statistical Planning and Inference*, 104:459–468, 2002.

C. Yang. Some designs for maximal $(+1, -1)$-determinant of order $n \equiv 2 \pmod 4$. *Mathematics of Computation*, 20:147–148, 1966.

C. Yang. On designs of maximal $(+1, -1)$-determinant of order $n \equiv 2 \pmod 4$. *Mathematics of Computation*, 22:174–180, 1968.

Y. H. Yang and T. Speed. Design issues for cDNA microarray experiments. *Nature Reviews Genetics*, 3:579–588, 2002.

F. Yates. Complex experiments. *Supplement to the Journal of the Royal Statistical Society*, 2:181–247, 1935.

K. Q. Ye. Indicator function and its application in two-level factorial designs. *Annals of Statistics*, 31:984–994, 2003.

K. Q. Ye. A note on regular fractional factorial designs. *Statistica sinica*, 14:1069–1074, 2004.

R. Zhang and R. Mukerjee. Minimum aberration designs for two-level factorials in N= 1 (mod 4) runs. *Statistica Sinica*, 23:853–872, 2013.

R. Zhang and D. Park. Optimal blocking of two-level fractional factorial designs. *Journal of Statistical Planning and Inference*, 91:107–121, 2000.

R. Zhang, F. K. Phoa, R. Mukerjee, and H. Xu. A trigonometric approach to quaternary code designs with application to one-eighth and one-sixteenth fractions. *Annals of Statistics*, 39:931–955, 2011.

Y. Zhang, Y. Lu, and S. Pang. Orthogonal arrays obtained by orthogonal decomposition of projection matrices. *Statistica Sinica*, 9:595–604, 1999.

W. Zhou, J.-F. Yang, and M.-Q. Liu. Optimal maximin L_2-distance Latin hypercube designs. *Journal of Statistical Planning and Inference*, 207:113–122, 2020.

Y. Zhou and B. Tang. Column-orthogonal strong orthogonal arrays of strength two plus and three minus. *Biometrika*, 106:997–1004, 2019.

Y. Zhou and H. Xu. Space-filling properties of good lattice point sets. *Biometrika*, 102: 959–966, 2015.

Y.-D. Zhou and H. Xu. Space-filling fractional factorial designs. *Journal of the American Statistical Association*, 109:1134–1144, 2014.

Y.-D. Zhou, K.-T. Fang, and J.-H. Ning. Mixture discrepancy for quasi-random point sets. *Journal of Complexity*, 29:283–301, 2013.

Index

Kolibri 360 GmbH
Ferdinand-Jühlke-Straße 7
99095 Erfurt, Deutschland
produktsicherheit@kolibri360.de